D1535236

The Aero- and Hydromechanics of Keel Yachts

J. W. Slooff

The Aero- and Hydromechanics of Keel Yachts

 Springer

J. W. Slooff
Uithoorn
The Netherlands

Every effort has been made to contact the copyright holders of the figures and tables which have been reproduced from other sources. Anyone who has not been properly credited is requested to contact the publishers, so that due acknowledgment may be made in subsequent editions.

ISBN 978-3-319-13274-7   ISBN 978-3-319-13275-4 (eBook)
DOI 10.1007/978-3-319-13275-4

Library of Congress Control Number: 2014956974

Springer Cham Heidelberg New York Dordrecht London

Printed on acid-free paper

Springer is part of Springer Science+Business Media (www.springer.com)

To my grandchildren:
Bente, Stan, Jouke, Mirte, Tessa and Laura

Preface

The intended reader will already have found out that this is not a book with glossy pictures by Beken of Cowes of the glamorous and fabulous sailing yachts of the world. It is a lot more sober and a little more scientific than that.

This book has come about as a follow-up on a course on the aero- and hydro-mechanics of sailing that was written by the author for the Heiner Sail Academy in the Netherlands in 2002/2003. The purpose of the course was to provide yachtsmen with (top-)amateur or (semi-) professional racing ambitions with a solid background of the aero- and hydromechanics of sailing. The ambition of the author was to do so in a manner that he could justify from his own background as a professional fluid dynamicist and a cruising yachtsman, while trying to keep the material accessible for attendees without profound academic schooling.

This, of course, is not the first book on the aero- and hydromechanics of sailing. A.C. Marchaj's "Aero- hydrodynamics of sailing" (ISBN 0-7136-5073-7) is generally considered as 'The Bible' on the science and technology of sailing. And rightly so; it contains an incredibly broad and deep treatment of almost all aspects of sailing technology. Many readers, however, may find it difficult to find their way in Marchaj's book, in particular if their objective is to find out in a relatively straightforward way about how and why sail boat performance depends on the configuration and trim of boat and sails.

While working on this book a new, very valuable volume on the science behind sailing yachts and their design appeared on the horizon: Fabio Fossati's "Aero-Hydrodynamics and the Performance of Sailing Yachts" (ISBN 978-0-07-162910-2). Fossati's book provides an excellent description of the state of the art (2007 level) of fluid dynamic technology as applied to sailing yachts. It also addresses the physical mechanisms of sailing in considerable depth.

At the other end of the spectrum there are numerous books on the subject of a more popular nature. Many of these are directed at the yachtsman that is interested in the 'what' and 'how' but not too much in the 'why' of boat configuration and boat trim. Few provide explanations of the mechanisms involved in a way that is scientifically justified. An excellent exception is formed by Frank Bethwaite's "High Performance Sailing" (ISBN 978-1-4081-2491-8).

It is the author's impression that there is a substantial gap between Marchaj's 'bible' as well as Fossati's 'volume a vela' and the more popular books on sailing. This book tries to bridge that gap. For this purpose the material is presented in a form that, on the one hand, is scientifically justified and consistent according to the author's best knowledge. On the other hand it also aims at facilitating on-board utilization of the knowledge that is acquired.

It is of course up to the reader to judge whether the author has succeeded in realising this objective. The author hopes, however, that his 50 years of experience in fluid dynamics and his 35 years of experience as a (cruising) yachtsman have contributed to the objective to create a 'New Testament' of sailing mechanics.

February 2004/2013   Joop Slooff
Uithoorn, The Netherlands

About the Author

Johannes ('Joop'[1]) W. Slooff is a (retired) fluid dynamicist by profession. He is also a cruising yachtsman (not yet retired).

In 1964 he obtained his degree in aeronautical engineering at Delft University of Technology in the Netherlands, also having followed the Diplome Course at the Von Karman Institute for Fluid Dynamics (VKI) in Rhode-St. Genèse, near Brussels in Belgium.

In 1965 he joined the National Aerospace Laboratory (NLR) in Amsterdam, The Netherlands, as a research scientist in aerodynamics, with airfoil and wing design and applied computational aerodynamic design methodologies as his main topics of research. Much of this research was done in support of and in close collaboration with the former Fokker Aircraft Company.

In 1976 he was appointed as head of the Theoretical Aerodynamics Department of NLR and in 1986 as chief of the Fluid Dynamics Division, a position he held until his retirement in 2001. From 1987 to 2001 he was also a part-time professor in applied computational aerodynamics at the Aerospace Faculty of Delft University of Technology.

He has served on several national and international advisory committees such as the Fluid Dynamics Panel of the former Advisory Group for Aeronautical Research and Development (AGARD) of NATO, the Research and Technology Organisation (RTO) of NATO, the Applied Aerodynamics Committee of the American Institute for Aeronautics and Astronautics (AIAA), the Group for Aeronautical Research and Technology in Europe (GARTEUR), the Technical Advisory Committee of the Von Kármán Institute for Fluid Dynamics in Belgium and the Scientific Advisory Council of the Maritime Research Institute of the Netherlands (MARIN). In 1996 he delivered the 36th Lanchester Lecture for the Royal Aeronautical Society in London. He is the 1997 recipient of the Von Kármán Medal of AGARD/NATO, in recognition of his outstanding contributions to that organisation. He is also a Fellow of the American Institute for Aeronautics and Astronautics.

His activities in sailing started in 1977 with the acquisition of a 28-foot cruising sloop. In 1980 he became involved, through MARIN, with the development

---

[1] Pronounced as "Yoap Sloaf".

of the 12-m yacht Australia II for the 1983 America's Cup campaign. He proposed the concept of the winged keel and was responsible for the computational fluid dynamic modelling, performed at NLR, that substantiated the performance potential of the concept[2]. In the 1980s he was also involved in keel research performed at Delft University of Technology by Gerritsma and Keuning.

After his (early) retirement in 2001 he wrote a course on the aero- and hydrodynamics of sailing for the Heiner Sail Academy. He also became involved, as a technical-scientific advisor, in the ABN-AMRO campaign for the 2005–2006 Volvo Ocean Race.

From 2008 through 2012 he has lectured on the aero- and hydrodynamics of sailing at an annual special course on sailing yacht technology at the faculty of Maritime Technology of Delft University.

He is (co)-author of about 40 conference papers and journal articles on fluid dynamic topics, including some on the hydrodynamics of sailing yacht keels.

Since 1986 he owns a 33-foot cruising sloop with a shallow draught winged keel and winged rudder of his own design.

...when tide and wind are gone: drying out on a winged keel...

---

[2] For an account of his involvement see: *"Who really designed Australia II?"* by Barbara Lloyd, Nautical Quarterly, Spring 1985.

Acknowledgements

No one works completely alone and, although I have done almost everything for this book by myself, there are a number of people that have played an essential stimulating or supporting role.

My addiction to the science of sailing probably started around 1980 when Peter van Oossanen, then project manager at MARIN, asked me to join him in the design research for the America's Cup winning, 12-m yacht Australia II.

About 30 years later Alexander Vermeulen, long-time friend from my years of studying in Delft, and once-a-year sailing mate, volunteered to, critically, proof-read what I had produced for this book. Sadly, he did not live long enough to witness the completion of our work. He died while scrutinizing Chap. 6.

Jouke van der Baan, another long-time friend from my years of studying in Delft, my co-skipper once a year, when we sailed a week or so around the Frisian Isles with Alexander Vermeulen, continued the proof-reading where Alex had stopped.

Lex Keuning, recognized expert in ship hydromechanics, with whom I shared a common interest in sailing and its technology since we first met over 35 years ago. He kept me up-to-date in sailing yacht technology during the decade when I was too busy to keep track of what was going on. More recently he gave me the opportunity to share my knowledge on the scientific aspects of sailing with the students and the other lecturers of the annual, so-called 'minor' course on the technology of sailing at Delft University of Technology.

Roy Heiner, professional racing yachtsman, gave me the opportunity to play a small role in the ABNAMRO Volvo Ocean Race campaign and to learn what ocean racing people consider to be important.

Last but not least my best and oldest friend Robbert van der Mije, sailing and travel companion for many, many years, always searching the internet for news on sailing that he thought I could use.

Finally my wife Lia, whom I should have mentioned first and before all. When I retired after a fairly busy professional career she must have thought that, finally, it was her turn to receive attention. Not quite so, I am afraid. After half a year of cleaning up and completing some technical/scientific work that I had started, but never finished, during my time at NLR, I was caught, once more, by the science of sailing. Again, my wife did not get the attention she deserved, without complaining. I beg her to forgive me for my addiction.

Contents

Symbols and Notation

A	Aspect ratio (of a lifting surface)
A_e	Effective aspect ratio, $= A\, b_e/b$
A_R	Propeller blade-area/disc-area ratio
$\underline{a}$	Acceleration (vector)
AC	Aerodynamic centre
B	Maximum width (beam) of a hull
B_{WL}	Maximum width (beam) of the waterline of a hull
BAD	Distance between the boom of the mainsail and the sheer line (Fig. 2.4.1)
BSF	Boat speed factor $= \sqrt{(C_T/C_R)}$
B	Damping ratio, Eq. (5.18.10)
b	Span of a lifting surface
bm	Distance between MC and CB
b0	Distance between the waterplane and CB at zero heel
b_e	Effective span of a lifting surface
b_w	Span of a winglet (at the tip of a lifting surface)
b'	'active' span of a lifting surface
b	Damping coefficient
CB	Center of buoyancy
CE	Centre of effort (sails)
CG	Centre of gravity
CLR	Centre of lateral resistance (underwater body)
CP	Centre of pressure
CSF	Chord length at the foot of a spinnaker
C_A	Coefficient of total aerodynamic force
C_B	Block coefficient of a sailing yacht hull
C_D	Drag coefficient
C_{D0}	Drag coefficient at zero lift
C_{D_F}	Friction drag coefficient
C_{D_f}	Form drag or 'viscous pressure' drag coefficient
C_{D_i}	Induced drag coefficient
C_{D_p}	Pressure drag coefficient
C_{Dprop}	Propeller drag coefficient

C_{D_v} $(=C_{D_F}+C_{D_f})$, 'viscous' drag coefficient (also called 'boundary layer' or 'profile' drag coefficient

C_{F_0} Friction drag coefficient of (one side of) a flat plate at zero angle of attack

C_f Local friction drag coefficient

C_H Heeling force coefficient

$C_{k\xi,\,k\eta,\,k\zeta}$ Dimensionless radius of gyration ('gyradius') about the ξ-, η-, ζ- axis, respectively

C_L Lift coefficient

$C_L{}'$ Coefficient of total lift under heel (in 'heeled' plane)

$C_{M¼}$ Moment coefficient with respect to quarter chord point of a foil section

C_{Mz} Yawing moment coefficient

C_m Hull maximum section coefficient

C_N Normal force coefficient of a lifting surface

C_{N90} Normal force coefficient at 90° angle of attack

C_P Prismatic coefficient

C_p Pressure coefficient

C_R Hydrodynamic resistance coefficient

C_{Raw} Coefficient of time-averaged added resistance due to waves

$C_{Rw\nabla}$ Coefficient of wave-making resistance due to displacement

C_{RwS} Coefficient of wave-making resistance due to side force

C_S Side force coefficient

C_{S_0} Hydrodynamic side force coefficient in the absence of free surface effects

C_T Driving force (thrust) coefficient

C_V Vertical fluid dynamic force coefficient

C_{WP} Waterplane coefficient = $(S_{WP}/(B_{WL}\,L_{WL})$

C_ζ Coefficient of fluid dynamic force component in ζ-direction

$C(k_f)$ Lift deficiency factor of an oscillating foil section

$C'(k_f)$ Lift deficiency factor of a 3D oscillating lifting surface

c Foil section chord length

c_m Foil mean chord length

c_{ma} Chord length of a mast section

c_R Chord length at the root or center section of a lifting surface

c_T Chord length at the tip of a lifting surface

c_t Chord length of trailing-edge flap or trimtab

c_{wD0} Constant in formula (7.6.4) for hull windage (drag)

c_{wD1} Proportionality constant in formula (7.6.4) for hull windage (drag)

c_{wL1} Proportionality constant in formula (7.6.3) for hull windage (lift)

c_β Coefficient representing the effect of fore-aft asymmetry on the aerodynamic lift of a sailing yacht hull

c_γ Coefficient representing the effect of fore-aft asymmetry on the effective wave incidence angle

$\boldsymbol{c_w}$ Wave propagation speed

D	Aerodynamic drag
D_i	Induced drag
D_F	Friction drag
D_P	Pressure drag
D_{par}	Parasite drag due to windage
D_v	'viscous' or profile drag
D	Hull depth, diameter of a circular cylinder or thickness of a mast section
D_F	Freeboard height (Fig. 2.2.1)
D_h	Hull draft
D_{max}	Maximum diameter of a body of revolution
D_{me}	Effective mast diameter/thickness, Eq. (7.4.1)
D_{prop}	Propeller diameter
D_T	Total draft of hull plus keel
DWL	Design waterline
$d...$	Difference operator
$\underline{d}$	Distance vector
d	Distance,
$d_{x, y, z}$	Distance between the centre of effort of the sails CE and the centre of lateral resistance CLR of the underwater body in the x, y, z direction, respectively
$\dfrac{dC_L}{d\alpha}, \dfrac{dC_D}{d\alpha}$	Rate of change of lift, drag coefficient with angle of attack
$\dfrac{dC_M}{d\alpha}$	Rate of change of moment coefficient with angle of attack
$\dfrac{dC_{MHz}}{d\beta}, \dfrac{dC_{MHz}}{d\lambda}$	Rate of change of hydrodynamic yawing moment coefficient with apparent wind angle or angle of leeway, respectively
$\dfrac{dC_S}{d\beta}, \dfrac{dC_S}{d\lambda}$	Rate of change of side force coefficient with apparent wind angle and angle of leeway, respectively
$\dfrac{dC_S}{d\delta_r}$	Rate of change of side force coefficient with rudder deflection angle
$\dfrac{dL}{d\alpha}, \dfrac{dD}{d\alpha}$	Rate of change of lift, drag with angle of attack
$\dfrac{dM_x}{d\dot{\phi}}$	Rate of change of rolling moment with roll rate
$\dfrac{dM_z}{d\lambda}, \dfrac{dM_z}{d\psi}$	Rate of change of yawing moment with angle of leeway, yaw, respectively
$\dfrac{dM_z}{d\dot{\psi}}$	Rate of change of yawing moment with the angular velocity of rotation in yaw (yaw rate)
E	Base of the mainsail,
E	'edge' factor determining the effective lift curve slope of a foil section (Eq. (5.15.14))

e	Base number of the natural logarithm (e=2.718...)
e	Efficiency factor of the spanwise lift distribution of a lifting surface
Fr	Froude number (Fr = $V/\sqrt{(L\,g)}$)
Fr_s	Froude number for resonant encounter with waves
F_{FS}	Free surface factor; ratio between lift in the presence of a free surface and lift without the pressure relief effect of the free surface (see Appendix E)
F_R	Ratio between the circulation at the root of a wing or fin and the circulation at the root for an elliptic distribution of circulation
F_w	Wave excitation factor (App. L)
F_0	Amplitude of an oscillatory force
$\underline{E}$, F	Force (vector)
F_e	External force per unit of fluid volume
F_F	Total friction force acting on a body
F_f	Frictional force per unit of fluid volume
F_g	Gravitational force per unit of fluid volume
F_i	Inertia force per unit of fluid volume
F_N	Normal force
F_{Np}	Normal force due to pressure
F_{Nt}	Normal force due to friction
F_P	Total pressure force acting on a body
F_p	Pressure force per unit of fluid volume
F_T	Tangential force
f	Frequency (periods per second)
f'	Factor modeling the effect of beam/draft ratio and displacement/length ratio on the yawing moment of a sailing yacht hull
f_c	Maximum camber of a foil section
f_e	Frequency of encounter with waves
$f_{s\varphi}$	Factor representing the effect of the downwash/sidewash from the keel on the rudder, including the effect of heel
f_0	Natural frequency of an object in periodic motion/deformation
$f_{\varepsilon\varphi}$	Factor representing the effect on the rudder of the downwash/sidewash from the keel, including the effect of heel
g	Constant of gravitational acceleration
g	Gap width
g_d	Distance between sail foot and deck
g_w	Distance between sail foot and water surface
g_{we}	Effective distance between sail foot and water surface
gb	Length of righting arm
gb(30)	Righting arm for lateral hydrostatic stability at 30° heel
gm	Metacentric height (lateral)
gm(0)	Metacentric height (lateral) at zero heel
gm_L	Longitudinal metacentric height
H	Heeling force
h	Height of a column of fluid,
h_e	Effective height of a sailing yacht rig (effective mast height)

h_m	Height of sail top above the water surface
h_{mast}	Height of a sailing yacht rig (between deck and sail top)
h_w	Wave height
$h_{w\frac{1}{3}}$	Significant wave height
h	Displacement (translation) of an object in heaving motion
h_m	Average displacement (translation) of an object in heaving motion
I	Height of fore triangle (foresail), mass moment of inertia
I_T	Geometric moment of inertia of the waterplane of a sailing yacht hull with respect to the centreline
I_{xx}, I_{yy}, I_{zz}	Mass moment of inertia around x, y, z-axis, respectively
$I_{\xi\xi}, I_{\eta\eta}, I_{\zeta\zeta}$	Idem, around ξ, η, ζ-axis, respectively
$I'_{xx}, I'_{yy}, I'_{zz}$	Moment of inertia around x, y, z-axis, respectively, of 'added mass'
IACC	International America's Cup Class
IMS	International Measurement System
IOR	International Offshore Rules
ι_0	Entry angle of a sail section (Fig. 7.4.8)
ι_1	Exit angle of a sail section (Fig. 7.4.8)
J	Base of the fore triangle (foresail)
K_i	Induced drag factor
K_S	Hydrodynamic drag-due-to-side force factor
K_w	Wave-drag-due-to-side-force factor, Eq. (6.5.32)
K_φ	Heel forcing factor, Eq. (7.9.19)
k	Form factor for profile drag
k^*	Critical roughness height for boundary layer transition
k^+	Admissible roughness height in a turbulent boundary layer
k_f	Reduced frequency $(=\pi f L/V)$
k_s	Equivalent sand roughness of a surface
k_{turb}	Constant $(\cong 0.4)$ in formula describing the velocity profile of the atmospheric boundary layer (vertical wind gradient)
k_w	Wave number, $= 2\pi/\lambda_w$
$k_{\xi, \eta, \zeta}$	Radius of gyration about the ξ-, η- or ζ-axis of a yacht, resp.
k_0	Form factor for friction drag
k_1	Form factor for viscous pressure drag
k	Spring stiffness factor in rotational motion
L	Fluid dynamic lift (horizontal component)
L_F	Lift due to friction forces
L_P	Lift due to pressure forces
L_0	Quasi-steady part of lift in unsteady flow
L_1	Part of lift in unsteady flow that depends on the rate of change with time of the motion (damping term)
L_2	Part of lift in unsteady flow that depends on the acceleration/deceleration of the fluid particles ('added mass' term)
L'	Total lift force acting on a sail under heel
$L(y), L(z)$	Local lift per unit span
L	Characteristic length of an object

L_{OA}	Length overall of a sailing yacht
L_{WL}	Length of the waterline of a sailing yacht
L_{CWL}	Circumferential length of the waterline
$L_{k\text{-}r}$	Longitudinal distance between keel and rudder
L_{ref}	Reference length
LCB	Longitudinal position of the centre of buoyancy, $=(0.5\,L_{WL}-x_{CB})/L_{WL}$
LP	Luff perpendicular (foresail, see Fig. 2.4.1)
lg	Girth length of a sail section
ln	Natural logarithm
M_0	Amplitude of an oscillatory moment of force
MC	Metacenter
$\boldsymbol{M}$	Moment (of force)
$\boldsymbol{M_{Ahz}}$	Aerodynamic yawing moment due to the hull
$\boldsymbol{M_{Hhz}}$	Hydrodynamic yawing moment due to the hull
$\boldsymbol{M_0}$	Quasi-steady part of moment in unsteady flow
$\boldsymbol{M_1}$	Part of moment in unsteady flow that depends on the rate of change with time of the motion (damping term)
$\boldsymbol{M_2}$	Part of moment in unsteady flow that depends on the acceleration/deceleration of the fluid particles ('added mass' term)
$\boldsymbol{M_{¼}}$	Moment with respect to quarter chord point of a foil section
$\boldsymbol{M_r}$	Righting moment
m	Mass (of an object), Meter (unit of length)
$\boldsymbol{m}$	Mass flux
$\boldsymbol{N}$	Normal force
O( )	Order of magnitude
P	Height of the mainsail
p	(Static) pressure
p'	Dynamical part of static pressure $(=p-p_g)$
p_a	Atmospheric pressure
p_g	Gravitational (static) pressure
p_w	Pressure at (the water side of) the water surface
p_0	Stagnation pressure
p_∞	Static pressure of undisturbed flow
q	Dynamic pressure $(=\frac{1}{2}\,\rho V^2)$
$\boldsymbol{R}$	Hydrodynamic resistance
$\boldsymbol{R_{aw}}$	Time-averaged added resistance in waves
$\boldsymbol{R_i}$	Induced hydrodynamic resistance
$\boldsymbol{R_{prop}}$	Propeller resistance
$\boldsymbol{R_{tot}}$	Total hydrodynamic resistance
$\boldsymbol{R_v}$	'viscous' resistance
$\boldsymbol{R_w}$	Wave-making resistance
$\boldsymbol{R_{w\nabla}}$	Wave-making resistance due to displacement
$\boldsymbol{R_{wS}}$	Wave-making resistance due to side force
R	Radius (of circle or circular cylinder)
R_{spec}	Specific gas constant

Re	Reynolds number ($Re = \rho\, V L/\mu$)
Re_{sep}	Critical Reynolds number for separated flow at the trailing edge of foil sections
Re_x	Local or running Reynolds number ($Re_x = V^* x/v$)
r	Radial coordinate
S	Side force
S_w	Local side force at the waterline due to deformation of the water surface
$S(z)$	Local side force (per unit spanwise coordinate)
$S_0(z)$	Local side force (per unit spanwise coordinate) without free surface effects
S	Surface area (lifting surfaces, hull, sails, appendages)
SF	Length of spinnaker foot
SLE	Length of spinnaker leech
SLU	Length of spinnaker luff
SMG	Spinnaker mid-girth length
SPL	Length of spinnaker pole
Sr	Strouhal number ($Sr = f\, L/V$)
S_H	Hydrodynamic reference area
$S_{f\Delta}$	Area of the foresail triangle
$S_{m\Delta}$	Area of the mainsail triangle
S_{proj}	Projected (planform) area of a lifting surface
S_{prop}	Frontal area of propeller
S_r	Rudder area
S_S	Sail area
S_{SS}	Spinnaker area
$S_{S\Delta}$	Total reference sail area
S_{wet}	Wetted area (of an object)
S_{WP}	Waterplane area
S_X	Cross-sectional area of a lifting surface
S_{Xmax}	Maximum cross-sectional area (submerged part of hull or bulb)
S_z	Projected area of a lifting surface
$S_\otimes$	Cross-sectional area (of a stream tube)
$S(\omega_w)$	Energy contents of waves with circular frequency ω_w
s	Seconds (unit of time)
s	Length of line (segment)
T	Aerodynamic thrust (driving force)
T	Temperature [Kelvin], Period (time) of an oscillation ($= 1/f$)
T_e	Period (time) of encounter with waves ($= 1/f_e$)
T_0	Period of the natural motion of an object
T_w	Wave period
T_{wa}	Average wave period
T'_{wa}	Normalized average wave period ($= T_{wa}\sqrt{g/L_{WL}}$)
T^*	Time it takes a flow particle to cover a characteristic distance L
TR	Taper ratio (c_T/c_R)
t	Time, (air)foil (maximum) thickness
U	Wind velocity at a certain height above the water surface

U_{10}	Wind velocity at 10 m above the water surface
U_e	Flow velocity at the edge of a boundary layer
u	Velocity variation due to surging
u_w	Horizontal component of velocity of the orbital motion of water particles in a wave
u_z	Rate of change of the flow velocity in the direction normal to the direction of the flow
$(u_z)_w$	Rate of change of the flow velocity in a boundary layer at the wall, in the direction normal to the wall
V	Volume of an object
VCB	Vertical position of the centre of buoyancy
VMG	$= V_{mg}$
V	Fluid velocity or flow speed
$\underline{V}$	Velocity vector
V_a	Apparent wind speed
V_A	Vertical aerodynamic force
V_b	Boat speed
V_b^*	Critical boat speed
V_{circ}	circulatory velocity (component)
V_H	Vertical hydrodynamic force
V_i	Induced velocity
V_{mg}	(VMG) Velocity-made-good
V_{mgl}	(Or downwind VMG), velocity-made-good to leeward
V_{mgw}	(Or upwind VMG), velocity-made-good to windward
V_{mg30}	Velocity-made-good in a direction at 30 degrees from the true wind
V_R	Resultant apparent wind velocity in rolling motion
V_{Rw}	resultant local fluid velocity in waves
V_{rot}	Velocity due to rotation of a fluid element
V_t	True wind speed
$V_{t\,ref}$	Reference true wind speed (at a height of 10 m above the water surface)
V_∞	Velocity of undisturbed (fluid) flow
V_n	Normal velocity component (perpendicular to a surface)
V_s	Tangential velocity component (parallel to a surface)
v_{if}	Velocity induced by foresail
v_{im}	Velocity induced by mainsail
v_r	Roll-induced apparent wind velocity component
v_w	Velocity of water particles in waves
X	Displacement (vector)
$\dot{X}$	Velocity of displacement
$\ddot{X}$	Acceleration of displacement
x, y, z	Hydrodynamic coordinate system (Fig. 3.1.6)
x_0, y_0, z_0	Coordinate system of sailing (Fig. 3.1.6)
x',y', z'	Aerodynamic coordinate system (Fig. 3.1.6)
x_a	X-coordinate of position of axis of rotation

x_f	Chordwise position of the maximum camber of a foil section
x_k	Longitudinal position of (the centre of pressure of) the keel of sailing yacht
x_{mast}	Longitudinal position of the (centre of the) mast
x_r	Longitudinal position of the (centre of pressure of the) rudder
x_{ref}	Longitudinal position of the reference point for the yawing moment
x_{tr}	Streamwise position of boundary layer transition
z_{CG}	Vertical position of the centre of gravity (zero heel)
z_H	Local water elevation
z_0	Roughness length (of atmospheric boundary layer)
$\|z'\|$	$= \|z\|/\zeta_w$ Amplitude of heaving motion, normalized by wave amplitude

Greek Symbols

α	Angle of attack (of a lifting foil or sail)
α'	Angle between spinnaker pole and foot of the spinnaker
α_{Dmax}	Angle of attack for which the drag of a sail attains its maximum
α_{Lmax}	Angle of attack for which the lift of a sail attains its maximum
$\alpha_{(L=D)}$	Angle of attack for which the lift and drag of a sail are equally large
$\alpha_{(L+D)max}$	Angle of attack for which the sum of lift and drag of a sail has its maximum
α_e	Effective angle of attack
α_f	Angle of attack at the foot of a sail
α_h	Angle of attack at the head of a sail
α_i	Induced angle of attack
α_{id}	'ideal' angle of attack of a sail section with sharp leading edge
α_m	Angle of attack of the mainsail
$\alpha_{\bar{m}}$	Mean angle of attack of a lifting surface in oscillatory motion
α_s	Angle of attack of (the foot of) a spinnaker
α_{Tmax}	Angle of attack for which the driving force has a maximum (dependent on apparent wind angle)
α_ε	Angle of the downwash (or side wash) behind the keel of a yacht
$\alpha_{\varepsilon Amin}$	Angle of attack for which the aerodynamic drag angle ε_A attains a minimum
$\alpha_{\varepsilon Pmin}$	Angle of attack for which the thrust angle ε_p attains a minimum
α_0	Zero lift angle of attack
$\dot{\alpha}$	Rate of change with time of angle of attack
$\dot{\alpha}_e$	Rate of change with time of effective angle of attack
$\ddot{\alpha}$	Angular acceleration of angle of attack
$\ddot{\alpha}_e$	Angular acceleration of effective angle of attack
β	Apparent wind angle (or apparent course angle, relative to the apparent wind vector)
β_{10}	Apparent wind angle at 10 m above the water surface

β_{min}	Minimum apparent wind angle
β_a	Apparent heading (angle between the longitudinal axis of the boat and the apparent wind vector)
β_a'	Indicated apparent heading under heel
Γ	Circulation
γ	True wind angle (true course angle, relative to the true wind vector)
γ_w	Wave incidence angle ($\gamma_w = 0°$ head waves, $\gamma_w = 180°$ stern waves)
Δ	As a prefix: difference or step width
$\mathit{\Delta}$	Hydrostatic (buoyancy) force
δ	Sheeting angle
δ^*	boundary layer displacement thickness
$\bar{\delta}$	Dimensionless geometrical twist of a sail (Eq. (7.3.4))
δ_{bl}	Boundary layer thickness
δ_f	Sheeting angle of (the foot of) a foresail, Flap (deflection) angle
δ_h	Sheeting angle at the head of a sail
δ_m	Sheeting angle of (the foot of) a mainsail
δ_p	Sheeting angle of spinnaker pole
δ_r	Rudder (deflection) angle
δ_{Tmax}	Sheeting angle for maximum driving force
$\delta_{\varepsilon Pmin}$	Sheeting angle for which the thrust angle ε_p attains a minimum
δ_0	Sheeting angle at the foot of a sail
δC_{FO}	Incremental friction drag coefficient due to surface roughness
$\delta C_{L\delta}$	Incremental lift coefficient due to flap deflection
$\delta C_L(\varphi)$	Incremental lift coefficient due to heel at zero leeway
δC_L	Incremental lift coefficient due to side edge or leading edge separation
δC_{Lmax}	Incremental maximum lift coefficient due to side edge or leading edge separation
$\delta C_{MHz}(\varphi)$	Incremental hydrodynamic yawing moment coefficient due to heel at zero leeway
$\delta F_{Hlat}(\varphi)$	Incremental lateral hydrodynamic force coefficient due to heel at zero leeway
$\delta M_{Hz}(\varphi)$	Incremental hydrodynamic yawing moment due to heel at zero leeway
δm	Element of mass
δy_{CG}	Lateral displacement of centre of gravity
δz	Sinkage
$\delta \alpha_r$	Incremental angle of attack due to rolling
ε_A	Aerodynamic drag angle ($= \arctan(D/L)$)
ε_{Amin}	Minimum aerodynamic drag angle
ε_H	Hydrodynamic drag angle ($= \arctan(R/S)$)
ε_{Hmin}	Minimum hydrodynamic drag angle
ε_P	Propulsion or thrust angle ($= \arctan(H/T)$)
ζ	Dimensionless distance along the mast measured from the foot of a sail, (Eq. (7.3.5))
ζ_w	Amplitude of orbital (wave) motion (halve wave height)

η_f	Flap effectiveness $(=\dfrac{dC_L}{d\delta_f}/2\pi)$
θ	Pitch angle
θ_w	Maximum wave slope $(=2\pi\zeta_w/\lambda_w)$
$|\theta'|$	$=|\theta|/(2\pi\,\zeta_w/L_{WL})$, Amplitude of pitching motion, normalized by maximum wave slope
$\dot{\theta}$	Angular velocity of pitching motion
$\ddot{\theta}$	Angular acceleration of pitching motion
ι_0	Entry angle of a sail foil section
ι_1	Exit angle of a sail foil section
κ	Factor governing the decay with depth of free surface effect
Λ	Sweep angle (usually of 25% chord line) of a lifting surface
Λ_e	effective angle of sweep of a sailing rig under heel, Eq. (7.9.12)
λ	Leeway angle
λ_0	Leeway angle at zero side force
λ_w	Wave length
μ	Dynamic viscosity of a fluid
ν	Kinematic viscosity of a fluid or gas $(=\mu/\rho)$
ξ, η, ζ	Ship coordinate system (Fig. 2.5.1, 3.1.6)
ξ_{luff}	Longitudinal position of the luff of a sail section
π	$=3.14\ldots$, Ratio between the circumference and the diameter of a circle
ρ	Mass density
σ	Velocity gradient angle in a boundary layer
Σ	Summation (mathematical operator)
τ	Shear stress in a fluid
τ_w	Shear stress at the wall or skin friction
φ	Heel (or roll) angle
φ_e	Effective angle of heel, Eq. (7.9.6)
φ_{k0}	Cant angle of a keel
φ^*	Critical angle of heel
$\dot{\varphi}$	Angular velocity of rolling motion (roll rate)
$\ddot{\varphi}$	Angular acceleration of rolling motion
ψ	Yaw angle
ψ_w	Wave angle
$\dot{\psi}$	Angular velocity of yawing motion
$\ddot{\psi}$	Angular acceleration of yawing motion
$\boldsymbol{\omega}$	Angular speed (of rotation)
$\dot{\boldsymbol{\omega}}$	Angular acceleration
ω	Circular (or angular) frequency $(=2\pi f)$
ω_e	Circular frequency of encounter with waves
ω_r	Circular frequency of rolling motion
ω_0	Natural circular frequency

Other Symbols

$\cong$	Approximately equal to
$\approx$	Asymptotically equal to
$\div$	Proportional to
$<$	Smaller than
$>$	Larger than
$\ll$	Much smaller than
$\gg$	Much larger than
$\lvert\cdot\rvert$	Amplitude of an oscillating quantity
∇	Displacement volume
∇	Gravitational force (weight)
∞	Infinite

Subscripts

A	Refers to air
ave	Average value
CB	Refers to centre of buoyancy
CE	Refers to centre of effort (of the sails)
CER	Refers to centre of effort (of the sails) in rolling motion
CLR	Refers to centre of lateral resistance
CP	Refers to centre of pressure (of a lifting surface)
f	Refers to foresail
H	Refers to water
hor	Refers to horizontal component
hull	Refers to hull
hv	Refers to heaving motion
k	Refers to keel
k-r	Refers to keel and rudder
lat	Refers to lateral component
lon	Refers to longitudinal component
L/Dmax	Refers to conditions for maximum lift/drag ratio
LWL	Refers to length of the waterline
m	Refers to mainsail
max	Maximum value
r	Refers to rudder
ref	Reference value
s	Refers to sail(s)
stat	Value in stationary flow
x, y, z	Refers to (component in) x, y, z-direction, respectively
ξ, η, ζ	Refers to (component in) ξ, η, ζ -direction, respectively

θ	Refers to pitching motion
φ	Refers to rolling motion
ψ	Refers to yawing motion
2D	Refers to two-dimensional flow

Chapter 1
Introduction

The art and science of sailing has intrigued mankind for many millennia. Not in the least, of course, because sailing was, until relatively recent, the most efficient way of transportation of men and goods over long distances over water. Pleasure craft, at least those of a small and improvised nature, may have existed already just as long or perhaps even longer ago.

Until, say, the middle of the nineteenth century, the design and construction of a sailing ship, and indeed sailing itself, were an art rather than a science, based on intuition, experience and craftsmanship. Progress in 'technology' was slow. Due to lack of understanding of the physical mechanisms involved the evolution of sailing and sailing ships was a Darwinian 'survival of the fittest'. An important reason for this is probably that the flow of water around the hull and the air flow around the sails are largely invisible for the naked eye and therefore difficult to fathom. With intuition, right or wrong, as a guide, innumerably different types of sailing vessels, in all places of the world, have come and gone over the ages. Those that have survived are probably the 'fittest' for the specific purpose for which they were used.

Sailing 'yachts' also serve a specific purpose and their development has, to a large extent, also been 'Darwinian'. They were introduced by the Dutch in the sixteenth century. The word, spelled as 'jacht' in Dutch, was used for relatively small and swift sailing vessels that were used for formal functions of the admiralty, communication between ships and shore and, since the seventeenth century, also for pleasure functions. Yacht racing probably also stems from this period, either as a contest between messengers ('who is the first to bring the message on shore') or as a game for the rich and mighty. Yacht racing as a sport evolved in the beginning of the nineteenth century. Inevitably this led to the formation of different classes of racing yachts, each with prescribed characteristics that leave little or no room for innovative developments (except, of course, for the so-called 'open classes').

Scientific approaches in sailing were non-existent until the second half of the nineteenth century when William Froude developed the first towing tank and published his similarity law on ship (hull) motion (1870) (Froude 1870). Froude's work

© Springer International Publishing Switzerland 2015
J. W. Slooff, *The Aero- and Hydromechanics of Keel Yachts,*
DOI 10.1007/978-3-319-13275-4_1

and that of Osborne Reynolds on frictional resistance and the similarity law of viscous fluids (1883) (Reynolds 1883) have formed the basis for the developments in ship hydrodynamics in the twentieth century.

Scientific methods in aerodynamics were not available until the beginning of the twentieth century. Here, much of the ground work, theoretical as well as experimental in wind tunnels, was done by Ludwig Prandtl[1] and his associates in Göttingen, Germany, and by Frederick Lanchester in the UK. Most of this ground work was associated with the dawning of the age of flight but an important basic law for the combined aero- and hydrodynamics of sailing was formulated by Lanchester (1907) in 1907.

With only a few exceptions, the main developments and wide-spread application of scientific methods in sailing technology did, however, not occur until after the second world war (see, for example, references Larsson 1990 and Milgram 1998). An important reason for this somewhat belated development is probably in the fact that commercial sailing vessels had more or less disappeared by the turn of the nineteenth to twentieth century. As a consequence funding of research and development in sailing technology had to come primarily from the yachting community and was necessarily scarce. The turning of the tide came when the commercial sponsoring of big yacht racing events such as the America's Cup, Witbread/Volvo Ocean Race and the like began to provide budgets for R&D of a different order of magnitude. Although much of this research was of a proprietary nature, the results were often vented some time later at the by then established international conferences and symposia on the technology of sailing such as the Chesapeake Sailing Yacht Symposium, the 'Ancient Interface' Symposium on the Aero/Hydronautics of Sailing of the American Institute of Aeronautics and Astronautics (AIAA) and the HISWA Symposium on Developments of Interest in Yacht Architecture in the Netherlands. The progress in sailing yacht technology was, of course, not limited to aero- and hydrodynamics. Developments in structural design and in particular of new materials have been equally if not even more important.

Several institutions and many individuals have contributed to the developments in sailing technology over, say, the past 40–50 years. Amongst those the groups at Southampton University in the UK (Marchaj et al.), Massachusetts Institute of Technology (MIT) in the US (Kerwin, Newman, Milgram), Delft University of Technology in the Netherlands (Gerritsma et al.) and Chalmers University in Sweden (Larsson) should be mentioned. As a result competitive racing yacht design is nowadays no longer possible without towing tank and wind tunnel tests, numerical flow simulations, race simulation models and velocity prediction programs (VPP). Even competitive sailing itself, at least in big ocean racing events, is no longer possible without 'high-tech' instrumentation and a VPP based performance database.

Yet, sailing is still much of an art, requiring knowledge, mental and physical capabilities and experience. The ever changing conditions of wind and waves and the large number (25 or so) of independent, partially controllable variables constitute probably the main reason for its charm.

---

[1] Much of Prof. Prandtl's work is reflected in the edited publications, by Tietjens (Prandtl and Tietjens 1934a, b), of his lectures.

As indicated in the Preface, the original primary objective of this book was to provide knowledge of the aero- and hydromechanics of sailing to the yachtsman who is interested in a proper understanding of the physical mechanisms that he is playing with. It has, however, evolved to a volume that may also be of interest for the yacht designer and other schooled professionals of sailing technology that wish to consult an overview of the fluid dynamic aspects of sailing.

Because of the primary objective of this book, the author has tried to avoid complex mathematical treatments and derivations. Where and when appropriate these are given in appendices. Nevertheless, it will help the reader if he is in command of the basic principles of physics, mathematics and general mechanics. For readers that wish to refresh their knowledge in this area Appendix A summarizes some basic mathematical notions.

It should also be mentioned that, as suggested by the title, the scope of this book is, in principle, limited to keel yachts. Furthermore, it is restricted to the most common type of keel yacht: single-masted mono-hulls with 'fore-and-aft', Bermuda-rigged sails. However, much of the material covered is also applicable to other types of sailing vessels such as multi-hulls, yachts with multiple masts, windsurf boards and the like.

In the chapters to follow, the reader is first familiarized (if required) with some basic notions about the main characteristics and geometry of sailing yachts (Chap. 2), the principles of mechanics and ship motions, and the forces acting on a sailing boat (Chap. 3). This is followed in Chap. 4 by an introduction to the general mechanics of sailing. Readers who are familiar with the principles of mechanics and sailing may wish to skip Chaps. 2 and 3.

Chapter 5 describes basic elements and phenomena of fluid mechanics (of both air and water) that, in the author's opinion, are required for understanding the more complex mechanisms of the flows about hulls, appendages and sails. For the underwater part these mechanisms are discussed in further detail in Chap. 6. Chapter 7 deals with the aerodynamic forces acting on the sails and the other air-exposed parts of a sailing yacht. Finally, several performance aspects of sailing are revisited in some more detail in Chap. 8.

References

Froude W (1870) The experiments recently proposed on the resistance of ships. Trans Inst Nav Archit 11:80

Lanchester FW (1907) Aerodynamics. A. Constable and Co., London

Larsson L (1990) Scientific methods in yacht design. Annu Rev Fluid Mech 22:349–385

Milgram JH (1998) Fluid mechanics for sailing vessel design. Ann Rev Fluid Mech 30:613–653

Prandtl L, Tietjens OG (1934a) Fundamentals of aero- & hydromechanics. McGraw-Hill, New York (also: Dover Publications, NY, 1957)

Prandtl L, Tietjens OG (1934b) Applied aero- & hydromechanics. McGraw-Hill, New York (also: Dover Publications, NY, 1957)

Reynolds O (1883) An experimental investigation of the circumstances which determine whether the motion of water will be direct or sinuous, and the law of resistance in parallel channels. Philos Trans R Soc Lond 35:84–99

Chapter 2
Sailing Yacht Geometry and Mass Properties

2.1 Introduction

The mechanics of sailing are almost entirely determined by the geometrical characteristics and mass properties of the sailing vessel. It is therefore useful to summarize
how the geometry and mass properties of a sailing yacht are, commonly, described
and to take notice of the nomenclature that is used for this purpose.

One problem in this respect is that there are many different types of sailing
yacht. It is an almost impossible job to address all of these. As already mentioned in
Chap. 1 we will limit ourselves to the most common type of sailing yacht: single-
masted mono-hulls with 'fore-and-aft', Bermuda-rigged sails. As a consequence the
list of characteristic parameters and notions to be discussed hereafter is not exhaus-
tive in the sense that it does not (fully) cover other types of sailing yacht such as
multi-hulls and multi-masted yachts.

When describing the geometry and mass properties of a sailing yacht it is useful
to distinguish three categories of quantities. For the hydro-mechanic forces acting
on a sailing yacht the geometrical characteristics of the under-water part of the hull
and the appendages are the most important. For the aerodynamic forces these are
the geometrical characteristics of the sailing rig and the above-the-water part of the
hull. The general mechanics and stability properties of a yacht are also determined
by the mass properties.

The three categories just mentioned are described in the following sections.

2.2 Hull Geometry

The geometry of the hull of a sailing yacht is usually described by means of traditional lines drawings and/or by computer aided mathematical descriptions (Larsson and Eliasson 1996; Claughton et al. 1999). The latter are sometimes known as
Lines Processing Programs (LPP). An example of a lines drawing is reproduced in

© Springer International Publishing Switzerland 2015
J. W. Slooff, *The Aero- and Hydromechanics of Keel Yachts,*
DOI 10.1007/978-3-319-13275-4_2

Fig. 2.2.1. A brief treatise of computer aided mathematical representations can be found in Claughton et al. (1999).

Although every detail of the geometry has, in principle, some influence on the hydro- or aerodynamics, there is a limited set of parameters or quantities that is commonly used to describe the most important characteristics. Here, we will largely follow the description given in Larsson and Eliasson (1996).

The two most important dimensions of a sailing yacht are the length and the displacement. Two different measures of the length of the hull are usually distinguished (see Fig. 2.2.1):

- The *length overall* (L_{OA})
- The *length of the waterline* (L_{WL})

L_{OA} is the distance between the most forward and the most rearward points on the hull. L_{WL} is the distance between the most forward and the most rearward points on the *design waterline* (DWL). The latter is the intersection of the plane, undisturbed water surface with the external surface of the hull under design weight, floating conditions and zero heel.

Measures of the displacement are:

- The volume *displacement* (∇) of the complete underwater part of the yacht. The corresponding weight displacement is denoted as V
- The volume displacement of the underwater part of the hull only (∇_h), without appendages[1]
- The *displacement/length ratio*. This is usually defined as

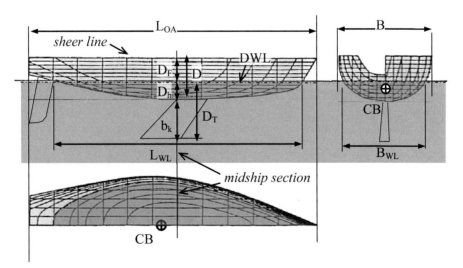

Fig. 2.2.1 *Example of traditional, three-view lines drawing of a sailing yacht. (From http://dsyhs. tudelft.nl, adapted)*

---

[1] In naval architecture it is common practice to indicate the hull by the subscript c (from canoe body). In this book we will use the subscript h because we have other use for the subscript c.

$$\nabla_h^{1/3}/L_{WL} \tag{2.2.1}$$

It is a dimensionless measure of the slenderness of the hull.

The most important lateral dimensions are

- The (maximum) *beam* (B) of the yacht, the maximum width of the hull
- The (maximum) *beam of the waterline* (B_{WL}), the maximum width of the design waterline (DWL)

The most important vertical dimensions are

- The *depth* D of the hull, the vertical distance between the deepest point of the hull and the sheer line (see Fig. 2.2.1)
- The *draft* D_h of the underwater part of the hull
- The *freeboard* D_F, the vertical distance between the sheer line and the waterline
- The *span* b_k of the keel and b_r of the rudder
- The total draft[2] $D_T = D_h + b_k$

Other quantities of importance are

- The *centre of buoyancy* CB, characterized by its longitudinal position LCB and its vertical position VCB
- The *midship section* is the cross-section at 50% of the length of the waterline.

LCB is usually measured from the midship section (positive forward) and expressed in fractions of L_{WL}. For example, LCB$=-0.05$ means that the centre of buoyancy is positioned 5% *aft* of the midship section.

- The *maximum area section* is the cross-section with the maximum submerged area. For sailing yachts it is usually positioned aft of amidships. The maximum section area is denoted as S_{Xmax} (see Fig. 2.2.2)
- The *section coefficient* C_m is the ratio between the maximum section area and the rectangle circumscribing the maximum area section:

$$C_m = S_{Xmax}/(B_{WL} * D_h) \tag{2.2.2}$$

- The *block coefficient* C_B. This is the ratio between the displacement volume ∇_h of the hull and the volume of the rectangular box circumscribing the underwater part of the hull:

$$C_B = \nabla_h/(L_{WL} * B_{WL} * D_h) \tag{2.2.3}$$

- The *prismatic coefficient* C_P. This is the ratio between the displacement volume and the volume of a cylinder with the length of the waterline circumscribing the submerged part of the maximum area section at zero heel:

---

[2] In naval architecture the total draft of a yacht is usually denoted as T. In this book we use D_T because we have other use for the letter T.

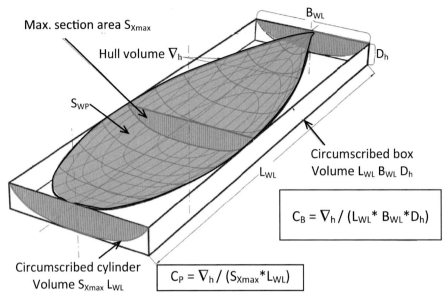

Fig. 2.2.2 *Defining hull areas and volumes*

$$C_P = \nabla_h/(S_{Xmax} * L_{WL}) = C_B/C_m \qquad (2.2.4)$$

The prismatic coefficient is a measure of the fullness of the bow and stern parts of a yacht. The fuller the bow and stern the larger the prismatic coefficient.

The ratio S_{Xmax}/L_{WL}^2 is another measure of the slenderness of the hull. From the expression (2.2.1) and Eq. (2.2.4) it follows that

$$S_{Xmax}/L_{WL}^2 = (\nabla_h/L_{WL}^3)/C_P \qquad (2.2.5)$$

- The *waterplane area* (S_{WP}) is the area enclosed by the design waterline
- The *waterplane coefficient* (C_{WP}) is the ratio between the waterplane area and the rectangle circumscribing the waterline:

$$C_{WP} = S_{WP}/(L_{WL} * B_{WL}) \qquad (2.2.6)$$

2.3   Appendages

The most important dimensions of the appendages (Fig. 2.3.1) are the span b_k, b_r and area S_k, S_r of the keel and rudder, respectively. The length of the intersection of the keel with the hull is called the root chord (c_R). The length of the keel at the tip is called the tip chord c_T. It follows that the area of a trapezoidal keel is given by

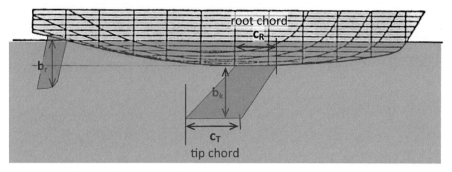

Fig. 2.3.1 *Definition of dimensions of keel and rudder*

$$S_k = b_k (c_R + c_T)/2 \qquad (2.3.1)$$

A similar expression holds, of course, for the rudder area.

Other quantities of importance are the aspect ratios A_k and A_r of keel and rudder. These are defined by

$$A_k = b_k^2 / S_k \qquad (2.3.2)$$

and

$$A_r = b_r^2 / S_r \qquad (2.3.3)$$

2.4   Rig and Sails

For a Bermuda-rigged yacht the dimensions of the rig and sails are usually described by the quantities indicated in Fig. 2.4.1. These follow the so-called IOR convention (International Measurement System 2011) of the Offshore Racing Council of the International Yacht Racing Union.

- 'I' is the height of the fore triangle
- J is the base of the fore triangle
- LP is the luff perpendicular
- P is the height of the mainsail
- E is the base of the mainsail
- BAD stands for Boom-Above-Deck (sometimes called BAS, Boom-Above-Sheerline), that is the distance between the sheer line and attachment point of the boom on the mast

With these definitions the area $S_{f\Delta}$ of the fore triangle is

$$S_{f\Delta} = 0.5\,I * J \qquad (2.4.1)$$

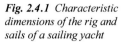

Fig. 2.4.1 Characteristic dimensions of the rig and sails of a sailing yacht

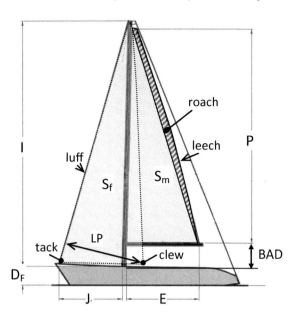

and the area $S_{M\Delta}$ of the mainsail triangle is

$$S_{m\Delta} = 0.5\, E * P \tag{2.4.2}$$

The total reference sail area is then given by

$$S_{S\Delta} = S_{m\Delta} + S_{f\Delta} \tag{2.4.3}$$

Note that this implies that any overlap between fore- and mainsail, defined as LP/J (%), is not accounted for in the definition of the reference sail area. This applies also to the 'roach' of the mainsail, that is the area between the triangle and the actual leech of the mainsail.

Note also that the mast height is about equal to or slightly larger than P+BAD and that the total height h_m of the sail top above the water surface is about

$$h_m \cong P + BAD + D_F \tag{2.4.4}$$

While for a masthead rig there holds $I \cong P + BAD$, one has $I < P + BAD$ for a fractional rig.

The dimensions of spinnaker type foresails (or head sails) are measured in a different way (See Fig. 2.4.2):

- SPL is the length of the spinnaker pole
- SF is the length of the spinnaker foot
- SLU is the length of the luff

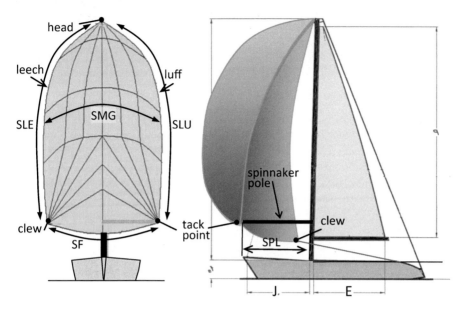

Fig. 2.4.2 *Describing the dimensions of spinnaker type foresails*

- SLE is the length of the leech
- SMG is the spinnaker mid-girth length (measured between the points at 50% of the luff and leech)

With these definitions the area S_{SS} of the spinnaker is calculated as (ORC Rating Systems 2011)

$$S_{SS} = \{(SLU + SLE)/2\} \; * \; (SF + 4SMG)/6 \tag{2.4.5}$$

It is further noted that for a symmetric spinnaker there holds $SLU = SLE$.

2.5 Mass Properties

In addition to the dimensions, the mass properties of a sailing yacht are also very important for the sailing performance as well as for other characteristics such as stability. The main mass properties are the following:

- The total mass m is given by

$$m = \rho \nabla \tag{2.5.1}$$

where ρ is the density of (sea)water and ∇ is the displacement volume

- The total weight V (a force) is then given by

$$V = g\,m\ =\rho g \nabla \tag{2.5.2}$$

  where g is the gravitational acceleration (9.82 m/s^2)

- The *centre of gravity* (CG) is the point where the total mass can be considered to be concentrated. That is, the action of gravity on the actual yacht is the same as if all its mass were concentrated in the centre of gravity.

  For a sailing yacht the centre of gravity is usually positioned somewhat aft of amidships. Under floating conditions the centre of gravity and the center of buoyancy are on the same vertical line.

  Depending on the type of yacht the vertical position of the CG can be above or below the water surface. For shallow draft, modern cruising yachts the CG is usually close to the water surface. For a racing yacht it can be a significant fraction of the total draft below the surface.

Measures of the distribution of mass are the so-called *mass moments of inertia*. They play an important role in the rotational motions of a ship in a seaway and in manoeuvering. Their definition requires the adoption of a coordinate system. The so-called ship-coordinate system of a sailing yacht is indicated in Fig. 2.5.1. The origin is usually taken in the centre of gravity (CG) or amidships in the intersection of the plane of symmetry with the waterplane. The longitudinal or ξ-axis is horizontal and in the plane of symmetry. The lateral or η-axis is perpendicular to the plane of symmetry. The vertical or ζ-axis is in the plane of symmetry. Note that this system is fixed to the yacht. It tilts with the yacht when the yacht heels.

The moment of inertia with respect to a certain, for example the ξ-, axis of a mass element (δm in Fig. 2.5.1) is given by

$$\delta I_{\xi\xi} = \delta m (d_\eta^{\,2} + d_\zeta^{\,2}), \tag{2.5.3}$$

where the position of the mass element is given by the distances d_η, d_ζ as indicated in Fig. 2.5.1. Note that the factor $(d_\eta^{\,2} + d_\zeta^{\,2})$ is nothing else but the square of the radial distance r_ξ between the mass element δm and the ξ-axis. According to Pythagoras' law:

$$r_\xi^2 = (d_\eta^{\,2} + d_\zeta^{\,2}) \tag{2.5.4}$$

The total mass moment of inertia around the ξ-axis is obtained by adding all mass elements that are part of the ship. This is expressed as

$$I_{\xi\xi} = \Sigma \delta m\,(d_\eta^{\,2} + d_\zeta^{\,2}) \tag{2.5.5}$$

where the symbol Σ stands for the summing operation over all mass elements.

The moments of inertia about the other axes are given by

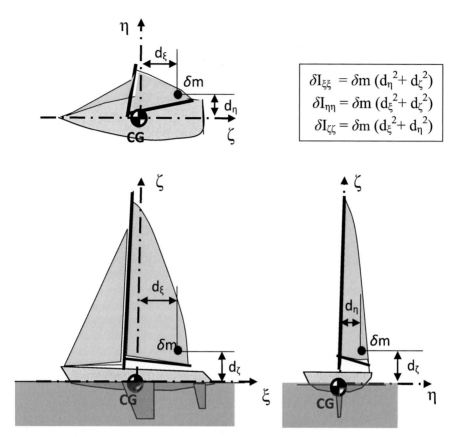

$$\delta I_{\xi\xi} = \delta m \ (d_\eta^2 + d_\zeta^2)$$
$$\delta I_{\eta\eta} = \delta m \ (d_\xi^2 + d_\zeta^2)$$
$$\delta I_{\zeta\zeta} = \delta m \ (d_\xi^2 + d_\eta^2)$$

Fig. 2.5.1 *Yacht coordinate system and mass moments of inertia*

$$I_{\eta\eta} = \Sigma \delta m \ (d_\xi^2 + d_\zeta^2) \qquad\qquad (2.5.6)$$

$$I_{\zeta\zeta} = \Sigma \delta m \ (d_\xi^2 + d_\eta^2) \qquad\qquad (2.5.7)$$

It is further customary to express a mass moment of inertia of a sailing yacht in terms of a so-called *radius of gyration* or *gyradius*. For the ξ-axis this is defined by

$$I_{\xi\xi} = mk_\xi^2 = \rho \nabla k_\xi^2 \qquad\qquad (2.5.8)$$

or

$$k_\xi = \sqrt{\{I_{\xi\xi} / (\rho \nabla)\}} \qquad\qquad (2.5.9)$$

with corresponding definitions for the other axes. The gyradius is the distance from the axis of rotation at which a mass equal to the total mass must be positioned in order to give the same moment of inertia.

The gyradius is sometimes expressed in terms of the length of the waterline of the yacht:

$$C_{k\xi} = k_\xi / L_{WL} \qquad (2.5.10),$$

etc. A typical value for the gyradius about the η-axis of a sailing yacht is $C_{k\eta} = 0.25$.

It will be clear from the definitions given above that the gyradius is a measure of the eccentricity in the distribution of mass. The gyradius is small when the mass is concentrated around the centre of gravity. Heavy loads in the bow and stern, a heavy mast and a deep keel with a heavy bulb lead to relatively large values of the gyradius.

References

Claughton AR, Shenoi RE, Wellicom JF (eds) (1999) Sailing yacht design: theory. Prentice Hall, Upper Saddle River. ISBN 0582368561

http://dsyhs.tudelft.nl/

International Measurement System IMS, Offshore Racing Congress (2011)

Larsson L, Eliasson RE (1996) Principles of yacht design. Adlard Coles Nautical, London. ISBN 0-7136-3855-9

ORC Rating Systems, Offshore Racing Congress (2011)

Chapter 3
Forces, Moments and Motions

3.1 Notions and Definitions

Vectors

The motion of a sailing yacht, that is its velocity and acceleration, is the result of the forces which act on it. Velocity, acceleration and forces are quantities that are characterized by two properties: magnitude and direction. They are also coupled to a specific location in space: their point of application. Such a quantity is called a *vector*.

The point of application of the velocity and acceleration vectors of an object is the center of mass, commonly referred to as the center of gravity. Forces may be acting in different locations on an object but are usually transferred to the center of mass. (As we will see later this transfer requires the introduction of another notion: that of the moment of a force).

Vectors require a coordinate system for their description. In this section they are denoted in ***bold*** underlined *italic* characters. In figures they are indicated by (fat) arrows. When only the magnitude of a vector is concerned we will use ***bold*** *italic* without underlining.

Coordinate systems are usually chosen to be orthogonal (rectangular) and right-handed, as in Fig. 3.1.1.

In an orthogonal (x, y, z) coordinate system like that of Fig. 3.1.1 a vector $\underline{F}$, as indicated by the full red arrow, can be described in two different ways:

- In *Cartesian coordinates* by its component vectors $\underline{a}$, $\underline{b}$ and $\underline{c}$ along the x-, y-, and z-axes
- In *polar coordinates* by its magnitude, usually indicated as $|\underline{F}|$ or just F, and the angles α between the direction of the vector and the axes of the coordinate system

Note that in polar coordinates two of the three angular coordinates suffice; the third follows from the mathematical relation that the sum of the angles is always 180°. Note also that Fig. 3.1.1 implies that vectors can be added and subtracted:

© Springer International Publishing Switzerland 2015
J. W. Slooff, *The Aero- and Hydromechanics of Keel Yachts,*
DOI 10.1007/978-3-319-13275-4_3

Fig. 3.1.1 *Description of a vec-*
tor in an orthogonal coordinate
system

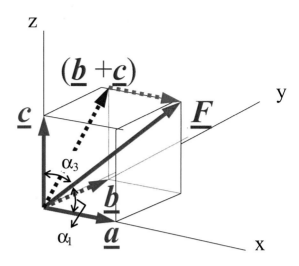

$$\underline{a} + (\underline{b} + \underline{c}) = \underline{F} \qquad (3.1.1)$$

$$\underline{F} - (\underline{b} + \underline{c}) = \underline{a} \qquad (3.1.2)$$

and that the vector $-(\underline{b} + \underline{c})$ is the opposite of $(\underline{b} + \underline{c})$.

The Moment of a Force

A force acting on an object is said to exert a *moment* (*of force*) $\underline{M}$ when it tends to set the object in a rotational motion about some point or axis (about which later). More specifically, the moment of a force $\underline{F}$ with respect to a point P is defined as the product of the force vector and the (shortest) distance d (the 'arm') between the point P and the *supporting line* on which $\underline{F}$ is positioned (See Fig. 3.1.2).

Introducing the distance vector $\underline{d}$, the moment $\underline{M}_P$ of the force vector $\underline{F}$ with respect to the point P is written as the (vector) product

$$\underline{M}_P = \underline{F} \times \underline{d} \qquad (3.1.3)$$

Note that the moment of a force is also a vector. The direction of the moment vector is perpendicular to the plane through the force vector $\underline{F}$ and the moment reference point P. Moment vectors are usually indicated by double or curved arrows. In the particular case of Fig. 3.1.2, where the force vector $\underline{F}$ and the point P are positioned in the x, y-plane, the moment vector is parallel to the z-axis. The magnitude is given by the product $F * d$ (or just Fd).

Fig. 3.1.2 *Defining the moment of a force*

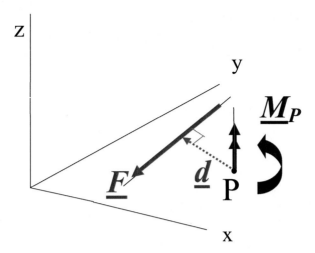

Note that the moment is independent of the position of the force vector $\underline{F}$ along its supporting line and that the supporting line of the moment vector is called the *moment axis*.

As already mentioned earlier, a force vector can be transferred from its supporting line to another, parallel supporting line by the introduction of a moment. For example, the force vector $\underline{F}$ in Fig. 3.1.3 can be replaced by a force vector $\underline{F}'$ in the point P of equal magnitude and direction plus the moment $\underline{M}_p$ of $\underline{F}$ with respect to P.

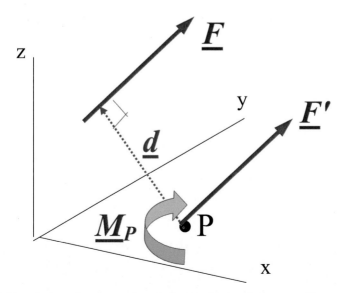

Fig. 3.1.3 *Transferring a force vector through the introduction of a moment of force*

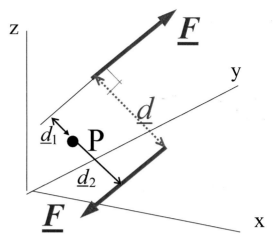

Fig. 3.1.4 *Illustrating a 'couple' (of forces)*

Note that $\underline{M}_P$ and $\underline{F}'$ cannot simply be 'added', because they have different *dimensions*; $\underline{F}'$ has the dimension of a force, while $\underline{M}_P$ has the dimension force∗ length. The equivalence is in the sense that the effect of a force $\underline{F}$ on the motion of an object is the same as the combined effects of the force $\underline{F}$ and the moment $\underline{M}_P$.

Two parallel forces of equal magnitude but opposite direction form a *couple* (*of forces*), (Fig. 3.1.4).

The moment of a couple of forces is equal to the product of the force and the distance between the supporting lines of the forces:

$$\underline{M} = \underline{F} \times \underline{d} \tag{3.1.4}$$

It is equal to the sum of the moments of the individual forces with respect to a common reference point:

$$\underline{M} = \underline{F} \times \underline{d} = \underline{F} \times \underline{d}_1 + \underline{F} \times \underline{d}_2 \tag{3.1.5}$$

Note that this is independent of the position of the common reference point.

Since a moment is a vector it can, similar to the vector $\underline{F}$ in Fig. 3.1.1, be decomposed into components along the axes of a coordinate system (Fig. 3.1.5). If F_x, F_y and F_z are the components of $\underline{F}$ along, respectively, the x-, y- and z-axis, then the components M_x, M_y and M_z of the moment (vector) of $\underline{F}$ with respect to the origin of the coordinate system can be expressed as:

$$M_x = -F_y d_z + F_z d_y$$
$$M_y = -F_x d_z + F_z d_x$$
$$M_z = -F_x d_y + F_y d_x \tag{3.1.6}$$

Fig. 3.1.5 *Illustrating the components of a moment vector*

$$M_x = -F_y d_z + F_z d_y$$
$$M_y = -F_x d_z + F_z d_x$$
$$M_z = -F_x d_y + F_y d_x$$

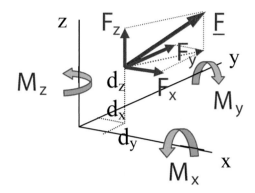

Static and Dynamic Forces and Moments

Forces and moments can be categorized according to the character of the mechanism through which they are generated. They are said to be of a *static* nature if they are independent of the motion of the object with respect to its surroundings. This is the case of a yacht floating in still water with zero boat speed and zero wind.

A force or moment is said to be of a *dynamic* nature if it is the result of the motion of the object with respect to the fluid(s) in which it is immersed. Hence, the forces and moments on a sailing yacht that are caused by the flows of air and water around the sails, hull and appendages are of a dynamic nature. The magnitudes of the dynamic forces depend, as we shall see later, on the relative velocities of air and water with respect to the yacht. In other words the dynamic forces and moments depend on wind speed and boat speed.

Coordinate Systems of Sailing

As indicated above, vectors require a coordinate system for their description. In the aero- and hydromechanics of sailing it is useful and common to distinguish several different coordinate systems. The most important are (Fig. 3.1.6):

- Ship coordinate system (ξ, η, ζ);
  already described in Sect. 2.3, fixed to the boat; ξ-axis longitudinal, in the plane of symmetry; η-axis perpendicular to the plane of symmetry, ζ-axis tilted with the boat. The ξ-η plane is parallel to or coincident with the calm-water level in the floating condition.
- Sailing coordinate system (x_0, y_0, z_0);
  attached to the direction of sailing; x_0-axis in the calm-water surface, parallel to the boat speed vector $\underline{V}_b$, positive in the direction of motion; y_0-axis in the calm-water surface, positive to port; z_0-axis vertical, positive upward.

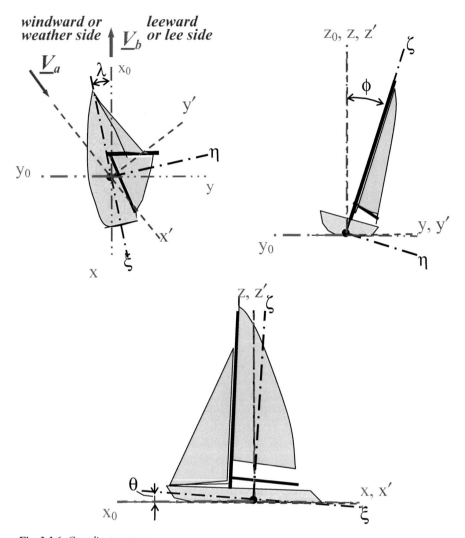

Fig. 3.1.6 *Coordinate systems*

- Hydrodynamic coordinate system (x, y, z);
  like the sailing coordinate system, but x-axis positive downstream and y-axis positive to starboard.
- Aerodynamic coordinate system (x', y', z');
  attached to the incoming wind; x'-axis parallel to the (relative) wind speed vector $\underline{V}_a$, y'-axis parallel to the calm-water surface, z'-axis vertical.

For all coordinate systems the origin is usually taken at amidships in the waterplane or at the most forward point of the waterline.

The side of a yacht that is exposed to the wind is called the weather side, the side in the direction whereto the wind is blowing is called the lee(ward) side. In Fig. 3.1.6 the weather side is the port (left) side.

The sailing coordinate system is used, in general, for the description of the mechanics of sailing performance and ship motion, the hydrodynamic system for the description of hydrodynamic forces and the aerodynamic coordinate system is used for the description of aerodynamic forces. The ship coordinate system is used for the description of boat characteristics and configuration such as centre of gravity, rudder deflection angles, sheeting angles of the sails, etc.

The attitude of a sailing yacht is described usually by means of the angles between the sailing, or the hydrodynamic, and the ship coordinate systems:

- The *angle of leeway* (or drift angle) λ between the x-axis and the intersection of the plane of symmetry (ξ-ζ plane) of the ship with the horizontal (x-y) plane
- The *angle of heel* φ between the vertical (z-)axis and the plane of symmetry of the boat (ξ-ζ) plane
- The *pitch angle* θ between the longitudinal (ξ-)axis of the ship and the horizontal plane

Because the geometry of sailing is symmetric with respect to the $(x_0$-$z_0)$ or (x-z) plane, it is convenient to define the drift angle λ and heel angle φ to be >0 when a yacht drifts and heels to leeward. The pitch angle θ is defined to be >0 when the attitude of the yacht is bow-up.

Under normal sailing conditions the angle of leeway and the pitch angle are usually small, that is smaller than $10°$. The angle of heel can be as much as $30°$.

The transformation of vector quantities from one coordinate system to another can be determined by means of the theory of rotation matrices (http://en.wikipedia.org/wiki/Rotation_matrix).

Ship Motions
The motions of a sailing yacht require also a coordinate system for their description. They are usually described in the coordinate system of sailing. The following types of motion are distinguished (Fig. 3.1.7):

Translational motions:

- *Surge*—longitudinal motion/acceleration, in the direction of the x_0-axis
- *Sway*—moving sideways, in the direction of the y_0-axis
- *Heave*—going up and down along the vertical (z_0-) axis

Rotational motions:

- *Roll*—rotation around the x_0-axis
- *Pitch*—rotation around the y_0-axis
- *Yaw*—rotation around the z_0-axis

The roll angle of a yacht is (about) equal to the instantaneous angle of heel during a rolling motion. The yaw angle (ψ) is (about) equal to the instantaneous angle of leeway during a yawing motion.

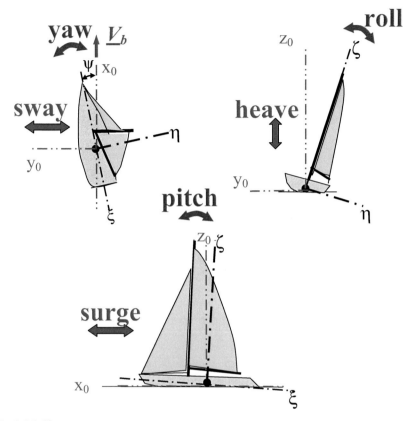

Fig. 3.1.7 *Ship motions*

Rotational motion can, like a moment of force, also be represented by a vector. Such a vector of rotation is directed along the axis of rotation. Its length is a measure of the angular velocity of the rotation.

Steady and Unsteady, Periodic and Time-Averaged Conditions
When the forces, moments and motion of a sailing yacht are constant, that is they do not vary with time, the conditions (of sailing) are said to be *steady*. When they do vary with time, which, in general, is the case for a sailing yacht, the conditions are said to be *unsteady*. When the unsteady moments and associated forces and moments repeat themselves in time in some regular way they are said to be *periodic* (Fig. 3.1.8). This is the situation of a yacht sailing in regular waves.

When the forces and motions of a sailing yacht are periodic it is possible and convenient to define *time-averaged* values of velocities, forces, etc. by averaging over a sufficiently large amount of time (Fig. 3.1.8). The time-averaged conditions can then be dealt with as in steady conditions. This is convenient because steady

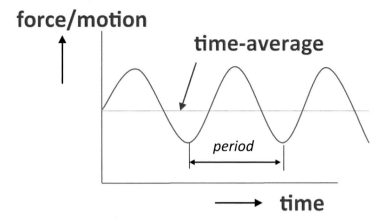

Fig. 3.1.8 *Periodic motion and time average*

conditions are much easier to deal with in sailing yacht mechanics than time-dependent conditions.

3.2   Forces Under Water

The forces and moments acting on the submerged parts of the hull and appendages of a sailing yacht are usually defined in the hydrodynamic (x, y, z) or the sailing (x_0, y_0, z_0) coordinate system. They are schematically depicted in Figs. 3.2.1, 3.2.4 and 3.2.5. For a yacht of given shape and size, the direction of the total hydrodynamic force vector $\underline{F}_H$ is determined primarily by the angle of leeway λ and the angle of

Fig. 3.2.1 *Hydrodynamic forces and moments in the horizontal plane*

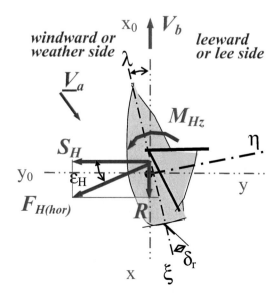

Fig. 3.2.2 *Variation (qualitatively) of side*
force and resistance with angle of leeway

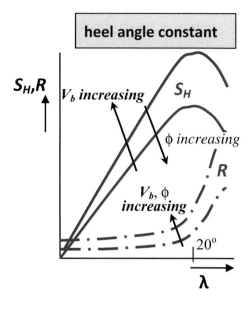

heel φ. The direction of $\underline{F}_H$ is usually such that it points to weather (windward). Its magnitude is mainly determined by the boat speed V_b.

In the horizontal (x, y) plane we have (Fig. 3.2.1):

- the horizontal component $F_{H(hor)}$ of the total hydrodynamic force
- the hydrodynamic resistance R, directed along the x-axis, that is opposite to the boat speed vector $\underline{V}_b$
- the hydrodynamic side force S_H, in the y-direction, that is perpendicular to the direction of sailing
- the hydrodynamic yawing moment M_{Hz} around the vertical (z-) axis.

Note that R and S_H are components of $F_{H(hor)}$ and are also positioned in a horizontal plane.

The yawing moment M_{Hz} is caused by $F_{H(hor)}$ and the distance between its supporting line and the center of gravity. It is strongly influenced by leeway and the deflection angle δ_r of the rudder. Its orientation is usually such that it tends to turn the boat to weather.

Because of the symmetry of sailing with respect to the (x, z) plane it is convenient to define S_H as >0 when it points to windward and M_{Hz} as >0 when it tends to turn the bow to windward (in other words: when it tends to increase leeway). The resistance R is considered >0 in the streamwise (x-) direction.

Figure 3.2.2 shows, qualitatively, how the side force and resistance vary with the angle of leeway for a constant heel angle. Note that for small angles of leeway, that is under normal sailing conditions, the side force S_H increases almost linearly with the angle of leeway λ and that it levels off and attains a maximum for some larger

*Fig. 3.2.3 Variation (qualitatively)
of hydrodynamic drag angle with angle
of leeway*

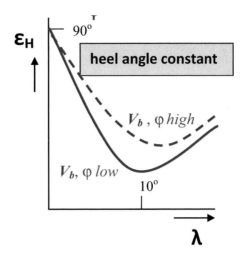

value of λ (usually somewhere around 20°). Note also that the slope of the curve increases with boat speed. As we will see later (Chap. 6), the slope decreases with increasing angle of heel.

The resistance **R** is usually much smaller than the side force. It also increases with leeway, but less rapidly so, at least initially, and continues to increase (more rapidly) beyond the angle of maximum side force. The level of the resistance increases with boat speed. It also increases with heel angle, as we will see in Chap. 6.

We will also see in Chap. 6 that the precise shapes and values of both the side force and resistance curves depend strongly on the configuration of the hull and the appendages, the latter in particular.

The angle ε_H between the side force S_H and the total horizontal force $F_{H(hor)}$ (Fig. 3.2.1) is called the *hydrodynamic drag (or resistance) angle.* It is defined through its (co)tangent:

$$\tan \varepsilon_H = R/S_H \tag{3.2.1a}$$

or

$$\cot an \varepsilon_H = S_H/R \tag{3.2.1b}$$

As illustrated by Fig. 3.2.3, the hydrodynamic drag angle varies strongly with the angle of leeway. To a lesser extent it is also a function of boat speed, in particular at high boat speeds, and of the angle of heel. Note that ε_H is 90° at zero leeway (because the side force **S** is zero at zero leeway) and that it has a minimum at some value of λ (usually somewhere around 10°). As we will see later (Chap. 6) this minimum and the corresponding angle of leeway are determined mainly by the configuration of the hull and appendages. We will also see later (Chap. 4) that ε_H is an important measure of the hydrodynamic efficiency of a sailing yacht.

Fig. 3.2.4 *Forces under water in the lateral plane*

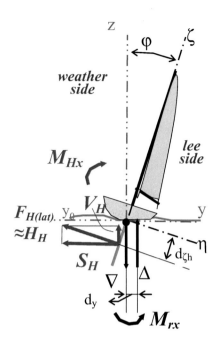

Figure 3.2.4 illustrates the forces and moments in the lateral (y, z-) plane. Here we have, in addition to the hydrodynamic side force S_H:

- the lateral component $F_{H(lat)}$ of the total hydrodynamic force
- the vertical component V_H of the hydrodynamic force (>0 upward)
- the hydrodynamic heeling moment M_{Hx} around the longitudinal axis, caused by $F_{H(lat)}$, defined as >0 when it tends to increase heel
- the gravitational force or weight ∇, acting downwards along the z-axis in the center of gravity
- the hydrostatic or buoyancy force Δ, acting upwards in the direction of the z-axis
- the lateral righting moment M_{rx}, caused by the non-alignment of $\underline{\nabla}$ and $\underline{\Delta}$, defined >0 when it tends to reduce heel

As we will see later, $F_{H(lat)}$ and its components S_H and V_H are generated mainly by keel and rudder. The direction of $F_{H(lat)}$ is therefore, roughly, perpendicular to the keel. This implies that for the heeling force H_H there holds

$$H_H \approx F_{H(lat)} \tag{3.2.2}$$

and that

$$S_H \approx H_H \cos\varphi \tag{3.2.3}$$

and

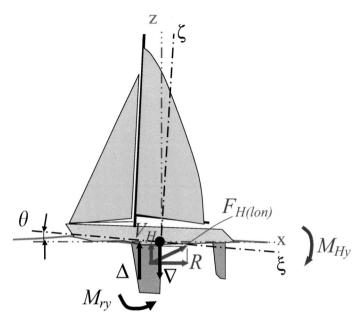

Fig. 3.2.5 *Forces under water in the longitudinal plane*

$$V_H \approx H_H \sin \varphi \tag{3.2.4}$$

The hydrodynamic heeling moment M_{Hx} around the longitudinal axis depends on the distance $d_{\varsigma h}$ between the point of application of the heeling force H_H and the ξ-η plane. It follows that

$$M_{Hx} \approx H_H d_{\varsigma h} \tag{3.2.5}$$

The point of application of the buoyancy force $\varDelta$ moves outboard when the boat is heeling. The lateral righting moment M_{rx} is caused by the non-alignment of the vectors $\underline{V}$ and $\underline{\varDelta}$ when the boat is under heel. It is given by

$$M_{rx} = \varDelta\, d_y \tag{3.2.6}$$

where d_y is the distance between the weight and buoyancy vectors.

In the longitudinal (x, z-) plane (Fig. 3.2.5) we have, in addition to the resistance R, the vertical hydrodynamic force V_H, the gravitational force V and the hydrostatic force $\varDelta$ described above, the following forces and moments:

- the total longitudinal component $F_{H(lon)}$ of the hydrodynamic force
- the hydrodynamic pitching moment M_{Hy}
- the longitudinal righting moment M_{ry},

The hydrodynamic pitching moment M_{Hy} is caused by the fact that the supporting line of $F_{H(lon)}$ does, in general, not pass through the origin of the ships axis system. Note that with the point of application of $F_{H(lon)}$ below the water surface, the hydrodynamic pitching moment M_{Hy} is 'bow down' (<0). The longitudinal righting moment M_{ry} is, similar to the lateral righting moment, caused by the non-alignment of the vectors ∇ and Δ when the boat is subject to an angle of pitch θ. In general, the orientation of M_{ry} is 'bow up' (<0) when the pitch attitude of the boat is 'bow down' ($\theta < 0$) and vice versa.

The physical mechanisms underlying the hydrostatic and hydrodynamic forces and moments will be discussed in Chaps. 5 and 6.

3.3   Wind Triangle

Before proceeding to the (aerodynamic) forces acting on the rig, sails, and the part of the hull of a sailing yacht that is exposed to the wind, it is useful and desirable to consider the relative motions of air and ship in some detail. Figure 3.3.1 serves to illustrate this purpose.

First of all we note that any vehicle moving with a velocity $\underline{V}_b$ in still air creates its own wind: it experiences an air stream with a velocity $-\underline{V}_b$ of equal magnitude

Fig. 3.3.1 *Wind triangle*

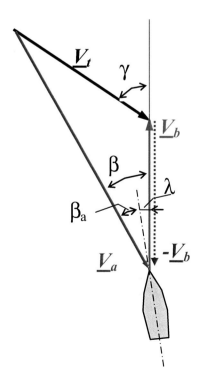

but direction opposite to $\underline{V}_b$. When the air stream itself is moving with respect to the surface of the earth, say with a (true) wind speed $\underline{V}_t$, then the resulting (apparent) wind speed $\underline{V}_a$ felt by the vehicle is the vector sum of $\underline{V}_t$ and $-\underline{V}_b$. Mathematically this is expressed as:

$$\underline{V}_a = \underline{V}_t + (-\underline{V}_b) = \underline{V}_t - \underline{V}_b \tag{3.3.1}$$

In the terminology of sailing V_t is the *true wind speed* and V_a is the *apparent wind speed*.

The *wind triangle* of Fig. 3.3.1 contains a number of characteristic angles:

The *true wind angle* γ is the angle between the boat speed vector and the true wind vector. The *apparent wind angle* β is the angle between the boat speed vector and the apparent wind vector. The *apparent heading* (angle) $β_a$ is the angle between the apparent wind speed vector and the longitudinal axis of the boat. As before, λ is the angle of leeway.

Figure 3.3.1 implies that the following relation holds between the apparent wind angle, the apparent heading and the angle of leeway:

$$β = β_a + λ \tag{3.3.2}$$

As indicated earlier, the hydrodynamic forces and moments are primarily determined by boat speed V_b and the leeway angle λ. The aerodynamic forces and moments are a function of the apparent wind speed V_a and the apparent heading $β_a$. The apparent wind angle β is determined by both aerodynamics and hydrodynamics.

In the art of sailing as expressed in terms of the wind triangle, the true wind vector $\underline{V}_t$ is an uncontrolled variable and the apparent heading $β_a$ is the only parameter that is under direct control of the yachtsman. For any given yacht configuration all the other quantities, i.e. apparent wind speed V_a, true wind angle γ, boat speed V_b and angle of leeway λ depend on V_t and $β_a$. How precisely depends on the type of yacht and its 'trim'. We can, however, obtain a qualitative impression of the dependence of the apparent wind speed and true wind angle on apparent wind angle and true wind speed without knowing the precise characteristics of the yacht.

For this purpose we consider trigonometric relations which can be derived from the wind triangle:

$$V_a/V_t = \cos(γ - β) + (V_b/V_t)\cos β \tag{3.3.3}$$

$$\tan γ = \{1 + (V_b/V_t)/\cos γ\}\tan β \tag{3.3.4}$$

From these two equations it is possible to calculate V_a/V_t and γ as a function of β for chosen values of V_b/V_t. The results of such calculations are presented in Fig. 3.3.2 for V_a/V_t and in Fig. 3.3.3 for γ.

Figure 3.3.2 tells us what we all know: the apparent wind speed decreases when the apparent wind angle is increased. It also decreases more strongly when the boat

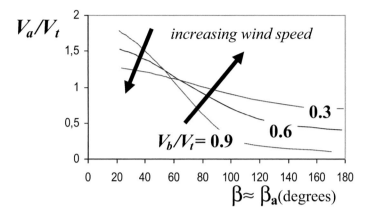

Fig. 3.3.2 *From the wind triangle: apparent wind speed V_a as a function of apparent wind angle β*

speed/wind speed ratio V_b/V_t is high. Note that when boat speed and true wind speed would be equal ($V_b/V_t = 1$) and the apparent wind angle 180°, the apparent wind speed would be zero. (This, of course, never happens in practice).

The figure also indicates the tendency of the relation with increasing true wind speed. Because, as we will see later, boat speed increases at a lower rate than true wind speed, the ratio V_b/V_t decreases with increasing (true) wind speed.

It is further useful to note that for apparent wind angles around 60° the apparent wind speed is almost independent of boat speed. This is due to the fact that, for β=60° and a usual range of boat speeds, the true wind angle γ is close to 90° (Fig. 3.3.3). This is when the true wind and boat speed vectors are (approximately) at right angles.

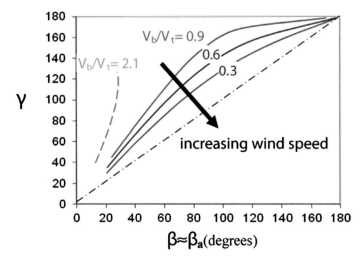

Fig. 3.3.3 *From the wind triangle: true wind angle γ as a function of apparent wind angle β*

Under normal conditions of sailing the angle of leeway λ is usually much smaller than the apparent wind angle β. Hence, it follows from Fig. 3.3.1 and Eq. (3.3.2) that equating β and β_a then implies only a modest error.

Figure 3.3.3 illustrates another phenomenon that all yachtsmen are familiar with: we can *track* closer to the direction of the true wind when the (true) wind speed increases. In terms of the figure it means that, for a given apparent wind angle β, the true wind angle γ decreases with decreasing V_b/V_t ratio. In the limiting case of $V_b/V_t \rightarrow 0$ the true wind angle γ becomes equal to the apparent wind angle β. Still another phenomenon reflected by Fig. 3.3.3 is that for very high boat speeds the apparent wind angle is always small, for almost all true wind angles (see the line for $V_b/V_t = 2.1$).

It is emphasized that Figs. 3.3.2 and 3.3.3 illustrate general trends only. They do not represent the behavior of any specific type of boat. The reason is, as we will see later, that boat speed, at a given true wind speed, varies with the apparent wind angle in a specific manner that depends on the type of yacht and its configuration. In addition, yachts cannot sail at apparent wind angles below a certain minimum value that is also dependent on the type of yacht.

For future reference it is further useful to note that in the terminology of sailing a number of notions is associated with sailing in certain ranges of direction relative to the wind:

- *Upwind sailing* is understood as sailing at true wind angles $<90°$
- *Downwind sailing* means true wind angles $>90°$
- *Sailing to windward* is understood as sailing in the direction of the true wind
- *Sailing close-hauled* is generally understood as sailing at small apparent wind angles (say $<45°$)
- *Reaching* is generally understood as sailing at apparent wind angles between, say, 45–135°, with *close-reaching* indicating the lower half of this range (45–90°), *beam reaching* meaning apparent wind angles of about 90° and *broad-reaching* indicating the upper half (90–135°) of the range
- *Running* means sailing in approximately the same direction as that where the wind comes from, i.e. (true) wind angles around 180°

In the daily way of speech the numbers in the definitions given above are not always strictly adhered to. It is often not even clear whether the wind angles that are referred to should be true or apparent wind angles. For qualitative discussion this does, however, not matter too much. When it does matter we will be more specific in this book.

3.4   Forces and Moments Above the Water Surface

The forces acting on a sailing yacht above the water surface are of an aerodynamic nature only. Because the density of the air inside the hull and other hollow parts of the yacht is equal to the density of the ambient air there are no net aerostatic forces.

The aerodynamic forces acting on a sailing yacht are, of course, mainly due to the sails. However, the effect of the hull, as we will see later, is not negligible.

As already mentioned in the preceding section, the aerodynamic forces on a sailing yacht are primarily determined by the apparent heading angle β_a and the apparent wind speed V_a. Other important parameters are the sheeting angle δ, (or rather sheeting angles, because mainsail and headsail will, in general, have different sheeting angles) and the angle of heel φ.

It is useful, at this point, to introduce the notion of *angle of attack*, usually denoted by the Greek symbol α. This is defined as the angle between the foot of a sail and the apparent wind vector. Since most sailing yachts have a mainsail and a headsail, each has, in principle, its own angle of attack. However, we will, at this stage, for convenience of simplicity, assume that the angles of attack of the head sail and the mainsail are coupled in the sense that there is a specific angle of attack of the head sail connected to each angle of attack of the mainsail. The latter is usually defined as the angle between the boom and the apparent wind vector.

As shown in Fig. 3.4.1 it follows that:

$$\beta_a = \alpha + \delta \qquad\qquad (3.4.1)$$

where δ is the sheeting angle of the mainsail. Note that, with the angle of attack of the head sail coupled to the angle of attack of the mainsail, the sheeting angle of the head sail is also coupled to that of the mainsail.

Then, more precisely, the direction of the total aerodynamic force vector $\underline{F}_A$ is primarily determined by the angle of attack α and the angle of heel φ. The magnitude of the aerodynamic forces is mainly determined by the apparent wind speed V_a.

The aerodynamic forces and moments in the horizontal (x, y) plane are schematically shown in Fig. 3.4.1.

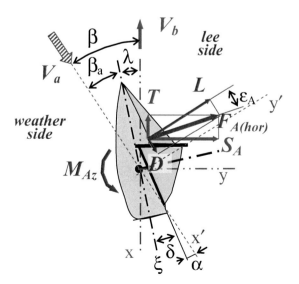

Fig. 3.4.1 *Aerodynamic forces and moments in the horizontal plane*

The following components are distinguished:

- the horizontal component $F_{A(hor)}$ of the total aerodynamic force F_A
- the aerodynamic *lift* force L, that is the component of $F_{A(hor)}$ perpendicular to the apparent wind speed vector V_a (i.e. in the direction of the y'-axis), defined as >0 when, as usual, it points to lee
- the aerodynamic *drag* D, that is the component of $F_{A(hor)}$ in the direction of the apparent wind speed vector V_a, (i.e. in the direction of the x'-axis)
- the aerodynamic yawing moment M_{Az} around the vertical (z- or z'-) axis, defined as >0 when it tends to turn the boat to weather

Note that L and D, like $F_{A(hor)}$, are located in a horizontal plane.

The aerodynamic yawing moment M_{Az} is caused by $F_{A(hor)}$ and the distance between its supporting line and the origin of the coordinate systems. For small apparent wind angles and small angles of heel its orientation is usually such that it tends to turn the boat to lee. When the apparent wind angle increases, M_{Az} tends to turn the yacht less to lee, or more to weather,. This is also the case with increasing angle of heel.

The dependence of the aerodynamic lift and drag of sails on angle of attack for two levels of the apparent wind speed is, qualitatively, shown by Fig. 3.4.2. The aerodynamic characteristics of two different sail configurations are depicted, one with an 'ordinary' (jib/genoa) type headsail and one with a spinnaker type of headsail.

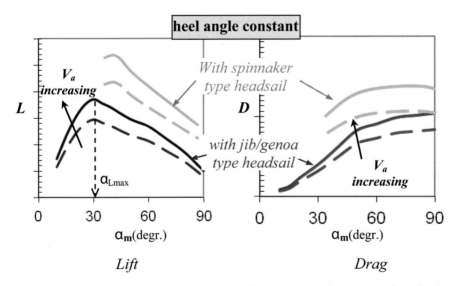

Fig. 3.4.2 *Variation (qualitatively) of aerodynamic lift L and drag D of sails (mainsail plus headsail) with angle of attack α_m of the mainsail*

The curves can be considered as 'typical' in the sense that they reflect an average of experimental data available to the author. The scales of the lift and drag figures are not indicated explicitly but are the same.

While the variation of side force and resistance of the underwater body is of interest mainly for small angles of leeway, the aerodynamics of the sails are important for the full range of angles of attack up to 90°. For small angles of attack the aerodynamic behaviour of the jib/genoa configuration is similar to the dependence of the hydrodynamic side force and resistance on angle of leeway and boat speed (Fig. 3.2.2). Note however, that, already for small angles of attack, the variation of the aerodynamic lift with α is much less linear than the variation of the hydrodynamic side force with leeway but that it also levels off and attains a maximum for some value of α (usually somewhere around 25–40°). For very large angles of attack the aerodynamic lift goes down again and the drag attains its maximum when α approaches 90°.

It is emphasized that the precise shape and values of both the lift and drag curves depend strongly on the configuration of the sails. For spinnaker type headsails lift and drag are usually substantially larger than for an ordinary headsail but they can be operated only at high angles of attack.The level and slope of the lift curve as well as the drag curve increase with apparent wind speed. The dependence of the aerodynamic forces on the angle of heel is, as we will see later, similar to that of the hydrodynamic forces. That is, the lift decreases but the drag increases with increasing heel angle.

The angle ε_A between the aerodynamic lift force L and the total horizontal aerodynamic force $F_{A(hor)}$ (Fig. 3.4.1) is called the *aerodynamic drag angle*. Like the hydrodynamic drag angle it is defined through its (co)tangent:

$$\tan \varepsilon_A = D/L \tag{3.4.2a}$$

or

$$\cot an \varepsilon_A = L/D \tag{3.4.2b}$$

It is an important measure of the aerodynamic efficiency of a sailing yacht. For a given sail configuration ε_A depends mainly on the angle of attack α. Its variation with α (Fig. 3.4.3) is qualitatively similar to the variation of the hydrodynamic drag angle ε_H with the angle of leeway λ (Fig. 3.2.3). For ordinary sail configurations the angle of attack for the minimum drag angle is of the order of 20°. The minimum drag angle itself is usually somewhere around 10–15°. For spinnaker type sails the drag angle at moderate angles of attack is appreciably higher than for ordinary sails.

For all types of sail the aerodynamic effectiveness and efficiency, like the hydrodynamic efficiency of keels, go down when the heel angle increases (Chap. 7).

For the purpose of considering sailing performance it is convenient to decompose the total horizontal aerodynamic force into components T and H that, respectively,

Fig. 3.4.3 *Variation (qualitatively) of aerodynamic drag angle with angle of attack*

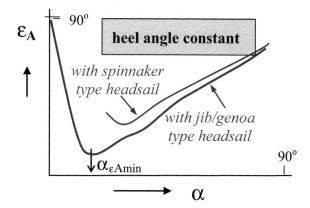

are parallel with and perpendicular to the boat speed vector (see Fig. 3.4.1). The component T in the direction of sailing is called the *driving force* or *thrust*. The lateral component S_A is called the aerodynamic side force. T and S_A are, of course, related to the aerodynamic lift L and drag D. These relations can be written as

$$T = L \sin\beta - D \cos\beta = F_{A(hor)} \sin(\beta - \varepsilon_A) \tag{3.4.3}$$

and

$$S_A = L \cos\beta + D \sin\beta = F_{A(hor)} \cos(\beta - \varepsilon_A) \tag{3.4.4}$$

Note that, while L and D are purely aerodynamic quantities, T and S_A depend also on the hydrodynamics of hull and appendages. In terms of Eqs. (3.4.3) and (3.4.4) the hydrodynamic dependence comes in through the apparent wind angle β, which is a function of the angle of leeway λ (see Eq. (3.3.2)).

Figure 3.4.4 illustrates the aerodynamic forces and moments in the lateral (y, z-) plane. Here we have, in addition to the aerodynamic side force S_A,

- the total lateral aerodynamic force or heeling force $F_{A(lat)}$
- the vertical component V_A
- the aerodynamic heeling moment M_{Ax}

Because the aerodynamic force is generated mainly by the sails, the direction of the total lateral force $F_{A(lat)}$ is roughly perpendicular to the top (ζ-) axis of the boat. This force component is generally identified as the heeling force H_A. It follows, see Fig. 3.4.4, that

$$S_A \approx H_A \cos\varphi \tag{3.4.5}$$

and

$$V_A \approx -H_A \sin\varphi \tag{3.4.6}$$

Fig. 3.4.4 *Aerodynamic forces in the lateral plane*

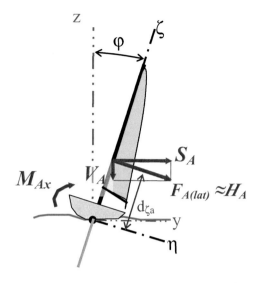

Fig. 3.4.5 *Aerodynamic forces in the longitudinal plane*

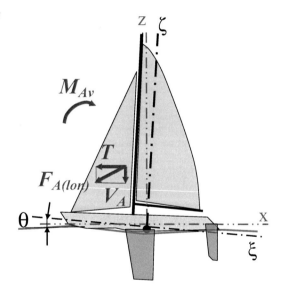

The aerodynamic heeling moment M_{Ax} is caused by H_A and the distance between its supporting line and the ξ-η plane. It follows that

$$M_{Ax} \approx H_A d_{\zeta a}$$ (3.4.7)

M_{Ax} is defined >0 when it tends to increase heel.

Figure 3.4.5 presents the aerodynamic forces and moments in the longitudinal plane.
    In this, (x, z), plane we have, in addition to the driving force T and the vertical aerodynamic force V_A:

* the total longitudinal component $F_{A(lon)}$ of the aerodynamic force
* the aerodynamic pitching moment M_{Ay}

M_{Ay} is defined >0 when it tends to rotate the boat bow-up. However, its orientation is usually bow-down, i.e. <0, due to the orientation of T and V_A.
    The physical mechanisms underlying the various aerodynamic forces and moments are discussed in more detail in Chaps. 5 and 7.

3.5 Newton's Law

The forces acting on an object and its resulting motion are coupled through the fundamental law of classical mechanics: Newton's law. What it says is that when an object of mass m is subject to a force F it will undergo an acceleration a of magnitude F/m in the same direction as the direction of the force (Fig. 3.5.1).
    In vector notation this can be written as

$$\underline{F} = m * \underline{a} \qquad (3.5.1)$$

The multiplication operator $*$ has been inserted here for clarity. When considering the product of two quantities it will usually be omitted in this book.
    The product $m * \underline{a}$, or rather $-m * \underline{a}$, is sometimes called *inertia force*. It expresses the virtual force that an object experiences when it accelerates or decelerates.
    When an object is subject to a force during a certain period of time Δt it experiences a change in velocity $\Delta \underline{V}$ that follows from multiplying both sides of Eq. (3.5.1) with Δt:

$$\underline{F} * \Delta t = m * \underline{a} * \Delta t = m * \Delta \underline{V} \qquad (3.5.2)$$

where

$$\Delta \underline{V} = \underline{a} * \Delta t \qquad (3.5.3)$$

Fig. 3.5.1 *Illustrating Newton's law*

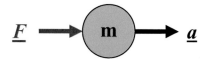

Fig. 3.5.2 *Illustrating the 'rotational version'*
of Newton's law

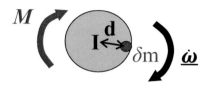

An important consequence of Newton's law is that when an object, like a sailing yacht, moves with a velocity of constant magnitude and direction, the net force acting on it is zero.

In many situations, including sailing, a 'rotational version' of Newton's law, expressing the relation between the moment of force acting on an object and its rotational motion is of equal importance (See Fig. 3.5.2). It can be derived from Eq. (3.5.1) that when an object is subject to a moment (of force) $\underline{M}$, it experiences an acceleration $\underline{\dot{\omega}}$ of angular velocity that is proportional to the magnitude of the moment, where the orientation of the angular acceleration is the same as that of the moment.

This can be written as

$$\underline{M} = I * \underline{\dot{\omega}} \tag{3.5.4}$$

The proportionality factor I is the mass moment of inertia of the object about the center of rotation. For a sailing yacht the latter is, in general, close to the center of gravity. We recall from Sect. 2.5 that a mass moment of inertia of an object like a sailing yacht is a measure of the eccentricity, in a certain direction, of the distribution of mass. In terms of Fig. 3.5.2 it can be expressed as

$$I = \Sigma \, (\delta m \, d^2) \tag{3.5.5}$$

where Σ represents the summation, over all mass elements 'δm' of the object, of the product of a mass element with the square of its distance 'd' from the axis of rotation (See also Sect. 2.4).

In general, a moment of inertia is defined with respect to an axis of a coordinate system. This implies that we should be a little more specific than Eq. (3.5.5). For example, if $\underline{M}$ is the moment of force around the x-axis (M_x), the moment of inertia should also be with respect to the x-axis. Indicating the latter as $I_{xx,}$ a more specific form of Eq. (3.5.4) is

$$M_x = I_{xx} * \dot{\omega}_x \tag{3.5.6}$$

where $\dot{\omega}_x$ is the acceleration of angular motion around the x-axis. Similar expressions hold, of course, for other axes.

An important consequence of Eq. (3.5.6) is that if the net moment of force acting on an object is zero, the object will not experience any acceleration or deceleration of angular motion. In other words the object will then rotate with a constant angular velocity (for example zero).

It is finally noted that Newton's law holds also in time-averaged situations (Sect. 3.1), as long as the variations with time of forces, moments and motions are sufficiently small. This means that we may use time-averaged values for forces, moments, velocity and acceleration in Eqs. (3.5.1) and (3.5.6).

Reference

http://en.wikipedia.org/wiki/Rotation_matrix

Chapter 4
Sailing: Basic Mechanics

4.1 Equilibrium of Forces and Moments

As we have seen in the preceding section Newton's law implies that the net forces
and moments acting on a sailing yacht in steady motion are zero. In other words:
the gravitational, hydrostatic, hydrodynamic and aerodynamic forces and moments
acting on a yacht that is sailing at a constant speed, in a fixed direction, in steady
wind and sea conditions, are in equilibrium. When the forces are not in equilibrium
the boat will accelerate or decelerate, change course and/or attitude until a (new)
equilibrium condition is established.

The process of obtaining equilibrium involves so-called independent and dependent variables. The *independent variables* are those over which the yachtsman has
no control at all such as wind and sea state and those that are set by the crew such
as sail configuration, sheeting angles and other trimming parameters and apparent
heading. The *dependent variables* are those that take their values as the result of
the equilibrating process, such as boat speed, angles of leeway, heel and pitch, over
which the yachtsman has no direct control. In steady sailing conditions, the rudder
angle, although controlled by the helmsman, is usually also considered as a dependent variable because it is set so as to realize directional equilibrium of moment in
the horizontal plane.

Equilibrium of Forces and Moments in the Horizontal Plane
Figure 4.1.1 illustrates the equilibrium of forces and moments in the horizontal
plane.

Equilibrium of force requires that the thrust T of the sails is equal in magnitude
but opposite in direction to the resistance R and that the same applies to the (aerodynamic) side force S_A and the hydrodynamic side force S_H.

This can be expressed as

$$T = R \qquad\qquad (4.1.1)$$

© Springer International Publishing Switzerland 2015
J. W. Slooff, *The Aero- and Hydromechanics of Keel Yachts,*
DOI 10.1007/978-3-319-13275-4_4

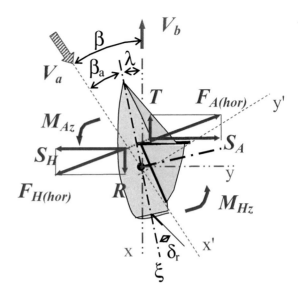

and

$$S_A = S_H \tag{4.1.2}$$

which also implies

$$F_{A(hor)} = F_{H(hor)} \tag{4.1.3}$$

Equilibrium of moment implies

$$M_{Az} + M_{Hz} = 0 \tag{4.1.4}$$

Note that the latter requires that the total horizontal components $F_{A(hor)}$ and $F_{H(hor)}$ of the aerodynamic and hydrodynamic forces are positioned on one and the same supporting line, as in Fig. 4.1.1. Note also that the net yawing moment, which is zero at equilibrium, is independent of the position of the centre of gravity; it depends only on the couple formed by $F_{A(hor)}$ and $F_{H(hor)}$ (or the couples formed by S_A and S_H, T and R). The condition (4.1.4) is controlled by the yachtsman through the trimming of the sails, determining M_{Az} and, in particular, the amount of helm that is the deflection (δ_r) of the rudder, which controls M_{Hz}.

Of the dependent variables of sailing, boat speed V_b and the angle of leeway λ are mainly determined by the conditions (4.1.1) and (4.1.2) of equilibrium in the horizontal plane. For a given wind speed, apparent heading β_a, sail configuration and sheeting angles, boat speed is mainly determined by the condition (4.1.1) of longitudinal equilibrium. The mechanism is illustrated by Fig. 4.1.2.

Fig. 4.1.2 *Variation (qualitatively) of thrust and resistance as a function of boat speed (upwind conditions)*

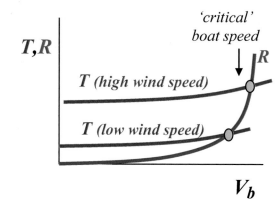

Figure 4.1.2 presents a qualitative picture of the variation of the hydrodynamic resistance **R** and the aerodynamic propulsive force (or thrust) **T** with boat speed. For conventional *displacement yachts*[1] a typical characteristic of the resistance curve is that beyond a certain 'critical' boat speed the hydrodynamic resistance increases much faster with boat speed than at low values of boat speeds (the mechanism of this phenomenon is described in Sect. 6.5). On the other hand, the aerodynamic thrust of the sails, for given apparent heading and sheeting angles, depends primarily on the apparent wind speed. As we have seen in Sect. 3.3, the apparent wind speed, for upwind sailing conditions, increases with boat speed (Fig. 3.3.2), but at a much lower rate than the hydrodynamic resistance. As a consequence the driving force **T** increases also much less rapidly with boat speed than the resistance **R**. (Note that for downwind sailing the apparent wind speed and, consequently, the driving force, decrease when boat speed is increased (also illustrated by Fig. 3.3.2)).

Equilibrium conditions of sailing are established when the thrust and resistance curves cross, that is, when **T** equals **R**. For high levels of (true) wind speed the level of the thrust **T** and the resulting boat speed are, of course, higher than at low wind speed. However, as illustrated by Fig. 4.1.2, the increase of boat speed as a result of an increase of wind speed is much smaller than proportional due to the rapid increase with boat speed of the hydrodynamic resistance. This phenomenon is typical for all conventional displacement yachts and a major determining factor in sailing yacht performance.

Another significant factor for boat speed is the angle of heel. As mentioned in Chap. 3, the lift and, hence, the driving force **T** of the sails decreases and the hydrodynamic resistance **R** increases with heel angle. As a consequence the longitudinal equilibrium of forces results in a lower boat speed when the heel angle is large.

As mentioned above, a second dependent variable that is mainly determined by the conditions of equilibrium in the horizontal plane is the angle of leeway λ. In this case the equilibrium of lateral forces, Eq. (4.1.2), is the governing condition. The mechanism is illustrated by Fig. 4.1.3.

---

[1] Yachts with a moderate to high displacement/length ratio.

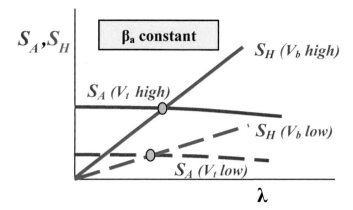

Fig. 4.1.3 *Variation (qualitatively) of aerodynamic and hydrodynamic side force as a function of leeway angle*

The hydrodynamic side force S_H is, as mentioned in Sect. 3.2, a function of, primarily, the angle of leeway and boat speed (that is, for a given heel angle). Under normal sailing conditions, when the angle of leeway is small, the dependence on leeway is practically linear (Fig. 4.1.3). The rate of increase depends on boat speed. At high boat speed the side force varies more strongly with leeway than at low boat speeds.

The aerodynamic side force S_A is, for a fixed heel angle, a function of, primarily, the apparent heading β_a, the sheeting angles δ and the apparent wind speed V_a. It is only weakly dependent on the angle of leeway.

Equilibrium conditions ($S_H = S_A$) are indicated in Fig. 4.1.3 for two situations: high wind speed (V_t)/high boat speed (V_b) and low wind speed/low boat speed. Note that, in general, the equilibrium angle of leeway varies much less than proportional with the aerodynamic heeling force. This is caused by the fact that a larger heeling force, due to, for example, a higher wind speed, is usually accompanied by a larger propulsive force and, hence, a higher boat speed. As a result the equilibrium angle of leeway at the higher wind speed is much smaller than it would be if boat speed would not have increased. Note also that the latter (no or little increase of boat speed) is the case when the yacht has attained its 'critical' boat speed beyond which the strong increase of hydrodynamic resistance (Fig. 4.1.2) prohibits a further significant increase of boat speed. In such conditions the angle of leeway increases much more strongly with increasing wind speed.

The effect of heel on leeway is of secondary importance. This is due to the fact that both the aerodynamic side force S_A and the hydrodynamic side force S_H decrease when the angle of heel increases.

Equilibrium of Forces and Moments in the Lateral Plane
The equilibrium of forces and moments in the lateral (vertical) plane is illustrated by Fig. 4.1.4. In addition to the condition (4.1.2) of the equilibrium of the hydrodynamic side force S_H and the aerodynamic side force S_A we have the equilibrium of vertical forces:

Fig. 4.1.4 *Equilibrium of forces and moments in the lateral (vertical) plane*

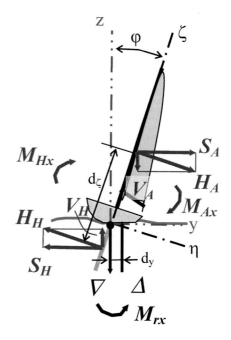

$$V_H + V_A + \varDelta = \nabla \qquad\qquad (4.1.5)$$

and the equilibrium of moment around the longitudinal (x−) axis:

$$M_{Hx} + M_{Ax} = M_{rx} \qquad\qquad (4.1.6)$$

The condition (4.1.5) of equilibrium of vertical forces governs the *sinkage* of the yacht. That is, for a given gravitational force or weight ∇, a difference between the magnitudes of V_A and V_H is balanced by a change of the (hydrostatic) buoyancy force or displacement $\varDelta$. When, with the yacht under heel, the absolute value of the vertical aerodynamic force V_A (<0) is larger than the vertical hydrodynamic force V_H, the yacht is pushed deeper into the water and the buoyancy force $\varDelta$ increases, and the other way around. However, under most sailing conditions, there is only a small difference between the magnitudes of V_A and V_H so that $\varDelta \approx \nabla$. At zero wind speed and zero boat speed (floating conditions) V_A and V_H are both zero so that Eq. (4.5) reduces to $\varDelta \approx \nabla$ precisely.

The condition (4.1.6) of equilibrium of lateral moment governs the angle of heel of a yacht. Figure 4.1.5 illustrates the mechanism. It shows (qualitatively) the heeling/rolling moments M_{Ax}, M_{Hx} and righting moment M_{rx} as a function of the angle of heel φ for a given apparent wind angle β and two different wind speeds.

As already mentioned before, the total lateral aerodynamic and hydrodynamic forces $F_{A(lat)}$ and $F_{H(lat)}$ are approximately perpendicular to the mast and keel of the boat (Sects. 3.2, 3.4). Therefore, the total heeling moment

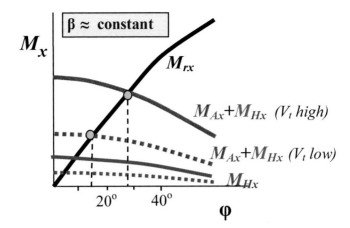

Fig. 4.1.5 *Variation of heeling and righting moments as a function of the angle of heel (close hauled conditions)*

$$M_x = M_{Ax} + M_{Hx} \qquad (4.1.7)$$

is approximately equal to the couple formed by $F_{A(lat)}$ and $F_{H(lat)}$:

$$M_x = (H_A + H_H) \, \mathrm{d}_\zeta / 2 \approx (F_{A(lat)} + F_{H(lat)}) \, \mathrm{d}_\zeta / 2 \qquad (4.1.8)$$

As shown by Fig. 4.1.5, the hydrodynamic heeling moment M_{Hx} is usually much smaller (typically by a factor 3–4) than the aerodynamic heeling moment M_{Ax}. This is due to the fact that the distance $\mathrm{d}_{\zeta h}$ between the hydrodynamic heeling force H_H and the longitudinal axis (Fig. 3.2.4) is much smaller than the distance $\mathrm{d}_{\zeta a}$ (Fig. 3.4.4) between the aerodynamic heeling force H_A and the longitudinal axis. We also see from Eq. (4.1.8) that the total heeling moment is independent of the position of the centre of gravity. The distance d_ζ between the points of application of the total aerodynamic and hydrodynamic lateral forces is the only geometrical quantity of direct importance.

Figure 4.1.5 also reflects that, as we will see later, the heeling moment(s) decreases when the angle of heel increases, (that is, for a given, constant wind speed and apparent wind angle) and that the general level of the heeling moment, like the aerodynamic forces, increases with wind speed.

Also shown in Fig. 4.1.5 is the lateral righting moment M_{rx} (Sect. 3.2, Eq. (3.2.7)). Under normal sailing conditions M_{rx} increases with the angle of heel because the buoyancy vector $\varDelta$ moves outboard progressively when the boat heels. Equilibrium in heel is established when the total heeling moment and the righting moment are equal and opposite, that is when the red and black curves in Fig. 4.1.5 cross. At high wind speeds, when the heeling moment is large, equilibrium is, of course, established at a larger angle of heel than at low wind speeds.

Equilibrium of Forces and Moments in the Longitudinal Plane
Figure 4.1.6 illustrates the forces and moments in the longitudinal (vertical) plane. In addition to the equilibrium of longitudinal (Eq. (4.1.1)) and vertical (Eq. (4.1.5)) forces, already dealt with above, we have here the condition of equilibrium of moment around the pitching (y−) axis:

$$M_{Ay} + M_{Hy} = M_{ry} \tag{4.1.9},$$

where M_{Ay} is the aerodynamic pitching moment, M_{Hy} is the hydrodynamic pitching moment and M_{ry} is the longitudinal righting moment. Equation (4.1.9) governs the trim-in-pitch angle θ of the boat through the longitudinal righting moment M_{ry}. The mechanism is similar to that of the equilibrium in heel illustrated by Fig. 4.1.5 and, therefore, does not require any further explanation.

Note that because, as mentioned above, $T = R$ (Eq. (4.1.1)) and $V_H \approx -V_A$, the total (aerodynamic plus hydrodynamic) pitching moment M_y is equal to the sum of the couples formed by T, R and V_A, V_H, respectively:

$$M_y \approx (T + R)\, d_z/2 + (V_H - V_A)\, d_x/2 = T d_z - V_A d_x \tag{4.1.10}$$

and, hence, independent of the location of the center of gravity.

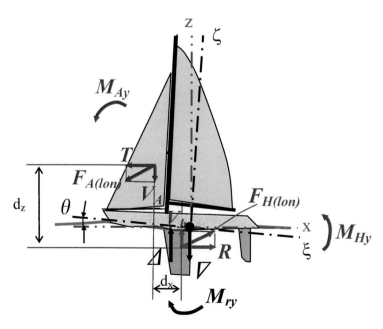

Fig. 4.1.6 *Equilibrium of forces and moments in the longitudinal (vertical) plane*

4.2   'Geometry' of the Mechanics of Sailing

The equilibrium of forces in the horizontal plane implies an important, general law relating the apparent wind angle β and the aerodynamic and hydrodynamic drag angles ε_A and ε_H introduced in the preceding chapter:

$$\beta = \varepsilon_A + \varepsilon_H \qquad (4.2.1)$$

This law is sometimes known as the *'geometry' of (the mechanics of) sailing* and is illustrated by Fig. 4.2.1. It was formulated in 1907 by the British aerodynamicist Frederick Lanchester (1907). We will refer to it as to *Lanchester's Law* in the remainder of this book.

For the purpose of understanding Lanchester's Law we first recall from Sect. 3.2 that the hydrodynamic drag angle ε_H is the angle between the side force S_H and the total horizontal component $F_{H(hor)}$ of the hydrodynamic force. It can be expressed as

$$\tan \varepsilon_H = R / S_H \qquad (4.2.2),$$

where R is the hydrodynamic resistance. Similarly, the aerodynamic drag angle ε_A is the angle between the lift L and the total horizontal component $F_{A(hor)}$ of the aerodynamic force. As already discussed in Sect. 3.4 it can be expressed as

$$\tan \varepsilon_A = D / L \qquad (4.2.3),$$

Fig. 4.2.1 *Illustrating the 'geometry' of the mechanics of sailing*

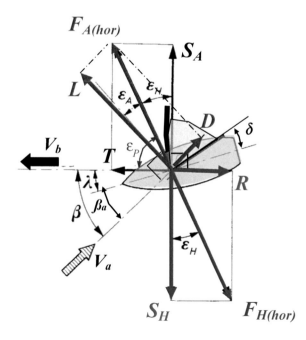

where D is the aerodynamic drag.

Note that because of the equilibrium of aero- and hydrodynamic forces the angle between the aerodynamic side force S_A and $F_{H(hor)}$ is equal to ε_H. The apparent wind angle β is the angle between the boat speed vector V_b and the apparent wind vector V_a and the side force S_A is perpendicular to V_b. Also, by definition, the lift L is perpendicular to V_a. It then follows that the angle between S_A and L is equal to β and Eq. (4.2.1) is satisfied. Note that this is the case for any sheeting angle δ.

While the general validity of Lanchester's Law is readily understood from Fig. 4.2.1, the significance of Eq. (4.21) for sailing practice may not be immediately evident. In the author's experience it may help to understand this significance by considering a yacht that is set to sail at a given apparent wind angle and wind speed. In terms of the basic mechanics of sailing the process of settling at equilibrium can, in a simplified way, be described as follows:

1. Setting of the sheeting angle(s) δ and the apparent heading β_a (through heading up or bearing off) determines, together with the apparent wind speed V_a, the aerodynamic lift L and drag D (Fig. 4.1.6, see also Sect. 3.4) and the aerodynamic drag angle ε_A (Eq. (4.2.3)).
2. The boat will then accelerate or decelerate and acquire a certain drift until the equilibrium conditions are met at a certain boat speed V_b and angle of leeway λ. Note that these two quantities determine the hydrodynamic resistance and side force (see Sect. 3.2) and, hence (Eq. (4.2.2)), the hydrodynamic drag angle ε_H.
3. When equilibrium has been reached, and only then, the angle of leeway λ and the hydrodynamic drag angle ε_H have attained such values that the condition (4.2.1), that is

$$\beta_a + \lambda = \varepsilon_A + \varepsilon_H \qquad (4.2.4),$$

is satisfied.

As already mentioned in Sect. 3.2, the direction of the hydrodynamic force is mainly determined by the angle of leeway λ and its magnitude mainly by boat speed V_b. This means that the ratio S_H / R between hydrodynamic side force and resistance, and thus the drag angle ε_H, is primarily a function of λ; the effect of boat speed on ε_H is of secondary importance, at least at low and moderate boat speeds.

Similarly, as mentioned in Sect. 3.4, the direction of the aerodynamic force is primarily determined by the apparent heading and sheeting angle(s); the apparent wind speed V_a determines the magnitude of the aerodynamic force. This means, that for a fixed sheeting angle, the aerodynamic drag angle ε_A is a function of β_a only.

Writing $\varepsilon_A = \varepsilon_A(\beta_a)$ to express that ε_A is a function of β_a and $\varepsilon_H \approx \varepsilon_H(\lambda)$ to express that ε_H is (primarily) a function of λ, and rewriting Eq. (4.2.4) as

$$\beta_a + \lambda = \varepsilon_A(\beta_a) + \varepsilon_H(\lambda) \qquad (4.2.5),$$

it follows that (4.2.5) is the dominating condition determining the angle of leeway λ for a given apparent heading and that this is, to a certain degree of approximation,

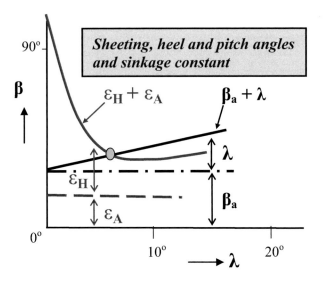

Fig. 4.2.2 *Directional balance of forces at a given apparent heading (qualitatively)*

independent of boat speed and wind speed. As such it can be considered as the condition governing the directional balance of aerodynamic and hydrodynamic forces in the horizontal plane, with the absolute level of forces governed by the condition (4.1.1) of longitudinal equilibrium.

In terms of the condition (4.2.5) the equilibrating or balancing process can be illustrated as in Fig. 4.2.2.

For a fixed sheeting angle and with the apparent heading β_a set at a given, constant value, the term $\beta_a + \lambda$ on the left-hand side of Eq. (4.2.5) is represented by the solid black line (note that vertical and horizontal scales in Fig. 4.2.2 are not the same; if they were, the slope would be 45°). With β_a fixed and constant, the aerodynamic drag angle ε_A is also fixed and constant. The hydrodynamic drag angle ε_H is 90° at zero leeway, when the side force S_H is also zero (see Eq. (4.2.2)), and attains a minimum value (Sect. 3.2) at some value of λ. The precise behaviour and values depend primarily on the keel configuration of the boat but for yachts with conventional keels the minimum value of ε_H occurs usually at values of λ around 5°–10°. The equilibrium condition is met when the blue curve, representing $\varepsilon_A + \varepsilon_H$, and the black line, representing $\beta_a + \lambda$, cross.

As already indicated, the equilibrating or balancing process as described above is based on certain simplifying approximations and assumptions. Real life is a little more complex. Apart from the fact that the hydrodynamic drag angle ε_H also depends on boat speed, at least for 'high' boat speeds, an important role is also played by the equilibrium in heel, and, but to a lesser extent, in pitch and heave. The reason is that the angle of heel has a significant influence on the aerodynamic and hydrodynamic drag angles (in the sense that both ε_A en ε_H increase with heel). This, however, does not change the general picture in a qualitative sense.

Important Implication of Lanchester's Law

Lanchester's Law has an important implication for the minimum sustainable value of the apparent wind angle that a sailing yacht can attain. Because, as mentioned earlier (Chap. 3), all yachts have minimum values for the attainable aero- and hydrodynamic drag angles ε_A en ε_H, it follows from Eq. (4.2.1), that there is also a minimum attached to their sum, that is, in equilibrium conditions, the apparent wind angle β. This, as illustrated by Fig. 4.2.3, has an important consequence for upwind sailing.

Figure 4.2.3 presents, qualitatively, the apparent wind angle β as a function of the apparent heading β_a under equilibrium conditions. The true wind speed and sheeting angle(s) are assumed to be constant. The different lines represent the different terms of Eq. (4.2.4).

The figure illustrates what happens if the apparent heading of a yacht is progressively, but stepwise, lowered to smaller and smaller values. Each step is supposed to last sufficiently long for equilibrium conditions to settle. If this process is started at a sufficiently large value of β_a, the apparent wind angle β (represented by the solid blue line) first decreases at almost the same rate as the apparent heading β_a. It will, however, gradually adopt a smaller slope because of the fact that the aerodynamic drag angle ε_A decreases at first until it attains its minimum value at some, sail configuration and sheeting angle dependent, value[2] of β_a. A similar behaviour is usually exhibited by the hydrodynamic drag angle ε_H, which is a function of λ (Fig. 3.2.3), although the minimum does not necessarily occur at the same value of β_a. As a consequence, the apparent wind angle $\beta(=\varepsilon_A+\varepsilon_H)$ also attains a minimum, for a value of β_a somewhere in the range where ε_A and ε_H have their minima. For still lower values of β_a the aero- and hydrodynamic drag angles increase again rap-

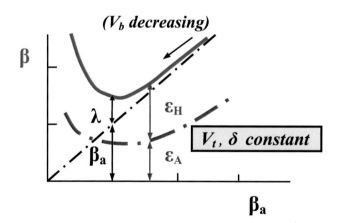

Fig. 4.2.3 *Illustrating that there is a lower limit to the attainable apparent wind angle*

---

[2] As we have seen in Sect. 3.4, Fig. 3.4.3, ε_A is a function of the angle of attack α of the sails, which, for a fixed sheeting angle δ, is directly linked to the apparent heading β_a.

idly and, consequently, the apparent wind angle β increases as well. (Note that for $\beta_a = 0$, when the boat points straight into the wind, there would only be drag and no lift or side force, so that both ε_A and ε_H would be 90°). Because of the fact that, in equilibrium, β is equal to $\varepsilon_A + \varepsilon_H$ as well as equal to $\beta_a + \lambda$, the implication of this is that the angle of leeway λ increases dramatically. In the terminology of sailing a yacht is said to be 'pinched' or 'squeezed' under such conditions.

As mentioned earlier both ε_A and ε_H also depend on the angle of heel in the sense that they both increase with heel. As a consequence the minimum sustainable apparent wind angle also increases with heel.

The hydrodynamic drag angle ε_H and the minimum sustainable apparent wind angle also increase rapidly when a yacht approaches its 'critical' boat speed (Fig. 4.1.2). Hence, the minimum sustainable apparent wind angle at high wind speeds (implying large heel angles and high boat speeds) is larger than at low wind speeds, in particular for 'tender' yachts that heel easily.

For conventional displacement yachts the minimum sustainable value of β is of the order of 20°–30°.

4.3 Driving Force, Side Force and the Efficiency of Propulsion

Driving Force
As indicated in Fig. 4.2.3 a simultaneous result of reducing the apparent heading is a decrease of boat speed, a phenomenon that every yachtsman is familiar with. This is a consequence of the fact that under normal sailing conditions, when the angle of leeway is small and there is not a large difference between the apparent heading and the apparent wind angle, the driving force T decreases when the apparent wind angle decreases. The latter is readily understood by (re)considering Eq. (3.4.3), which expresses the driving force as a function of the lift and drag of the sails and the apparent wind angle:

$$T = L\sin\beta - D\cos\beta \qquad (4.3.1)$$

This expression is readily derived from Fig. 4.2.1. It can be rewritten as

$$T = L(\sin\beta - \tan\varepsilon_A \cos\beta) \qquad (4.3.2a)$$

or

$$T/L = \sin\beta - \tan\varepsilon_A \cos\beta \qquad (4.3.2b)$$

Figure 4.3.1 shows how the driving-force/lift ratio T/L varies with the apparent wind angle β for three different levels of the aerodynamic drag angle ε_A. Note that

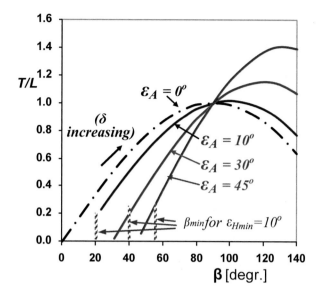

Fig. 4.3.1 *Driving-force/ lift ratio as a function of apparent wind angle*

these curves do not represent equilibrium conditions. They merely illustrate the general trend that, for small apparent wind angles, the driving force/lift ratio increases rapidly with β and that a small drag angle is favourable for $\beta < 90°$. The latter trend is reversed for $\beta > 90°$. This, of course, is due to the fact that, conforming to Eq. (4.3.1), the contribution of the drag to the driving force is negative for $\beta < 90°$, zero for $\beta = 90°$ and positive for $\beta > 90°$.

Note that the driving force (and, hence, boat speed!) goes to zero for small apparent wind angles and more rapidly so for large drag angles. More precisely, *T/L* goes through zero for $\beta = \varepsilon_A$. However, the latter cannot occur in sailing practice, because in equilibrium conditions we have $\beta = \varepsilon_A + \varepsilon_H$ and there is a lower limit to the attainable value of ε_H. Assuming, for example, that the lower limit ε_{Hmin} of ε_H is 10°, the lower limit of β, as indicated in Fig. 4.3.1, is equal to $\varepsilon_A + 10°$ for each value of ε_A.

Figure 4.3.1 also indicates that the driving-force/lift ratio attains a maximum for an apparent wind angle somewhere beyond 90°, depending on the level of ε_A. In fact, it can be shown that the maximum occurs for $\beta = 90° + \varepsilon_A$.

Note also that the (hypothetical) curve for $\varepsilon_A = 0$, when there is lift only and no drag (which never happens in practice), represents the contribution of the lift to the driving force.

The driving force *T* itself is, for a given apparent wind angle, a function of both the lift *L* and the drag angle ε_A of the sails. As we have seen in Sect. 3.4, the lift *L*, for a constant apparent wind speed, increases with angle of attack (or, for a fixed sheeting angle, with the apparent heading β_a) until it attains its maximum at, say, $\alpha = \alpha_{Lmax}$ (Fig. 3.4.2). However, the aerodynamic drag angle at maximum lift is, in general, far from its minimum value. As a consequence, the driving force *T* itself will, for $\beta < 90°$, attain its maximum for an angle of attack somewhere between $\alpha_{\epsilon min}$ and α_{Lmax}. The higher the lift in this range of angles of attack, the larger the driving force. For apparent wind angles around 90°, when the relative effect of ϵ_A on the driving force is small (Fig. 4.3.1), the angle of attack for the maximum driving force will be close to α_{Lmax}. For small apparent wind angles, when the relative effect of ϵ_A is large, it will be closer to $\alpha_{\epsilon min}$.

It has been mentioned already that for apparent wind angles beyond 90° the contribution of the aerodynamic drag to the driving force becomes increasingly important and the lift contribution diminishes. The driving-force/lift ratio T/L then loses its significance as a propulsion parameter. It then makes more sense to consider the driving-force/drag ratio T/D. It follows from Eq. (3.4.3)/(4.3.1) that this can be written as

$$T/D = \cotan\epsilon_A \sin\beta - \cos\beta \qquad (4.3.3)$$

Figure 4.3.2 illustrates the variation of T/D with β for different levels of ϵ_A. Note that for $\epsilon_A = 45°$, that is for $L = D$, the T/D curve is identical to the T/L curve in Fig. 4.3.1 and that the curve for $\epsilon_A = 90°$, when there is drag only and no lift, represents the contribution of the aerodynamic drag to the driving force. For $\beta < 90°$ this contribution is negative.

For $\beta = 180°$, that is when running downwind, the contribution of the aerodynamic lift is always zero (because sin 180° = 0) and the driving force is equal to the aerodynamic drag. Maximizing the driving force then requires that the sails are set for maximum drag. As already indicated in Sect. 3.4, Fig. 3.4.2, this usually implies an angle of attack of the sails close to 90°.

From the preceding discussion it follows, that for apparent wind angles between 90° and 180°, the maximum driving force is obtained for angles of attack between,

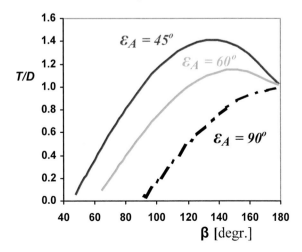

Fig. 4.3.2 *Driving-force/drag ratio as a function of apparent wind angle*

respectively, α_{Lmax} and 90°. It also follows that for $\varepsilon_A = 45°$, that is for $L = D$, the maximum driving force occurs for $\beta = 135°$. The other way around is not necessarily the case. Because $\sin 135° = -\cos 135° (= \frac{1}{2}\sqrt{2})$, the contributions of lift and drag to the driving force are equally important (see Eq. (4.3.1)). This implies that for $\beta = 135°$ the maximum driving force is obtained when the sum $L + D$ of lift and drag has its maximum. The corresponding angle of attack $\alpha_{(L+D)max}$ is, in general, not the same as the angle of attack $\alpha_{(L=D)}$ at which lift and drag are equally large.

An instructive way of considering the role of the lift and drag of the sails for the driving force is in the form of a so-called *lift-drag polar diagram*. The lift-drag polar of a sail is constructed by plotting the lift versus the drag for a given angle of attack and connecting the points for all angles of attack through a line. As an example Fig. 4.3.3 gives the lift-drag polar for the generic sail configurations considered in Fig. 3.4.2 for a constant apparent wind speed.

Considering Fig. 4.3.3, we first note that a line between the origin and a point on the polar curve (shown for the jib/genoa type head-sail) represents the total aerodynamic force vector F_A for the angle of attack that belongs to the point on the polar. For the sake of simplicity it has been assumed that the heel angle is zero. For non-zero angles of heel the total aerodynamic force vector F_A is to be replaced by its horizontal component $F_{A(hor)}$ (see Fig. 4.1.1). Also indicated in Fig. 4.3.3 is a line from the origin of the polar indicating the direction of motion of the yacht. The angle β between this line and the horizontal, 'drag'-axis represents the apparent wind angle.

A useful feature of a 'polar' diagram of the type under discussion is that it is relatively easy to indicate, for a given apparent wind angle, the sail setting that provides the maximum driving force. In a diagram like that of Fig. 4.3.3 this condition is

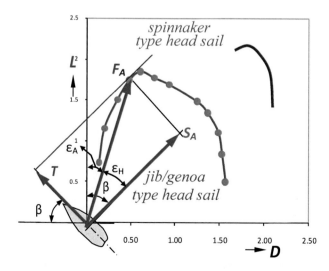

Fig. 4.3.3 *Lift-drag polar and decomposition of lift and drag into driving force and side force*

obtained by constructing the line that is tangent to the polar curve and perpendicular to the direction of the course of the boat. Decomposition of the total aerodynamic force vector F_A to the point of contact into the driving force T and the side force S_A then provides the maximum driving force and the corresponding heeling force for the apparent wind angle considered.

Note, that Lanchester's law is also visualized in Fig. 4.3.3: the angle between the heeling force vector and the vertical ('lift') axis is equal to the apparent wind angle β, the angle between F_A and the lift axis is, by definition, equal to the aerodynamic drag angle ε_A, and the angle between S_A and F_A is, in equilibrium conditions, equal to the hydrodynamic drag angle ε_H (see Sect. Equilibrium of Forces and Moments in the Lateral Plane). It is easily verified that when the apparent wind angle has its minimum value ($\beta_{min} = \varepsilon_{Amin} + \varepsilon_{Hmin}$), the total aerodynamic force vector F_A is tangent to the polar curve.

Figure 4.3.4 gives a qualitative impression of how the maximum driving force varies with the apparent wind angle in the absolute sense. The figure was also constructed on the basis of the 'generic' aerodynamic characteristics for a constant apparent wind speed shown in Fig. 3.4.2. For all apparent wind angles the angle of attack and, hence, the sheeting angle, was chosen so as to produce the maximum driving force.

It can be shown, through application of the calculus of variation, that for a fixed angle of attack of the sails, i.e. for a fixed value of ε_A, the driving force attains a maximum when the following condition is satisfied:

$$L\cos\beta + D\sin\beta = 0 \qquad (4.3.4a)$$

This is equal to the condition

$$\beta = 90 + \varepsilon_A \text{ (degrees)} \qquad (4.3.4b)$$

Fig. 4.3.4 *Variation (qualitatively) of maximum driving force as a function of apparent wind angle*

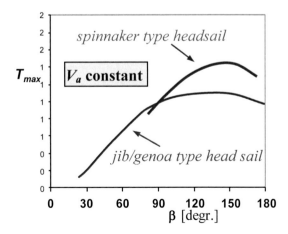

The latter is, of course, what we have already seen in Figs. 4.3.1 and 4.3.2.

It can also be shown that for a given value of the apparent wind angle β, the maximum driving force is obtained when the sails are set such that the following condition is satisfied:

$$\frac{dL}{d\alpha}\sin\beta - \frac{dD}{d\alpha}\cos\beta = 0 \qquad (4.3.5),$$

where $\dfrac{dL}{d\alpha}$ and $\dfrac{dD}{d\alpha}$ are the rates of change of lift and drag with the angle of attack (see Fig. 3.4.2). This can also be written as

$$\frac{dL}{d\alpha} / \frac{dD}{d\alpha} = \cotan\beta = \tan(90-\beta) \qquad (4.3.6)$$

or

$$dD / dL = \tan\beta \qquad (4.3.7)$$

The latter is precisely what is illustrated by Fig. 4.3.3: the tangent to the lift-drag polar is perpendicular to the line indicating the apparent course.

It is also clear from Fig. 4.3.3 that the absolute maximum driving force is realized when the total resulting aerodynamic force F_A has its maximum absolute value and acts in the same direction as the direction of sailing (hence, no side force). This is when the conditions (4.3.4) and (4.3.7) are both satisfied. It appears (Fig. 4.3.4) that this is usually the case for an apparent wind angle of about 135°, as already suggested by Figs. 4.3.1 and 4.3.2. However, if the tangency condition (4.3.7) occurs at a high value of the lift/drag ratio L/D (or a low value of the aerodynamic drag angle ε_A) the driving force will, because of condition (4.3.4b), attain its absolute maximum at a lower value of the apparent wind angle. The opposite is the case when the total aerodynamic force F_A attains its maximum for a low value of L/D.

Side Force

The aerodynamic side force S_A is, like the driving force, also a strong function of the apparent wind angle. Recalling, from Sect. 3.4, Eq. (3.4.2), see also Fig. 4.2.1, that

$$S_A = L\cos\beta + D\sin\beta, \qquad (4.3.8)$$

and rewriting this as

$$S_A/L = \cos\beta + \tan\varepsilon_A \sin\beta \qquad (4.3.9),$$

we can plot S_A/L as a function of β for different levels of ε_A (Fig. 4.3.5).

What Fig. 4.3.5 shows is that the aerodynamic side force decreases when the apparent wind angle is increased and goes through zero for some value of β > 90°,

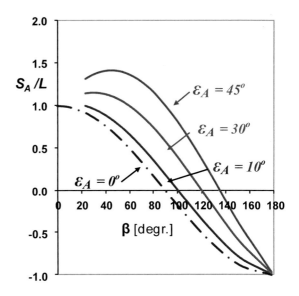

Fig. 4.3.5 *Side force/lift ratio
as a function of apparent wind
angle*

depending on the level of ε_A. It can be derived from Eq. (4.3.9) that the ratio S_A/L attains its maximum when $\beta = \varepsilon_A$. It can also be shown that the value of β for which S_A/L becomes zero is the same as the value of β for which T/L attains its maximum, that is at $\beta = 90° + \varepsilon_A$. For still larger values of β S_A/L becomes negative, that is the side force points to weather. The (hypothetical) curve for $\varepsilon_A = 0$, when there is lift only and no drag represents the contribution of the lift to the heeling force. Note that the latter is zero for $\beta = 90°$ but large for small and large apparent wind angles.

The decrease of the side force (and heeling force) with increasing apparent wind angle implies that there is an equally rapid decrease with β of the aerodynamic heeling moment M_{Ax} (Sect. 3.4). This, of course, is in agreement with every yachtsman's experience that the heel angle of a yacht is small when the apparent wind angle is large.

Figure 4.3.5 also illustrates that the level of the side force is low when the aerodynamic drag angle ε_A is small, at least for low and moderate apparent wind angles. It is also clear that the relative effect of the drag angle on the side force is large at apparent wind angles around 90°. For such conditions, when the contribution of the aerodynamic lift to the side force is small, it makes sense to consider the side-force/ drag ratio S_A/D, which can be expressed as

$$S_A/D = \cotan\varepsilon_A \ \cos\beta + \sin\beta \qquad (4.3.10)$$

The variation of S_A/D with β, for different levels of ε_A is illustrated by Fig. 4.3.6. The figure shows that:

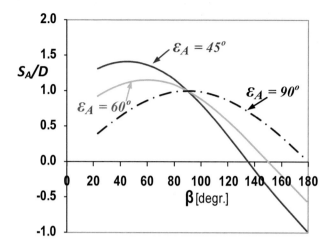

Fig. 4.3.6 *Side-force/drag ratio as a function of apparent wind angle*

- the effect of the drag angle ε_A on S_A/D is zero for $\beta = 90°$ when the aerodynamic lift does not contribute to the side force.
- the curve for $\varepsilon_A = 45°$, when lift and drag are equally large, is the same as the corresponding curve in Fig. 4.3.5 for S_A/L
- the curve for $\varepsilon_A = 90°$, when there is drag only and no lift, represents the contribution of the aerodynamic drag to the side force

The figure also illustrates that the contribution of the aerodynamic drag to the side force has its maximum for $\beta = 90°$ and is zero for $\beta = 0°$ and $180°$.

A large side force and heeling force affects boat performance in two (or three) different ways. First of all it follows from the condition (4.1.2), $S_A = S_H$, of lateral equilibrium that the hydrodynamic side force S_H is equal to the aerodynamic side force S_A. This implies that a large aerodynamic side force leads automatically to a large hydrodynamic side force. The latter, as mentioned in Sect. 3.2 and shown in Fig. 3.2.2, implies a large angle of leeway λ. Since the hydrodynamic resistance R also increases when the angle of leeway λ increases (Fig. 3.2.2), this implies also that a large side force leads to a high(er) hydrodynamic resistance, and, as a consequence, low(er) boat speed.

The side force also affects boat performance through the angle of heel. As we have seen in Sects. 3.2 and Equilibrium of Forces and Moments in the Horizontal Plane, a large side force implies a large heeling moment and, thus, a large angle of heel (Fig. 4.1.4). As mentioned in Sects. 3.4 and 3.2, and as to be discussed in more detail in Chaps. 6 and 7, the aerodynamic lift and driving force decrease and the hydrodynamic resistance increases with increasing angle of heel. This, of course, also implies a low(er) boat speed.

Efficiency of Propulsion

An interesting quantity to consider is the ratio T/S_A between driving force and side force. As we have seen above, a large driving force contributes directly to a high boat speed and a small aerodynamic heeling force contributes indirectly through a lower hydrodynamic resistance and a smaller heel angle. Recalling that the ratio L/D is associated with the notion of aerodynamic efficiency and, similarly, the ratio S_H/R with hydrodynamic efficiency, we can associate T/S_A with the notion of efficiency of propulsion.

It follows directly from Eqs. (4.3.2) and (4.3.9) that

$$T/S_A = (\sin\beta - \tan\varepsilon_A \cos\beta)/(\cos\beta + \tan\varepsilon_A \sin\beta) \qquad (4.3.11)$$

It also follows from the equilibrium conditions (4.1.1) and (4.1.2) and the definition (3.2.1) of the hydrodynamic drag angle, see also Fig. 4.2.1, that

$$T/S_A = R/S_H = \tan\varepsilon_H \qquad (4.3.12)[3]$$

In ordinary language this means that, in equilibrium sailing conditions, the hydrodynamic efficiency of the underwater body is equal to the inverse of the propulsive efficiency of the sails. This, perhaps surprising, result is already contained in the description of aerodynamic forces in Sect. 3.4. In fact, it follows directly from Eqs. (3.4.3) and (3.4.4) upon substitution of the equilibrium condition $\beta = \varepsilon_A + \varepsilon_H$. It can be understood in a broad sense if it is realized that while the aerodynamic efficiency L/D of the sails is determined by the apparent heading β_a, the propulsive efficiency T/S_A is determined by the apparent wind angle β (Fig. 4.2.1). The latter depends also on the angle of leeway λ (Eq. (3.3.2)), which determines ε_H.

Figure 4.3.7 presents T/S_A as a function of β—according to the relation (4.3.12). The figure indicates, not surprisingly after Figs. 4.3.1 and 4.3.5, that the propulsive efficiency increases rapidly with the apparent wind angle and that a small aerodynamic drag angle is favourable, at least for apparent wind angles below, say, 90°.

Note that T/S_A becomes infinitely large for apparent wind angles somewhere beyond 90° when the heeling force S_A goes to zero (Fig. 4.3.5), that is for $\beta = 90° + \varepsilon_A$.

Figure 4.3.7 does, like Figs. 4.3.1 and 4.3.5, *not* represent equilibrium conditions. Rather, it illustrates the potential of the efficiency of propulsion. However, if we introduce horizontal lines representing levels of ε_H (or rather levels of tan ε_H), then the intersections with the curves representing T/S_A for different values of ε_A do represent equilibrium conditions because of Eq. (4.3.8). It then follows from Fig. 4.3.7 that low values of ε_H are required for equilibrium at small apparent wind angles and high values of ε_H for equilibrium at large apparent wind angles. It is also clear that equilibrium conditions cannot exist below the line representing ε_{Hmin}, that is the minimum hydrodynamic drag angle, or for values of $\beta < \varepsilon_A + \varepsilon_{Hmin}$. It

---

[3] The result (4.3.12) is, of course, also obtained from Eq. (4.3.11) upon substitution of $\beta = \varepsilon_A + \varepsilon_H$.

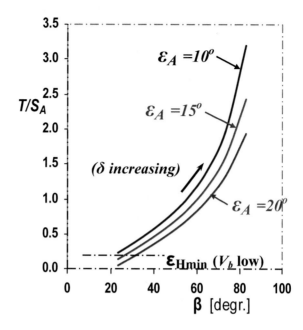

Fig. 4.3.7 *Efficiency of propulsion as a function of apparent wind angle*

is further noted that a high level of the (minimum) hydrodynamic drag angle ε_{Hmin} (high boat speeds) implies a large minimum apparent wind angle but a relatively high T/S_A ratio at equilibrium conditions and that the opposite is the case for a low value of ε_{Hmin} (low boat speeds).

Because T/S_A becomes infinitely large for $\beta = 90° + \varepsilon_A$, a more convenient quantity for expressing the efficiency of propulsion, at least for apparent wind angles beyond 90°, is one defined by

$$\cotan \varepsilon_P = T/S_A \tag{4.3.13}$$

The quantity ε_P may be called the *propulsion or thrust angle*. As indicated in Fig. 4.2.1 ε_P is the angle between the horizontal component $F_{A(hor)}$ of the total aerodynamic force vector and the direction of motion of the yacht. Note that ε_P is 0° when the aerodynamic force vector is in line with the direction of motion, that is when $S_A \to 0$ and $T/S_A \to \infty$. Also, in equilibrium conditions, Eq. (4.3.4) implies that

$$\varepsilon_H = 90° - \varepsilon_P \tag{4.3.14}$$

If ε_P is plotted as a function of β for different levels of ε_A one obtains straight lines as depicted in Fig. 4.3.8. For each value of the aerodynamic drag angle ε_A, the lines cross the $\varepsilon_P = 0°$ level at $\beta = 90° + \varepsilon_A$. Figure 4.3.8 illustrates that for $\beta < 90° + \varepsilon_{Amin}$ the thrust angle ε_P attains its minimum when the sails are set for the minimum drag angle ε_{Amin}. For $\beta \geq 90° + \varepsilon_{Amin}$ it is, in principle, always possible

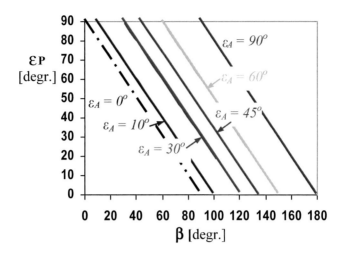

Fig. 4.3.8 *Propulsion or thrust angle as a function of apparent wind angle*

to realize the maximum efficiency of propulsion (or a thrust angle of 0°), provided that the sails are set so as to realize an aerodynamic drag angle ε_A equal to $\beta - 90°$.

   Note that for $\beta = 180°$ ('dead run') the thrust angle zero situation is obtained for $\varepsilon_A = 90°$, that is when the sails have drag only and no lift.

Effectiveness of Propulsion

A key question in relation to boat performance is how the sails should be set in order to realize maximum boat speed for a given apparent wind angle. Unfortunately, the answer to this question is not simple. Maximizing the driving force is probably the first thing that comes to one's mind. This, however, is not necessarily the right answer. The reason is that maximizing the driving force may lead to a large side force and heeling force, in particular for small apparent wind angles. A large side force and heeling force implies, as we have seen in the preceding paragraphs, a large hydrodynamic resistance and reduced aerodynamic as well as hydrodynamic efficiency due to a large heel angle. Hence, it could be advantageous to give up some driving force for the benefit of a smaller heeling force, in particular for high wind speeds when the aerodynamic forces are large and/or for yachts with a high sensitivity of the hydrodynamic resistance to side force and/or for 'tender' yachts with low lateral (static) stability that heel easily. Aiming for the maximum efficiency of propulsion or zero thrust angle may not be the best solution either because the ratio T/S_A does not say much about the absolute level of the driving force. If, for example, for an apparent wind angle of 100°, the sails would be set for $\varepsilon_A = 10°$ (say the minimum drag angle) so as to realize $S_A = 0$ and a thrust angle of 0° (Figs. 4.3.5 and 4.3.8), then the lift of the sails is probably significantly less than its maximum. In addition Fig. 4.3.1 shows that for $\beta > 90°$ the driving force/lift ratio T/L is higher for a higher value of ε_A. Hence, it is very likely that accepting some heeling force and setting the sails for a higher value of ε_A is more effective for $\beta > 90°$.

This in particular for low wind speeds and/or or for yachts with a low sensitivity of the hydrodynamic resistance to side force and/or 'stiff' yachts with high lateral (static) stability that do not heel easily.

From the discussion just given it follows that the most effective 'mode' of propulsion of a yacht depends on the characteristics of the yacht as well as on the apparent wind angle and wind speed. Figure 4.3.9 summarizes the preceding discussion in a qualitative sense. The figure was constructed on the basis of the generic aerodynamic characteristics depicted in Figs. 3.4.2 and 3.4.3.

Considering Fig. 4.3.9 we first recall that for $\beta < 90° + \varepsilon_{Amin}$ the angle of attack $\alpha_{\varepsilon Pmin}$ for the minimum thrust angle is equal to the angle of attack $\alpha_{\varepsilon Amin}$ for the minimum drag angle. Hence, the lines representing the conditions of the minimum thrust angle ($\alpha_{\varepsilon Pmin}$) or the maximum propulsive efficiency have a kink for $\beta = 90° + \varepsilon_{Pmin}$. This is due to the fact that the minimum thrust angle decreases with increasing β (Fig. 4.3.8) until it hits the zero level for $\beta = 90° + \varepsilon_{Amin}$ and stays at that level for larger values of β. For the configuration with jib/genoa the kink occurs at $\beta \approx 100°$. For the spinnaker configuration with its much larger value of ε_{Amin} it occurs at a higher value of β. Note that with spinnaker type headsails the angle of attack α_{Tmax} for the maximum driving force is somewhat larger than for jib/genoa type headsails. Note also that the lines representing the angle of attack $\alpha_{\varepsilon Pmin}$ for the maximum propulsive efficiency (or minimum thrust angle intersect the line for $\alpha_{L=D}$ at about $\beta \cong 135°$. This is due to the fact that the side force is zero and changes sign at this condition (see Fig. 4.3.6).

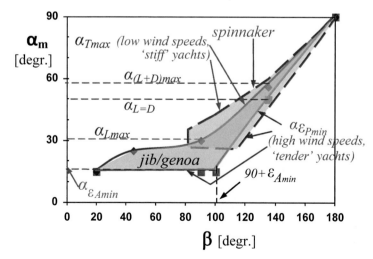

Fig. 4.3.9 *Range (qualitatively) of the most effective angle of attack as a function of apparent wind angle*

Maximizing boat speed for a given apparent wind angle requires in general that the sails are set for an angle of attack somewhere in the range between the angle of attack α_{Tmax} for the maximum driving force and the angle of attack $\alpha_{\varepsilon Pmin}$ for the maximum propulsive efficiency (or minimum thrust angle). For 'stiff' yachts and/or for low wind speeds, the most effective propulsion mode will be close to the condition for maximum driving force. For 'tender' yachts and/or for high wind speeds, the sails should be set so as to be close(r) to the maximum efficiency of propulsion.

4.4   The Role of the Sheeting Angle

Until now we have hardly paid any specific attention to the sheeting angle(s) of the sails. The sheeting angle(s), together with the apparent heading, is, however, an important variable for controlling the aerodynamics of the sails, and, through that, for controlling boat performance.

For the purpose of describing the essence of the role of sheeting angle in boat performance we first recall from Sect. 3.4 that the aerodynamics of the sails are primarily determined by the angle of attack α and the apparent wind speed and that α is defined by Eq. (3.4.1), i.e. $\beta_a = \alpha + \delta$, where δ is the sheeting angle of the mainsail (Fig. 3.4.1). It was also indicated, that the aerodynamic drag angle ε_A is a function of α rather than β_a.

For the sake of simplicity we will still assume, as in Sect. 3.4, that the sheeting angle of the head sail is coupled to that of the mainsail. A more in-depth treatment of the interaction between head sail and mainsail and their respective sheeting angles will be given in Chap. 7.

Figure 4.4.1 illustrates qualitatively how the sheeting angle δ can be used to control the directional equilibrium of forces for a given, constant value of the angle of attack α of the sails. The figure is basically the same as Fig. 4.2.2 with different scales for the two axes.

We note first of all that, with the angle of attack α constant, the aerodynamic drag angle ε_A is also constant. Because ε_H is an function of λ, the blue curve representing $\varepsilon_A + \varepsilon_H$ is then independent of the sheeting angle δ. With the angle of attack α fixed, the apparent heading β_a varies like δ. The black straight lines in Fig. 4.4.1 represent three different values of δ and the points where the lines representing $\beta_a + \lambda$ intersect the blue line representing $(\varepsilon_A + \varepsilon_H)$ represent the equilibrium conditions corresponding with the three sheeting angles.

Alternatively, one may use the sheeting angle to control the angle of attack and the aerodynamics of the sails for a fixed apparent heading. This situation is depicted in Fig. 4.2.2. Because β_a is now constant, there belongs a different angle of attack α to each sheeting angle δ and vice versa. As a consequence the aerodynamic drag angle ε_A has a different value for each sheeting angle and the curves representing $\varepsilon_A + \varepsilon_H$ have three correspondingly different levels. Again, we have three different equilibrium conditions; one for each sheeting angle. What Figs. 4.4.1 and 4.4.2 try to say is that the sheeting angle can be used to control the apparent wind angle β

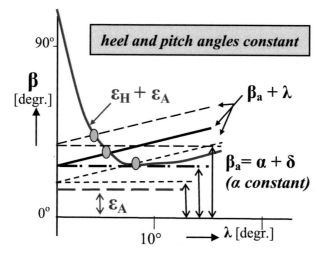

Fig. 4.4.1 *Effect (qualitatively) of sheeting angle on the directional equilibrium of forces at constant angle of attack of the sails*

independently of the apparent heading β_a through control of the angle of attack α of the sails. This, of course, within the limits set by the aerodynamic characteristics of the sails (ε_A as a function of α) and the hydrodynamic characteristics of the hull and appendages (ε_H a function of λ). For example, the sheeting angle can be adjusted to keep the angle of attack at the value ($\alpha_{\varepsilon Amin}$) for which the aerodynamic drag angle

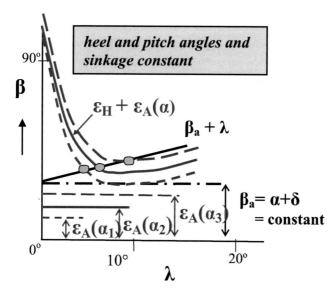

Fig. 4.4.2 *Effect (qualitatively) of sheeting angle on the directional equilibrium of forces at constant apparent heading*

ε_A has its minimum (Fig. 3.4.2) while the apparent heading β_a is set such that the apparent wind angle $\beta = \varepsilon_A + \varepsilon_H$ also attains its minimum value.

Figure 4.4.1 illustrates this situation. If α is set such that ε_A attains its minimum value, β attains its minimum for the smallest of the three sheeting angles δ considered. Note that a further reduction of β_a and sheeting angle would lead to a larger value of β. Note also that the angle of leeway at the minimum apparent wind angle is not necessarily small. It depends on the hydrodynamic characteristics of hull and appendages, the hydrodynamic drag angle as a function of leeway (Fig. 3.2.3) in particular.

As mentioned already in Sects. 3.2, 3.4 and 4.2, the aerodynamic and hydrodynamic drag angles both increase with increasing angle of heel. The hydrodynamic drag angle also increases with boat speed; in particular when a yacht approaches its 'critical' boat speed (Fig. 4.1.2). Since both the angle of heel as well as boat speed increase with increasing wind speed, the aerodynamic and hydrodynamic drag angles also increase with wind speed. Figure 4.4.3 illustrates the consequences for the minimum apparent wind angle and the associated apparent heading and sheeting angle. It has been assumed that the angle of attack of the sails is set at the value that gives the minimum aerodynamic drag angle and that this value is independent of wind speed. The figure shows that for a higher level of $\varepsilon_H + \varepsilon_{Amin}$ the apparent heading β_a must be larger in order to realize equilibrium at the point where the curve $\varepsilon_H + \varepsilon_{Amin}$ has its minimum. Since $\beta_a = \alpha + \delta$, and α has been assumed constant, this implies a larger sheeting angle δ. We can conclude that sailing at the minimum apparent wind angle requires a larger sheeting angle at high wind speeds than at low wind speeds, in particular for 'tender' yachts that heel easily.

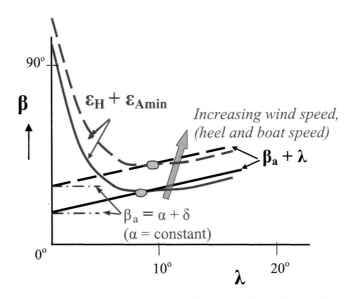

Fig. 4.4.3 *Effect of wind speed (qualitatively) on the directional equilibrium of forces*

The sheeting angle is of course also of major importance for the driving and heeling forces of the sails. Referring to Figs. 4.3.1, 4.3.2, 4.3.3, 4.3.4, 4.3.5, 4.3.6, 4.3.7, 4.3.8 we recall, again, that, because the aerodynamic drag angle ε_A depends on the angle of attack α of the sails, the lines of constant ε_A are also lines of constant α. This means that, since $\beta = \beta_a + \lambda$ and $\beta_a = \alpha + \delta$, and the angle of leeway λ is relatively small, at least under normal sailing conditions, the sheeting angle increases with β along the lines of constant ε_A, as indicated in Fig. 4.3.7.

We have also seen (Fig. 4.3.9) that the most effective 'mode' of propulsion requires an angle of attack somewhere between α_{Tmax} (maximum driving force) and $\alpha_{\varepsilon Pmin}$ (maximum propulsive efficiency), depending on wind speed and the lateral hydrodynamic and static stability characteristics of the yacht. The corresponding 'optimal' sheeting angles can, assuming that the angle of leeway is small, be obtained from the approximate relation $\beta \approx \alpha + \delta$.

Figure 4.4.4 presents, qualitatively, the range of the optimal sheeting angle as a function of the apparent wind angle. The figure is based on the angle of attack data contained by Fig. 4.3.9. As before, we have, for the time being, assumed that the sheeting angles of the mainsail and headsail are coupled so that it suffices to consider the sheeting angle of the mainsail only.

It follows from the discussion in Sect. 4.3 and Fig. 4.3.9 that for 'stiff' yachts and/or for low wind speeds, the optimal sheeting angle will be close to the condition for the maximum driving force (δ_{Tmax}). On the other hand, for 'tender' yachts and/or for high wind speeds, the sails should be set to operate close(r) to the condition for the maximum efficiency of propulsion (ε_{Pmin}, defined by Eq. (4.3.13)).

It is emphasized again that the sheeting angle characteristics of Fig. 4.4.4 are generic and may vary significantly between one yacht and sail configuration and

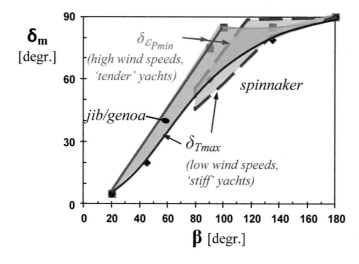

Fig. 4.4.4 *Range (qualitatively) of optimal sheeting angle of the sail(s) as a function of apparent wind angle*

another. Different types of foresail and mainsail in particular may require different sheeting angles for maximum thrust, as will be discussed in more detail in Chap. 7. The main message of Fig. 4.4.4 is that, for ordinary sail configurations, the optimal sheeting angle (of the mainsail) increases almost linearly with the apparent wind angle from $\delta \approx 5°$ at $\beta = 20°$ to $\delta \approx 75°$ at $\beta \approx 100-120°$ and, from there, increases slowly to 90° at $\beta \approx 180°$. For sail configurations with spinnaker type headsails smaller sheeting angles of the mainsail may be better for apparent wind angles below 140° (to be discussed in some detail in Chap. 7).

It is further noted that for most rigs the sheeting angle is mechanically limited to a maximum of about 80° when the boom is stopped by the shrouds.

4.5   Upwind Sailing

Sailing to Windward
The purpose of sailing to windward is usually to reach a *waypoint* with a position in the direction from which the true wind comes in the shortest possible time. When the direction towards the waypoint and the direction from which the wind blows differ less than a certain amount it is not possible to reach the waypoint in one tack; the waypoint can only be reached through successive port and starboard tacks. The reason for this is that, as we have seen above, the sails do not generate a propulsive or driving force for apparent wind angles below a certain, boat dependent, value. It is easily understood that the fastest way to reach a waypoint to windward is to maximize, during each successive tack, the so-called *Velocity-Made-Good-to-windward* (*upwind VMG*, or V_{mgw}) which is defined as

$$V_{mgw} = V_b \cos\gamma \qquad (4.5.1)$$

Here, as before, V_b is boat speed and γ is the true wind angle. As illustrated by Fig. 4.5.1 the velocity-made-good is nothing else but the (magnitude of the) component of the boat speed vector $\underline{V}_b$ in the direction of the true wind vector $\underline{V}_t$.

Equation (4.5.1) might suggest that maximizing V_{mgw} is 'simply' a matter of maximizing V_b and minimizing the true wind angle γ (Fig. 4.5.2). This, unfortunately, is not quite so simple. The reason is that, for any given type of boat, boat speed and true wind angle cannot be varied independently. Besides, there is in general, as mentioned in Sect. 4.1 (Fig. 4.1.2), an upper 'limit' to boat speed (the 'critical' boat speed of conventional displacement yachts), as well as a, boat and wind speed dependent, lower limit to the attainable true wind angle. The latter is a consequence of the lower limit to the sustainable apparent wind angle (Sect. Equilibrium of Forces and Moments in the Lateral Plane see also the relation between γ and β from the wind triangle, Fig. 3.3.3).

Fig. 4.5.1 *Defining Velocity-Made-Good to windward (V_{mgw})*

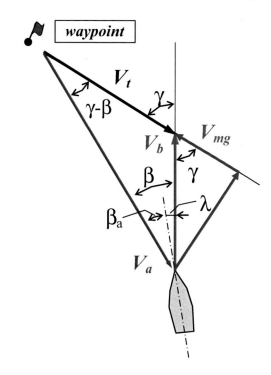

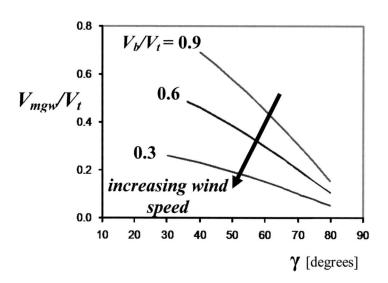

Fig. 4.5.2 *From the wind triangle: Velocity-Made-Good to windward as a function of true wind angle*

A little more insight into the conditions for the best VMG can be obtained from an alternative formula[4] that can also be derived from the wind triangle:

$$V_{mgw}/V_t = \cfrac{1}{\cfrac{\tan\gamma}{\tan(\gamma-\beta)} - 1} \qquad (4.5.2)$$

If we combine Eq. (4.5.2) with Eq. (3.3.4), i.e. with

$$\tan\gamma = \{1 + (V_b/V_t)/\cos\gamma\}\tan\beta,$$

which expresses the relation between γ, β and boat speed V_b, it is possible to calculate the ratio V_{mgw}/V_t as a function of the apparent wind angle β and the ratio V_b/V_t of boat speed to true wind speed. The result of such calculation is visualized in Fig. 4.5.3.

Figure 4.5.3 illustrates that, as one might expect, the VMG increases with decreasing apparent wind angle and that the rate of this increase is higher for high ratios V_b/V_t of boat speed to wind speed. The latter implies, that the VMG is more sensitive to the apparent wind angle at low wind speeds (when the ratio V_b/V_t is high) than at high wind speeds (when the ratio V_b/V_t is low). It is caused by the fact that the true wind angle is more sensitive for variation of the apparent wind angle at high values of V_b/V_t (Fig. 3.3.3).

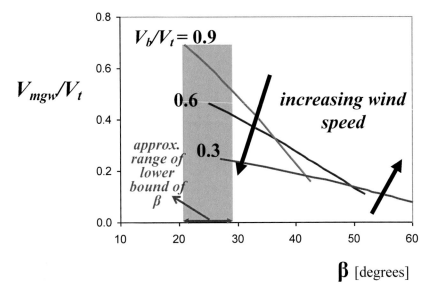

Fig. 4.5.3 *From the wind triangle: Velocity-Made-Good to windward as a function of apparent wind angle*

---

[4] Reportedly (Marchai 2000), this expression was first published by H. Barkla (1965).

Also indicated in Fig. 4.5.3 is the approximate range in which the minimum sustainable apparent wind angle usually occurs. As already discussed in the preceding sections it is impossible to sail (steadily) at values of β below this minimum. Additionally, we have seen that the driving force and efficiency of propulsion decrease when the apparent wind angle is decreased (Figs. 4.3.1, 4.3.7). This implies that boat speed also decreases when the apparent wind angle is decreased. As a consequence, the conditions for the best VMG are a compromise between apparent wind angle and boat speed: the apparent wind angle for the best VMG is usually somewhat larger than the minimum apparent wind angle for the benefit of a higher boat speed. In terms of Fig. 4.5.3 it means that the increase of V_b has a larger benefit for the VMG than the decrease due an increase of β.

The key question is how much larger than $β_{min}$ should β be? Unfortunately the answer depends intricately on the specific aerodynamic, hydrodynamic and lateral stability characteristics of the yacht. It is possible, however, to identify some general trends.

Obviously the rate of increase of boat speed with increasing apparent wind angle is an important factor. When this rate is high it pays (in terms of VMG) to increase the apparent wind angle. When this rate is low the best VMG will be obtained for an apparent wind angle close to $β_{min}$.

'Tender' yachts, that heel easily, will, in principle, profit a little more (in a relative sense) from an increase of apparent wind angle than 'stiff' yachts. The reason is that the more rapid reduction of heel with apparent wind angle of 'tender' boats causes the driving force to *in*crease faster and the hydrodynamic resistance (at constant boat speed) to *de*crease a little more than with 'stiff' boats. As a consequence the margin between the apparent wind angle for the best VMG and $β_{min}$ for 'tender' yachts will be a little larger then for 'stiff' yachts. However, the rate of decrease of the side force with β is rather small for small apparent wind angles (Fig. 4.3.5) so that the difference will be small also.

It is recalled in this context, that the hydrodynamic resistance of conventional displacement yachts increases sharply with (absolute) boat speed when the so-called 'critical boat speed' is approached (see Sect. 4.1, Fig. 4.1.2). Under such conditions (which usually imply high wind speed) boat speed can hardly increase any further when the apparent wind angle is increased. This means that for a yacht approaching its 'critical boat speed' (that is at high wind speeds), a low value of the apparent wind angle is even more important for the VMG than at low absolute boat speeds (and low wind speeds). In other words, the best VMG at high wind speeds is realized at an apparent wind angle that is closer to the minimum apparent wind angle than at low wind speeds.

We have seen in the preceding sections that the minimum attainable apparent wind angle increases with wind speed (Fig. 4.4.3). This, as mentioned, is due to increasing heel and increasing hydrodynamic resistance, in particular for 'tender' yachts that heel easily. The implication in terms of Fig. 4.5.3 is that, for a given yacht, the minimum attainable $β$ is smaller for high values of the ratio V_b/V_t (usually corresponding with low wind speeds and low absolute boat speeds) than for low values of the ratio V_b/V_t (high wind speeds and high absolute boat speeds).

We have also seen (Sect. 4.3), that the most effective 'mode' of propulsion for a given apparent wind angle requires in general that the sails are set for an angle of attack somewhere in the range between the angle of attack α_{Tmax} for the maximum driving force and the angle of attack $\alpha_{\varepsilon Pmin}$ for the maximum propulsive efficiency (or minimum thrust angle ε_{Pmin}). For 'stiff' yachts and/or for low wind speeds, the most effective propulsion mode was found to be close to the condition for the maximum driving force. For 'tender' yachts and/or for high wind speeds, the sails should be set so as to be close(r) to the maximum efficiency of propulsion.

For small apparent wind angles, that is maximum VMG conditions, we can be a little more specific. We have seen in Sect. 4.3, Fig. 4.3.9, that α_{Tmax} is then close to, and $\alpha_{\varepsilon Pmin}$ is equal to the angle of attack $\alpha_{\varepsilon Amin}$ for the minimum aerodynamic drag angle. It follows that the most effective 'mode' of propulsion for maximum VMG conditions requires that the sails are set for an angle of attack close to $\alpha_{\varepsilon Amin}$, in particular for 'tender' yachts and/or for high wind speeds.

It is also clear from Sects. 4.2 and 4.4 that for all yachts the minimum apparent wind angle can be realized only when the sails are set for the minimum aerodynamic drag angle. This is another indication that the best VMG is always obtained with the angle of attack of the sails set close to $\alpha_{\varepsilon Amin}$.

It also appears from Fig. 4.5.3 that the VMG increases with increasing V_b/V_t for $\beta < \cong 35°$ but decreases with boat speed at larger apparent wind angles. For $\beta \cong 35°-50°$ the VMG is almost independent of boat speed.

Surprising as this may seem at first sight, it is readily understood when it is realized that, for a constant apparent wind angle, the true wind angle increases when boat speed increases (see Fig. 3.3.3). This means that the factor $\cos\gamma$ in Eq. (4.5.1) decreases when boat speed increases. As it happens, the positive effect of the increase of boat speed is nullified by the decrease of the factor $\cos\gamma$ when the apparent wind angle is about 35°–50°.

What Fig. 4.5.3 seems to say, is that there is no point in trying to improve VMG by improving boat speed for apparent wind angles above about 35°–50°. Rather, one should then try to reduce the apparent wind angle at the cost of some boat speed. This situation, however, hardly occurs in practice and then only for boats with very poor aerodynamic as well as poor hydrodynamic characteristics (minimum drag angles of the order of 20° or more).

Figure 4.5.4 gives an example of the upwind performance of a real sailing yacht. Shown is the (maximum) VMG as a function of β for a 12-Metre type yacht for three different wind speeds. The figure is based on data from Marchaj (2000b).

Note that the shape of the curves is similar to those of the generic curves in Fig. 4.5.3, except at small apparent wind angles. It can be seen that the best VMG is obtained for an apparent wind angle that is about 5° higher than the lowest values of β as contained by the data (left end of the curves).

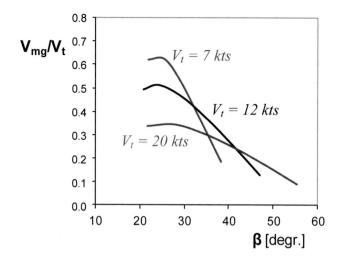

Fig. 4.5.4 *Example of variation of VMG with apparent wind angle (12-Metre type yacht (Marchai2000b))*

Maximizing Boat Speed for a Given True Wind Angle
This is usually the objective if a waypoint can be reached in one tack. In general, boat speed will increase when the driving force T is increased. Maximizing the driving force requires, as we have seen in Sects. 4.3 and 4.4, an angle of attack of the sails somewhere between the angle for the minimum drag angle ($\alpha_{\varepsilon Amin}$) and the angle for maximum lift (α_{Lmax}). For small apparent wind angles α should be close to $\alpha_{\varepsilon Amin}$. For apparent wind angles around 90° it should be close to α_{Lmax}. However, because of the fact that the relation between the true wind angle and the apparent wind angle depends on the ratio V_b/V_t between boat speed and true wind speed (Fig. 3.3.3), the condition for the maximum driving force for a given true wind angle depends also on V_b/V_t. We can see from Fig. 3.3.3 that, for a given true wind angle, the apparent wind angle decreases when V_b/V_t increases. This implies that, for a given true wind angle, the angle of attack for the maximum driving force should be smaller (and the sheeting angle larger) for a high value of V_b/V_t (that is, usually, for low wind speeds) than for low values of V_b/V_t (that is, usually, for high wind speeds).

An example of the dependence of (maximum) boat speed on the apparent wind angle is given in Fig. 4.5.5. The data are for the same 12-Metre type yacht (Marchai 2000b) as in Fig. 4.5.4.

In agreement with the behaviour of the driving force and efficiency of propulsion (Figs. 4.3.1 and 4.3.6), boat speed is seen to increase steadily with increasing apparent wind angle for the higher wind speeds.

However, for the lowest wind speed indicated (7 kts), boat speed appears to reach a maximum at an apparent wind angle of about 35°. The reason for this is

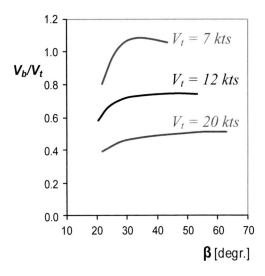

Fig. 4.5.5 *Example of variation of boat speed with apparent wind angle (12-Metre type yacht)*

the dependence of the aerodynamic forces, the lift force in particular, on the apparent wind speed. As we have seen in Fig. 3.3.2 the apparent wind speed decreases rapidly with increasing apparent wind angle for high values of V_b/V_t. For a true wind speed of 7 kts the decrease of the aerodynamic lift L as a consequence of the decreasing apparent wind speed overrules the increase of the driving force/lift ratio with increasing apparent wind angle.

4.6   Reaching, Running and Downwind Sailing

Reaching
If beam reaching is defined as sailing at an apparent wind angle of about 90°, it follows from Sect. 3.4, Eqs. (3.4.3) and (3.4.4), that for $\beta = 90°$, the driving force T is equal to the lift L of the sails and the aerodynamic heeling force S_A equal to the drag D. (See also Fig. 4.6.1). As already discussed in Sect. 4.3, the maximum driving force is obtained when the sails are operating at maximum lift. This, usually, requires angles of attack of the order of 30°–40° and corresponding sheeting angles of the order of 60°–50°. As also illustrated by Fig. 4.3.1, the level of the aerodynamic drag is irrelevant for the driving force under such conditions. There is, however, as already discussed in Sect. 4.3, an indirect effect of the aerodynamic heeling force on boat speed through the effect of side force S_H on the hydrodynamic resistance R and through the heel angle. Because in equilibrium conditions we have $S_H = S_A = D$, a low aerodynamic drag at maximum lift is, indirectly, favourable for boat speed in reaching conditions. The most effective 'mode' of propulsion at beam reaching conditions is therefore in general with the sails set at an angle of attack a little below

Fig. 4.6.1 *Orientation of forces in beam reaching conditions*

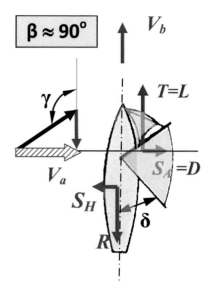

α_{Lmax} (see also Fig. 4.3.9). As argued in Sect. 4.3 the angle of attack should be close to α_{Lmax} for 'stiff' yachts and/or for low wind speeds. For 'tender' yachts and/or for high wind speeds, it should be (a little) closer to $\alpha_{\varepsilon Amin}$. See also Fig. 4.4.4 for a qualitative indication of the corresponding sheeting angles.

Running

Running is usually defined as sailing in the same direction as the wind, that is at a true wind angle of about 180°. The apparent wind angle is then also 180°.

   As indicated in Fig. 4.6.2 the driving force is then equal to the drag of the sails and the heeling force equal to the lift. It follows, as already discussed in Sect. Equilibrium of Forces and Moments in the Longitudinal Plane, that the maximum driving force is obtained when the sails are set for maximum drag. Because of the indirect effect of the heeling force on boat speed it is now advantageous to have a small aerodynamic lift force. The most effective propulsion 'mode' in running conditions is therefore obtained when the sails are set at an angle of attack somewhere between the value (α_{Dmax}) for maximum drag and the value $\alpha_{\varepsilon Pmin}$ for the minimum thrust angle (see also Fig. 4.3.9). It should be close to α_{Dmax} ($\approx 90°$) for yachts with a low sensitivity of the hydrodynamic resistance to side force and/or 'stiff' yachts and/or for low wind speeds. For yachts with a high sensitivity of the hydrodynamic resistance to side force and/or 'tender' yachts and/or for high wind speeds, it should be closer to $\alpha_{\varepsilon Pmin}$. There is, however, in general not a large difference between α_{Dmax} and $\alpha_{\varepsilon Pmin}$ for $\beta = 180°$; for most sail configurations the angle of attack for maximum drag is close to 90° and the corresponding lift relatively small (Fig. 3.4.2). As a consequence the angle of attack for a zero thrust angle at $\beta = 180°$ (zero lift situation) is also about 90°.

Fig. 4.6.2 *Orientation of forces when running dead downwind*

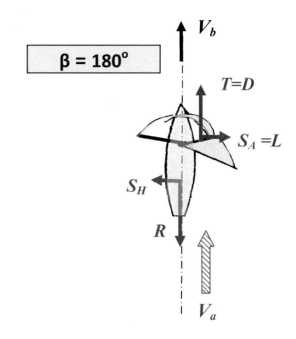

Downwind Sailing

When the purpose of sailing is to reach a waypoint that is positioned in the same direction as the true wind in the shortest possible time a *dead run* may not be the best solution. There are several reasons for this.

First of all we have seen in Sect. 4.3, Figs. 4.3.1 and 4.3.2, that the driving force/lift ratio and driving force/drag ratio both have a maximum at $\beta = 135°$ for $\varepsilon_A = 45°$, that is when lift and drag are equally large. This means, that for a given wind speed, the driving force potential at $\beta = 135°$ will be larger than at $\beta = 180°$. More specifically we have seen in Fig. 4.3.4 that, for a given apparent wind speed, the driving force itself has a maximum for an apparent wind angle of the order of 130°–150°, depending on the sail configuration. We have also seen that for a given true wind speed the apparent wind speed increases if the apparent wind angle is moving away from 180° (Fig. 3.3.2). As a result of both factors the driving force and, consequently, boat speed will be higher at apparent wind angles smaller than 180° than on a dead run. How much higher depends, of course, on the apparent wind angle, but also on the level of boat speed. If the boat speed is already close to the 'critical' boat speed (which would be the case at high wind speed) there is not much room for increasing boat speed by heading away from $\beta = 180°$. The shortest road ($\beta = 180°$) is then also the fastest road. At low boat speeds, however, the rate of increase of boat speed with decreasing apparent wind angle can be so large that the component of boat speed in the true wind direction is larger than on a dead run.

Under such conditions, that is at low wind speeds, the faster road downwind is by alternating port and starboard tacks, through gybing, keeping the apparent wind angle somewhere well below 180°. The optimum apparent wind angle depends on wind speed (and boat speed). For low wind speeds it will be in the 130°–150° range, or even smaller. For high wind speeds it will be closer to 180°. The precise value depends not only on wind speed but also on the characteristics of the yacht, hull as well as sails. For heavy displacement yachts, which usually have a pronounced 'drag rise' at the 'critical' boat speed, it does not pay very much to sail at $\beta < 180°$ at high wind speeds. For light displacement yachts that have a less pronounced 'drag rise' tacking downwind may also pay off at high wind speeds.

It is appropriate at this point to introduce the notion *Velocity-Made-Good-to-lee-ward* (V_{mgl}) or *downwind VMG*. This is nothing else but the component of boat speed in the true wind direction (see Fig. 4.6.3). One can write

$$V_{mgl} = V_b \ (-\cos \gamma) \qquad\qquad (4.6.1),$$

where $90° < \gamma < 180°$. This expression is, of course, basically the same as Eq. (4.5.1). In fact, one could use Eq. (4.5.1) irrespective of the sign of $\cos \gamma$ and call the value of the result the Velocity-Made-Good-to-leeward when it is negative.

Fig. 4.6.3 *Defining Velocity-Made-Good to leeward (* V_{mgl} *)*

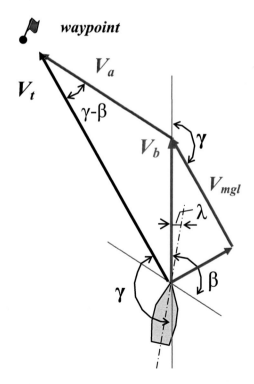

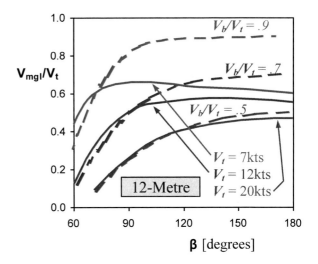

Fig. 4.6.4 *Downwind VMG*
as a function of apparent
wind angle for a 12-Metre
type of yacht

We can also use Eqs. (4.5.2) and (3.3.4) to calculate V_{mgl} as a function of β and the ratio V_b/V_t of boat speed to true wind speed. The dashed lines in Fig. 4.6.4 give the result of such calculations. Also given in Fig. 4.6.4 is the downwind VMG of a 12-Metre type yacht at three different true wind speeds (solid lines, data from Ref. Marchai 2000b). It appears, in agreement with the discussion given above, that for a light, 7 kts true wind speed the downwind VMG has its maximum at an apparent wind angle of about 100°. For a 12 kts wind this is at $\beta \approx 140°$. For a 20 kts wind it does not pay anymore for a heavy displacement yacht like a 12-Metre to sail at an apparent wind angle below 180°.

Figure 4.6.5 presents similar information in terms of the true wind angle. In a 7 kts wind the optimal downwind true wind angle of a 12-Metre is about 145°. In a

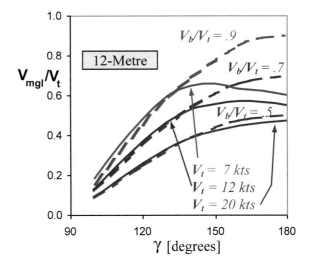

Fig. 4.6.5 *Downwind VMG*
as a function of true wind
angle

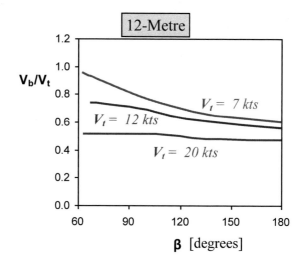

Fig. 4.6.6 *Downwind boat speed of a 12-Metre yacht as a function of true wind angle*

12 kts wind this is about 160°. Note also from Figs. 4.6.4 and 4.6.5 that the curve for $V_t = 20$ kts almost coincides with the curve for $V_b/V_t = 0.5$. This means that in a 20 kts wind a 12 m runs at its maximum ('critical') boat speed for all apparent wind angles above, say, 70° and that this 'critical' boat speed is about $0.5*20 = 10$ kts. The latter is confirmed by Fig. 4.6.6 which gives downwind boat speed as a function of apparent wind angle for the same 12-Metre yacht: for a true wind speed of 20 kts V_b/V_t is almost constant and about 0.5 for all apparent wind angles.

4.7 Speed Polar Diagram

A convenient and common way to summarize the speed potential of a sailing yacht is to do so in the form of a *speed polar diagram*. A polar diagram in general is a representation of a vector quantity in so-called polar coordinates. That is, the (relative) direction of the vector is represented by the angle (or azimuth) between the vector and a reference direction. The magnitude of the vector is represented by its length in the radial direction. In the speed polar diagram of a sailing yacht the vector quantity is ('the best') boat speed and the azimuth is the true wind angle. The data on which a speed polar is based can be from full scale sailing tests or computations (about which later, in Chap. 8) or from a combination of both.

Figure 4.7.1 gives an example of a speed polar. Shown is the maximum boat speed as a function of the true wind angle of a 63 ft light/medium displacement cruising yacht (Gerritsma and Keuning 1984). The colored lines represent the envelopes of the boat speed vectors as a function of the true wind angle for five different levels of the true wind speed.

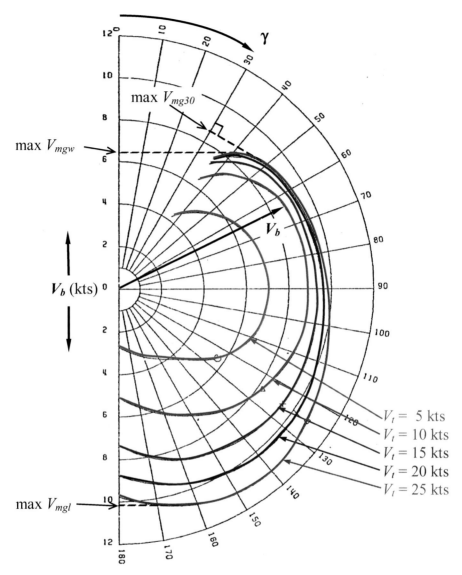

Fig. 4.7.1 *Example of a speed polar diagram (63 ft light/medium displacement cruising yacht, Gerritsma and Keuning (1984))*

Also indicated in the figure are the maximum upwind and downwind VMG. For any level of true wind speed these can be obtained readily by projection of the envelope on the vertical axis (dashed lines).

Incidentally, the figure illustrates the point discussed in the preceding section that for light displacement yachts tacking downwind may also pay off at high wind speeds (usually not the case for heavy displacement yachts). For completeness we remark that the notion of Velocity-Made-Good can be generalized so as to cover also directions to waypoints not coinciding with the direction of the true wind. For example, the VMG towards a waypoint in a direction at 30° (V_{mg30}) from the true wind can be obtained by projection of the polar curve on the line for $\gamma = 30°$. In the case of Fig. 4.7.1 it appears that in a 25 kts wind sailing at $\gamma \approx 50°$ gives the best V_{mg30} (≈ 8.6 kts) while sailing at the best VMG to windward, at $\gamma \approx 40°$, gives $V_{mg30} \approx 8.3$ kts. This is a difference of almost 4%. In a 5 kts true wind the corresponding figures are $\gamma \approx 60°$ with $V_{mg30} \approx 5.7$ kts and $\gamma \approx 45°$ with $V_{mg30} \approx 5.3$ kts, respectively, a difference of 7%. The example illustrates that, even upwind, sailing towards a waypoint in one tack (or as close to the wind as possible) is not always the fastest way, in particular at low wind speeds. In general the shortest course to a waypoint in terms of distance and the fastest course in terms of time will differ significantly when boat speed varies rapidly with the true wind angle. In terms of Fig. 4.7.1 that is when the polar curves intersect the semi-circle(s) of constant boat speed at a large angle. The latter is the case at the downwind as well as at the upwind end of the polar curves. For true wind angles between, say, 80°–100° at low wind speeds and 130°–160° for high wind speeds, there will be very little difference between the shortest and fastest course. Heading directly for the waypoint is then the fastest option, not in the least because tacking or gybing also takes time.

References

1. Barkla H (1965) The best course to windward, yachts and yachting, Feb 19
2. Gerritsma J, Keuning JA (1984) Experimental analysis of five keel-hull combinations, Delft University of Technology, Ship Hydromechanics Lab, Report no. 637
3. Marchai CA (2000a) Aero-hydrodynamics of sailing. Adlard Coles Nautical, London, p 48
4. Marchai CA (2000b) Aero-hydrodynamics of sailing. Adlard Coles Nautical, London, p 71

Chapter 5
Elements of Fluid Mechanics (Air and Water)

5.1 Introduction

There are many good books on fluid mechanics. Classical examples are the books by Prandtl and Tietjens (1934a in Chap. 1), already mentioned in the Introduction, and *'The Theory of Wing Sections'* by Abbott and Von Doenhoff (1949). More recent examples are *'Fundamentals of Aerodynamics'* by John D. Anderson (2001), and *'Low-Speed Aerodynamics'* by Katz and Plotkin (1991).

This chapter is, of course, not intended to replace such books. Rather, the objective is to provide the reader with a certain level of basic knowledge of general fluid mechanics. This level, it is believed, will be sufficient to serve the purpose of facilitating the description and understanding of the more specific phenomena and mechanisms that are underlying the aero- and hydrodynamic forces acting on a sailing yacht. The latter will be discussed in more detail in Chaps. 6 and 7.

In most if not all of the books mentioned above we can find that the basic laws of physics governing the motion of liquids (like water) and those governing the motion of gasses (like air) are the same. Fluid mechanics, in fact, is the science of the forces acting in fluids, where everything that can flow is considered a fluid. Both, liquids like water and gasses like air belong to this category.

In a subdivision of fluid mechanics distinction is made between forces acting in still fluids, that is *fluid statics*, and forces acting in moving fluids, that is *fluid dynamics*.

Because air and water have different properties, in particular in terms of density, and are separated by a boundary surface, there are also phenomena in water that do not occur in air and vice versa. The most obvious example of this is, of course, the generation of waves and associated phenomena in the boundary surface between air and water. This will be addressed later, in Chaps. 6 and 7.

In the following sections we will discuss the laws and the associated phenomena that are common in the mechanics of air and water.

© Springer International Publishing Switzerland 2015
J. W. Slooff, *The Aero- and Hydromechanics of Keel Yachts,*
DOI 10.1007/978-3-319-13275-4_5

5.2   Pressure and Friction Forces

In the preceding chapters we have considered the overall forces acting on a sailing yacht. That is, the aerodynamic as well as the hydrodynamic forces were 'added-up' to form resulting, overall forces and moments in air and water respectively. In doing so the resulting force vectors and moments were concentrated in the center of gravity of the yacht (or some other appropriate location) and the factors determining the performance of a sailing yacht were discussed in terms of what is called *'point mechanics'*. If we want to consider the forces on hull and sails and the underlying physical mechanisms in some detail we must zoom in on the surfaces where the forces are generated.

Any object that is fully or partly submerged in a moving fluid (or any object that moves in and relative to a fluid) is subject to forces acting on the external surface of the object. In terms of the orientation of the forces two types can be distinguished: pressure forces and friction forces (Fig. 5.2.1):

Pressure forces are acting 'normal', that is perpendicular to the surface. Friction forces act 'tangential' to the surface, that is their orientation is (locally) parallel to the surface.

Pressure forces are already present in a still fluid. In still air they are the result of the internal, atmospheric pressure, which is due to the 'weight' of the column of air above the object. In still water they are the result of the weight of the column of water plus the column of air above the object. This, as already mentioned above, is the subject of *fluid statics* (Sect. 5.3).

Fig. 5.2.1 *Illustrating pressure and friction forces*

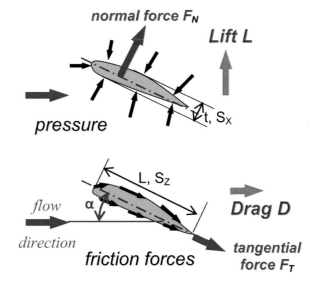

Friction forces require that there is 'friction' between the object and the surrounding fluid. This is the case only when the object and the fluid in which it is submerged are in relative motion. In the latter situation the pressure around the object also is no longer constant. Like the friction forces the pressure varies along the surface. Why and how they vary along the surface is the subject of *fluid dynamics.*

Added-up ('integrated') around the surface the pressure and friction forces give rise to resulting overall forces like lift L and drag D (recall from Chaps. 3 and 4 that the lift force acts normal to the direction of the incoming flow and the drag parallel). For relatively thin bodies, such as wings, keels and sails, it is useful to introduce the notions normal force F_N and tangential force F_T (see Fig. 5.2.1). Lift and drag can be expressed in terms of the normal force and tangential force:

$$L = F_N \cos\alpha - F_T \sin\alpha \qquad (5.2.1a)$$

$$D = F_N \sin\alpha + F_T \cos\alpha, \qquad (5.2.1b)$$

where α is the angle of attack between the main axis or main plane of the body and the direction of the incoming flow.

Distinguishing forces due to pressure and forces due to friction one can write:

$$F_N = F_{Np} + F_{Nf} \qquad (5.2.2a)$$

and

$$F_T = F_{Tp} + F_{Tf} \qquad (5.2.2b)$$

It will be clear from Fig. 5.2.1 that the part F_{Np} due to pressure of the normal force is, in two dimensions, proportional to the length L and, in three dimensions, to the area S_Z of the main plane of the body. Similarly, the part F_{Nf} due to friction is proportional to the thickness 't' or cross-sectional area S_X. This means that F_{Np} and F_{Nf} can be expressed as, respectively,

$$F_{Np} = f_{np}\, S_Z \qquad (5.2.3a)$$

and

$$F_{Nf} = f_{nf}\, S_X \qquad (5.2.3b)$$

Similarly

$$F_{Tp} = f_{tp}\, S_X \qquad (5.2.4a)$$

and

$$F_{Tf} = f_{tf}\, S_Z \qquad (5.2.4b)$$

The factors f_{np}, f_{nf}, f_{tp} and f_{tf} are functions of the distribution of pressure and friction forces around the body.

Combining the various expressions gives

$$L = (f_{np}\ S_Z + f_{nf}\ S_X)\cos\alpha - (f_{tp}\ S_X + f_{tf}\ S_Z)\sin\alpha \qquad (5.2.5a)$$

$$D = (f_{np}\ S_Z + f_{nf}\ S_X)\sin\alpha + (f_{tp}\ S_X + f_{tf}\ S_Z)\cos\alpha \qquad (5.2.5b)$$

The lift L_P due to pressure and L_F due to friction can be written as

$$L_P = f_{np}S_Z\cos\alpha - f_{tp}S_X\ \sin\alpha \qquad (5.2.6a)$$

$$L_F = f_{nf}S_X\cos\alpha - f_{tf}S_Z\ \sin\alpha \qquad (5.2.6b)$$

The pressure drag D_P and friction drag D_F are given by

$$D_P = f_{np}\ S_Z\ \sin\alpha + f_{tp}\ S_X\ \cos\alpha \qquad (5.2.7a)$$

$$D_F = f_{nf}\ S_X\ \sin\alpha + f_{tf}\ S_Z\ \cos\alpha \qquad (5.2.7b)$$

It follows from Eqs. (5.2.3) and (5.2.4) that for thin or slender bodies, that is when $S_X \ll S_Z$, the normal force is built-up mainly by pressure forces and the tangential force mainly by friction forces. This means that for thin or slender bodies at small angles of attack, ($\sin\alpha \ll 1, \cos\alpha \approx 1$), Eq. (5.2.5) reduce to

$$L \approx f_{np}\ S_Z \approx L_P, \quad (\sin\alpha \ll 1) \qquad (5.2.8a)$$

$$D \approx f_{tf}\ S_Z \approx D_F, \quad (\sin\alpha \ll 1) \qquad (5.2.8b)$$

This means that, then, the lift consist mainly of pressure forces and, in principle, the drag mainly of friction forces.[1] The opposite is the case for a thin body at (very) high angles of attack ($\cos\alpha \ll 1, \sin\alpha \approx 1$). Equation (5.2.5) then reduce to

$$L \approx f_{tf}\ S_Z \approx L_F, \quad (\cos\alpha \ll 1) \qquad (5.2.9a)$$

$$D \approx f_{np}S_Z \approx D_P, \quad (\cos\alpha \ll 1) \qquad (5.2.9b)$$

Then, the drag will consist mainly of pressure forces and the lift mainly of friction forces. When the friction forces are small, the total force on a thin body is approximately normal (perpendicular) to the plane of the body.

For thick bodies ($S_X \approx S_Z$) such arguments do not hold. Based on relative dimensions only, pressure and friction forces are then equally important for lift and drag, irrespective of the angle of attack.

---

[1] We will see later, in Sect. 6.5, that an exception is formed by the wave-making resistance, which consists of pressure drag.

*Fig. 5.3.2 Still fluid static pressures
on a (partially immersed) sailing
yacht*

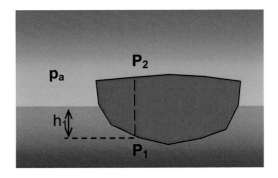

that the pressure at the air/water interface is equal to the atmospheric pressure p_a caused by the weight of the column of air above. Secondly, the height of the column of air above a yacht is very much larger that the vertical dimension of the yacht and the density of air is very small (Sect. 5.5). It then follows (Eq. (5.3.1), that the static air pressure acting on the yacht can, for all practical purposes, be taken as constant and equal to the atmospheric pressure at sea level.

In a submerged point (P_1) of the hull, the static pressure is the result of the weight of the column of water plus the weight of the column of air above it. It follows that

$$p_1 = p_a + \rho\, g\, h_1 \qquad (5.3.4)$$

For a point P_1 on the bottom and a point P_2 on the top of the hull that are positioned directly above each other the pressure difference Δp is now equal to

$$\Delta p = p_1 - p_2 = p_a + \rho\, g\, h_1 - p_a = \rho\, g\, h_1 \qquad (5.3.5)$$

Note that the atmospheric pressure has cancelled out in the final expression for the pressure difference and, hence, is of no consequence for the vertical force.

It now follows that the total vertical lift or buoyancy force is equal to

$$\Delta = \rho\, g\, \nabla, \qquad (5.3.6)$$

where ∇ is the volume of the submerged part (in water) of the hull. ∇ is also called the displacement volume. At this point we can conclude that aerostatic forces do not play a role of any significance in sailing yacht mechanics but that hydrostatics do. We will come back to hydrostatic forces in some more detail in Chap. 6.

5.4   The Conservation Laws of Fluid Motion

The motion of fluids and the associated forces acting on a body in a moving fluid are governed by a number of fundamental laws: the so-called *conservation laws of physics*. For every elementary volume occupied by the fluid there holds:

Fig. 5.4.1 *Mass conservation for an elementary volume of fluid*

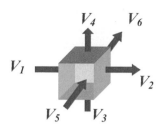

- *conservation of mass,* expressing that mass cannot just appear or disappear
- *conservation of momentum,* expressing that the momentum (or quantity of motion) of a fluid particle in equilibrium is invariant
- *conservation of energy,* expressing that energy cannot appear or disappear spontaneously

A consequence of the mass conservation law is that the quantity of fluid matter that flows into a volume element must be equal to the quantity that flows out of it. For a volume element as depicted in Fig. 5.4.1 this means that

$$V_1 - V_2 + V_3 - V_4 + V_5 - V_6 = 0,$$

provided that the sides of the volume element are of equal size and small enough to justify the assumption that the variation of the normal velocity components V_1 to V_6 over the sides of the volume element is negligible.

The conservation law for momentum is nothing else but Newton's Law applied to a fluid particle. In the form of Sect. 3.5 it describes the balance between the forces acting on an object and the motion of the object in a fixed frame of reference. This was written as $\underline{F} = m * \underline{a}$ (Eq. 3.5.1). As mentioned already in Sect. 3.5 the object itself and anything attached to it seems to experience a virtual, so-called *'inertia'* force F_i equal to $-m * \underline{a}$ in the direction opposite to the direction of acceleration.

In a frame of reference that moves with the fluid particle Newton's Law can therefore be written as

$$\underline{F}_e + \underline{F}_i = 0, \qquad\qquad (5.4.1)$$

expressing that the external forces $\underline{F}_e$ and the inertia forces $\underline{F}_i$ are in balance.

Distinguishing the external forces as pressure forces $\underline{F}_p$ and friction forces $\underline{F}_f$ one can also write

$$\underline{F}_p + \underline{F}_f + \underline{F}_i = 0, \qquad\qquad (5.4.2)$$

expressing that the pressure forces, friction forces and inertia forces acting on a fluid particle are in balance.

Two further remarks are in place at this point. The first is that a fluid particle (or any object, for that matter) is also subject to the gravitational force. However, the gravitational force is constant throughout a homogeneous fluid of constant density and independent of the fluid motion. It then decouples from the dynamics of fluid motion (Prandtl and Tietjens 1934a in Chap. 1) and plays a role only in fluid statics (Sect. 5.3). Since, as we will see in Sect. 5.8, water is practically incompressible and the density of air moving at low speed is practically constant, the gravitational force is, in general, not important in the hydro- and aerodynamics of sailing yachts. There is an exception, however. Because 'the fluid' around a sailing yacht is not homogeneous at the air/water interface, gravitational forces do play a role near the water surface, (about which later, in Sect. 5.19).

The second remark is that Eq. (5.4.2) is a vector equation. Decomposing the vectors in components along the axes of an orthogonal coordinate system (Sect. 3.1), Eq. (5.4.2) is equivalent with three algebraic equations, one for each direction.

The energy conservation law is not of great importance in the aero- and hydrodynamics of sailing yachts. The reason is that it can be shown (Prandtl and Tietjens 1934a in Chap. 1), that the flow of an incompressible fluid[2] with negligible temperature effects is already completely determined by the conservation laws for mass and momentum.

It is worth noting at this point that the conservation laws discussed above also form the basis of the relatively new discipline of Computational Fluid Dynamics (CFD). In CFD the space around an object is divided into a very large number (of the order of millions) of elementary volumes and the conservation laws are satisfied through numerical approximation for each element. Solving the resulting set of millions of coupled algebraic equations requires very large computer power. The final result consists of the velocity components, pressure, friction, etc. in all of the volume elements. These can be integrated (added-up) to determine the forces acting on the object. CFD has been applied to the aero- and hydrodynamics of sailing yachts since the 1980s and is nowadays a valuable alternative to wind tunnel and towing tank testing (Larsson 1990; Milgram 1998).

5.5   A Consequence of Mass Conservation: The Venturi Effect

Streamlines and Stream Tubes
In any (steady) fluid flow it is possible, in principle, to draw curves such that the direction of the curve in each point agrees with the direction of the velocity at that point. Such curves are called *streamlines*. Streamlines can be used to map out the direction of the fluid velocity at every point in the space occupied by the fluid. Figure 5.5.1 gives an example of streamlines of the two-dimensional flow about the front part of a

---

[2] As already mentioned and as we will see later (Sect. 5.8), both water and air at low speeds can be considered as incompressible for all practical purposes.

Fig. 5.5.1 *Streamlines of a steady flow*

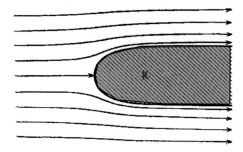

Fig. 5.5.2 *Stream tube*

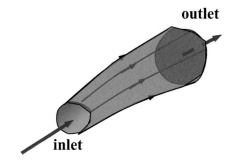

body. For a steady flow it can be shown (Prandtl and Tietjens 1934a in Chap. 1) that the streamlines are identical with the paths or trajectories of fluid particles.

A bunch of adjacent streamlines form a *stream tube* or a *stream filament* (Fig. 5.5.2).

Because the outer mantle of a stream tube is made up of streamlines there can be no mass flowing through it. Hence, fluid matter can enter or leave the stream tube only through the front and rear inlet/outlet faces. It follows that the stream tube acts like a real tube with a rigid wall inside which fluid flows.

The Venturi Effect

Let us now consider a stream tube or converging-diverging channel as in the upper part of Fig. 5.5.3. It follows from the mass conservation law that the mass flux m through the channel is constant and equal to

$$m = \rho V S_{\otimes} = \text{constant} \tag{5.5.1}$$

This equation is also known as the continuity equation. The implication is that the fluid flow velocity V in a channel with varying cross-section is inversely proportional to the cross-sectional area $S_{\otimes}$. As a consequence the fluid flow velocity V varies along the length (x) of the channel as indicated by the red curve in the middle graph of Fig. 5.5.3: the flow velocity is high in the middle, narrow part of the channel and low in the wider parts.

This phenomenon is sometimes called the *Venturi effect*, after the Italian physicist Venturi (1746–1822) who experimented with these kind of channels and tubes.

Fig. 5.5.3 *Illustrating the Venturi effect*

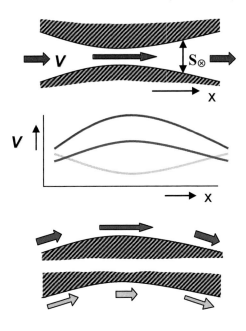

In case the flow is bounded by a curved wall on one side only, such as illustrated by the lower sketches of Fig. 5.5.3, similar phenomena occur with respect to the fluid flow velocity. This can be understood by imagining that the top wall in the upper part of Fig. 5.5.3 moves further and further away from the bottom wall. The variation of the flow velocity along the bottom wall will then be qualitatively similar to the variation in the case of the confined channel. However, the variation will be less pronounced because the fluid has 'more room' away from the wall (see the blue curve in the middle graph of Fig. 5.5.3).

In the case just described, the wall is said to have convex curvature (locally narrowing channel). If we would have started out from a diverging-converging channel rather than a converging-diverging one, the lower wall would have concave curvature (locally widening channel (see the bottom sketch in Fig. 5.5.3). The velocity distribution would then be like the yellow curve in the middle graph of Fig. 5.5.3, that is the flow velocity would be low where the indentation of the wall is large.

5.6   A Consequence of Conservation of Energy: Bernoulli's Equation

For non-viscous, that is frictionless, steady fluid flows it is possible to derive from the energy conservation law (or, alternatively, for incompressible fluids, from the momentum conservation law) an equation that expresses a relation between the pressure and the local flow velocity along a streamline (Prandtl and Tietjens 1934a in Chap. 1). This equation is known as Bernoulli's equation or Bernoulli's law (after the Swiss mathematician and physicist Daniel Bernoulli (1700–1782). In the general case that a boundary surface (e.g. between air and water) is present and gravitational forces are important it can be written as:

$$p + \tfrac{1}{2}\rho V^2 + \rho g z = \text{constant}, \qquad (5.6.1)$$

where, as before, p is the (static) pressure, ρ the density, g the constant of gravitational acceleration, z is the vertical position of the streamline above some chosen reference height (see also Fig. 5.3.1, where z=−h) and V is the local flow velocity.

A simpler form of Bernouilli's equation is obtained by replacing the gravitational term $\rho g z$ (= −$\rho g h$) by −p_g (see Eq. 5.3.1):

$$p - p_g + \tfrac{1}{2}\rho V^2 = \text{constant} \qquad (5.6.2)$$

Replacing the difference $p - p_g$ between the static pressure and the gravitational pressure by p' Eq. (5.6.2) takes the form

$$p' + \tfrac{1}{2}\rho V^2 = \text{constant} \qquad (5.6.3)$$

Here p' is called the *dynamical part of the static pressure* and the product $\tfrac{1}{2}\rho V^2$ is called the *dynamic pressure*.

Bernoulli's law can also be interpreted in terms of the energy contents of an elementary fluid particle. Associating the pressure with the internal or potential energy and the term $\tfrac{1}{2}\rho V^2$ with the kinetic energy, Eq. (5.6.3) then says that the sum of the internal energy and the kinetic energy of a fluid particle is constant along a streamline.

Bernoulli's equation is of great practical importance because it relates the local pressure in a fluid flow to the local fluid velocity. More specifically, Eq. (5.6.3) tells us that the local pressure is low when the local flow velocity is high and vice versa. It also tells us that the pressure adopts a maximum, called the *total pressure*, when the flow stagnates, that is when the local flow velocity V is zero. The total pressure is usually denoted as p_0. Eq. (5.6.3) can then also be written as

$$p' + \tfrac{1}{2}\rho V^2 = p_0 \qquad (5.6.4)$$

Obviously, the total pressure is the sum of the (dynamical part of) the static pressure and the dynamic pressure and Bernoulli's law implies also that in a non-viscous flow the total pressure is constant along a streamline.

It is instructive at this point to consider once more the converging-diverging channel flow of Fig. 5.5.3 that was introduced to illustrate the Venturi effect. As we have seen in Sect. 5.5 the mass conservation law implies that the local flow velocity is high in the narrow, middle part of the channel or tube and low in the wider sections at the inlet and outlet. Bernoulli's law teaches, as illustrated by Fig. 5.6.1, that this means that the pressure is low in the middle of the channel where the flow velocity is high and that the pressure is higher at the ends where the flow velocity is lower.

*Fig. 5.6.1 Illustrating Bernoulli's law for
a Venturi tube: the pressure is low when the
velocity is high
and vice versa*

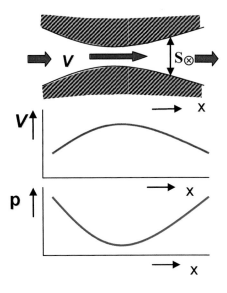

5.7   Consequences of Conservation of Momentum: Scaling Laws for Inertia Forces, Friction Forces and Pressure Forces

From the conservation law for momentum (or Newton's law applied to an elementary fluid particle, Sect. 5.4) it is possible to derive a number of scaling laws that indicate how the different types of forces acting on a body in a fluid flow depend, qualitatively, on specific characteristics of the fluid flow and the size of the body. For example, it can be shown (Prandtl and Tietjens 1934b in Chap. 1) that the inertia forces per unit of volume acting in a fluid flow about a body scale like

$$F_i \div \rho V^2/L \tag{5.7.1}$$

That is, the magnitude of the inertia forces per unit volume is proportional to the fluid density ρ, the square of the velocity and inversely proportional to a measure of the size of the body or object, such as a characteristic length L. Equation (5.7.1) implies amongst others, that for the flow about a body of fixed size, the inertia forces become four times larger when the fluid velocity is doubled (Fig. 5.7.1).

It can also be shown (Prandtl and Tietjens 1934b in Chap. 1) that the friction forces per unit volume in a fluid flow scale like

$$F_f \div \mu V/L^2, \tag{5.7.2}$$

where μ is the dynamic viscosity or 'stickiness' of the fluid and L is, again, a measure of the size of the object. As before, V is the flow velocity. Note that the friction forces scale with the viscosity rather than the density and with the ratio V/L^2 rather than V^2/L.

Fig. 5.7.1 *Illustrating the scaling law for inertia forces*

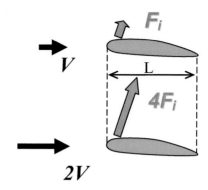

Figure 5.7.2 illustrates the scaling law for friction forces for a body of given, fixed size: the friction forces are proportional with the flow velocity.

Unfortunately it is not possible to derive a general, simple scaling law for pressure forces. However it is possible to arrive at some conclusions by considering Newton's Law applied to a fluid particle (Eq. 5.4.2). In words Eq. (5.4.2) says that the inertia, friction and pressure forces acting on a fluid particle are in equilibrium. From this it follows that the pressure forces scale like something as the sum of the inertia forces and the friction forces. This could be written as

$$F_p \div c_i \rho V^2/L + c_f \mu V/L^2, \tag{5.7.3}$$

where c_i and c_f are constants of proportionality.

The 'general' scaling law (5.7.3) for pressure forces is fairly complex. It can only be simplified by assuming a certain relation between the inertia and the friction forces. It is useful for this purpose to consider the ratio F_i/F_f between inertia forces and friction forces. It follows from Eqs. (5.7.1) and (5.7.2) that

$$F_i/F_f \div (\rho V^2/L)/(\mu V/L^2) = \rho V L/\mu \tag{5.7.4}$$

Fig. 5.7.2 *Illustrating the scaling law for friction forces*

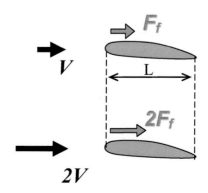

The last quantity is called the *Reynolds number* (Re), named after the British physicist Osborne Reynolds (1842–1912). It is a dimensionless number given by

$$Re = \rho V L / \mu \tag{5.7.5}$$

The Reynolds number is an important, so-called *similarity parameter* for all viscous fluid flows. It implies that the flows about similarly shaped but different size bodies are similar when the Reynolds numbers are the same. This is true even for flows of different types of fluid.

In many if not most fluid flows of practical interest, including the fluid flow about sailing yachts, the Reynolds number is very large, of the order of several millions (see Sect. 5.9). For such conditions it follows from Eq. (5.7.4) that the inertia forces are much larger than the friction forces. It also follows, from Eq. (5.7.3) that the pressure forces then scale like the inertia forces. In other words, for (very) high Reynolds numbers the pressure forces per unit volume scale, approximately, like

$$\boldsymbol{F}_p \div \rho V^2 / L \tag{5.7.6}$$

It must be emphasized that the scaling laws (5.7.1), (5.7.2) and (5.7.6) are applicable for the forces per unit of fluid volume. Scaling laws for the total forces acting on an object can be obtained by multiplying the expressions with a factor representing the total volume of the fluid that is influenced by the flow about the object. Because a volume has the dimension of a length to the third power, the factor representing the total volume can be taken as L^3. This means that the total pressure force $\boldsymbol{F}_p$ acting on an object at very high Reynolds numbers approximately satisfies the scaling law

$$\boldsymbol{F_P} \div \rho (V^2 / L) L^3 \tag{5.7.7a}$$

This can also be written as

$$\boldsymbol{F_P} \div \rho V^2 L^2 \tag{5.7.7b}$$

Since the factor L^2 in (5.7.7b) has the dimension of the area of a surface we may replace (5.7.6) by

$$\boldsymbol{F_P} \div \rho V^2 S \tag{5.7.7c}$$

where S is a characteristic area of the object. From the discussion in Sect. 5.2 it follows further that, in Eq. (5.7.7c), S should be taken as the area of the longitudinal section through the main plane of the body if the lift is concerned. For the pressure drag S should be taken as the cross-sectional area.

A scaling law for the total frictional force $\boldsymbol{F}_F$ acting on an object can be derived in the same way. The result is

$$\boldsymbol{F_F} \div \mu (V / L) S \tag{5.7.7d}$$

Here, S is to be taken as the wetted surface S_{wet} of the body.

For completeness and further reference it is noted that the scaling laws given above can also be expressed in terms of the Reynolds number. For example the scaling law for the total friction force can be written as

$$F_F \div \rho V^2 S \, Re^{-1} \tag{5.7.8}$$

and the scaling law for the pressure force at high Reynolds number can be written as

$$F_P \div \rho V^2 S \, Re^0 \tag{5.7.9}$$

In the latter expression we have used the mathematical identity that any number to the power zero is equal to 1.

5.8   Physical Properties of Air and Water

As already mentioned before, the conservation laws of fluid mechanics and their consequences are valid for all kinds of fluids, liquids as well as gasses, air as well as water. The only thing we have to keep in mind is that we have to insert the proper values for physical quantities like density, viscosity, etc. into the various expressions and equations describing the various kinds of forces.

Water is, of course, a liquid. That is, it is an incompressible fluid for all practical purposes. The mass density depends a little bit on the temperature and a little more on the salinity (Lewis 1988). For fresh water the density ρ_H is 1000 kg/m³ at a temperature of 4 °C and 998 kg/m³ at a temperature of 20 °C. For seawater the density is a little larger, depending on the salinity. At a 'normal' salinity level of 3.5 % the density of seawater varies between 1028 kg/m³ at a temperature of 0 °C and 1025 kg/m³ at 20 °C, a difference less than a third of a percent.

The viscosity or 'stickiness' of water also depends on the temperature and the salinity. It is usually expressed in terms of the kinematic viscosity ν which is the ratio between the dynamic viscosity μ and the density ρ:

$$\nu = \mu/\rho \tag{5.8.1}$$

Note that this is precisely the ratio appearing in the definition (5.7.4) of the Reynolds number. In terms of the kinematic viscosity ν the Reynolds number can be expressed as

$$Re = VL/\nu \tag{5.8.2}$$

As indicated by Fig. 5.8.1 the dependence of the kinematic viscosity on the temperature (Lewis 1988) is quite significant. At 25 °C the value of ν is only about half that at 0 °C.

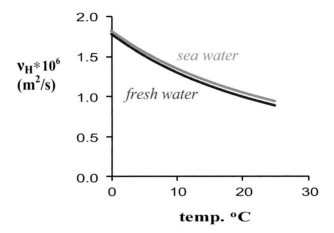

Fig. 5.8.1 *Kinematic viscosity of water as a function of temperature*

Air, on the other hand, is a mixture of gasses (International Standard Atmosphere
n.d.). As such, it is, in principle, compressible. However, it can be shown, and it is
the general experience (Prandtl and Tietjens 1934a in Chap. 1), that air flows behave like the flow of an incompressible fluid as long as the speed or flow velocity
is smaller than about 50 m/s (or about 100 kts). This means that the air flow about
sails, rig and hull of a sailing can be considered as incompressible for all practical
purposes.

The mass density of air is, of course, several orders of magnitude smaller than
that of water. Like the density of water it depends on the temperature. However it
also depends on the atmospheric pressure and, to a lesser extent, on the humidity.
The dependence on temperature and atmospheric pressure is governed by the general gas law. When expressed in terms of density the latter can be written as

$$p = \rho R_{spec} T \qquad (5.8.3a)$$

or

$$\rho = p_a /(R_{spec} T) \qquad (5.8.3b)$$

Here, p_a is the atmospheric pressure, T the temperature in Kelvin (i.e. °C+273)
and R_{spec} is the specific gas constant. For dry air $R_{spec} = 287$ J/kg/K. As indicated by
Eq. (5.8.3) the density increases with increasing atmospheric pressure and decreasing temperature. The variation of the mass density with atmospheric conditions is
significant, as illustrated by Fig. 5.8.2. The difference between high pressure/cold
air and low pressure/hot air conditions can be as much 20%. The effect of humidity,
also indicated in Fig. 5.8.2 for a low pressure condition, comes in through the spe-

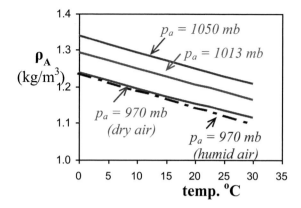

Fig. 5.8.2 *Mass density of air at sea level as a function of atmospheric conditions*

cific gas constant R_{spec} but is small. Humidity is here to be interpreted as waterdamp and not as the (small) fluid water particles that form mist or fog.

The dynamic viscosity μ of air is a (weak) function of the temperature only (International Standard Atmosphere n.d.). At 0 °C its value is 1.7×10^{-5} kg/(m · s) and at 35 °C it is 1.9×10^{-5} kg/(m · s). For all practical purposes it can be taken as 1.8×10^{-5} kg/(m · s) and constant in the temperature range 0–35 °C. However, the kinematic viscosity ν ($= \mu/\rho$) of air, like the mass density, varies significantly with the atmospheric conditions.

As indicated by Fig. 5.8.3 'ν' varies between about 1.3×10^{-5} m²/s at high-pressure/low-temperature and about 1.7×10^{-5} m²/s at low pressure/high temperature, a difference of about 25 %. Note that the kinematic viscosity of air is about a factor 10 larger than that of water and that the dependence on temperature is of opposite sign. This is mainly due to the variation in density.

Fig. 5.8.3 *Kinematic viscosity of air as a function of atmospheric conditions at sea level*

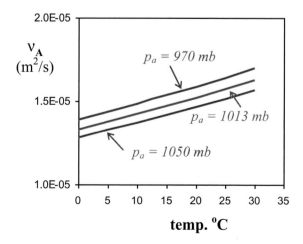

5.9 Sailing Yacht Reynolds Numbers

Having established the ranges of values of the kinematic viscosity of air and water it is now also possible to estimate the range of Reynolds numbers of the flow of water about the under body and the flow of air about the sails of a sailing yacht. We recall that the Reynolds number is defined by

$$\mathrm{Re} = VL/\nu, \qquad (5.9.1)$$

where V is the flow velocity, L is a measure of the length of the (part of the) yacht that is considered and ν is the kinematic viscosity. It is then clear that for the Reynolds number of the water flow about the hull and the appendages the flow velocity V is equal to the boat speed V_b and that for the Reynolds number of the air flow about the sails the flow velocity is equal to the apparent wind speed V_a. It is also clear from the factor L in the numerator of (5.9.1) that the Reynolds numbers increase with the size of a yacht. Boat speed, as we will see later, usually also increases with the length of a yacht.

Unlike classical sailing ships, modern sailing yachts have keels and rudders that are not an integrated part of the hull, although they are still attached to it. As a consequence the flow about these 'appendages' is distinctly separate from the flow about the hull. This means that the Reynolds numbers of the flows about keel(s) and rudder(s) should also be based on the longitudinal dimension of the appendage rather than the length of the hull.

For sails the viscous flows about the foresail and mainsail are also separated, so that each has it own characteristic length. The length of a sail also varies from some minimum value at the mast top to a maximum at the foot of the sail. This means that, effectively, the Reynolds number of the flow about the sails varies along the length of the mast.

Figure 5.9.1 gives an indication of 'typical' Reynolds numbers of the flows about hull, keel and rudder for three different sizes of sailing yachts; a 5 m yacht/dinghy, a 10 m yacht and a 20 m yacht. Note that both scales of the figure are logarithmic.

It appears that for a small, 5 m yacht/dinghy, the Reynolds number of the rudder is of the order 10^5–10^6, that of the keel about 10^6 and that of the hull about 10^7. For a large, 20 m yacht the corresponding values are about 10^6, 10^7, and 10^8, respectively, depending, of course, also on boat speed.

Similar information for the Reynolds number of the air flow about the sails is given in Fig. 5.9.2. As indicated, the Reynolds number of the sails of a small, 5 m yacht/dinghy ranges, roughly, between 10^5 and 10^6. For a 20 m yacht this is 10^6–10^7 and for a 10 m yacht it is somewhere in between.

The general conclusion from Figs. 5.9.1 and 5.9.2 is that the Reynolds number of the airflow about the sails of a sailing yacht and the Reynolds numbers of the water

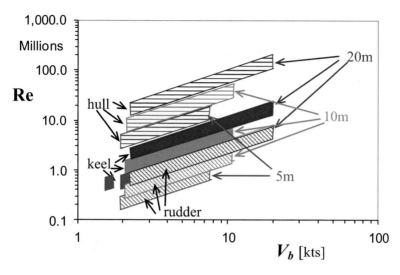

Fig. 5.9.1 *Range of Reynolds numbers for flows about hull, keel and rudder of different size yachts*

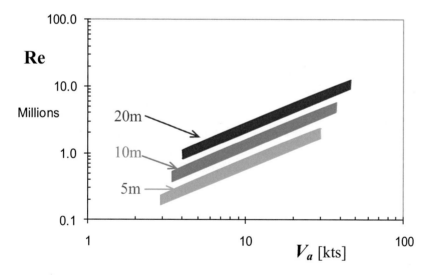

Fig. 5.9.2 *Range of Reynolds numbers of the flow about the sails of different size yachts*

flow about the appendages are of the same order of magnitude. Both are very large, of the order of 10^5–10^7. Hence, it can be concluded that both the aerodynamics and hydrodynamics of sailing yachts are characterized by (very) high Reynolds numbers.

5.10   High Reynolds Number Flows

Now that we have established that the flows of air and water around sailing yachts are characterized by high Reynolds numbers, it is appropriate to consider high Reynolds number flows in some more detail.

As already discussed in Sect. 5.7, one characteristic of high Reynolds number flows is that, in general, the friction forces are much smaller than the pressure forces. We have also seen, in Sect. 5.2, that for thin bodies at small angles of attack the lift consists mainly of pressure forces and the drag mainly of friction forces (see also Fig. 5.2.1). This means that in high Reynolds number flows about thin bodies at small angle of attack, the drag is, in general, much smaller than the lift (or side force). The opposite is the case for very high angles of attack .

Note that the generic descriptions of aero- and hydrodynamic forces in Chap. 3 are in agreement with these general conclusions.

Because, in high Reynolds number flows, the pressure forces are, in general, much larger than the friction forces, it is common practice to express the aero- and hydrodynamic forces acting on a body in a moving fluid on the basis of the (approximate) scaling law (5.7.6) for pressure forces at high Reynolds numbers (Sect. 5.7). More precisely the lift force L is expressed as

$$L = \frac{1}{2}\,\rho V^2 S C_L \qquad\qquad (5.10.1)$$

and the drag force D as

$$D = \frac{1}{2}\,\rho V^2 S C_D \qquad\qquad (5.10.2)$$

In Eqs. (5.10.1) and (5.10.2) S is a suitably chosen reference area such as the area of the wetted surface, a cross-sectional area or a related quantity. The coefficients C_L and C_D are called the *lift coefficient* and *drag coefficient*, respectively.

At small angles of attack, when the lift consists mainly of pressure forces, the lift coefficient C_L will be almost independent of the Reynolds number or the flow speed V. In other words, for most practical purposes C_L is a function of the shape and attitude of the body only.

The same applies to the drag coefficient at very high angles of attack. At small angles of attack, when the drag is mainly determined by friction forces, there is a significant dependence of C_D on Reynolds number or flow speed.

Another characteristic of high Reynolds number flows is that the flow field adopts a certain, layered structure. It appears (Prandtl and Tietjens 1934a in Chap. 1), that sufficiently far away from a body the flow behaves as if it were frictionless (inviscid). This outer layer of the flow field is usually called the *outer inviscid flow*. At a very large distance from the body the flow direction and velocity are the same as that of the undisturbed, incoming flow far upstream of the body.

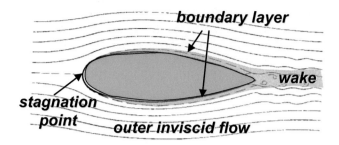

Fig. 5.10.1 *Layered structure of high Reynolds number flows*

Phenomena related to friction are felt only in a thin layer adjacent to the body. This layer is called the *boundary layer* (see Fig. 5.10.1). The boundary layer origi-nates from the *stagnation point* on the front side of the body where the external flow comes to a standstill. The streamline through the stagnation point is called the *stagnation streamline*. The stagnation streamline branches around the body from the stagnation point and initiates the development of the boundary layer. Due to the action of viscosity the fluid in the boundary layer sticks to the surface of the body, but the flow velocity in the boundary layer increases rapidly in the direction nor-mal to and away from the surface (Fig. 5.10.2). In the downstream direction more and more fluid material is dragged into the boundary layer and the thickness of the boundary layer increases in the downstream direction. At the rear of the body the thickness of the boundary layer at small angles of attack is usually of the order of a few percent of the length of the body. Behind the body the boundary layers from top and bottom surface merge and continue as the *viscous wake*.

Fig. 5.10.2 *Velocity profile in a bound-ary layer and wake*

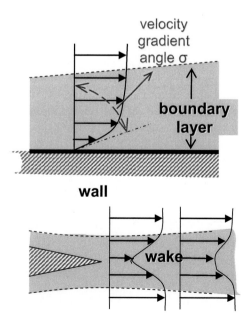

While the frictional forces are much smaller than the inertia and pressure forces in the outer inviscid flow, they are of the same order as the inertia forces in the boundary layer. Because the thickness δ_{bl} of the boundary layer is much smaller than the characteristic length of the body the scaling of the frictional forces per unit volume in the boundary layer is given by (Prandtl and Tietjens 1934b in Chap. 1)

$$F_f \doteq \mu V / \delta_{bl}^2 \qquad (5.10.3)$$

rather than by (5.7.1). With the magnitude of the inertia forces given by Eq. (5.7.1) and equating them with the magnitude of the frictional forces it follows that

$$\rho V^2 / L \doteq \mu V / \delta_{bl}^2 \qquad (5.10.4a)$$

Relating the viscosity with the Reynolds number by means of Eq. (5.7.4) this can also be written as

$$\rho V^2 / L \doteq (\rho V L / Re) V / \delta_{bl}^2 \qquad (5.10.4b)$$

From this it follows that

$$\left(\delta_{bl} / L \right)^2 \doteq 1/Re \qquad (5.10.5a)$$

or

$$\delta_{bl} / L \doteq Re^{-1/2} \qquad (5.10.5b)$$

In words this means that the (relative) boundary layer thickness decreases with increasing Reynolds number like $1/\sqrt{Re}$.

As in Sect. 5.7 we can derive a scaling law for the total frictional force acting on a body by multiplying the expression (5.10.3) for the frictional force per unit volume with a factor representing the total volume of fluid material that is influenced by the viscous flow about the object. Because we are considering friction forces and the friction phenomena are now limited to the boundary layer, the total volume of the boundary layer is now the appropriate quantity. Hence, the multiplication factor can be taken as $\delta_{bl}.S$ where S is, again, the area of the wetted surface. The scaling law for the total friction force for high Reynolds numbers can then be written as

$$F_F \doteq (\mu V / \delta_{bl}^2) \delta_{bl} S \qquad (5.10.6)$$

Relating the viscosity factor μ and boundary thickness δ_{bl} with the Reynolds number Re by using (5.7.4) and (5.10.5b), respectively, Eq. (5.10.6) can also be written as

$$F_F \doteq \rho V^2 S Re^{-1/2} \qquad (5.10.7)$$

Comparing the scaling law (5.10.7) for high Reynolds numbers with the original scaling law (5.7.7), which was derived without any assumption on the magnitude of the Reynolds number, we can see that the friction force decreases less rapidly with Reynolds number when the latter is (very) large. Note that the friction force still goes to zero for vanishing viscosity, when the Reynolds number becomes infinitely large.

If we compare the scaling law (5.10.7) with the general expression (5.10.2) for the drag of a body it follows that at small angles of attack, when the drag consists mainly of friction forces, the drag coefficient C_D varies with Reynolds number like $Re^{-\frac{1}{2}}$.

5.11   Boundary Layers, Wakes and Friction Drag

As already mentioned in the preceding section, the fluid in the boundary layer sticks to the surface of a body due to the action of viscosity. However, the flow velocity increases rapidly in the direction normal to and away from the surface until it becomes more or less constant at the edge of the boundary layer. It is said that the boundary layer has a certain *velocity profile* (Fig. 5.10.2).

It can be shown (Prandtl and Tietjens 1934b in Chap. 1) that the rate of change of the velocity or the *'velocity gradient'* at the wall or surface of the body is coupled to the *shear stress* at the wall or *skin friction*. Denoting the velocity gradient at the wall as $\tan\sigma_w$, where σ_w is the angle between the tangent to the velocity profile and the normal at the wall (Fig. 5.10.2), the relation between the shear stress τ_w and the velocity gradient σ_w at the wall can be written as:

$$\tau_w = \tan\sigma_w, \qquad (5.11.1)$$

expressing that the local shear stress or skin friction is proportional to the rate of change of the velocity at the wall.

It has also been found (Prandtl and Tietjens 1934b in Chap. 1) that the pressure in the boundary layer is constant in the direction normal or perpendicular to the wall (but can vary along the wall, depending of the shape and attitude of the body).

The pressure is also constant across the wake behind the body, at least in the normal direction. The velocity profile in the wake reflects the merging of the boundary layers from the upper and lower surfaces of the body (Fig. 5.10.2). The velocity defect in the wake decreases in the downstream direction and disappears at a very large distance behind the body.

Laminar and Turbulent Flows
Until now we have assumed implicitly that the flow in the boundary layer is steady and neatly layered or stratified. It appears that this is indeed the case as long as the Reynolds number is not too large. The boundary layer is then said to be *laminar*. However, it is found that if the Reynolds number is sufficiently high, the flow in the boundary layer becomes unstable and acquires an unsteady, time-variant character.

The boundary layer is then said to be *turbulent. Transition* from laminar to turbulent flow usually takes place when the local or 'running' Reynolds number, defined by

$$\mathrm{Re}_x = Vx/\nu, \tag{5.11.2}$$

is larger than about 2 to 5×10^5, where x is the distance along the surface stream-line measured from the stagnation point. Other quantities that have an influence on boundary layer transition (Young 1989) are the roughness of the surface, the stream wise pressure gradient, the amount of cross flow in the boundary layer, and the level of turbulence in the free stream. Boundary layer transition takes place at lower Reynolds numbers when the surface is rough, when the pressure along the surface increases in the downstream direction, when there is a large amount of cross flow and when the incoming flow has already a high level of turbulence.

We have seen in Sect. 5.9 (Fig. 5.9.1) that at low boat speeds, the Reynolds numbers of the rudder and keel of small to medium size yachts and the rudder of large yachts is of the order of 10^5–10^6. Hence, there is, depending on the other conditions, of surface roughness, etc., a chance that the flow is partly laminar under these conditions. For medium size yachts at high boat speeds, and large yachts at all but the lowest of boat speeds, this is out of the question.

As indicated by Fig. 5.9.2, the Reynolds numbers of the sails of small and medium yachts at low apparent wind speeds would also allow some laminar flow, depending also on the other conditions of surface roughness, etc.

In a turbulent boundary layer the flow velocity fluctuates in time and space in a chaotic way. In general, the range of the time- as well as the length scales of these fluctuations or 'eddies' is very large. That is, the length scale can be microscopic or of the size of the thickness of the boundary layer and anything in between. Similarly, the time scales can range between microseconds and the time ($\div L/V$) it takes for a disturbance to travel along the length of the body. Because of the fluctuations the interaction between adjacent layers in a turbulent boundary layer is much stronger than in a laminar one. This causes an additional, virtual or 'eddy' viscosity. As a consequence the total, effective viscosity and resulting shear stresses in a turbulent boundary layer are also larger. At the same time, the time-averaged velocity profile of a turbulent boundary layer is 'fuller' than that of a laminar boundary layer (Fig. 5.11.1).

Close to the surface, the velocity fluctuations are constrained by the presence of the wall. Because of this, there is always a thin, viscous sub-layer adjacent to the wall in which the flow is more or less laminar.

Because turbulence is a chaotic phenomenon there is no generally valid theory for turbulent boundary layers. However, it has been found from experiments (Prandtl and Tietjens 1934b in Chap. 1) that the scaling laws governing turbulent boundary layer flow are quite different from those that are applicable in laminar boundary layers. For example, while theory predicts that the thickness of a laminar boundary layer scales like

$$\delta_{bl}/L \div \mathrm{Re}^{-1/2}$$

Fig. 5.11.1 *Velocity profiles of a laminar and a turbulent boundary layer*

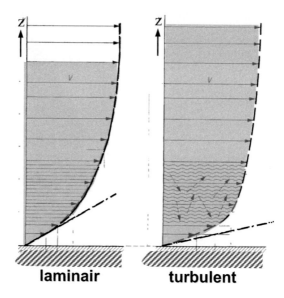

laminair   turbulent

(see Eq. 5.10.5b), it has been found that there holds, approximately

$$\delta_{bl}/L \div Re^{-1/5} \tag{5.11.3}$$

for a turbulent boundary layer. In other words, the thickness of a turbulent boundary layer decreases less rapidly with Reynolds number than a laminar one.

In terms of the local or 'running' Reynolds number (5.11.2), the boundary layer thickness scales like

$$\delta_{bl}/x \div Re_x^{-1/2} \tag{5.11.4a}$$

or

$$\delta_{bl} \div x^{1/2} \tag{5.11.4b}$$

for laminar flow and like

$$\delta_{bl}/x \div Re_x^{-1/5} \tag{5.11.5a}$$

or

$$\delta_{bl} \div x^{4/5} \tag{5.11.5b}$$

for turbulent flow. This means that a turbulent boundary layer grows more rapidly in the downstream direction than a laminar one.

Fig. 5.11.2 *Converging and diverg-ing boundary layer flow*

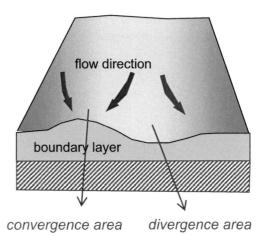

convergence area   divergence area

Other factors that have an effect on the rate of growth of the thickness of a boundary layer are streamwise pressure gradient and *convergence/divergence* of the boundary layer flow. When the pressure increases in the downstream direction the boundary layer grows more rapidly. When the pressure decreases the boundary layer grows less rapidly. Convergence or confluence occurs when the streamlines in the boundary layer are forced towards each other, causing the boundary layer to thicken more rapidly in the streamwise direction (Fig. 5.11.2). In case of divergence it thickens less rapidly. Convergence also promotes transition while transition is postponed in case of divergence of the boundary layer flow.

Friction Drag and Surface Roughness
In analogy with Eq. (5.10.7), the total frictional force acting on a body with fully turbulent boundary layers is found to scale like

$$F_F \div \rho V^2 S \mathrm{Re}^{-1/5} \tag{5.11.6}$$

rather than (5.10.7). It then follows from Sect. 5.10 that for turbulent flow, the drag coefficient of a body at small angle of attack scales also like $\mathrm{Re}^{-1/5}$.

For an infinitely thin flat plate at zero angle of attack there can be no pressure drag so that the drag consist of friction drag only. The drag can then be expressed as

$$D_F = \tfrac{1}{2}\,\rho V^2 S_{\mathrm{wet}} C_{F_0} \tag{5.11.7}$$

(see also Eq. (5.10.2)).

Figure 5.11.3 presents the friction drag coefficient C_{F_0} of (one side of) a flat plate at zero angle of attack as a function of Reynolds number for laminar and turbulent flow. For full laminar flow on a smooth surface C_{F_0} satisfies the exact theoretical expression (due to Blasius 1908)

$$C_{F_0} = 1.33\,\mathrm{Re}^{-1/2} \tag{5.11.8}$$

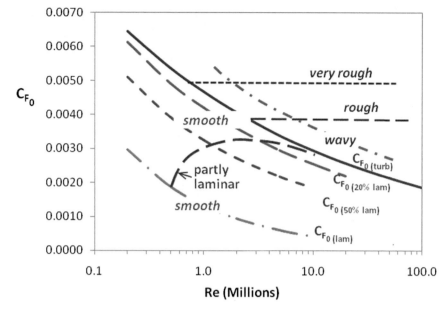

Fig. 5.11.3 *Friction drag of a flat plate at zero angle of attack as a function of Reynolds number*

A similar expression for turbulent flows (due to Prandtl 1927 and Von Kármán 1921) is

$$C_{F_0} = 0.074 \, \text{Re}^{-1/5} \qquad\qquad (5.11.9a)$$

It should be mentioned that Eq. (5.11.9a) represents only one of several alternative, empirical formulae for the friction drag of turbulent boundary layers that can be found in the literature (Young 1989). It is particularly applicable for Reynolds numbers between 5×10^5 and 10^7 and is generally used in the aeronautical community. For Reynolds numbers between 5×10^6 and 10^8 slightly better results are obtained by the formula

$$C_{F_0} = 0.045 \, \text{Re}^{-1/6} \qquad\qquad (5.11.9b)$$

Figure 5.11.3 shows clearly that the friction drag caused by a turbulent boundary layer[3] is much larger than that of a laminar boundary layer (on a smooth surface). The figure also indicates that on a smooth flat plate full laminar flow is not possible for Reynolds numbers above about 0.5×10^6 and that for Reynolds numbers above about 5×10^6 the flow will be practically fully turbulent. At Reynolds numbers in between the flow will be partly laminar (on the front part of the plate) and partly turbulent, depending on the other factors that govern transition. For a constant position of the location of transition x_{tr}/L the friction drag as a function of Reynolds number will follow a line in between those for full laminar and full turbulent flow.

---

[3] Shown is the friction drag according to Eq. (5.11.9a).

Fig. 5.11.4 *Examples of surface roughness*

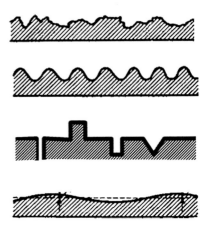

Also indicated in Fig. 5.11.3 is the effect of *surface roughness*. It is not uncommon
to distinguish between two types of surface roughness : irregular types of roughness
such as scratches, grooves, dents, cracks, ridges, slots, marine fouling, uneven sail
tissue, etc. and regular, wave-like disturbances (see Fig. 5.11.4).

As already mentioned above one effect of surface roughness, irregular roughness
in particular, is to cause early transition from laminar to turbulent flow. Another effect is that, even in a turbulent boundary layer, (irregular) roughness causes a higher
friction drag above a certain, critical Reynolds number that depends on the level of
the surface roughness. Moreover, the friction drag then becomes almost independent of Reynolds number.

When the dimension of the roughness is small compared to the thickness of the
boundary layer, the boundary layer 'will not notice the roughness' until at fairly
high Reynolds numbers when the boundary layer is thin(ner). For large(r) values of
the ratio between the size of the roughness and the thickness of the boundary layer
the friction drag coefficient will already be higher at lower Reynolds numbers (as
indicated by the line labeled 'very rough' in Fig. 5.11.3). What this means in practice is that it is important to keep the body surface as smooth as possible, in particular where the boundary layer is thin, that is on the front part of the body.

The effect on friction drag of regular, 'wavy wall' types of roughness, such as
in the bottom figure of Fig. 5.11.4, is somewhat different. It usually causes a more
or less parallel shift of the friction drag coefficient curve for a smooth surface to
a higher level, rather than becoming independent of Reynolds number (see also
Fig. 5.11.3).

The friction related drag of bodies of finite thickness such as the hull, keel, rudder
and rig of a sailing yacht depends in a similar way on Reynolds number. However,
there are differences. First of all there will, in general, be a difference in the position
of transition due to pressure gradients and three-dimensional effects like cross-flow
and/or convergence or divergence of the boundary layer flow . Secondly, the (average) fluid velocity outside the boundary layer on a body with thickness is higher

than on a flat plate at zero angle of attack because of the 'Venturi-like' effects (mass flow conservation) described in Sect. 5.5. As a consequence the friction drag is also higher. This is usually accounted for by expressing the friction drag coefficient of a body with thickness like

$$C_{D_F} = (1 + k_0) \, C_{F_0}, \tag{5.11.8}$$

where k_0 is a so-called *form factor* (Young 1989). This form factor k_0 depends on the shape of the body, the thickness/length ratio and the associated 'supervelocity' in particular. For airfoil type shapes k_0 is of the order of 0.2. For a flat plate $k_0 = 0$, of course.

It is useful to note that the friction drag coefficient of a wavy wall in Fig. 5.11.3 behaves like Eq. (5.11.8).

The friction drag of a body depends, in principle, also on the lift. Under lifting conditions the flow velocity on one side of the body is higher than on the other. As a consequence the effect of lift on friction drag is relatively small and usually neglected.

Unlike a flat plate at zero angle of attack, a body with finite thickness also experiences pressure drag, even at zero angle of attack. This kind of pressure drag is due to the fact that the boundary layer, through its thickness, changes the effective shape of the body (Young 1989). It is useful at this point to introduce the notion of *displacement thickness*. Because of the velocity defect in the lower part of the boundary layer, the mass flow through the part of space occupied by the boundary layer is smaller than when there would be no boundary layer, that is when the outer inviscid flow would continue right up to the wall or surface.

As illustrated by Fig. 5.11.5 the mass flow defect can be modelled by adding a layer with thickness δ^* to the surface of the body, where the product $U_e \, \delta^*$ is equal

Fig. 5.11.5 *Illustrating displacement thickness (δ^*)*

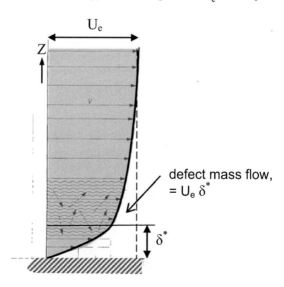

to the mass flow defect in the boundary layer (assuming the density equal to 1). Here, δ^* is the displacement thickness and U_e is the flow velocity at the edge of the boundary layer.

The displacement thickness causes changes in the velocity and pressure distributions in the sense that lower pressures occur on the rear part (Young 1989) of a body. The resulting pressure drag is usually referred to as *form drag* or *'viscous pressure drag'*. At small angles of attack the form drag of a thin or slender, streamlined body is relatively small, of the order of, say, 10% of the friction drag. The thicker the boundary layer and the larger the displacement thickness, the higher the form drag. Thick bodies have thicker boundary layers towards the rear end and a higher pressure drag than thin bodies, at least at small angles of attack.

The viscous pressure drag of a body depends also on lift. This is already the case for a flat plate at an angle of attack. As discussed in Sect. 5.2, the drag of a body at very high angles of attack consists even mainly of pressure drag.

The form drag of a body is usually also accounted for by a form factor of the type k_0 introduced above. That is, the sum of form drag and friction drag, called boundary layer drag (Young 1989), viscous drag or profile drag, is expressed as

$$C_{D_v} = (1+k)\, C_{F_0}, \qquad\qquad (5.11.9)$$

where C_{F_0} is again the flat plate friction drag coefficient and the form factor k now includes the form drag. 'k' depends primarily on the thickness of the body profile but is also a function of the lift and angle of attack.

5.12   Boundary Layer Separation and Flow Detachment

In layered, high Reynolds number flows there is a sort of hierarchy between the layers (Young 1989) in the sense that the outer inviscid flow determines what happens to the boundary layer. When the outer flow accelerates, which, according to Bernoulli's law (Sect. 5.6), means that the pressure decreases in the stream wise direction, the flow in the boundary layer will accelerate also. When the outer flow decelerates and the pressure increases in the stream wise direction, the flow in the boundary layer will decelerate. When the deceleration of the outer flow and the stream wise increase of pressure are strong enough, they may cause reversed flow in the lower part of the boundary layer, close to the wall, where the flow velocity was already small (see Fig. 5.12.1). The reversed flow causes a local re-circulatory flow pattern that is separated from the original boundary layer by what is called the *separation streamline*, (or, in three dimensions, the separation stream surface). The separation streamline is the continuation of the stagnation streamline (Sect. 5.10) aft of the *separation point* where the boundary layer is said to separate from the surface.

When the boundary layer on a body separates, the displacement thickness increases rapidly and the form drag is much higher than when the flow is fully at-

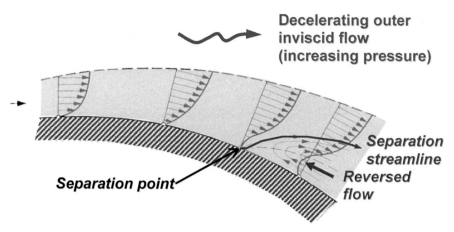

Fig. 5.12.1 *Illustrating boundary layer separation*

tached. It has been found (Young 1989) that thick, low Reynolds number boundary layers are more susceptible to separation than thin, high Reynolds number boundary layers and that a laminar boundary layer separates much earlier than a turbulent one. The latter is caused by the more intensive vertical exchange of energy in turbulent boundary layers (Sect. 5.11) which tends to make the time-averaged velocity profile more uniform.

On a smooth surface, a boundary layer becomes resistant to separation when the Reynolds number becomes infinite large. However, a boundary layer, laminar or turbulent, always separates when it encounters a sharp edge (Fig. 5.12.2).

This is independent of the Reynolds number. Even in the limit that the Reynolds number becomes infinitely large there is still separation at a sharp edge. The reason is that the momentum conservation law excludes that the flow (or any flow for that matter) can change its direction discontinuously.

A separated boundary layer can, under certain conditions, reattach to the surface to form a *separation bubble* (Fig. 5.12.3). Inside a separation bubble the flow veloc-

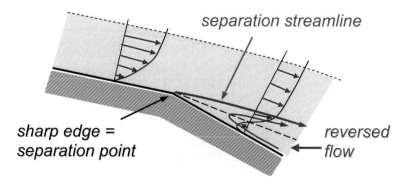

Fig. 5.12.2 *Flow separation at a sharp edge*

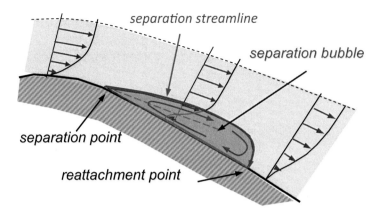

Fig. 5.12.3 *Sketch of a separation bubble*

ity is usually very low and the pressure is almost constant in all directions (Young 1989). This is sometimes called a 'dead water' region. Whether the formation of a separation bubble happens or not depends on the shape and attitude of the body and the position of the separation point. When separation occurs near the front of a body the separating boundary layer will often be laminar. However, transition to turbulent flow always takes place around reattachment.

When there is no reattachment the separation is said to be of the 'open' type and the flow is said to detach at the separation point. The separation point is then also a detachment point.

An important example of the latter is the sharp trailing-edge of a keel, rudder or sail (Fig. 5.12.4). When the boundary layer has not separated in the sense described above it will in any case leave the surface of the body when it arrives at the sharp trailing-edge. In this situation the flow is said to be attached on both upper and lower surfaces and there is flow detachment at the trailing-edge only. It is also

Fig. 5.12.4 *Separation and detachment at a sharp trailing edge*

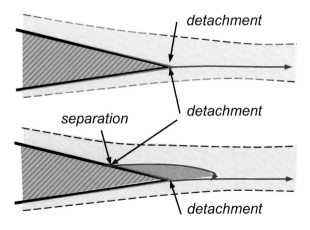

possible that the boundary layer on one surface separates before it arrives at the trailing-edge (see bottom figure of Fig. 5.12.4). In that case there is one detachment point upstream of the trailing-edge and one at the trailing-edge.

5.13   Rotation, Circulation and Vortices

Rotation and Circulation
Under the action of the frictional forces, the fluid particles in the boundary layer acquire a *rotation*al motion, in addition to the translational motion they already had before they became part of the boundary layer. Figure 5.13.1 illustrates, for a rectangular element of fluid volume, that the translational and rotational motion together form the kind of boundary layer velocity profile that was pictured in Fig. 5.10.2. (The 'vertical' components of the rotational velocity of a fluid element are cancelled by the components of the adjacent fluid elements on the left and on the right).

Rotation, like velocity, is a vector quantity with, in a general three dimensional situation, three components. In the case of Fig. 5.13.1 the axis of rotation is perpendicular to the paper.

It can be shown (Prandtl and Tietjens 1934a in Chap. 1) that the total amount of rotation within a fluid element in a certain plane is equal to the product of the tangential velocity (component) at a segment of the circumference of the fluid element in that plane and the length of the segment, integrated ('added-up') over all segments.[4] For example, if the fluid element is a square box with four segments of equal length s, as in the case of Fig. 5.13.1, and the tangential components of the rotational velocity all have the same magnitude V_{rot}, then the total amount of rotation Γ within the box is equal to

$$\Gamma = 4V_{rot} \cdot s \qquad\qquad (5.13.1)$$

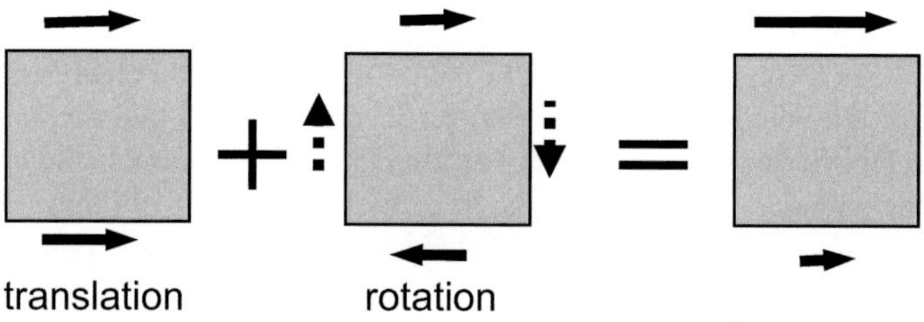

translation   rotation

Fig. 5.13.1 *Boundary layer flow is translation plus rotation*

---

[4] This is known as Stokes' theorem (1854).

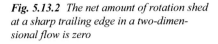

Fig. 5.13.2 *The net amount of rotation shed at a sharp trailing edge in a two-dimensional flow is zero*

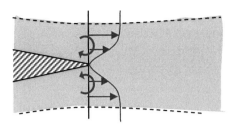

The quantity Γ is also called the *circulation* around the fluid element. More in general one speaks of the circulation around an arbitrary area of fluid when the summation process described above is performed around a closed curve surrounding that area.

Flow Detachment and the Shedding of Rotation

At the location(s) where the flow detaches from a body, an amount of rotation is shed or 'dumped' into the flow field. In a two-dimensional, attached flow about a body with a sharp trailing-edge the net amount of rotation that is shed at the trailing-edge is zero. This is caused by the fact that the amounts of rotation contained by the boundary layers on upper and lower surface at the trailing-edge are equal but of opposite sign (Fig. 5.13.2).

The explanation of this is found by realizing that the continuity laws of physics requires that the pressure at the trailing-edge is the same at the upper and lower surface. In turn this implies that the velocities at the edge of the boundary layer at the upper and lower surfaces are equal. Together with the 'zero slip' condition that the velocity on the surface of the body is zero, this implies that the total amounts of rotation contained by the upper and lower surface boundary layers at the trailing-edge are equal but of opposite sign. In two-dimensional flow this is always the case, irrespective of, for example, the angle of attack.

In a three-dimensional flow the situation is different, at least in the case with lift. As an example, illustrated by Fig. 5.13.3, the direction of the detaching flow at the trailing-edge of a body on the upper surface (solid black arrow) will, in general, not be the same as the direction of the detaching flow on the lower surface (dashed black arrow). As a consequence the net rotation (coupled to the blue, 'cross flow' component of the velocity profiles in Fig. 5.13.3) that is shed into the wake is no longer zero. This because the amounts of rotation due to cross flow in the upper and lower surface boundary layers are of the same sign. The result is a thin layer containing stream-wise rotation, that is the axis of rotation has the same direction as the mean flow in the wake. As in the two-dimensional case there is also *stream-normal* rotation in the wake of which the axis of rotation is normal (perpendicular) to the red, 'mean flow' components of the velocity profile. The net amount of the stream-normal rotation is again zero because of the opposite signs of the rotation from the upper and lower surface boundary layers.

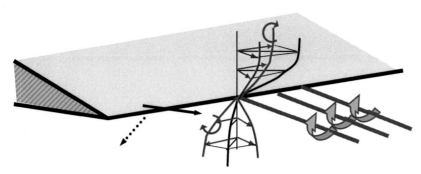

Fig. 5.13.3 ... *In a three-dimensional attached flow the net rotation 'dumped' at the sharp trailing edge of a body is $\neq 0$ and leads to the formation of a thin layer of stream-wise rotation*

Vorticity and Vortices

At this point it is convenient to introduce the notions of a *vortex* and *vorticity*. While rotation is a kind of mathematical quantity, the word 'vorticity' is used to indicate the associated physical phenomenon of rotating elements of fluid.

A vortex is a filamentary, cylindrical region or tube of fluid with concentrated rotation/vorticity and a circulating fluid motion around it. The axis of rotation coincides with the direction of the filament (See Fig. 5.3.4). Inside the filament the flow rotates as if it were a solid body. This is called the vortex core. Outside the filament the law of conservation of mass flow dictates that the velocity in the direction of circulation behaves like (Prandtl and Tietjens 1934a in Chap. 1)

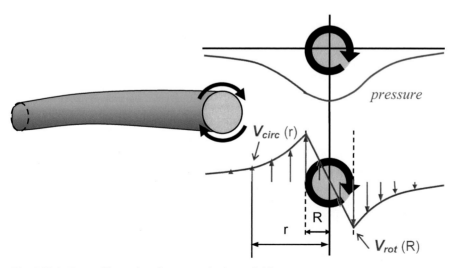

Fig. 5.13.4 *Vortex (filament) and associated velocity field*

$$V_{circ}(r) = \Gamma/(2\pi r), \quad r > R, \tag{5.13.2}$$

where r is the distance to the axis of the filament, R is the radius of the vortex core and Γ is the strength of the vortex. The latter is equal to the (amount of) circulation around the vortex which is given by

$$\Gamma = 2\pi R V_{rot}(R) \tag{5.13.3}$$

It is noted that it follows from Stokes' theorem (see p. 152) that the circulation around any closed curve surrounding the vortex is the same.

Because the velocity at the edge of the vortex core is high, it follows from Bernoulli's equation (Sect. 5.6) that the pressure in and near the core is low. In other words the characteristics of a vortex closely resemble those of the meteorological phenomena known as tornados, hurricanes and typhoons.

Vortices in an 'ideal' that is inviscid fluid and the associated fluid motion satisfy a number of theorems (Prandtl and Tietjens 1934a in Chap. 1) formulated around 1860 by Thompson (alias Lord Kelvin) and Helmholtz:

- Once formed, the circulation around a vortex filament is the same everywhere along its length
- Fluid particles which at any time are part of a vortex filament always belong to that same vortex filament
- Vortex filaments must either be closed in themselves or disappear into the bounding surface of the fluid volume

The second of these theorems implies that vortex filaments follow streamlines. The first and the third theorem are related and can be combined into what could be called a vorticity conservation law: once formed, vortices in an ideal fluid are fully preserved. In a real, that is viscous fluid, vortices dissipate under the influence of viscosity and friction. However, because of the fact that the friction in high Reynolds number flows is very small (except near the surface of a body) the rate of dissipation of vortices is also very small. As a consequence vortices in high Reynolds number flows of fluids with small viscosity are very persistent (like tornados and hurricanes).

Bound and Trailing Vorticity

We now return to the detaching flow at the sharp trailing-edge of a lifting surface (Fig. 5.13.5). It will be clear that the thin layer of nett, stream-wise rotation in the wake behind the trailing-edge can be interpreted as a thin layer with stream-wise vorticity, that is continuously distributed elementary vortices with their axis of rotation in the direction of the mean flow in the wake. The stream-wise vorticity in the wake is usually called *trailing vorticity*. As we have seen above, the stream-wise

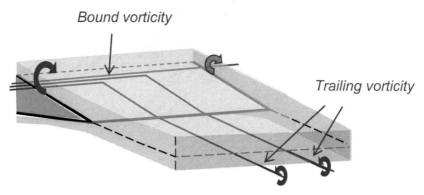

Fig. 5.13.5 *Bound and trailing vorticity on a lifting surface*

rotation in the wake, that is the trailing vorticity, is coupled to the rotation in the boundary layers on the upper and lower surface of the body. The rotation in the boundary layers (as in Fig. 5.13.2), with its axis of rotation normal to the direction of the flow ('stream-normal '), can be represented by a thin layer of stream-normal or *bound vorticity* attached (or 'bound') to the body surface. The trailing vorticity behind the body is then coupled to this bound vorticity on the body as indicated schematically in Fig. 5.13.5 for the bound vorticity contained by the upper surface boundary layer. The same happens at the lower surface and the net trailing vorticity in the wake is the sum of the vorticity shed by the upper and lower surface boundary layers.

More in general it can be shown (Van der Vooren 2006) that stream-wise vorticity can be formed only out of stream-normal vorticity in three-dimensional boundary layers and wakes with a non-planar, 'twisted' velocity profile (such as in the wake, just after the trailing-edge). It can also be shown that such 'twisted' velocity profiles can develop only when the fluid particles are subject to forces that are perpendicular to the streamlines. Van der Vooren (2006) shows further that the rate of development of stream-wise vorticity along a streamline is equal to the rate of change of the stream-normal vorticity. This, of course, is in agreement with the Kelvin-Helmholtz theorem of conservation of vorticity.

It is further noted that Fig. 5.13.5 depicts only a limited segment of a body. Because of the Kelvin-Helmholtz theorem on conservation of vorticity, mentioned above, the trailing vorticity in Fig. 5.13.5 must continue to downstream infinity and the bound vorticity must either run sideways to infinity or trail-off downstream in some other segment of the body. The latter will be the case on bodies of finite dimensions. The vorticity associated with flow detachment on bodies of finite dimensions can therefore be represented schematically by a system of distributed 'horseshoe' vortices of the kind sketched in Fig. 5.13.6.

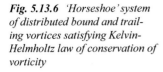

Fig. 5.13.6 'Horseshoe' system of distributed bound and trailing vortices satisfying Kelvin-Helmholtz law of conservation of vorticity

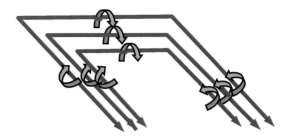

In discussing flow detachment and vorticity shedding we have up till now only considered the case of flow detachment at a sharp trailing-edge. It will be clear, however, that vorticity shedding of a similar kind takes places wherever there is boundary layer separation and flow detachment on a three-dimensional body.

Vorticity Layers and Vortex Formation

Thin layers containing distributed stream-wise vorticity as shed from a three-dimensional body are subject to deformation. As illustrated by Fig. 5.13.7 they 'roll up' under the influence of their self-induced circulatory motion and form discrete vortices of the type depicted in Fig. 5.13.4. This rolling-up takes place around the location where the vorticity density has a maximum. During the rolling-up process the distributed vorticity is collected in the developing discrete vortex and the original vorticity layer thins and, eventually, disappears.

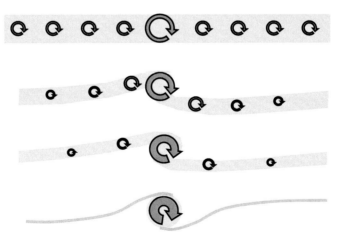

Fig. 5.13.7 Deformation of a thin layer with distributed stream wise vorticity and formation of a discrete vortex

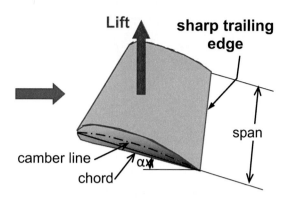

Fig. 5.14.1 *Lifting surface*

5.14   Lifting Surfaces, Foil Section Characteristics

Lifting surfaces are flat bodies intended to generate a force (lift or side force) perpendicular to the direction of the undisturbed, incoming flow. Usually the lift is to
be generated for the lowest possible drag. In general, lifting surfaces are thin and
have a sharp trailing-edge (Fig. 5.14.1). Aircraft wings are probably the best known
examples of a lifting surface. Rotating examples are propellers and wind turbines.
On a sailing yacht the sails and appendages (keels, rudders, centre boards, dagger
boards) are lifting surfaces.

Notions and Definitions
It is useful at this point to introduce a number of notions and definitions in connec-
tion with lifting surfaces. A lifting surface is also called a *foil*; in air an airfoil or
aerofoil and in water a *hydrofoil*.

The projection of a lifting surface on a suitably chosen flat reference plane, approximately parallel to the lifting surface, is called the *planform*. The lateral dimension is the span and the streamwise dimension is called the chord length. The latter
usually varies in the spanwise direction. The *angle of* attack α of a lifting surface is
the angle between the plane of the planform and the incoming flow.

A streamwise cross-section of a lifting surface is called a *(hydro/air) foil section*.
A foil section has a *profile* shape characterized by a thickness distribution and camber. The *camber* line or *mean* line of a foil section is the collection of points halfway
between the upper and lower surface. The chord line or *chord* is the straight line
connecting the leading- and trailing-edge points of the camber line. The *incidence*
of a section is the angle between the chord and the plane of the planform. The variation of the incidence in the spanwise direction is called the *(geometric) twist*.

Mechanism of Lift Generation
As illustrated by Fig. 5.14.2 for a streamwise section of a lifting surface or (air/
hydro)foil, the sharp trailing-edge is essential for the generation of lift at high Reyn-

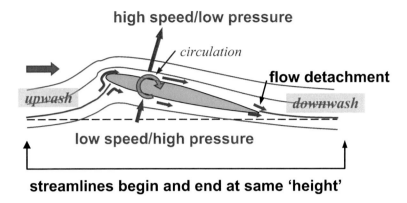

high speed/low pressure

circulation

flow detachment

upwash   *downwash*

low speed/high pressure

streamlines begin and end at same 'height'

Fig. 5.14.2 *Lift mechanism of a foil section*

olds numbers. We have seen in Sect. 5.12 that at sufficiently high Reynolds numbers the flow on both upper and lower surface will detach at a sharp trailing-edge. When the sharp trailing-edge points downward it forces the flow to detach smoothly with a downward velocity component. This is known as the Kutta condition (1902) or the condition of Kutta-Youkovski (Abbott and Von Doenhoff 1949). As a result there is *downwash* behind the trailing-edge. Conservation of mass then requires that there is an equal amount of *upwash* in front of the foil. In a 'horizontal' mean flow, streamlines must necessarily begin and end at the same 'height'. If this were not the case the velocities far downstream and far upstream of the foil could not be the same.

The upwash and downwash imply that, in terms of the 'Venturi' effect described in Sect. 5.5, the flow passing over the foil experiences the foil as a convex wall and the flow passing under it experiences it as concave wall. As a consequence the velocity on the upper surface is higher and the velocity on the lower surface is lower that the velocity of the mean flow. Bernoulli's equation (Sect. 5.6) then teaches that the pressure on the upper surface is lower than the pressure on the lower surface and this means that the body experiences lift.

Because of the different (average) velocities on the upper and lower surface there is circulation around the foil section. This means (see also Sect. 5.13) that, along any closed contour surrounding the foil section, the product of the tangential velocity (component) at a segment of the closed contour and the length of the segment, integrated ('added-up') over all segments of the contour, gives a quantity (circulation) unequal to zero.[5] From the discussion in the preceding section it follows that the circulation around the foil is equal to the net bound vorticity in the boundary layers on the upper and lower surfaces. It can also be shown (Prandtl and Tietjens 1934b

---

[5] In mathematical terms this can be written as, where V is the velocity and ds the length of an elementary contour line segment. See also p. 152.

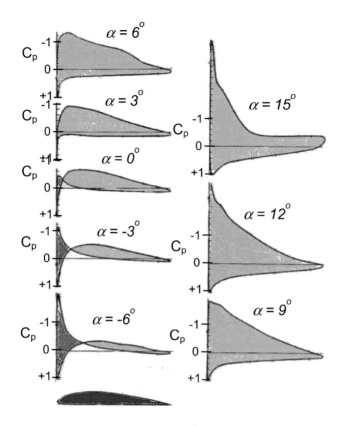

Fig. 5.14.3 *Pressure distributions on an airfoil at several angles of attack*

in Chap. 1) that the lift per unit span **L(y)** is proportional to the local circulation according to the relation

$$L(y) = \rho V_\infty \Gamma(y) \qquad (5.14.1)$$

This relation is known as the law of Kutta-Youkowski (1902/1906). For a positive lift, the sign/orientation of the circulation is as indicated in Fig. 5.14.2.

An example of pressure distributions on an airfoil at several angles of attack is presented in Fig. 5.14.3. The figure, for a cambered airfoil, has been adapted from (Prandtl and Tietjens 1934b in Chap. 1). As is customary in aerodynamics, the pressure has been expressed in terms of the *pressure coefficient* C_p, given by

$$C_p = (p - p_\infty) / (\tfrac{1}{2}\rho V_\infty^2) \qquad (5.14.2)$$

Note that this means that C_p expresses the difference between the local static pressure and the static pressure of the undisturbed flow at infinity upstream in terms of the dynamic pressure. This implies that the stagnation pressure corresponds with $C_p = +1$ and that $C_p = -1$ implies a suction level equal to the dynamic pressure.

The pressure distributions in Fig. 5.14.3 are given for both the upper and lower surface of the airfoil. At positive angles of attack the upper surface has the highest suction while the pressure on the lower surface is higher than the undisturbed static pressure ($C_p > 0$). The blue area between the curves for the upper and the lower surface is a measure for the (positive) lift on the airfoil. The red areas at the lower angles of attack indicate negative lift.

Note also that with increasing angle of attack a suction peak develops on the upper surface at the leading-edge. This means locally high velocities followed by a strong deceleration of the flow. As discussed in Sect. 5.12 this usually leads to separation of the boundary layer. In the case of Fig. 5.14.3 boundary layer separation is reflected in the levelling-off of the pressure distribution on the upper surface towards the trailing-edge ($\alpha = 12°$ and $15°$).

Maximum Lift

As discussed above, a downward pointing sharp trailing-edge is essential for the generation of lift. It is also found that the more downward the trailing-edge is pointing, for example by a higher angle of attack or by more camber, the higher the lift is. More precisely, and as illustrated by Fig. 5.14.4, the lift increases linearly with the angle of attack, at least as long as there is attached flow. For two-dimensional foils,

Fig. 5.14.4 *Variation of lift with angle of attack for symmetrical foil sections (qualitatively)*

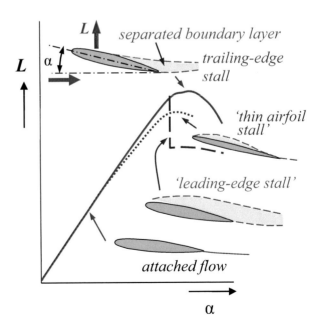

that is foils of infinite span, the slope of the linear part of the variation of the lift coefficient with angle of attack is about 0.1/degree, or a little less than 2π/radian.[6]

For every kind of foil the lift attains a maximum at a certain angle of attack. At and beyond maximum lift the foil is said to be *stalled*. The mechanism that determines maximum lift can vary between foil sections. Three different mechanism are usually distinguished (Woodward and Lean 1992): trailing-edge stall, leading-edge stall and thin airfoil stall.

If the leading-edge of the foil section is sufficiently round with a sufficiently large radius, which is usually the case when the foil is sufficiently thick, the boundary layer begins to separate ahead of the trailing-edge beyond a certain angle of attack. The point of separation moves forward with increasing angle of attack and the slope of the lift curve decreases (see the solid blue line in Fig. 5.14.4). The lift usually attains a maximum when the separation point has, approximately, reached the mid-chord position of the foil. (For the airfoil of Fig. 5.14.3 this is the case for $\alpha = 15°$). In this case, with boundary layer separation moving forward from the trailing-edge, one speaks of *trailing-edge stall*.

Some foils, with a relatively blunt leading-edge followed by a local area of relatively high curvature causing a high suction peak, can develop a short laminar separation bubble just aft of the leading edge that bursts at a certain angle of attack without reattachment. This is accompanied by a sudden loss of lift. When this is the mechanism determining maximum lift it is said that the foil experiences *leading-edge stall* (see the dashed red line in Fig. 5.14.4).

Thin foils with a small or zero leading-edge radius develop a laminar separation bubble at the leading-edge of finite extent, with turbulent reattachment further downstream on the foil. The length of the separation bubble increases when the angle of attack is increased. At the same time there is a (slight) kink in the lift curve (see the purple dotted line in Fig. 5.14.4). This is indicative of a reduction of the lift curve slope due to the onset of leading-edge separation. Maximum lift is attained, roughly, when the reattachment point of the separation bubble reaches the trailing-edge. This kind of stall pattern is known as *thin-airfoil stall*.

Leading-edge stall and thin-airfoil stall can also occur with some trailing-edge separation already present.

Surface roughness, in particular at the leading-edge, has a negative effect on maximum lift. The reduction can be as much as 35%, depending on the type of roughness (Prandtl and Tietjens 1934b in Chap. 1; Abbott and Von Doenhoff 1949).

Profile (or 'viscous') Drag
As already discussed in Sect. 5.11 the drag of a body with thickness is somewhat higher than that of a flat plate and is a function of the angle of attack. Figure 5.14.5 gives a qualitative example, derived from Abbott and Von Doenhoff (1949), of the variation of the profile drag of a foil section as a function of the lift. The figure

---

[6] The radian is the basic, dimensionless measure of angle. The 360° contained by a full circle are equal to 2 radians.

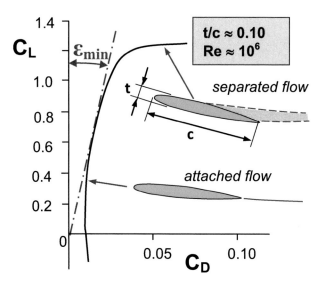

Fig. 5.14.5 *Force polar diagram showing the variation of profile drag with lift for a foil section with thickness/chord (t/c) ratio 0.10*

reflects that in aerodynamics it is common practice to present such information in the form of a *force polar diagram*. That is, the lift coefficient is plotted versus the drag coefficient[7]. Note that the scales of C_L and C_D are almost a factor 10 different.

Figure 5.14.5 illustrates that the variation of the profile drag with lift is small as long as the flow is attached but increases rapidly when boundary layer separation occurs. As discussed in Sect. 5.11 some 80–90 % of the profile drag (or 'viscous' drag) is friction drag when the angle of attack and the lift is small. The remaining 10–20 % is form drag (pressure drag). The rapid increase of the drag with lift at high lift coefficients when the flow is separated is almost entirely pressure drag.

Also indicated in Fig. 5.14.5 is the minimum drag angle ε_{min}. (Recall that $\tan\varepsilon = D/L = C_D/C_L$). As we have seen in Chap. 4, the drag angles of the sails and the under-water body are very important quantities in the mechanics of sailing.

Effect of Profile Shape on Lift, Drag and Moment
The lift and drag characteristics of a foil section, that is the variations of lift and drag with angle of attack and Reynolds number, depend on the profile shape. A qualitative picture of the influence of foil thickness and camber on lift and drag for an 'average' level of Reynolds number is given in Fig. 5.14.6. The figure is based on information contained by Prandtl and Tietjens (1934b in Chap. 1), Marchai (2000 in Chap. 4) and Abbott and Von Doenhoff (1949). It can be seen that, as already discussed in Sect. 5.11, thick symmetrical foil sections have a higher drag at low and moderate lift coefficients than thin symmetrical sections. However, they also

---

[7] See Eqs. (5.10.1) and (5.10.2) in Sect. 5.10 for the definition of the lift and drag coefficients.

Fig. 5.14.6 *Effect of foil section
shape on lift and drag (qualitatively)*

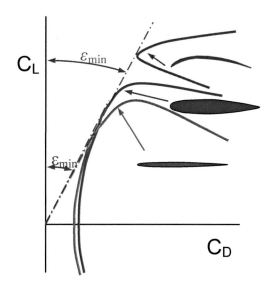

have a somewhat higher maximum lift coefficient, at least as long as the thickness
is not excessively large. This is caused by the fact that for thick foils, which usually
have a larger leading-edge radius, the suction peak at the leading-edge develops less
rapidly than for thin foils.

Thin sections with a large amount of camber, as is characteristic for sails, have
a significantly higher maximum lift coefficient ; this, however, at the expense of a
much higher drag level.

The minimum drag angles of thin and thick symmetrical foils are usually about
the same but the minimum drag angle of thick sections is found at a slightly higher
value of C_L. The minimum drag angle of thin, highly cambered sections occurs at
much higher values of lift. However, the minimum drag angle itself is usually not
better than that of thick symmetrical sections.

Other geometrical foil section quantities with significant influence on the fluid
dynamic characteristics are the chord-wise positions of maximum thickness and
maximum camber. It has been found (Abbott and Von Doenhoff 1949) that the minimum drag of a foil section usually increases and the maximum lift decreases when
the position of maximum thickness is moved aft. Foils with the position of maximum camber far aft also have a high minimum drag. However, the maximum lift
also increases when the position of maximum camber is moved aft, except for low
Reynolds numbers.

For a symmetrical foil section the point of application of the lift force, also called
the *centre of pressure*, is close to the point on the foil chord at 25 % of the chord
length from the nose. (Fig. 5.14.7). This means that the moment $M_{1/4}$ with respect to
the 'quarter chord point ' is almost zero. For symmetrical sections this is the case at

Fig. 5.14.7 *Effect of profile shape on the moment of a foil section (qualitatively)*

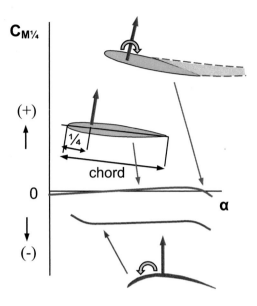

all angles of attack as long as the flow is attached. When boundary layer separation develops the quarter chord moment becomes negative (nose down).

Note that Fig. 5.14.7 presents the quarter chord moment coefficient $C_{M\frac{1}{4}}$ rather than the moment itself as a function of the lift coefficient. The moment coefficient is defined by

$$M_{\frac{1}{4}} = \frac{1}{2}\,\rho V^2 S\ c\ C_{M\frac{1}{4}} \tag{5.14.3}$$

Here, as before, S is the foil surface area and 'c' the chord length. For sections with positive camber the point of application of the lift force is positioned aft of the quarter chord point. This means that the quarter chord moment is negative (nose down). However, the change of the lift force due to a change in angle of attack still takes place in a point at or near the quarter chord point. This point is called the *aerodynamic center, (AC)*. As a consequence the centre of pressure of a section with positive camber moves forward with increasing angle of attack and rearward with decreasing angle of attack. The opposite is the case for sections with negative camber.

The description just given does not hold for situations around and beyond maximum lift, when there is a large amount of separated flow. Wind tunnel data indicate that the centre of pressure moves rearward when the point of separation of the boundary layer on the upper surface moves forward.

Effects of Reynolds Number (Scale Effects)
The most important effects of Reynolds number are those on drag and maximum lift. A general rule is that the effects are small as long as boundary layer separation does not occur or does not vanish as a result of a variation in Reynolds number. The

variation of profile drag then follows the trend of the variation of the friction drag with Reynolds number discussed in Sect. 5.11 and depicted in Fig. 5.11.3. Due to the higher value of the form factor k in Eq. (5.11.9) the variation of profile drag with Reynolds number is a little larger for thick sections than for thin sections. When the thickness and/or the lift are sufficiently large and the Reynolds number is sufficiently low for boundary layer separation to occur, then there is, in general, a rapid change of drag with Reynolds number. This in the sense that the drag coefficient increases strongly when boundary layer separation begins or moves forward rapidly.

Because maximum lift is always associated with boundary layer separation, the effect of Reynolds number on maximum lift can be considerable, in particular for thick sections exhibiting trailing-edge stall. In general, the maximum lift of thick sections increases with increasing Reynolds number. The variation with Reynolds number of the maximum lift of thin sections is usually much smaller (Marchai 2000 in Chap. 4), in particular in the case of a sharp leading-edge. This is caused by the fact that a sharp leading-edge always forces the flow to separate, irrespective of the Reynolds number. An exception is formed by foils with a large amount of upper surface curvature, such as a highly cambered sail section. In this case secondary separation can develop towards the rear of such foils, that worsens for low Reynolds numbers.

Although the fluid dynamic characteristics of most foil sections conform to the general trends described above there is considerable variation within these general trends. The precise characteristics of foil sections depend very much on their specific profile shape. Abbott and Von Doenhoff (1949) contains a wealth of information for 'standard', NACA airfoil sections for Reynolds numbers ranging from 3×10^6 to 6×10^6. Marchai's book (2000 in Chap. 4) contains data and references for lower Reynolds numbers and thin, sharp edged sections. Data for thin, sail-like sections can also be found in Milgram (1971).

   It is further noted that in the current practice of yacht and sail design the utilization of 'standard' sections for foil design is gradually being replaced by 'custom' design and optimisation for specific requirements by means of Computational Fluid Dynamics.

5.15   Lifting Surfaces in Three Dimensions; Downwash and Induced Drag

Effects of Finite Span

In the preceding section we have considered the characteristics of foil sections, that is lifting surfaces of infinite span with identical geometric and flow characteristics in all streamwise sections along the span. Lifting surfaces of finite span exhibit additional phenomena.

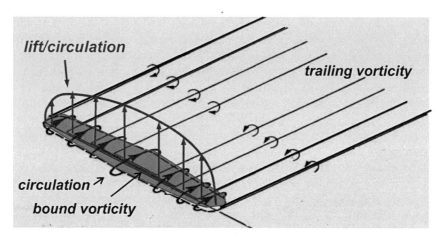

Fig. 5.15.1 *Lift, circulation, bound and trailing vorticity for a lifting surface of finite span*

As discussed in Sect. 5.13 a lifting body with a sharp trailing-edge exhibits bound and trailing vorticity. We have seen also that the trailing vorticity is linked to the net bound vorticity contained by the upper and lower surface boundary layers, that the net bound vorticity is equal to the circulation around a streamwise section and that the lift in a section is proportional to the circulation. It then follows that, as illustrated by Fig. 5.15.1, the strength of the trailing vorticity is equal to the rate of change of the spanwise distribution of circulation which is proportional to the span-wise distribution of lift. Because the lift drops to zero towards the tips of the lifting surface, the span-wise rate of change of the distribution of circulation and, hence, the strength of the trailing vorticity is the largest at the tips.

We have seen in Sect. 5.13 that layers of distributed stream-wise vorticity de-form and roll-up around the location where the vorticity density has a maximum, to form a discrete vortex. Because the density (or strength) of the trailing vorticity behind a three-dimensional lifting surface has its maximum at the tips, the trailing vortex sheet rolls-up around its edges to form discrete *tip vortices* (Fig. 5.15.2). At

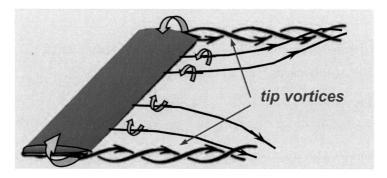

Fig. 5.15.2 *Formation of tip vortices behind a lifting surface*

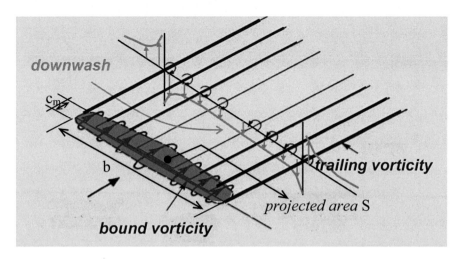

Fig. 5.15.3 *Downwash induced by trailing vorticity*

high angles of attack the fluid tends to flow around the tips with separation at the side-edges (as well as at the trailing-edge). The tip vortices then already begin to form at the side-edges.

The circulatory motion induced by the trailing vorticity and tip vortices causes downwash at and behind the lifting surface (Fig. 5.15.3) and upwash outside of the tip. As a consequence the angle of attack of a three-dimensional lifting surface is, effectively, smaller than in the case of two-dimensional flow. Introducing the notion of *effective angle of attack*, α_e, one can write

$$\alpha_e = \alpha - \alpha_i, \tag{5.15.1}$$

where α is the geometric angle of attack or *incidence* and α_i is the *induced angle of attack* due to the downwash caused by the trailing vorticity (see also Fig. 5.15.4). Because the effective angle of attack is smaller than the geometric angle of attack, the lift on a three-dimensional lifting surface is smaller than the lift in two-dimensional flow. As already indicated in Fig. 5.15.1, the loss of lift due to the induced downwash is, in general, the largest at the tips.

As a further consequence of the induced downwash the lift vector cants backwards over an angle equal to the induced angle of attack (Fig. 5.15.4). This causes an additional, *induced drag, D_i,* that is absent in two-dimensional flow.

It has been found (Prandtl and Tietjens 1934b in Chap. 1) that the induced downwash and induced drag increase with:

- decreasing span 'b' (Fig. 5.15.3) of the lifting surface
- increasing lift
- decreasing velocity of the undisturbed onset flow

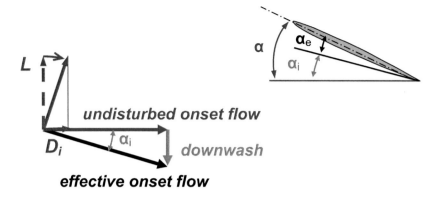

Fig. 5.15.4 *Induced angle of attack and induced drag*

More specifically it can be shown (Prandtl and Tietjens 1934b in Chap. 1), that when the span 'b' is much larger than the mean chord length c_m, there holds, for the (mean) induced angle of attack

$$\alpha_i \approx \frac{L}{(\frac{1}{2}\, \rho V^2 b)(\pi b)}, \text{(in radians)} \tag{5.15.2}$$

and for the induced drag

$$D_i \approx L\alpha_i \approx \frac{L^2}{(\frac{1}{2}\, \rho V^2 b)(\pi b)} \tag{5.15.3}$$

In ordinary language this means that the induced downwash of a lifting surface is proportional to the lift and inversely proportional to the square of the span. The induced drag is proportional to the square of the lift and also inversely proportional to the square of the span.

Using the formulae (5.10.1) and (5.10.2) for expressing lift and drag in terms of dimensionless coefficients C_L and C_D, Eqs. (5.15.2) and (5.15.3) can also be written as

$$\alpha_i \approx C_L /\pi A, \quad (\text{radians}) \tag{5.15.4}$$

and

$$C_{D_i} \approx C_L{}^2 /\pi A \tag{5.15.5}$$

Here, A is the aspect ratio of the lifting surface, defined by

$$A = b^2/S, \tag{5.15.6}$$

where S is the projected area (Fig. 5.15.3). In terms of the mean chord length c_m the aspect ratio can also be expressed as

$$A = b/c_m \tag{5.15.7}$$

The total drag of a lifting surface is the sum of the profile drag (due to viscosity) introduced in Sect. 5.11 and the induced drag given above. One can write:

$$C_D = C_{D_v} + C_{D_i} \approx C_{D_v} + C_L^2/\pi A \tag{5.15.8}$$

It is noted that while the profile drag has both friction and pressure components (Sect. 5.11) the induced drag consists of pressure drag only. The latter follows from the fact that the induced drag is the result of a rotation of the lift vector and that lift was found to consist of pressure forces only (at least at high Reynolds numbers and at small angles of attack, Sect. 5.2).

It can also be shown (Abbott and Von Doenhoff 1949) that the lift coefficient of a thin lifting surface of high aspect ratio in inviscid flow (that is for Reynolds numbers approaching infinity), approximately satisfies the relation

$$C_L = 2\pi\alpha_e = 2\pi(\alpha - \alpha_0 - \alpha_i), \quad (\text{all } \alpha\text{' s in radians}) \tag{5.15.9}$$

Here, α_0 is the angle of attack at which the lift is zero and the factor 2π represents the lift curve slope per radian of the two-dimensional foil section. It is easily verified that this is about equal to the value of 0.1/degree mentioned in Sect. 5.14.

After substitution of Eq. (5.15.4) into (5.15.9) the latter can also be written as

$$C_L \approx \frac{2\pi(\alpha - \alpha_0)}{1 + 2\pi/\pi A} \tag{5.15.10}$$

From this it follows that, for high aspect ratio foils, the lift curve slope $\dfrac{dC_L}{d\alpha}$ (rate of change of the lift coefficient with angle of attack) is equal to

$$\frac{dC_L}{d\alpha} \approx \frac{2\pi}{1 + 2\pi/\pi A} = \frac{2\pi}{1 + 2/A}, \quad (\text{per radian}) \tag{5.15.11}$$

For two-dimensional flow, that is for $A \rightarrow \infty$, this becomes equal to the foil section lift curve slope of 2π/radian. If required, a more accurate result can be obtained by replacing the factor 2π in Eqs. (5.15.10) and (5.15.11) by the actual lift curve slope of the foil section in viscous flow.

Fig. 5.15.5 *Lift versus angle of attack for rectangular wings of various aspect ratios*

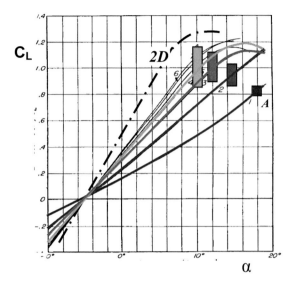

Figure. 5.15.5 shows the dependence of the lift versus angle of attack curve on aspect ratio. The figure is based on classical experimental data (Prandtl and Tietjens 1934b in Chap. 1) for aircraft wing-like lifting surfaces with rectangular planform of aspect ratios ranging from 1 to 7. All wings have the same airfoil section, with a thickness/chord ratio of about 10% and a small amount (about 2.5%) of camber. Added to the figure is an estimate of the lift curve for two-dimensional (2-D) flow, that is for $A \to \infty$.

The figure illustrates that the dependence of lift on aspect ratio is appreciable. Bearing in mind that many sailing yachts have keels and sail plans with an aspect ratio of 1–4 (or less) it is clear that, for a given angle of attack, the reduction of lift due to finite aspect ratio may be as much as 30–70%.

The effect of aspect ratio on the lift-drag polar of a lifting surface is illustrated by Fig. 5.15.6. The data are for the same wing-type lifting surfaces as those of Fig. 5.15.5. We can see that, in agreement with Eq. (5.15.5), the effect of aspect ratio on drag is even larger than that on lift. For example, the total drag at $C_L = 0.8$ for an aspect ratio of 3 is more than four times as large as in two-dimensional flow. In other words, the induced drag at $C_L = 0.8$ is three times as large as the drag in two-dimensional flow, that is the profile drag C_{D_v}.

It is also evident from Fig. 5.15.6 that the effect of aspect ratio on the minimum drag angle is quite large. For two-dimensional flow the minimum drag angle is about equal to arctan (0.02), that is about 1°. For an aspect ratio of 3 this has increased to about arctan (0.07) which is about 4°!

Fig. 5.3.1 *Still fluid static*
pressure—proportional to the height
of the column of fluid

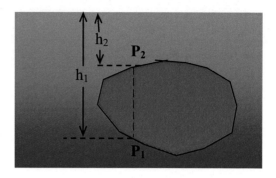

5.3   Static Forces

As already mentioned in the preceding section the pressure forces acting on the external surface of an object immersed in a still fluid are the result of the weight of the column of fluid above the object. It can be shown (Prandtl and Tietjens 1934a in Chap. 1) that the pressure p_g at a point P on the surface of an object immersed in a still fluid of constant mass density ρ is proportional to the height h of the column of fluid above it (See Fig. 5.3.1). The precise relation is:

$$p_g = \rho g h, \tag{5.3.1}$$

where g is the constant of gravitational acceleration.

If a point P_1 on the bottom and a point P_2 on the top of the object are positioned directly above each other there will a difference Δp in pressure equal to

$$\Delta p = \rho g(h_1 - h_2) \tag{5.3.2}$$

Because $h_1 > h_2$ the pressure difference between top and bottom will cause an upwards directed force. It can be shown (Prandtl and Tietjens 1934a in Chap. 1) by adding-up ('integrating') the pressure differences all around the surface that the object as a whole experiences a vertical lift or buoyancy force equal to

$$\Delta = \rho g V, \tag{5.3.3}$$

where V is the volume of the object. This, of course, is nothing else but Archimedes' law (Sicily, about 250 BC), which states that a submerged body experiences a vertical lift force equal to the weight of the displaced volume of fluid.

Note that if the points P were positioned on different sides but at the same depth h, there would be no difference in pressure. This is the reason why an object immersed in a still fluid does not experience a side force.

For a sailing yacht, being an object that is partially immersed in air and partially in water, the situation is a little bit more complicated (Fig. 5.3.2). First of all we note

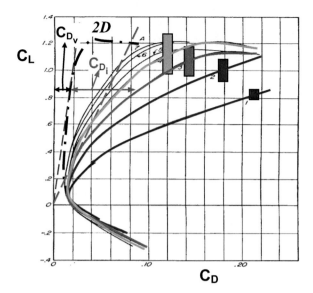

Fig. 5.15.6 *Effect of aspect ratio on drag polar for rectangular wings*

It can also be seen that the lift coefficient at which the drag angle has its minimum drops from about 0.9 in two-dimensional flow to about 0.35 for an aspect ratio of 3.

We can now summarize the effects of aspect ratio on the fluid dynamic characteristics of lifting surfaces as follows:

- For aspect ratios of the order of 1–4, as commonly applied in sailing yacht sail plans and hull appendages, the three-dimensional (aspect ratio) effects dominate at high lift coefficients but the foil section characteristics are important at low lift.
- For very high aspect ratios, say >10, the section characteristics become also important at high lift.
- Foil section maximum lift is important for the maximum lift of 3-D lifting surfaces of all (except very small) aspect ratios.

Effects of Planform Shape
We have seen above that the induced downwash and induced drag of a lifting surface are determined by the level of lift and the span or aspect ratio. It has been found (Prandtl and Tietjens 1934b in Chap. 1; Abbott and Von Doenhoff 1949), that, to a lesser extent, they also depend on the distribution of circulation (or lift) in the spanwise direction. The latter is not only determined by angle of attack and aspect ratio

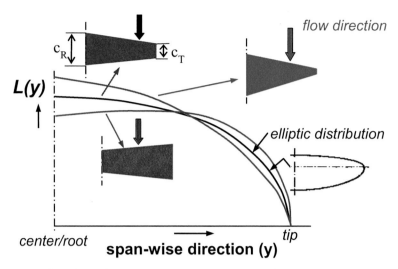

Fig. 5.15.7 *Effect of plan-form shape (taper) on the span-wise lift distribution of a lifting surface*

but also by the shape of the planform and span-wise variations of foil section shape and incidence ('*twist*'), if applicable. It has also been found that the induced drag of a planar lifting surface of given span or aspect ratio acquires a minimum value when the spanwise distribution of circulation is elliptic. According to a theorem of Munk (Munk 1923) this is the case when the induced downwash is constant along the span.

For lifting surfaces with symmetrical foil sections and without twist, such as keels and rudders, an elliptical span-wise distribution of circulation is obtained when the plan-form shape is elliptic (Fig. 5.15.7). For small to moderate aspect ratios $(A <\cong 6)$ a distribution of circulation close to the elliptical one is obtained for a straight-tapered plan-form with a taper ratio $TR = c_T/c_R$ of about 0.45–0.5.[8] Here, c_T is the chord length at the tip and c_R is the chord length at the centre section of the lifting surface. However, it has also been found that for very small aspect ratios $(A <\cong 1)$ the distribution of circulation is always almost elliptic, independent of the taper ratio, except at high angles of attack.

 As indicated qualitatively in Fig. 5.15.7, the spanwise lift distribution is, in general, over-elliptic (more lift near the centre, less near the tip) for plan-form shapes with more taper, i.e. a lower value of c_T/c_R. For plan-forms with inverse taper, that is taper ratios >1, the lift distribution is under-elliptic, with more lift near the tip and less near the centre.

 An elliptical spanwise lift distribution can also be obtained for non-elliptical planforms and planforms with arbitrary taper. It requires that the non-elliptical spanwise lift distribution due to angle of attack is corrected into an elliptical one by

---

[8] Note that only the starboard halves of the lifting surfaces are shown in Fig. 5.15.7.

applying spanwise variation of foil section shape or incidence (twist). Figure 5.15.7 indicates, that for an over-elliptic (or 'triangular') lift distribution due to angle of attack, the twist should be such that the incidence is lowered at the centre and increased at the tip. For an under-elliptic lift distribution due to angle of attack it is the other way around. Note that in this situation the elliptical lift distribution can be obtained for one angle of attack only.

For completeness it is mentioned that the lift distribution due to angle of attack is also called the *additional lift distribution* and the distribution at zero total lift is called the *basic lift distribution*. When the latter is not zero due to twist or spanwise variation of foil section there will be trailing vortices and already some induced downwash/drag at zero total lift.

Another parameter affecting the span-wise lift distribution and associated fluid dynamic characteristics of a lifting surface is the *angle of sweep*. This is usually defined as the angle between a line perpendicular to the direction of the undisturbed flow and the line connecting the quarter chord points of the root and tip sections (the angle in Fig. 5.15.8). The effect of sweep-back (Küchemann 1978) is to move the centre of pressure, that is the point of application of the resulting lift force, in the direction of the tip (red curve in Fig. 5.15.8). Sweep-forward moves the centre of pressure towards the root (blue curve).

It will also be clear from the discussion given above that taper and sweep can be combined in order to realize certain required spanwise load (lift) distributions. For example, it is possible to realize a near-elliptical spanwise load distribution giving minimum induced drag by combining sweep-back with a large amount of taper $(TR \ll 1)$, or inverse taper $(TR > 1)$ with sweep-forward. Figure 5.15.9, derived

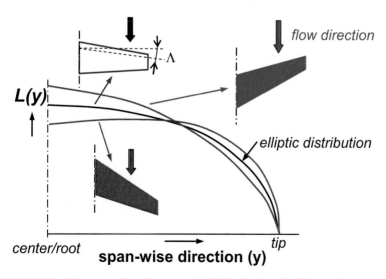

Fig. 5.15.8 *Effect of sweep angle on the span-wise lift distribution of a lifting surface*

Fig. 5.15.9 *Required taper ratio TR for a (near-) elliptic span-wise load distribution as a function of sweep angle*

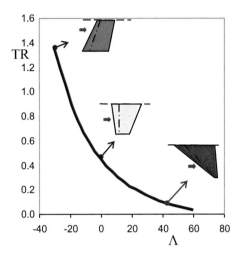

from DeYoung and Harper (1948), illustrates, for moderate aspect ratios, for which combinations of sweep and taper this is the case. Under-elliptic distributions of circulation are obtained to the right of this curve and over-elliptic distributions to the left.

It should be noted that Fig. 5.15.9 has little meaning for foils of very small aspect ratios (A < 1, say) because these always assume an elliptical spanwise load distribution, irrespective of sweep and taper.

The shape of the planform of a lifting surface is also of importance for the lift-curve slope. For high aspect ratios the effect of sweep on the lift-curve slope can be approximated through the expression (Slooff 1984)

$$\frac{dC_L}{d\alpha} \approx \frac{2\pi\cos\Lambda}{1+2\cos\Lambda/A} \qquad (5.15.12)$$

Equation (5.15.12), which reduces to

$$\frac{dC_L}{d\alpha} \approx \frac{2\pi}{1+2/A}$$

(Equation (5.15.11)), for $\Lambda = 0$, implies that the lift-curve slope decreases with increasing angle of sweep Λ. It also implies that the effect of sweep is more pronounced for high aspect ratios, when the term 1 in the denominator dominates over the term $2\cos\Lambda/A$. For very small aspect ratios, say A < 1, it is the other way around.

Low Aspect Ratio Effects

It has been found (Katz and Plotkin 1991) that, for small aspect ratios, the lift-curve slope is overestimated by the expression Eq. (5.15.12). The reason for this is that the foil section characteristics become less and less significant for vanishing aspect ratio. According to Ref. Abbott and Von Doenhoff (1949) this can be taken into account by multiplying the factor 2π in Eq. (5.15.11), that is the foil section lift-curve slope, with an 'edge correction factor' E:

$$\frac{dC_L}{d\alpha} = \frac{2\pi E}{1 + 2E/A} \tag{5.15.13}$$

In its original form, proposed by R. T. Jones (1941) for wings of elliptical planform, E, or rather $1/E$, is taken as the ratio of the semi-perimeter to the span of the wing. For more general wing shapes, E is a function of aspect ratio, angle of sweep and taper ratio. An empirical approximation (*USAF* n.d.) that is often used is:

$$E = (A/2)\cos\Lambda / \sqrt{\{\cos^2\Lambda + (A/2)^2\}} \tag{5.15.14}$$

It is easily verified that

$$E \to A/2 \to 0 \quad \text{for } A \to 0 \tag{5.15.15}$$

and

$$E \to \cos\Lambda \quad \text{for } A \to \infty \tag{5.15.16}$$

Note that Eq. (5.15.16) implies that Eq. (5.15.13) reduces to Eq. (5.15.12) for high aspect ratios.

Effective Aspect Ratio
It is often convenient to express the effects of planform shape and the associated spanwise lift distribution on the induced downwash and induced drag in terms of what is called the *effective span* or *effective aspect ratio*. The effective span is usually defined as the span of a fictitious lifting surface with an elliptic span loading that, for the same lift, has the same (average) downwash as the actual lifting surface with a non-elliptic span loading. The expression (5.15.2) for the induced downwash is then written as

$$\alpha_i \approx \frac{L}{(\frac{1}{2}\rho V^2 b)(\pi b_e)} \quad \text{(in radians)} \tag{5.15.17}$$

where b_e is the effective and b the geometric span. The expression for the lift curve slope then becomes

$$\frac{dC_L}{d\alpha} = \frac{2\pi E}{1 + 2E/(A\frac{b_e}{b})} \tag{5.15.18}$$

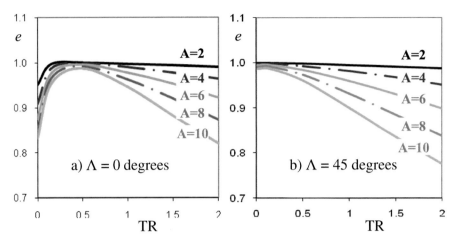

Fig. 5.15.10 *Dependence of the effective span on taper ratio and aspect ratio, for two angles of sweep*

and the expression for the induced drag is written as

$$C_{D_i} = C_L^2 / (\pi A \, b_e / b) \qquad (5.15.19)$$

The product $A \, b_e / b$ is sometimes called the effective aspect ratio A_e:

$$A_e = A \, b_e / b \qquad (5.15.20)$$

and the ratio b_e / b is often referred to as the aerodynamic efficiency factor 'e':

$$e = b_e / b \qquad (5.15.21)$$

Obviously, $e = 1$ for an elliptic span loading and there holds $e < 1$ for any other, non-elliptic spanwise lift distribution.

Figure 5.15.10, derived from Abbott and Von Doenhoff (1949) and DeYoung and Harper (1948) (see also Appendix B), shows the dependence of the aerodynamic efficiency factor on taper ratio and aspect ratio for two different angles of sweep. The figure illustrates, in agreement with Fig. 5.15.9, that a taper ratio of about 0.5 leads to a maximum effective span for a plan-form with zero sweep, but that smaller taper ratios are required for high angles of sweep. The figure also illustrates that the effects of taper and sweep become insignificant for (very) small aspect ratios. As mentioned above, all lifting surfaces adopt an elliptic spanwise load distribution (i.e. $e = 1$) for vanishing aspect ratio ($A \ll 1$), at least at small angles of attack.

Not shown in Fig. 5.15.10, but indicated by Fig. 5.15.9, is the fact that taper ratios > 1 are required to maximize the effective span for negative angles of sweep.

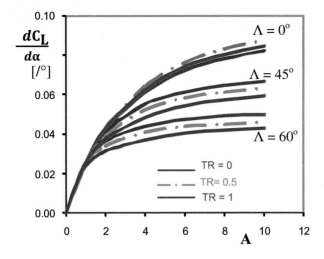

Fig. 5.15.11 *Lift curve slope as a function of aspect ratio for several angles of sweep Λ and several taper ratios TR*

Figure 5.15.11 presents the lift curve slope $\frac{dC_L}{d\alpha}$ as a function of aspect ratio A for various angles of sweep Λ and taper ratios TR. The figure is based on data from DeYoung and Harper (1948). It confirms that the effect of aspect ratio on the lift curve slope is large, as already evident in Fig. 5.15.5. It can also be seen that the effect of the angle of sweep is quite large for high and moderate aspect ratios, but that the effect of taper is relatively small.

It is appropriate at this point to mention that the lift-curve slope $\frac{dC_L}{d\alpha}$ is a measure of the effectiveness of a lifting surface. When $\frac{dC_L}{d\alpha}$ is large, the surface (sail, keel, rudder) area required to produce a given level of lift is small. When $\frac{dC_L}{d\alpha}$ is small, the area must be large, implying, amongst others, a large level of friction drag.

It should also be mentioned that the precise shape of the tip of a lifting surface has some effect on the lift and the effective span. As discussed in the next paragraph these effects are most pronounced at high angles of attack and for low aspect ratios.

Low Aspect Ratio Phenomena at High Angles of Attack
A small aspect ratio also has an effect on the high-lift characteristics of a lifting surface (Katz and Plotkin 1991). This is because the flow about the tip(s) becomes increasingly important for small aspect ratios, in particular at high angles of attack. Figure 5.15.12 illustrates what happens on a thin lifting surface with rectangular planform. At high angles of attack the flow about a thin lifting surface separates at the side-edge (as well as at the leading-edge when the latter is sufficiently sharp).

Fig. 5.15.12 *Side-edge/tip separation at*
high angle of attack of a flat, low aspect ratio
lifting surface

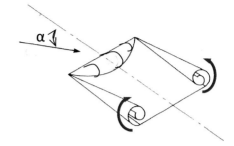

When the angle of sweep is sufficiently small and the angle of attack not too large, the leading-edge separation usually forms a closed separation bubble. The separated boundary layer at the side-edge rolls up to form a vortex above and just inboard of the tip. This side-edge vortex induces additional suction on the upper surface which results in some additional lift as compared to the situation in which there is only flow detachment/separation at the trailing-edge. The additional suction also causes an additional, nose-down pitching moment .

When the span of the wing is small as compared to the chord, the side-edge vortices cover a relatively large part of the span. For this reason the relative magnitude of the additional lift increases when the aspect ratio decreases.

Figure 5.15.13 gives an impression of the magnitude of the additional lift δC_L for flat, rectangular lifting surfaces of aspect ratios 1 and 0.5 (based on data from Ref. Winter (1937). The straight lines denoted by *'linear theory'* indicate what the lift would be without the side-edge separation (Eq. 5.15.18). Note that the additional lift δC_L for an aspect ratio A=0.5 is almost twice as large as for an aspect ratio A=1. Note also that the total lift attains a maximum at an angle of attack of about 35–40°. This is caused by the fact that beyond a certain, high angle of attack, that depends on aspect ratio, the side vortices loose their concentrated structure, a phenomena known as 'vortex burst'.

Fig. 5.15.13 *Lift characteristics*
of flat, low aspect ratio lifting
surfaces

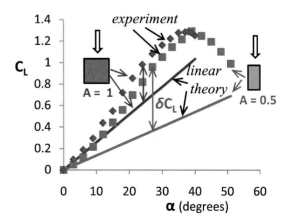

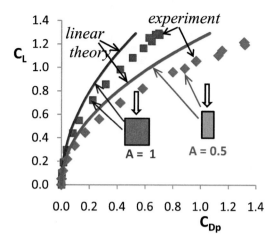

Fig. 5.15.14 Drag characteristics of flat, low aspect ratio lifting surfaces

Figure 5.15.14 compares drag polars with and without side-edge separation with the latter again indicated by *'linear theory'*. This can be interpreted as the induced drag (Eq. 5.15.19). Note that the figure reflects only pressure drag. The friction drag has not been taken into account but is, practically, negligible on the scale considered. The figure shows that the side-edge separation also causes a substantial amount of additional pressure drag.

A similar phenomenon, involving a leading-edge vortex, takes place at the leading-edge of highly swept, low aspect ratio delta wings (Fig. 5.15.15). In this case the additional lift is even larger (see Fig. 5.15.16), because the leading-edge vortices cover a larger portion of the span than the side-edge vortices in the case of the rectangular wing. In both cases the effects are most pronounced when the lifting surface /wing has sharp edges. When the side- or leading-edge is rounded, the separation starts at a higher angle of attack . As a consequence both the additional lift and the associated additional drag of a lifting surface with rounded edges are smaller than in the case of a foil with the same planform but with sharp edges at the same angle of attack.

The additional lift due to side-edge or leading-edge separation also causes an increase of the maximum lift. Figure 5.15.17 presents the additional maximum lift δC_{Lmax} as a function of aspect ratio for rectangular and delta planforms. The figure is based on experimental data given in Etkin and Reid (1996) and Winter (1937). As shown, the additional maximum lift attains a maximum for an aspect ratio $A \cong 1$. Given the fact that the maximum lift of a flat plate in two-dimensional flow is about

Fig. 5.15.15 Leading edge vortices on a flat, low aspect ratio delta wing

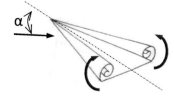

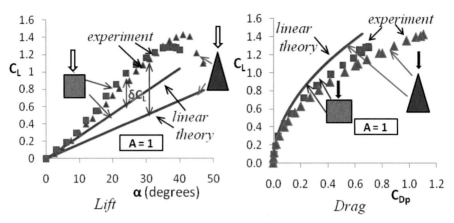

Fig. 5.15.16 *Aerodynamic characteristics of flat, low aspect ratio lifting surfaces with rectangular and delta planforms*

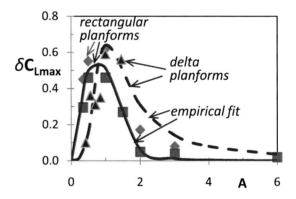

Fig. 5.15.17 Additional maximum lift due to side edge or leading edge separation for thin, flat, rectangular and delta planforms, as a function of aspect

0.8, (as we will see later, in Chap. 7), it can be concluded that the maximum lift is almost doubled by side-edge/leading-edge separation for $A \cong 1$. The additional lift is approximately proportional to the taper ratio TR and the (tangens of) the angle of sweep (see Appendix C.2). It is therefore much smaller for trapezoidal planforms with small leading-edge sweep and a small taper ratio.

Centre of Pressure

For the purpose of determining structural loads and the stability and control of air and sea vehicles the position of the centre of pressure of the lifting surfaces of the vehicle is of importance. Figure 5.15.18 presents the spanwise position y_{CP} of the centre of pressure of plane lifting surfaces (that is the aerodynamic centre) as a function of the taper ratio and the angle of sweep for different aspect ratios. The figure is based on the information given in DeYoung and Harper (1948). Note that

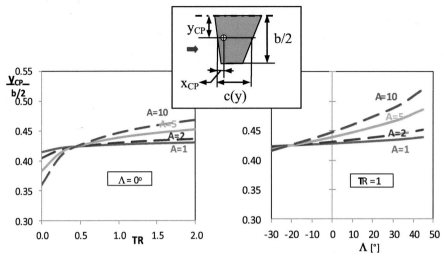

Fig. 5.15.18 *Span-wise position of the centre of pressure of a plane lifting surface as a function of taper ratio and angle of sweep, for different aspect ratios*

the lines cross at $y/(b/2) = 4/(3\pi) \cong 0.424$, which is the position of the centre of pressure when the spanwise distribution of lift is elliptic.

For moderate and high aspect ratios the chordwise position (Fig. 5.15.19) of the centre of pressure of a plane lifting surface is close to the 25 % chord line (DeYoung and Harper 1948). However, there is theoretical (Katz and Plotkin 1991; Letcher 1965 in Chap. 6) and experimental (Whicker and Fehlner 1958 in Chap. 6) evidence that it moves forward for low aspect ratios, in particular for large taper ratios and small angles of sweep. This is illustrated by Fig. 5.15.19. The two sets of curves are based on the data contained by Katz and Plotkin (1991), DeYoung and Harper (1948), Letcher (1965 in Chap. 6) and Whicker and Fehlner (1958 in Chap. 6). They suggest that the chordwise position of the centre of pressure of a plane lifting

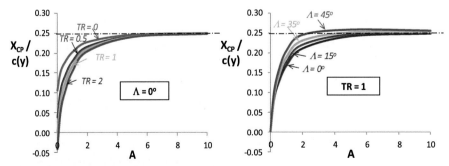

Fig. 5.15.19 *Chord-wise position of the centre of pressure of a plane lifting surface as a function of aspect ratio, for different taper ratios and angles of sweep (small angles of attack)*

surface is significantly forward of the 25 % chord line when the (effective) aspect ratio is smaller then, say, 2. Although this is undoubtedly correct for small angles of attack, the position of the centre of pressure is probably further backward at high angles of attack due to the effects of side-edge separation.

Non-planar Lifting Surface Configurations

In the preceding paragraphs we have only considered planar lifting surfaces, that is lifting surfaces that, approximately, fit into a flat plane. It has been found (Blackwell 1976), that if the latter restriction is lifted it is possible to realize efficiency factors 'e' significantly > 1.

For sailing yacht applications the most important non-planar lifting surface configurations are those with *winglets*. Winglets are fin-like extensions at the tip, approximately perpendicular to the mean plane of a lifting surface (Fig. 5.15.20). They may be pointing upward or downward, or both. Their main purpose is to reduce the induced drag for a given span. The concept was developed in the aerospace sector in the early 1970s by Richard T. Whitcomb at the NASA Langley Research Centre(Whitcomb 1976) but can and has also been applied to sailing yacht keels (Slooff 1984, 1985). In principle it could also be applied to sails.

Figure 5.15.20b) illustrates the mechanism of the reduction of the induced drag. When designed properly, that is when there is no boundary layer separation in the junction between winglet(s) and the main wing, the winglet(s) move(s) the tip vortex away from the plane of the wing to the tip(s) of the winglet(s). (In case of double winglets the original wing tip vortex is split into two winglet tip vortices). Because of the larger distance between the tip(s) of the winglet(s) and the main wing, the downwash on the main wing induced by the tip vortex/vortices is smaller than in the case without winglets. This can be seen from, for instance, the expression (5.13.2) for the circulatory flow velocity induced by a vortex. The latter is seen to be inversely proportional to the distance 'r' between the vortex core and a point where the induced circulatory flow is felt. In addition, the direction of the induced velocity

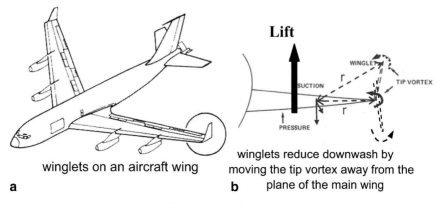

winglets on an aircraft wing

a

winglets reduce downwash by moving the tip vortex away from the plane of the main wing

b

Fig. 5.15.20 Example and downwash reducing mechanism of winglets **a** *winglets on an aircraft wing* **b** *winglets reduce downwash by moving the tip vortex away from the plane of the main wing*

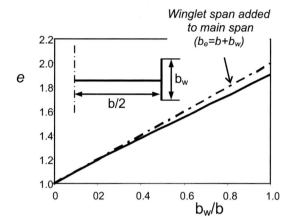

Fig. 5.15.21 *Effective span of lifting surfaces with winglets as a function of winglet-span/main-span ratio*

is such that there is a significant component parallel to the main wing and a smaller normal component (Fig. 5.15.20b).

We have seen earlier that for planar lifting surfaces an elliptic spanwise loading leads to minimum induced drag. Spanwise lift distributions giving minimum induced drag can also be determined for non-planar configurations of lifting surfaces (Lundry 1968). In general, the optimal spanwise lift distribution for a non-planar lifting surface is non-elliptic. The maximum efficiency that can be realized in this way for configurations with winglets is summarized in Fig. 5.15.21. Shown is the efficiency factor 'e' ($= b_e/b$, Eq. (5.15.21)) as a function of the winglet-span/wing-span ratio b_w/b. Note that only half of the configuration is shown in the picture and that winglets have been applied at both tips. With winglets attached to one tip only the effect would be less than half of that shown in Fig. 5.15.21.

Also shown in the figure is the (maximum) efficiency factor of planar lifting surfaces with a span equal to that of the basic wing plus half the span of the winglets. Note that this always leads to a better efficiency than with the winglets 'vertical'. It illustrates the point that, when there is no constraint on the span, increasing the span of the main lifting surface is always better than adding winglets.

The application of winglets has negative aspects also. It implies a larger 'wetted' surface and, hence, a larger frictional or profile drag. The total (profile plus induced) drag of configurations with winglets will therefore be lower above a certain level of lift only (Fig. 5.15.22). The 'break-even' level of lift is determined mainly by the relative amount of added frictional resistance and the winglet-span/main-span ratio. This means that the level of drag of other configuration parts, such as a fuselage or hull, also plays a role. For a sailing yacht this means, for example, that the break-even lift for a small lifting surface attached to a large hull will be lower than for a large lifting surface mounted on a small hull. Because the added frictional drag of winglets will be roughly proportional to the winglet span, the break-even lift for small winglets will be lower than for big ones.

As a final remark with respect to winglets it is emphasized that they should not be confused with *endplates*. Endplates, that is plane, non-profiled fences mounted

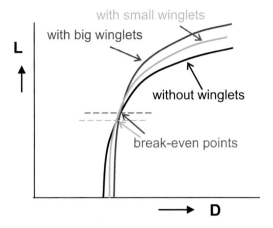

*Fig. 5.15.22 Effect of winglets
on the lift-drag polar of a lifting
surface (qualitatively)*

at the tip of a lifting surface, are based on the erroneous concept of reducing the loss
of lift at the tip by eliminating the tip vortex and the associated induced downwash.
That the concept is erroneous is inferred immediately by the Kelvin-Helmholtz law
of conservation of vorticity introduced in Sect. 5.13. Elimination of a tip vortex is,
unfortunately, impossible without eliminating lift. Tip vortices can only be moved
to a location where they are less harmfull and this, precisely, is the background of
the winglet concept.

5.16   'Non-lifting' Bodies

Slender Bodies
In the science of fluid dynamics cigar-like objects, sailing yacht hulls, and bodies
like keel-bulbs, etc. are usually classified as *non-lifting bodies.* The reason is that
they carry little or no lift and are usually not intended to do so either. The fluid dynamics of non-lifting bodies differ significantly from the fluid dynamics of lifting
surfaces. The main reason is that hulls, bulbs, etc. are usually relatively slender in
the longitudinal direction. That is, the lateral dimensions are small in comparison
with the length of the body. In the terminology of lifting surfaces this means that the
'span' is relatively small. Another difference is the absence, in general, of a sharp
trailing-edge. As a consequence of both the small 'span' and the absence of a sharp
trailing-edge the 'lift' or side force of slender objects is usually very small, at least
when they 'operate' in isolation. (We will see later that the situation is different
when a non-lifting body is attached to a lifting surface).

As illustrated qualitatively by Fig. 5.16.1, the slope of the lift curve of slender
bodies is practically zero at small angles of attack when the flow is more or less attached. Then, there is positive lift on the front half of the body but almost equally
large negative lift on the rear half. With increasing , a net lift develops slowly as a

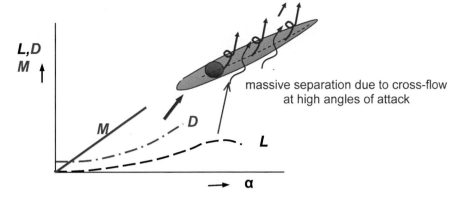

Fig. 5.16.1 *Lift, moment and flow pattern of slender*

result of cross-flow separation and vortex formation, as on a low-aspect-ratio wing, until the flow breaks down completely by massive separation due to excessive cross-flow. Bodies with a blunt base develop a little more lift than bodies with a pointed (rear) end (Katz and Plotkin 1991).

The drag at small angles of attack is mostly viscous drag. However, with increasing angle of attack, when lift develops as a result of cross-flow separation and vortex formation, there is also a rapid development of induced drag. At very high angles of attack, when the lift has collapsed due to vortex bursting, most of the drag is form drag.

While the net lift of a slender body is negligible at small angles of attack, the moment around a pitching (y-) axis is substantial due to the positive lift on the front half and the negative lift on the rear half. It can be shown (Katz and Plotkin 1991) that for slender, ellipsoidal bodies of revolution there holds

$$M_y \approx \tfrac{1}{2}\,\rho V^2 2\alpha \nabla, \qquad (5.16.1)$$

where α is the angle of attack is in radians and ∇ is the volume of the body. A slightly more general expression (Ashley and Landahl 1985/1965) for the moment of ellipsoidal bodies of revolution can be written as

$$M_y \approx \tfrac{1}{2}\rho V^2 2\alpha f(D/L)\nabla \qquad (5.16.2)$$

The factor $f(D/L)$ is a function of the maximum diameter/length ratio. For slender bodies, i.e. $D/L \ll 1$, the value of $f(D/L)$ approaches 1. For a sphere, i.e. $D/L = 1$, it takes the value 0. The author has found that a suitable approximation for $f(D/L)$ is given by

$$f(D/L) \cong 1 - D/L \qquad (5.16.3)$$

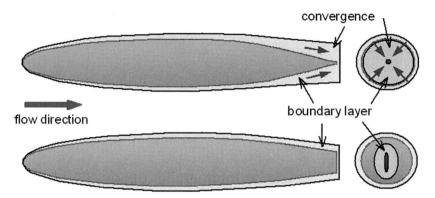

Fig. 5.16.2 *Slender bodies with flattened tail: thinner boundary layer and less form drag through less convergence of boundary layer flow*

Friction is the dominant source of resistance of slender bodies of revolution, at least at small and moderate angles of attack. As discussed before (Sects. 5.2, 5.11, 5.12), the pressure drag (form drag and induced drag) becomes more important at high angles of attack.

As already mentioned in Sect. 5.11 the growth of the boundary layer thickness and associated form drag is co-determined by convergence or divergence of the boundary layer flow. On slender bodies of revolution with pointed rear ends there is considerable convergence (confluence) of boundary layer material towards the rear, leading to a relatively high form drag. It is the author's impression that a reduction of convergence and form drag can be realized by flattening out the tail part of the body, (Fig. 5.16.2) so that the boundary layer flow detaches from a short, lateral trailing-edge ('beaver tail' or 'fish tail'). Although such flattening increases the wetted surface, the total drag can be lower due to a reduction of the form drag.

In aeronautics, the concept has been applied to, e.g., the fuselage of the Mc-Donnell-Douglas MD-80 series of aircraft (Henne 1989) and the Boeing 777. It is, occasionally, also applied to keel bulbs, although not necessarily for the purpose of reducing the viscous drag in all cases.

Sailing Yacht Hulls
Sailing yacht hulls are usually also fairly slender but the flow about them is complicated by the fact that they are partly immersed in water and partly in air. The part of a sailing yacht hull that operates in air has many sharp edges and discontinuities that force the flow to separate, detach and shed vortices at all apparent wind angles (Fig. 5.16.3). As a consequence the aerodynamic pressure drag (form drag and induced drag) of a hull is quite large and, in general, larger than the friction drag. The relatively large induced drag is caused by the small lateral dimensions ('span') of the hull and the associated relatively large downwash.

The under-water part of the flow about a hull is usually much better 'organized'. Flow separation and vortex formation is, in general, restricted to the stern section

Fig. 5.16.3 *Typical 'messy'*
air flow about a sailing yacht
hull

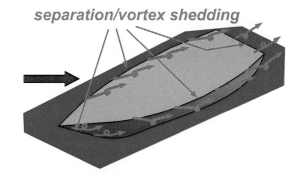

Fig. 5.16.4 *Under-water flow*
about a hull

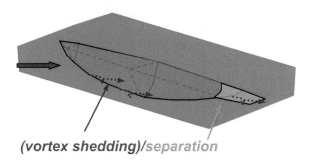

of the hull. Exceptions are hulls with a sharp V-bottom or cross-sections with kinks or chines (Fig. 5.16.4). In such cases vortices are generated at the sharp edges that may generate a significant amount of induced resistance.

In the absence of sharp edges, the profile drag (friction drag plus form drag) constitutes the larger part of the total resistance, at least at low boat speeds. At higher boat speeds, when the effects of the air/water interface become dominant, this is no longer the case (about which later in Chap. 6).

5.17   Fluid Dynamic Interference Between Lifting Surfaces and 'Non-lifting' Bodies

Lifting surfaces usually do not operate in isolation (except in the case of flying wing types of aircraft). In general they are attached to a fuselage or hull type of 'non-lifting' body. In the case of sailing yachts the appendages are fully attached to the hull and the sails operate in close proximity to it.

The close proximity of a fuselage or hull type of body interferes with the flow about a lifting surface and vice versa. How precisely depends primarily on the relative dimensions of, and distance between the lifting surface and the other body.

Fig. 5.17.1 *Lifting surface attached to a flat wall of infinite extent*

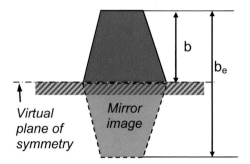

A special example is when a lifting surface is attached to a plane wall of infinite extent (Fig. 5.17.1). This situation is representative of the case of a (very) small fin attached to a (very) large hull. It has been found that, as long as there is no boundary layer separation at the junction, the wall acts as a reflection plane that creates a virtual, mirror image of the lifting surface at the other side of the wall. The induced downwash and the induced angle of attack experienced by the actual lifting surface are then only half of that corresponding with the actual geometrical span. It is said that in this case the effective span b_e (Sect. 5.15) is twice the geometrical span b.

Matters differ considerably when there is a gap of finite width (g) between the wall and the lifting surface (Fig. 5.17.2). The reason is that a gap allows the formation of a tip vortex at the inner tip (root) close to the wall. As a consequence the lifting surface experiences much more downwash. This, however, is partly compensated by upwash induced by the mirror-image counterpart of the root vortex. When the lifting surface is attached to the wall, that is when the gap width is zero, the effect of the inner (root) vortex is fully cancelled by its mirror-image counterpart.

Figure 5.17.3, derived from Marchai (2000, p. 398), Lakshminarayana (1964), presents the effective span b_e as a function of the width g of the gap. The associated spanwise lift distributions for $g/b = 0, 0.1$ and ∞ are shown in Fig. 5.17.4.

Fig. 5.17.2 *Lifting surface in proximity of a flat wall of infinite extent*

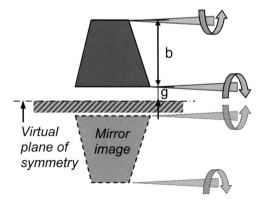

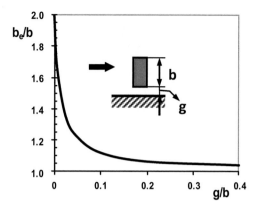

Fig. 5.17.3 Effect on effective span of a gap between a lifting surface and a wall

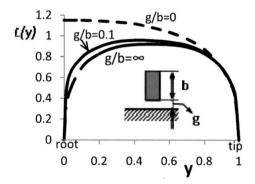

Fig. 5.17.4 Effect on the spanwise distribution of lift of a gap between a lifting surface and a wall

It appears that, while the effective span is twice the geometric span when the gap width is zero, the ratio b_e/b decreases very rapidly with increasing value of the ratio g/b. For a gap width of about 10% of the span, the effective span has already dropped to only about 110% of the geometric span. The precise form of the curve shown in Fig. 5.17.3 will depend slightly on the taper ratio and sweep angle of the lifting surface. Because of the effect of taper and sweep on the spanwise lift distribution (see Figs. 5.15.7 and 5.15.8), it is to be expected that the drop of the effective span as a function of the gap width will be a little faster for taper ratios <1 and a little slower for TR>1. It will also be a little slower for positive angles of sweep.

A special (but not unusual) case arises when the lifting surface is attached to the wall, but suffers from boundary layer separation in the foil-wall junction. The configuration then behaves like there is a (small) gap.

Fin-Body Configurations

When attached to a 'non-lifting' body of finite dimensions, a lifting surface induces considerable lift on that body through a mechanism called 'lift carry-over' (Schlichting and Truckenbrodt 1979). Figure 5.17.5 illustrates the mechanism for the case of a non-lifting body fitted, symmetrically, with fins or wings. When there is no boundary layer separation at the fin-body junctions, the rotation in the fin boundary layer and

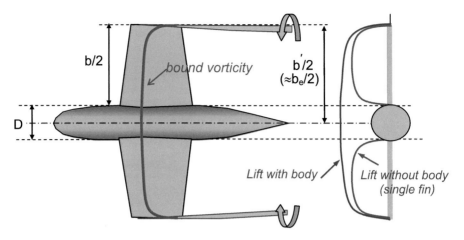

Fig. 5.17.5 *Fins or wings attached to a 'non-lifting body'*

the associated circulation or bound vorticity (Sect. 5.13) do not go to zero and there
is no trailing vortex at the junction. This means that the circulation and bound vorticity are carried over onto the body (Slooff 1984, 1985). Note that this a consequence
of the Kelvin-Helmholtz theorem of conservation of vorticity (Sect. 5.13).

At this point it is useful to introduce the notion '*active span*'. This is defined
as the lateral distance b' over which the configuration carries a load. As shown in
Fig. 5.17.5 it follows that $b = b + D$. The *active span* is not to be confused with the
effective span b_e which was defined, in Sect. 5.15, as the span of an (equivalent) lifting surface with an elliptical span loading having the same lift and the same induced
drag. Because the induced drag depends on the efficiency factor e of the spanwise
lift distribution, introduced in Sect. 5.15, the effective span can be expressed as
$b_e = be + D$. With $e \leq 1$ this means that b_e is approximately equal to, but in general
slightly smaller than, the distance between the tip vortices, i.e. $b_e \lesssim b'$.

It will be clear that the active span and the effective span are related in the sense
that a larger active span will also lead to a larger effective span. The interested
reader can find further details in Appendix C.1. Here, we will suffice by showing
the lift curve slope $\dfrac{dC_L}{d\alpha}$ as a function of the ratio D/b for fin-body configurations
with a body of circular cross-section (Fig. 5.17.6[9]). The figure illustrates clearly that
the effect of inserting a body at the root of a fin or wing is quite large and increases
almost linearly with D/b. Note that the range of D/b considered is $0 \leq D/b \leq 1$. As
explained in Appendix C.1 the effect of the presence of the body disappears again
for $D/b \rightarrow \infty$.

Because the amount of lift carry-over at the root of the fin or wing depends on
the circulation at the root, it also depends on the taper ratio and angle of sweep of
the fin. As indicated in Figs. 5.15.7 and 5.15.8, a high level of circulation at the root,

---

[9] In this figure span, area and aspect ratio are those of the exposed wing/fin.

Fig. 5.17.6 *Lift curve slope of a wing-body configuration as a function of the ratio of the body diameter to the span of the exposed wings*

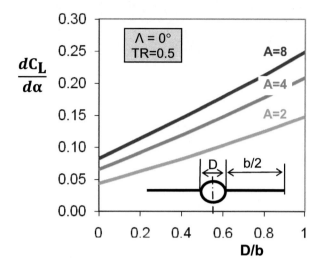

that is more lift carry-over on the body, is obtained for low taper ratios and negative angles of sweep. The opposite is the case for high taper ratios and positive angles of sweep. The ratio D/b is, however, the most important parameter.

Although the lift carry-over from the wing or fin on the body forms the most important aspect of wing/fin-body interference, there is also an effect of the fin on the moment of the body.[10] This is caused by the fact that the wing or fin changes the effective incidence of the body. Although the wing or fin induces upwash on the part of the body ahead of the wing, it forces the flow in the direction of the axis of the body at and aft of the position of the wing or fin. The net effect is that the moment acting on the body with a wing or fin attached is a little smaller (roughly by a factor around 3/4, depending on the configuration) than that of the isolated body at the same angle of attack.

Effect of a Bulb

Examples of 'non-lifting' bodies attached to a lifting surface (or the other way around) in keel yacht fluid dynamics are fin keels attached to sailing yacht hulls, bulbs attached to a fin keel and a mainsail attached to a boom. The case of a semi-submerged hull with a fin keel is special and will be discussed in Chap. 6. (The case of a mainsail on a boom is also special, but in a different way.)

The case of a bulb-type non-lifting body attached to the tip of a fin-type lifting surface (or an aircraft wing with a tip tank), attached to a wall or hull is illustrated in Fig. 5.17.7. In terms of fluid dynamics the case of a fin plus bulb is similar to that of a fin-body configuration. In both cases the main mechanism is an increase of the active span and the effective span. When the bulb has a pointed tail, the tip vortex is moved outboard over a distance approximately equal to half the diameter D of the bulb.

---

[10] Here we consider the moment with respect to an axis in the span-wise direction of the fin.

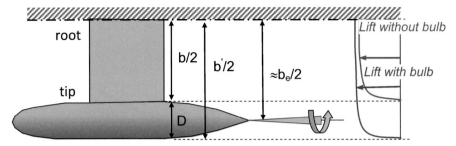

Fig. 5.17.7 *Fin with bulb (or wing with tip-tank)*

When the bulb is fitted with a vertical, flattened ('fish') tail, as in Fig. 5.16.2, the tip vortex is moved further away from the root. Hence, in case of a bulb with a pointed tail, we have $b_e \approx be + D/2$ and with a 'fish' tail we have $b_e = be + D \approx b'$. It will also be clear that in case there is also a body at the wing centre, the increase of the active span and the effective span of the two separate bodies will have to be added.

Furthermore, when the bulb has a 'beaver tail', that is a flat tail perpendicular to the fin, the bulb will act somewhat like a winglet. Then, the effective span will be somewhere between those of the pointed bulb (with $b_e = be + D/2$) and the fishtail bulb (with $b_e = be + D$).

The general conclusion is that adding a bulb to a fin increases the effective span with about one half to one diameter of the bulb, depending on the configuration of the bulb and provided there is no boundary layer separation in the fin-bulb intersection. Note however, that this implies that when compared with a plane fin (without bulb) with the same span $(b+D)$ as the fin plus bulb configuration, the effective span of the fin plus bulb will, in general, be somewhat smaller (see also Ref. Tinoco et al. (1993).

The span-wise lift distribution on the fin with and without bulb is indicated qualitatively in the right part of Fig. 5.17.7. Two phenomena are to be noted here. Firstly, there is the lift carry-over from the fin onto the bulb. This by itself causes an increase of lift. A second effect of the bulb is that, due to the increased effective span, the lift on the fin is also somewhat higher than in the case without bulb (which has a smaller effective span).

Other, perhaps more secondary aspects of the mutual interference, are that the cross-flow about the bulb influences the spanwise lift distribution of the fin and that the upwash and downwash generated by the fin create something like a curved onset flow for the bulb. The latter implies an increase of the cross-flow on the part of the bulb ahead of the fin and a reduction of the cross-flow on the rear part, aft of the fin. While these effects are usually of secondary importance, at least at small angles of attack, they become more pronounced when the angle of attack is large.

Local Viscous Interference at the Junction of a Body with a Wall
As already indicated above, it is important to avoid boundary layer separation at the fin-body junction in order to avoid reduction of the effective span and to minimize the induced drag and form drag. Usually, that is without special measures, a blunt body

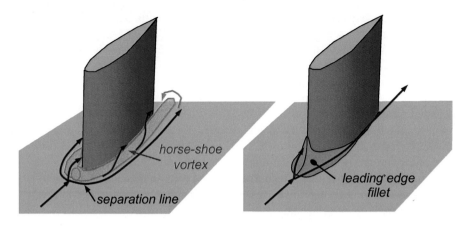

a   *without leading edge fillet*   **b**   *with leading edge fillet*

Fig. 5.17.8 *Flow patterns at the junction of a blunt body with a wall* **a** *without leading edge fillet* **b** *with leading edge fillet*

or lifting surface (fin), attached to a wall provokes separation of the boundary layer on the wall. This separation is caused by rapidly increasing pressure just in front of the body and leads to the formation of a *'horse-shoe' or 'necklace' vortex* that wraps around the body at the junction (Fig. 5.17.8a). When the body or fin carries lift, the boundary layer separation and resulting vortex formation is a-symmetric and concentrated on the suction side. The circulation around the fin is then (partially) lost which manifests itself in the appearance of a trailing vortex aft of the junction. The separation of the wall boundary layer and associated vortex formation can, in general, be avoided by reducing the pressure gradient through application of a leading-edge fillet at the wall-body junction (Fig. 5.17.8b). Leading-edge fillets of this kind, at keel-hull, rudder-hull and keel-bulb junctions, are a prerequisite for low drag.

5.18   Unsteady, Periodic Flow Phenomena

Introduction
As mentioned earlier (Sect. 5.11) the (turbulent) flows about the hull, appendages and sails of a sailing yacht are unsteady, that is time dependent, by nature. We have also seen that for many practical purposes, such as performance analysis, the flows can be treated as if they were steady through the introduction of a time-averaging process (see also Sect. 3.1). There are, however, a number of flow phenomena in sailing yacht fluid dynamics of which the unsteady nature is essential for the behaviour of the yacht and its structure. Most of these unsteady phenomena are of a periodic nature, that is they repeat themselves in time. In general, they are also coupled to the periodic motion and/or the deformation of the (part) of the yacht that is being considered.

One, obvious, example of periodic, unsteady flow conditions is the flow about hull, appendages and sails of a yacht sailing in waves. Waves bring the hull, appendages and sails in periodic, unsteady motion. As a consequence the flows of air and water about a yacht sailing in waves are also unsteady. The effects of waves on the flow about hull and appendages will be discussed in Chap. 6. The effects of periodic, unsteady ship motion on the aerodynamic performance of the sails is a subject of Chap. 7.

Other examples of unsteady flows and motions are the rolling behaviour of sailing yachts in downwind conditions (to be discussed in Chaps. 7 and 8), the 'whistling' of the shrouds of the rig in (strong) winds and 'mast fluttering '. Since every yachtsman will encounter such phenomena at some point in time it is appropriate to pay some attention to them. However, we will first consider some general aspects of unsteady, periodic phenomena.

Fluid-Structure Interactions and Their Relation with Frequency
For the purpose of bringing some order in the various unsteady phenomena it is useful to consider three different levels of interaction between the flows about and the motion of an object:

a. Flow about a rigid body in forced, periodic motion with little or no effect of the flow on the motion of the body. An example is the flow about a yacht sailing in waves
b. A still, rigid body, but unsteady flow as a result of periodic flow phenomena. Example: 'whistling' of the shrouds
c. Mutual interaction between the motion and/or deformation of the body and the unsteady flow. Examples are downwind rolling and mast fluttering.

A well-known phenomenon underlying the second of the three categories listed above is the so-called *Von Kármán vortex trail* (or *vortex street*).[11] It is an unsteady, periodic, asymmetric boundary layer separation and vortex formation phenomenon that, for a certain range of flow conditions, occurs at and downstream of blunt bodies. It is illustrated by Fig. 5.18.1 for the case of a circular cylinder in a crossflow. The mechanism can be described as follows:

At a given point in time and for a certain range of Reynolds numbers, the boundary layer on either the upper or the lower surface of the cylinder will separate because it cannot cope with the strong pressure gradient downstream of the point of maximum thickness. This leads to the formation of a discrete stream-normal vortex that is washed downstream. This vortex induces a change in flow angle at the cylinder in such a way that in the next phase the separation and vortex formation takes place at the other side. This alternating process then repeats itself periodically in time. The resulting oscillatory flow causes pressure fluctuations that are experienced as sound waves.

---

[11] The phenomenon is named after the fluid dynamicist Theodore von Kármán, who was the first to describe its mechanism (1911).

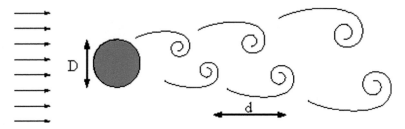

Fig. 5.18.1 *Von Kármán vortex trail or vortex street*

It can be shown (Prandtl and Tietjens 1934b in Chap. 1), that a periodic vortex pattern of the kind that is characteristic for a Von Kármán vortex street can exist and is stable only for a certain value (about 0.25) of the ratio D/d between the diameter D of the cylinder and the distance 'd' between two successive vortices with the same sense of rotation. It is easily verified that the distance d between two successive vortices is a function of the flow speed V and the frequency f of vortex shedding:

$$d = V/f \qquad (5.18.1)$$

Hence, the ratio D/d can be expressed as

$$D/d = D/(V/f) = f\,D/V \qquad (5.18.2)$$

Because this expression contains the flow velocity V and the basic mechanism (boundary layer separation) is governed by viscosity, the phenomenon depends also on Reynolds number. It has been found to occur only at Reynolds numbers VD/v between 2000 and 3000.

An important dimensionless parameter for unsteady flows in general is the *Strouhal number* Sr defined by

$$Sr = f\,L/V, \qquad (5.18.3)$$

where f is the frequency of the unsteady flow phenomenon, L is a characteristic dimension of the body and V is the flow velocity. Obviously, the expression (5.18.3) for the Strouhal number is the same as (5.18.2) if we take the diameter D of the circular cylinder as the characteristic length L. This means that the ratio D/d between the diameter of the cylinder and the distance between two successive vortices in a Von Kármán vortex street is the same as the Strouhal number. We have seen above that $D/d \cong 0.25$ is a prerequisite for the existence of a Von Kármán vortex street. In other words, Von Kármán vortex streets can occur only at a Strouhal number (=D/d) of about 0.25.

From Eq. (5.18.3) it follows that

$$f = Sr\, V/D \tag{5.18.4}$$

Because Sr is more or less constant (0.25), this implies that the frequency of the sound generated by Von Kármán vortex streets at the shrouds of a sailing yacht rig goes up with increasing wind speed and decreasing diameter of the shroud. The range of Reynolds numbers indicated above, $2000 < VD/\nu < 3000$, implies that the sound generation is taken over by shrouds with smaller diameters D when the wind speed V increases.

Another dimensionless parameter that is often used to characterize unsteady flow conditions is the so-called *reduced frequency* (Theodorsen 1935), usually defined as

$$k_f = \omega L/(2V) \tag{5.18.5}$$

(or, sometimes, as $k_f = \omega L/V$).

Here, $\omega \equiv 2\pi f$ is the so-called *circular or angular frequency*. Comparing (5.18.5) with (5.18.3) it is obvious that the Strouhal number and the reduced frequency differ only by a factor π (or 2π). While 'natural' unsteady flow phenomena like Von Kármán vortex streets are usually described in terms of Strouhal number, the terminology of reduced frequency is generally used in the case of flows about objects in forced oscillatory motion.

Because the ratio L/V in Eq. (5.18.5) can be considered as the time (T^*) it takes the flow to cover a distance L and f is the reciprocal of the period T of an oscillatory motion, the reduced frequency can also be expressed as

$$k_f \div T^*/T \tag{5.18.6}$$

T^* can also be interpreted as the time it takes for the flow to respond to an unsteady motion such as a (sudden) change in angle of attack. The reduced frequency is therefore a (relative) measure of the rapidity of response of the flow to an unsteady, periodic motion:

1. For $k_f \ll 1$ we have the situation that the (oscillatory) motion of the object is so slow that the flow can follow the motion as if it were steady at each point in time. Such conditions are called *quasi-steady*.
2. For $k_f \approx 1$ the response time of the flow is of the same order of magnitude as the period of oscillation of the object. As a consequence there is a time-lag between the periodic variation of the flow and the motion of the object.
3. For $k_f \gg 1$ the flow can no longer follow the very rapid oscillatory motion of the object and hardly responds anymore. This does not mean, however, that the unsteady forces become negligible.

Many of the unsteady, periodic phenomena in sailing yacht fluid dynamics, such as the flows (both under and above the water surface) when sailing in (long) waves or in downwind rolling conditions, are of the quasi-steady kind ($k_f \ll 1$). However, for high frequencies (short waves) and low boat/wind speeds the time lag between motion and flow response can become significant ($k_f \cong O(1)$). Von Kármán vortex street phenomena and sailing in relatively short waves at low wind and boat speeds belong to this category. Conditions corresponding with the third category ($k_f \gg 1$) hardly occur in sailing yacht fluid dynamics.

Dynamic Systems

All objects in periodic motion or periodic deformation are subject to periodic forces. When the object is rigid, that is, it cannot deform, it can, in addition to inertia and gravitational forces, only be subject to external forces acting at its surface, such as fluid dynamic forces. Objects with a flexible structure can also be subject to internal forces associated with spring-like forces due to bending, torsional stresses and forces associated with internal friction.

An important parameter in unsteady motions and the dynamic behaviour of objects and their structures is the *'natural'* or *'eigen'frequency* of the motion or the vibration modes of the object. An object that is triggered into some kind of motion or vibration by some instantaneous external force (initial disturbance) usually adopts its own frequency of motion or vibration. This is called the natural frequency of the motion or vibration. The type or mode of vibration (for example bending or torsion) or motion (for example translation or rotation) depends on how and where the force is applied and on the distribution of mass and stiffness of the object. Mass and stiffness also determine the natural frequency (f_0) or period ($T_0 = 1/f_0$). Damping of the motion by, for example, friction, causes the amplitude of the natural motion to decrease with time after the initial disturbance (Fig. 5.18.2).

It is often instructive to consider the mechanisms and properties of objects in oscillatory motion or vibration in terms of a generalized, so-called mass-spring-damper system (http://en.wikipedia.org/wiki/Harmonic_oscillator). Figure 5.18.3 depicts such a *'dynamic system'*. The mass 'm' is positioned in between and attached to a spring with stiffness 'k' and a damping cylinder 'b'. Two situations are of interest here. The first is the case when there is some initial disturbance of the position

Fig. 5.18.2 *Illustrating the effect of damping on the natural motion of an object*

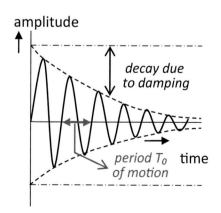

Fig. 5.18.3 *Schematic of a 'dynamic system'*

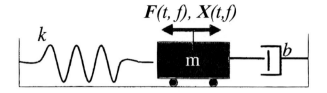

$$F(t, f), X(t,f)$$

X of the mass but no further external force (F). In this situation the equilibrium of forces acting on the mass 'm' can be expressed as

$$m\ddot{X} + b\dot{X} + kX = 0 \tag{5.18.7}$$

Here, $\ddot{X}$ is the acceleration in the X-direction of the mass m, $m\ddot{X}$ is the inertia force, $\dot{X}$ is the velocity of displacement, b is the *damping coefficient*, $b\dot{X}$ is the damping force opposing the direction of motion and $k\,X$ is the restoring force opposing the direction of the displacement. It can be shown through the calculus of differentiation that when the displacement X varies with time like sinωt, its rate of change with time, $\dot{X}$, varies like cosωt and the corresponding acceleration $\ddot{X}$ like $-$sinωt.

It is easily understood that due to compression and expansion of the spring ('k') (*restoring force*) and the 'dragging' motion of the piston in the damping cylinder 'b' (*damping force*), the mass 'm' will move back and forth periodically with a certain frequency f. In general, the system then behaves as in Fig. 5.18.2; the mass 'm' adopts a damped oscillation with a certain, the 'natural', frequency. When there is no damping the (undamped) natural frequency f_0 can be shown (http://en.wikipedia. org/wiki/Harmonic_oscillator) to depend on the spring stiffness 'k' (dimension N/m) and the mass 'm' like

$$2\pi f_0 \ (\equiv \omega_0) = \sqrt{(k/m)} \tag{5.18.8}$$

Equation (5.18.8) expresses that the natural frequency f_0 of the system is low when the mass 'm' is large and the spring stiffness k is small, and that f_0 is high for stiff systems with a small mass.

The decay of the amplitude of the motion is determined by the magnitude of the *damping coefficient* 'b' (dimension (N/(m/sec)) in Eq. (5.18.7). When the damping is large, the (undamped) natural frequency f_0 shifts to a lower, the so-called damped natural frequency f_1 which is given by

$$f_1 / f_0 = \sqrt{(1 - B^2)}, \tag{5.18.9}$$

where

$$B = b / (2m\omega_0) = b / (4\pi f_0 m) \tag{5.18.10}$$

is the *damping ratio*. For $B < 1$ the system is said to be underdamped and the motion is periodic as in Fig. 5.18.2. When $B = 1$, the system is said to have critical damping.

The natural response of the system to an initial disturbance is then such that the motion decays in precisely one period. For $B>1$ the system is said to be overdamped and the motion is no longer periodic. The mass 'm' then returns to its neutral position without oscillation.

In the other case of interest the mass 'm' is subject to a continuous, external, oscillatory force F that varies in time with frequency 'f'. When the external force is sinusoidal (like $F = F_0 \sin\omega t$, where $\omega = 2\pi f$), the equilibrium of forces acting on the mass 'm' can be expressed as

$$m\ddot{X} + b\dot{X} + kX = F_0 \sin\omega t \qquad (5.18.11)$$

In general, the system then adopts an oscillatory motion with, eventually, a constant amplitude. This is called a limit cycle oscillation (see Fig. 5.18.4). A sailing yacht pitching in waves is an example of this. Note that the limit cycle stage is preceded by a transient stage from the time $t=0$ at which the external force is initiated. The transient stage is determined by the properties of the natural motion of the object. In mathematical terms this is determined by the solution of the 'homogeneous' Eq. (5.18.7).

It can be shown (http://en.wikipedia.org/wiki/Harmonic_oscillator) that the limit cycle amplitude is given by

$$|X| = (F_0/m)/\sqrt{\left[\left\{\omega_0{}^2 - \omega^2\right\}^2 + \left\{2\,\omega_0\,\omega\,B\right\}^2\right]}, \qquad (5.18.12a)$$

which can also be written as

$$|X| = (F_0/k)/\sqrt{\left[\left\{1 - \omega^2/\omega_0{}^2\right\}^2 + \left\{2\,(\omega/\omega_0{}^2)B\right\}^2\right]} \qquad (5.18.12b)$$

Here, F_0 is the amplitude of the external force, m is the mass, k is the spring stiffness and B is the damping ratio. As before, 'ω' is the circular frequency $2\pi f$ and $\omega_0 = 2\pi f_0$ is the natural circular frequency of the system.

Fig. 5.18.4 *Limit cycle oscillation*

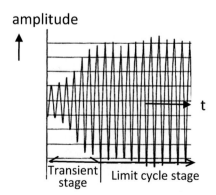

Fig. 5.18.5 *Amplitude of motion of a driven 'dynamic system' as a function of frequency for different levels of damping (see Eq. (5.18.12)*

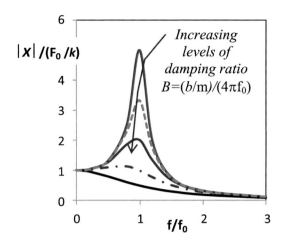

Equation (5.18.12) implies that the amplitude $|X|$ of the limit cycle motion becomes large when the circular frequency ω of the external force is close to the natural circular frequency ω_0 of the system This is called the *resonance* condition. The amplitude becomes particularly large when the *damping ratio B* is small. This is illustrated by Fig. 5.18.5 which gives the normalized amplitude $|X|/(F_0/k)$ as a function of the frequency ratio f/f_0 for different values of the damping ratio $B = b/(4\pi f_0 m)$. Resonance is seen to occur for $f/f_0 \cong 1$. In case of heavy damping the maximum normalized amplitude of the motion moves to lower frequencies.

It is further useful to know that there is, in general, a phase lag between the exciting force and the responsive motion. The larger the damping the larger the phase lag. The phase lag is also a function of the frequency of the external force and the natural frequency. Near the resonance condition the phase lag is about a quarter period.

For sailing yachts the 'rotational version' of a dynamic system of the type sketched above is equally important. In that case the rectilinear motion is replaced by an angular motion with the external force *F* replaced by an external moment *M*, the mass 'm' replaced by the mass moment of inertia 'I' around the axis of rotation and the rectilinear displacement *X* by an angular displacement . The equilibrium of moment can then be expressed as

$$I\ddot{\theta} + b_\theta \dot{\theta} + k_\theta \theta = M_0 \sin\omega t \qquad (5.18.13)$$

and the amplitude is given by

$$|\theta| = (M_0/I)/\sqrt{\left[\left\{\omega_0^2 - \omega^2\right\}^2 + \left\{2\,\omega_0\omega B_\theta\right\}^2\right]} \qquad (5.18.14)$$

Here, I is the mass moment of inertia and B_θ is the damping ratio, now given by $B_\theta = b_\theta/(2I\omega_0)$. The dimension of the torsional spring constant k_θ is now Nm/radian and that of the damping coefficient b_θ is Nm/(rad/sec). The natural frequency f_0 of the rotational motion is now given by $\omega_0 (= 2\pi f_0) = \sqrt{(k_\theta/I)}$. As for a rectilinear displacement X, it can be shown that when the angular displacement varies with time like sinωt, its rate of change with time, $\dot\theta$, varies like cosωt and the corresponding angular acceleration $\ddot\theta$ like −sinωt.

Depending on the magnitude and the duration, limit cycle oscillations may lead to structural fatigue problems. It is the author's impression that the structural fatigue problems with the keel-hull attachments of most of the yachts participating in the 2005–2006 Volvo Ocean Race form an example of this.

   Another, more common but less severe example of limit cycle oscillations for sailing yachts is mast vibration or fluttering. It is caused by a Von Kármán vortex street shed by the mast when the wind speed is such that the frequency of vortex shedding is about equal to the frequency of the natural bending vibration of the mast. The periodic lateral fluid dynamic forces due to the vortex shedding may then enhance the mast vibration and vice-versa. Mast fluttering usually happens at the dock, when the sails are down. Fortunately it is almost never catastrophic due to damping in the rigging and due to the fact that the wind speed and direction is constantly varying.

We have seen above that a 'dynamic system' adopts a stable, oscillatory motion with decaying amplitude or a limit cycle oscillation when there is a positive restoring force and positive damping. A potentially more dangerous situation arises when the restoring force and/or the damping force have the wrong sign.

   When the spring constant k, or its equivalent, is <0, the 'restoring' force is not restoring, but destabilizing. This situation is called statically unstable. The motion is then non-periodic and diverges immediately.

   When the damping coefficient b, or its equivalent, is <0, the motion is still periodic. However, the amplitude increases rather than decreases in time (Fig. 5.18.6). In this case the amplitude of oscillation increases rapidly during the transient stage and a limit cycle stage is never established. In this situation one speaks of a *'dynamic*

Fig. 5.18.6 *Dynamic instability: the amplitude of an oscillatory motion or vibration increases with time*

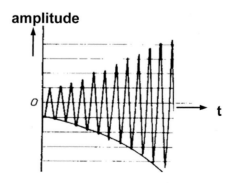

Fig. 5.18.7 *Lifting surface in pitching and heaving motion*

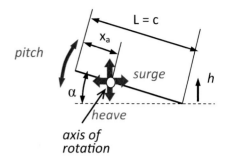

instability'. Needless to say that the consequences of a static or a dynamic instability are, potentially, catastrophic. This applies to the stability and control of a yacht as a whole as well as to the structural dynamics of its components.

Fluid Dynamic Forces and Moments in Unsteady Flow
It has been found (Katz and Plotkin 1991; Ashley and Landahl 1985/1965) that the fluid dynamic forces acting on a rigid body in sinusoidal oscillatory motion can be distinguished into three components that differ in their dependence on time. For a general lifting surface, oscillating in sinusoidal heaving and/or pitching motion,[12] see Fig. 5.18.7, the lift force L can be written as

$$L = L_0(\alpha_e) + L_1(\dot{\alpha}_e) + L_2(\ddot{\alpha}_e) \tag{5.18.15}$$

Note that Eq. (5.18.15) implies a dependence of lift on an effective angle of attack α_e. We have seen in Sect. 5.15 that in steady flow α_e is a function of the aspect ratio of the lifting surface. Here, it is also a function of the frequency of the motion.

The first term $L_0(\alpha_e)$ is due to the instantaneous angle of attack α. It represents the usual (quasi-steady) lift force as a function of the (effective) angle of attack. We consider sinusoidal oscillations with the displacement h in heave varying like $h = h_m + |h|\sin\omega t$, where h_m is the mean vertical position of the lifting surface and $|h|$ the amplitude. The angle of attack α due to pitching varies like $\alpha = \alpha_m + |\alpha|\sin\omega t$, where $|\alpha|$ is the amplitude of the variation and α_m the mean angle of attack. The velocity (surge) varies like $V = V_0 + |u|\sin\omega t$, where V_0 is the average velocity and $|u|$ is the amplitude of the variation of the velocity due to surging. L_0 can then be approximated as

$$L_0(\alpha_e) = \frac{1}{2}\,\rho V^2 S\left\{ C_L(\alpha_m) + \frac{dC_L}{d\alpha} C'(k_f) |\alpha|\sin\omega t \right\} \tag{5.18.16}$$

---

[12] For a sailing yacht this would, for example, be a keel or a sail oscillating in yaw, roll, or pitch.

Fig. 5.18.8 *Lift deficiency factor of a lifting surface as a function of reduced frequency. (Theodorsen 1935)*

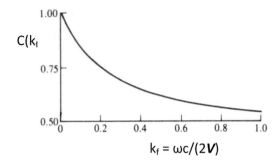

Here, $C_L(\alpha_m)$ is the lift coefficient in steady flow at the mean angle of attack α_m, $\dfrac{dC_L}{d\alpha}$ is the lift-curve slope and

$$C'(k_f) = 1 + \{C(k_f) - 1\} A/(1 + A) \tag{5.18.17}$$

is a factor representing the effect of frequency on the amplitude of the lift variation. The *lift deficiency factor* (Theodorsen 1935) $C(k_f)$ in Eq. (5.18.17) is a function of the reduced frequency k_f (see Eq. 5.18.5). It represents the unsteady effects of the time-varying vortex wake on the downwash and the effective angle of attack. Figure 5.18.8 shows the lift deficiency factor $C(k_f)$ as a function of the reduced frequency k_f. Note that $C(k_f)$ and $C'(k_f)$ both take the value 1 for $k_f \to 0$.

The factor $A/(1 + A)$ in Eq. (5.18.17) models the effect of aspect ratio A on the unsteady effects. Eq. (5.18.17) implies that the lift deficiency effect vanishes also for vanishing aspect ratio, irrespective of the reduced frequency. For a lifting surface with an aspect ratio of 1 the lift deficiency is only half of one with a very high aspect ratio.

In terms of a 'dynamic system' the quasi-steady part $L_0(\alpha_e)$ has the character of a restoring force. It follows from Eq. (5.18.16) that the restoring force is zero in a pure heaving motion.

The second term $L_1(\dot{\alpha}_e)$ of (5.18.15) is due to the rate of change with time of the translation (surging and heaving motion) and/or the angle of attack (rotation) of the body. It can, to first order, be expressed as

$$L_1(\dot{\alpha}_e) = \tfrac{1}{2}\,\rho V_0^{\,2} S \times k_f \cos\omega t \times$$
$$\left\{ C_L(\alpha_m)\frac{|u|}{2V_0} + \frac{dC_L}{d\alpha}\left[2C'(k_f)\left\{ -\frac{|h|}{c} + \left(\frac{3}{4} - \frac{x_a}{c}\right)|\alpha|\right\} + \frac{|\alpha|}{2}\right]\right\}, \tag{5.18.18}$$

where x_a is the position of the axis of rotation (see Fig. 5.18.7).

Equation (5.18.18) implies that $L_1(\dot{\alpha}_e)$ is <0 for a pure heaving motion. The direction of the force than opposes the direction of the motion. For this reason the L_1 term is associated with the notion of damping. For a pure pitching motion $L_1(\dot{\alpha}_e)$ is >0 when the centre of rotation is located forward of the trailing edge of the lifting surface. $L_1(\dot{\alpha}_e)$ is also >0 for a pure surging motion.

The term $L_2(\ddot{\alpha}_e)$ is associated with the acceleration of the fluid particles due to the rate of change in time of the translational and/or rotational velocity of the body. This can be expressed as

$$L_2(\ddot{\alpha}_e) = \tfrac{1}{2}\,\rho V_0^{\,2} S \frac{dC_L}{d\alpha} f_L k_f^{\,2} \sin\omega t \qquad (5.18.19)$$

The factor f_L is a function of the amplitudes of the heaving and pitching motions and the position of the axis of rotation. It has been found (Katz and Plotkin 1991) that $f_L > 0$ for a pure heaving motion. For a pure pitching motion it is found that $f_L = 0$ when the axis of rotation is at the $\tfrac{1}{2}$-chord point, <0 when the axis is more rearward and >0 when the axis is forward of the $\tfrac{1}{2}$-chord point. $L_2(\ddot{\alpha}_e)$ is generally known as the *added mass* term. It was given this name because the force component that it represents is in phase ($\div \sin\omega t$) with the inertia force acting on the oscillating body. This means that it acts as if the body has additional, virtual mass m that is to be added to the real mass of the body (see the paragraph on *'dynamic systems'*).

Considering the expressions (5.18.16–5.18.19) it is immediately clear that $L_1(\dot{\alpha}_e)$ and $L_2(\ddot{\alpha}_e)$ increase with frequency but vanish when the frequency goes to zero. It can further be seen from the time dependency (factor $\cos\omega t$) that the $L_1(\dot{\alpha}_e)$ term is $90°$ out of phase with the motion ($\cos\omega t = \sin(90° - \omega t)$), see also Fig. 5.18.9. This means that the lift force develops a time lag relative to the motion with increasing

Fig. 5.18.9 The lift acting on a lifting surface in sinusoidal heaving motion as a function of time during one period

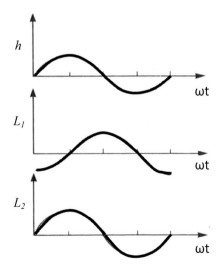

frequency. For very high frequency ($k_f \gg 1$) the time lag disappears again because the in-phase added mass term (with factor k_f^2, Eq. 5.18.19) then begins to dominate.

What has been described above for the lift force is, qualitatively, also applicable to other forces and moments. The details are different, however.

For a fully submerged 'non-lifting' body such as a slender ellipsoid, the fuselage of an airplane, a submarine, or the bulb of a sailing yacht, oscillating in heave, sway, pitch or yaw, the unsteady lift or side force (damping) is usually small but the unsteady moments and added mass effects are not. The added mass in particular can be substantial, being of the order of 30% of the mass of the displaced volume. For a non-lifting' body in rolling oscillation all the unsteady components of the fluid dynamic forces and moments are usually small, in particular if the shape of the cross-section is close to a circle.

For semi-submerged non-lifting bodies, such as a ship hull, both damping and added mass can be quite large (Ref. Lewis (1989). The added mass, for example, can be of the same order as the real mass of the ship. The main reason for this behaviour is the generation of additional waves due to plunging (heaving) and pitching motion. We will come back to this in Chap. 6.

Propulsion Potential of Heaving Motion
An interesting point to mention is that a lifting surface in a heaving motion has the potential to generate thrust. The mechanism is illustrated in Fig. 5.18.10.

Shown is the path of the lifting surface and the lift vector during a down-stroke of the foil. Because the direction of the lift vector is perpendicular to the path of the foil, it is tilted forward. This means that the lift vector has a, propulsive, component in the direction of the average motion. During the up-stroke, the lift is negative but has also a forward component. When the drag is sufficiently small, the net effect can be a time-averaged propulsive force component.

Dynamic Stall
Figure 5.18.11 gives an impression of the variation in time of the total lift for a NACA 0012 airfoil oscillating in pitch (McCroskey 1981) around $\alpha = 3°$ with a fairly large amplitude, $|\alpha| = 10°$. The reduced frequency is fairly low, $k_f = 0.1$. Shown is

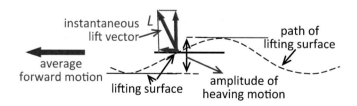

Fig. 5.18.10 *Illustrating the propulsive potential of a lifting surface in heaving motion*

Fig. 5.18.11 *Histogram for lift on a*
NACA 0012 airfoil in pitching oscillation
$(x_d/c = 1/4, k_f = 0.1)$. *(McCroskey 1981)*

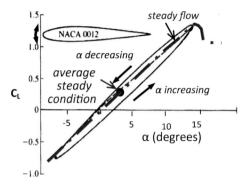

the lift coefficient C_L as a function of the angle of attack as it varies with time.[13] The lift is seen to lag behind below the level for steady flow when the angle of attack increases. When the angle of attack decreases the lift is higher than in steady flow. This hysteresis phenomenon is mainly determined by the damping term $L_1(\dot{\alpha}_e)$ and increases with increasing reduced frequency.

As long as the flow is attached during an oscillation cycle, the maximum lift is not much different from that in steady conditions and the time-averaged lift is about equal to the lift in steady flow at the mean angle of attack. This because the average over a cycle of the factors $\sin\omega t$ and $\cos\omega t$ in Eqs. (5.18.17–5.18.19) is zero. However, the average over a cycle is no longer zero when there is separated flow over part of the cycle. This because $\frac{dC_L}{d\alpha}$ is no longer constant. It has been found (McCroskey 1981) that there can be a substantial 'overshoot' of the maximum lift beyond the steady flow value when the airfoil oscillates through or beyond stall, see Fig. 5.18.12. The mechanism involved is periodic detachment at the leading-edge of a vortex that spills downstream and creates local additional suction (lift) on the surface of the airfoil.

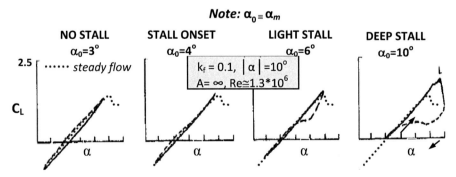

Fig. 5.18.12 *Histogram for lift of a NACA 0012 airfoil in pitching oscillation before and beyond stall. (McCroskey 1981)*

---

[13] A diagram of this kind is known as a histogram.

It has also been found (McCroskey 1981) that, even in a pure heaving oscillation, there can be negative damping over part of the cycle when there is separated flow.

The phenomena described above are known (McCroskey 1981) to be more pronounced for large amplitudes and high reduced frequencies. We will see in Sect. 6.8 that it can possibly play a role in the flow about the keel of a yacht in large-amplitude rolling oscillations at low boat speeds.

Structural Dynamics
Lifting surfaces type structures can also suffer from a form of structural dynamic instability known as (classical) *flutter* (Bisplinghoff et al. 1996). This kind of structural instability can occur when the mass and stiffness properties of the structure are such that the natural frequencies of the bending and torsional vibration modes of the structure are sufficiently close. When this is the case and the flow speed is sufficiently high, the fluid dynamic forces due to the bending motion may then enhance the torsional vibration, and vice versa.

Flutter is a phenomenon that requires careful investigation for aircraft wings and bridges. To the author's knowledge it is not, in general, a problem of sailing yachts; except, probably, for extreme canting keel configurations such as adopted in the Open 60 and Volvo 70 class racing yachts (Broekhuijsen 2006).

5.19   The Air-Water Interface, Surface Waves

Discontinuities, Jump Conditions and Their Consequences
As already discussed in Sect. 5.3 the air-water interface or boundary surface plays a crucial role in hydrostatics. It is also important for the flow about a sailing yacht (or any sailing vessel for that matter).

In physics a boundary between volumes of different fluids is called a *contact discontinuity*. Contact discontinuities are subject to the same conservation laws as those discussed in Sect. 5.4. When these conservation laws are applied to a small volume element enclosing (part of) the surface of discontinuity (Fig. 5.19.1) it is found that, if we ignore the effects of viscosity, the flow quantities satisfy the following jump conditions across the boundary:

- zero pressure difference
    $$(p_A = p_H)$$
- jump in density
    $$(\rho_A \neq \rho_H)$$
- zero normal velocity (normal to the discontinuity)
    $$(V_{n,A} = V_{n,H} = 0)$$
- jump in direction and magnitude of the tangential (or shear) velocity
    $$(V_{s,A} \neq V_{s,H})$$

Fig. 5.19.1 *Jump conditions at the air-water interface*

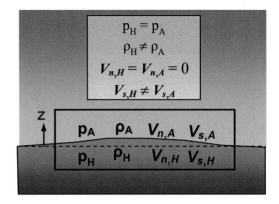

For the air-water interface this means of course that both p_A and p_H are equal to the atmospheric pressure (at least in still air), ρ_H is the mass density of water and ρ_A that of air. The zero normal velocity condition implies that the air-water interface is a stream surface.

The condition for the tangential velocity is a little bit more complex if the effects of viscosity are taken into account. Viscosity prevents the occurrence of 'slip' at interfaces between fluids and solids as well as at the interface between different fluids. This means that at the surface of contact the velocities of air and water must be equal. Viscosity also causes the formation of boundary layers and associated friction or shear stresses on both sides of the interface (Fig. 5.19.2). We have seen in Sect. 5.11 that the shear stress τ is proportional to the slope of the velocity profile in the boundary layer. This means that on the air side it satisfies the relation

$$\tau_A = \mu_A (u_z)_A \tag{5.19.1}$$

and on the water side

$$\tau_H = \mu_H (u_z)_H, \tag{5.19.2}$$

Fig. 5.19.2 *Effect of viscosity on the tangential velocity at the air-water interface*

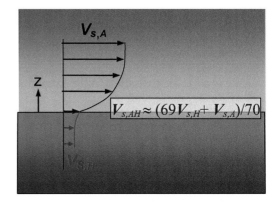

where μ is the dynamic viscosity coefficient and u_z represents the rate of change of the tangential velocity in the vertical or 'normal' (z-) direction. (Note that in this notation the tangential velocity 'u' would be equal to V_s at the edge of the boundary layer and that the 'normal' direction is positive when it points away from the interface).

It also follows, from Newton's law, that in steady conditions the friction forces must be in equilibrium at the interface. This means that the shear stresses on both sides must be equal, i.e., there holds $\tau_A = \tau_H$ or

$$\mu_A (u_z)_A = \mu_H (u_z)_H \tag{5.19.3}$$

Because the dynamic viscosity of water μ_h is much larger than that of air μ_a (about a factor 70, see Sect. 5.8), Eq. (5.19.3) implies that the rate of change of the tangential velocity in the boundary layer on the water side is about a factor 70 smaller than that on the air side. Assuming that the velocity profiles in air and water are similar this also implies that the common velocity $V_{s,ah}$ at the interface differs very little from the water velocity at the edge of the boundary layer at some distance below the surface. More precisely there must hold

$$V_{s,AH} \approx (69V_{s,H} + V_{s,A})/70 \tag{5.19.4}$$

It should be mentioned that in the description given above it has been assumed for the sake of simplicity that $V_{s,H}$ and $V_{s,A}$ have the same direction. In general this is not necessarily the case. The boundary layers then acquire twisted velocity profiles and Eq. (5.19.4) should be interpreted as a vector equation.

The conditions at the air-water interface have also consequences for some other flow properties. One consequence of the condition that the interface is a stream surface, is that free vortex filaments cannot pierce through it, because they follow streamlines (see the Kelvin-Helmholtz vortex theorems of Sect. 5.13). It was also indicated in Sect. 5.13 that stream-wise vorticity can only develop out of stream-normal vorticity in viscous layers when the velocity profile of the viscous layer is non-planar ('twisted'). In the case that there is only stream-normal vorticity perpendicular to the water surface this means that no stream-wise vorticity can develop in the air-water interface as long as this is sufficiently flat. However, it would seem that stream-wise vorticity can develop when there is stream-normal vorticity *in* the water surface. The latter would be the case when a body intersects the water surface at an angle different from 90°.

Surface Deformation
Perhaps the most important consequence of the jump conditions given above is that the water surface deforms under the influence of velocity and pressure disturbances. We can see this by applying Bernoulli's law (Sect. 5.6) to both sides of the air-water interface. In the situation of a sailing vessel moving steadily through water and

wind at a boat speed V_b and an apparent wind speed V_a we have, on the water side, in a frame of reference that moves with the vessel:

$$p_H + \tfrac{1}{2}\, \rho_H\, V_{s,H}{}^2 + \rho_H\, g\, z_H = \text{constant} = p_a + \tfrac{1}{2}\, \rho_H\, V_b{}^2 \tag{5.19.5}$$

and on the air side

$$p_A + \tfrac{1}{2}\rho_A\, V_{s,A}{}^2 + \rho_A\, g\, z_H = \text{constant} = p_a + \tfrac{1}{2}\rho_A\, V_a{}^2 \tag{5.19.6}$$

Note that in obtaining these relations we have already made use of the condition of zero normal velocity $(V_{n,H} = V_{n,A} = 0)$ and that z_H is the height of the water surface above some reference level. The value of the constants are determined by the conditions far ahead of the vessel, where the water surface is not yet deformed (i.e. $z_H = 0$), the pressure is equal to the atmospheric pressure p_a and $V_{s,H}$, $V_{s,A}$ are equal to V_b, V_a, respectively. Subtracting (5.19.6) from (5.19.5) and inserting the jump condition $p_H = p_A$ we get

$$\tfrac{1}{2}\rho_H\, V_{s,H}{}^2 + \rho_H\, g\, z_H - \left(\tfrac{1}{2}\, \rho_A\, V_{s,A}{}^2 + \rho_A\, g\, z_H\right) = \tfrac{1}{2}\rho_H\, V_b{}^2 - \tfrac{1}{2}\rho_A\, V_a{}^2 \tag{5.19.7}$$

This can be rearranged to read

$$(\rho_H - \rho_A)\, g\, z_H = \tfrac{1}{2}\rho_H\left(V_b{}^2 - V_{s,H}{}^2\right) + \tfrac{1}{2}\rho_A\left(V_a{}^2 - V_{s,A}{}^2\right) \tag{5.19.8}$$

Because the density ρ_A of air is very much (about a factor 1000) smaller than that of water (ρ_H) we can safely neglect terms with ρ_A in Eq. (5.19.8). If we do so and divide by ρ_H we get

$$g\, z_H \approx \tfrac{1}{2}\left(V_b{}^2 - V_{s,H}{}^2\right) \tag{5.19.9}$$

In words, Eq. (5.19.9) expresses that the water surface deforms under the influence of velocity disturbances (variations in $V_{s,H}$) caused by the moving vessel. The water level sinks locally if the local flow velocity $V_{s,H}$ is higher than the boat speed V_b and it rises when $V_{s,H} < V_b$. The static pressure at the surface is, for most practical purposes, equal to the atmospheric pressure because the mass density of air is so much smaller than that of water.

Because the water surface is 'free' to deform it is also called a free surface.

Froude Number

The terms $\rho\, g\, Z_h$ appearing in the equations of the preceding paragraph reflect that gravity forces play an essential role in the flow at and near the water surface. It can be shown (Prandtl and Tietjens 1934a in Chap. 1) that these gravity forces scale, per unit volume, like

$$F_g \div \rho g \tag{5.19.10}$$

We have also seen in Sect. 5.5.3 that inertia forces per unit volume scale like

$$F_i \div \rho V^2 / L \tag{5.19.11}$$

where L is a characteristic length of the object moving through or near the air-water interface.

It follows from Eqs. (5.19.10) and (5.19.11) that the ratio of inertia and gravity forces scales like

$$F_i / F_g = V^2 / (L\, g) \tag{5.19.12}$$

This ratio, or rather the square root of it, is known as the *Froude number* [14]

$$\mathrm{Fr} = V / \sqrt{(L\, g)} \tag{5.19.13}$$

The Froude number, like the Reynolds number, is an important similarity parameter in ship hydrodynamics. It was shown by William Froude (1870 in Chap. 1) that the flows about surface ships of similar shape but different size are only similar under conditions of equal Froude number.

It is further useful to note that it follows from Eq. (5.19.9), that the surface elevation z_H can be expressed as

$$z_H / L \approx \tfrac{1}{2}\, \mathrm{Fr}^2 (1 - V_{s,H}{}^2 / V_b{}^2), \tag{5.19.14}$$

if the Froude number is based on boat speed V_b.

Surface Waves
Contact discontinuities with a jump in tangential velocity, like the air-water interface, are known to be dynamically unstable. That is, running surface waves are formed under the influence of velocity disturbances, pressure, friction and gravity forces. Two types of surface waves can be distinguished: those generated by moving vessels and those driven by the wind.

In the following wind driven waves are considered in some detail. Waves generated by moving vessels will be discussed in Sect. 6.5.

(Part of) the mechanism of wind driven waves is illustrated by Fig. 5.19.3. Assuming that there is an initial disturbance in the form of a small ridge of water, the wind will accelerate locally over this ridge due to the Venturi effect described in Sect. 5.5. The higher wind velocity V_{crest} at this ridge causes a lower pressure p (see Bernoulli's law, Sect. 5.6), as well as a higher friction force τ at the water surface. Due to the lower pressure the water level is raised to form a wave crest. When the water goes up in one place the mass conservation law requires that an equal amount

---

[14] Named after the British scientist William Froude, mentioned earlier, in Chap. 1.

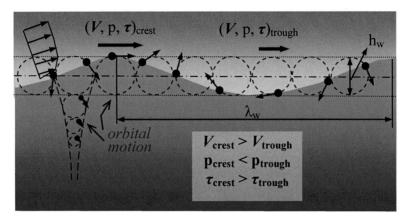

Fig. 5.19.3 *Illustrating the mechanism of surface waves*

goes down somewhere else, with a smooth transition region somewhere in between. As a consequence a trough is formed at the same time as a crest. At a trough, the wind velocity and surface friction are locally lower and the pressure higher than the mean, undisturbed values. The increased friction at a crest causes the water particles to flow in the direction of the wind and mass conservation requires that an equal amount of water flows in the opposite direction (at the trough).

It turns out (Lewis 1989) that the net kinematic result of all this motion is that the water particles at the surface move approximately in circles, as also indicated in Fig. 5.19.3. The diameter of the circles is equal to the wave height. Away from the surface the water particles also move along circles, but the diameter decreases with increasing depth. The whole motion pattern is known as the *orbital motion* of water waves.

A consequence of the orbital motion is that the water particles experience a centrifugal force. On the surface, at a wave crest, the centrifugal force is opposite to the gravitational force while in a trough the centrifugal and gravitational forces act in the same direction. As a result, the apparent weight of the water particles at a wave crest is smaller and in a trough larger than their weight in calm water. It can be shown (Lewis 1989) that the apparent weight w_w of a water particle of unit mass at a wave crest is given by

$$w_{wc} = \left\{ 1 - \pi h_w/(2\lambda_w) \right\} g, \tag{5.19.15}$$

while in a wave trough it is equal to

$$w_{wt} = 1 + \pi h_w/(2\lambda_w)\} g \tag{5.19.16}$$

Equation (5.19.15) implies that for a given ratio of wave height h_w to wavelength λ_w a water particle at a wave crest becomes weightless. This is the case for

$$h_w/\lambda_w = 2/\pi \qquad (5.19.17)$$

The condition (5.19.17) is a rough indication of when waves will begin to break.[15]

Equation (5.19.15) also offers an explanation for another phenomenon. Water particles at wave crests can easily be blown away by the wind to form spray or foam when their apparent weight becomes sufficiently small.

In deep water, the amplitude ζ_w of the orbital motion (the diameter of the circles) is known to decrease exponentially with depth:

$$\zeta_w(z) = \zeta_w(z=0)\, e^{k_w z} = \frac{h_w}{2} e^{k_w z}, \qquad (5.19.18)$$

In the exponential function $e^{k_w z}$, k_w is the wave number

$$k_w = 2\pi/\lambda_w \qquad (5.19.19)$$

and z, (<0), is the distance below the (undisturbed) water surface, $(e=2.718\ldots)$.

A similar dependence on depth is found for the velocity of the water particles due to the orbital motion. The maximum values u_{wmax} of the horizontal component of the orbital velocity is given by

$$u_{wmax} = \pm k_w (h_w/2)\sqrt{(g/k_w)e^{k_w z}}, \qquad (5.19.20)$$

where the $+$ sign refers to a crest and the $-$ sign to a trough.

While the water particles move only along relatively small circles, the visual appearance of the whole motion pattern is one of a traveling wave or, rather, a traveling wave form. It can be shown (Lewis 1989), that the propagation speed c_w of a traveling surface wave depends only on the wavelength λ_w:

$$c_w = \sqrt{(g\,\lambda_w/2\pi)} \qquad (5.19.21)$$

The wave period T_w, that is the time it takes a wave to travel over a distance equal to the wavelength, satisfies the relation

$$T_w = \sqrt{(2\pi\lambda_w/g)} \qquad (5.19.22)$$

Equations (5.19.21) and (5.19.22) imply that the propagation speed and wave period increase with wave length.

Both wave height h_w and wavelength λ_w and, hence, the maximum wave slope θ_w, which is equal to

$$\theta_w = \pi h_w/\lambda_w = 2\pi\zeta_w/\lambda_w, \qquad (5.19.23)$$

---

[15] The indication is a very rough one because waves of significant height are no longer sinusoidal.

depend on wind speed in a rather complex way. In general both increase with wind speed. However, because time is needed to transfer energy from the wind to the sea and to build-up a wave, they depend also on the duration of the wind and the 'age' of, or the distance ('fetch') traveled by the wave. Moreover, new waves are being created continuously in all places and already existing waves may disappear by dissipation and/or interference with other waves. Dissipation is caused by internal friction forces. However, the rate of dissipation of surface waves is very low, except for short wavelengths. This is the reason why long waves may propagate almost without damping over a long time and over large distances.

Waves may also run in different directions as a result of changes in wind direction. 'Old' waves, in particular those with long wavelengths, may persist a long time in the form of what is called a 'swell'. A *sea state* is, therefore, characterized by a whole spectrum of wave lengths and wave heights and, possibly, wave directions. The irregular waves of a seaway can be considered to be made up of many regular ('sinusoidal' or 'harmonic') waves superimposed upon each other (Fig. 5.19.4). Between 10 and 15 wavelengths are generally sufficient to describe a realistic sea.

The wave spectrum of a sea is usually considered in terms of a so-called variance or energy density spectrum (Fig. 5.19.5). It is characterized by the distribution of the energy S contained by waves of a given wavelength or frequency/period as a function of the wave length or wave period/frequency. In the example of Fig. 5.19.5 it is given as a function of the so-called angular or circular frequency $\omega_w = 2\pi/T_w$, where T_w is the wave period.

The energy density is defined by the relation (Lewis 1989)

$$S(\omega_{wi})\delta\omega_w = \tfrac{1}{2}\zeta_{wi}^2, \tag{5.19.24}$$

where ζ_{wi} is the wave amplitude (half the wave height) and $\delta\omega_w$ the bandwidth of a narrow frequency band. Eq. (5.19.24) implies that the energy contents of a wave is proportional to the square of the amplitude or wave height.

A commonly adopted family of (semi-) empirical wave spectra is the so-called Bretschneider Spectrum. It has the form (Lewis 1989)

Fig. 5.19.4 *Irregular, composite waves are made up of many regular waves superimposed upon each other*

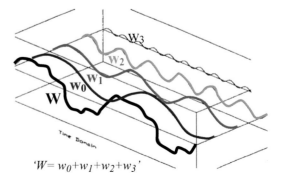

'$W = w_0 + w_1 + w_2 + w_3$'

Fig. 5.19.5 *Example of wave spectrum*

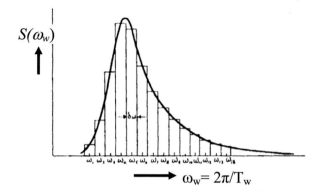

$$S(\omega_w) = (A/\omega_w^5)\, e^{-B/\omega_w^4} \qquad (5.19.25)$$

The quantities 'A' and 'B' are parameters that allow to distinguish between several spectra.

For engineering purposes that are aiming at average conditions it is recommended to take (Lewis 1989)

$$A = 173\, h_{w_{1/3}}^{\;2} / T_{wave}^{\;4} \qquad (5.19.26)$$

$$B = 691 / T_{wave}^{\;4} \qquad (5.19.27)$$

Here, $h_{w_{1/3}}$ is the *significant wave height* and T_{wave} the *average wave period*. The significant wave height is defined as the average 1/3 highest waves. It correlates (Lewis 1989) closely with the estimate of an experienced observer of the average wave height. The average wave period correlates closely with the corresponding estimate of the average wave period.

As an example Fig. 5.19.6 gives an impression of the significant wave height and the average wave period as a function of wind speed for different environments. Note that the curves level-off at high wind speeds. For the open ocean (North Atlantic) this is caused by the fact that the wave propagation speed for long wavelengths (long periods) begins to approach the wind speed, which limits the transfer of energy from the wind to the sea. For bounded stretches of water such as the North Sea and coastal, estuary type of waters, when the waves do not have sufficient time and distance ('fetch') to build-up fully, the curves level-off earlier at lower values of wave height and wave period.

It should also be mentioned that in the description of waves given above it has been assumed that the water depth is much larger than the wave length. This, of course, is the case for the open ocean, but only marginally so for the North Sea. In

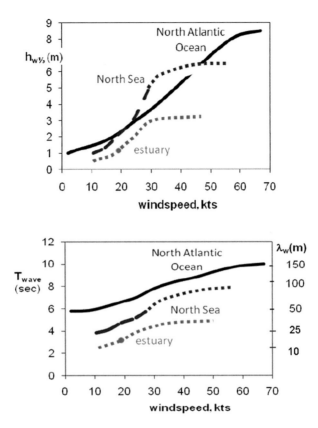

Fig. 5.19.6 *Estimated significant wave height and average wave period as a function of wind speed for different sailing environments*

shallow water, that is when the depth is (much) smaller than about half the wave length, the propagation speed, for a given wave period, is lower than in deep seas due to interference with the bottom. As a consequence waves tend to be shorter and steeper in shallow water (Lewis 1989). This, as illustrated by Fig. 5.19.7, is the case in coastal, estuary type environments.

An associated phenomenon is that waves steepen-up and eventually break when they run into shallow water.

Waves are also known to be affected by currents (Lewis 1989). When the current is flowing in the same direction as the direction of propagation of the waves, the wavelengths can increase while heights decrease. The opposite is the case when the current flows in the opposite direction.

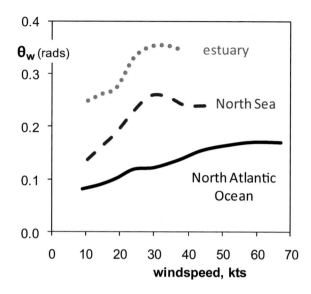

Fig. 5.19.7 Estimated maximum wave slope as a function of wind speed

References

Abbott IH, Von Doenhoff AE (1949) Theory of wing sections. McGraw Hill, New York

Anderson Jr, John D (2001) Fundamentals of aerodynamics. McGraw-Hill, New York. ISBN 0-07-118146-6

Ashley H, Landahl M (1985/1965) Aerodynamics of wings and bodies. Dover Publications, New York. ISBN 0-486-64899-0

Bisplinghoff RL, Ashley H, Halfman H (1996) Aeroelasticity. Dover Science, Mineola. ISBN 0-486-69189-6

Blackwell JA (1976) Numerical method to calculate the induced drag or optimum loading for arbitrary non-planar aircraft, NASA SP 405

Blasius H (1908) Boundary layers in fluids of small viscosity. Z Math Physik 56:1

Broekhuijsen TF (2006) KEEL FLUTTER a new hazard on the high sea, Leonardo Times, December 2006

DeYoung J, Harper C (1948) Theoretical symmetrical span loading at subsonic speeds for wings having arbitrary planform. NACA Rept 921

Etkin B, Reid LD (1996) Dynamics of flight: stability and control. John Wiley and Sons, New York

Flax AH (1973) Simplification of the wing-body problem. J Aircraft 10(10):640

Greeley DS, Cross-Whiter JH (1989) Design and hydrodynamic performance of sailing yacht keels. Marine Technol 26(4):260–281

Henne PA (1989) Private communication, McDonnell-Douglas Aircraft Company

Hoerner SF (1965) Fluid dynamic drag. Hoerner Fluid Dynamincs

http://en.wikipedia.org/wiki/Harmonic_oscillator

International Standard Atmosphere. http://www.ae.su.oz.au/aero/atmos/atmos.html

Jones RT (1941) Correction of the lifting line theory for the effect of the chord. NACA TN No. 617

Katz J, Plotkin A (1991) Low-speed aerodynamics. McGraw-Hill, New York. ISBN 0-07-05040446-6

Küchemann D (1978) The aerodynamic design of aircraft. Pergamon Press, Oxford

Lakshminarayana B (1964) Effect of a chordwise gap in an aerofoil of finite span in a free stream. J R Aeronaut Soc 68:276–280

Larsson L (1990) Scientific methods in yacht design. Annu Rev Fluid Mech 22:349–385

Lewis EV (ed) (1988) Principles of naval architecture, vol II. SNAME, Jersey City. ISBN 0-9397703-01-5

Lewis EV (ed) (1989) Principles of naval architecture, vol III—motions in waves and controllability. SNAME, Jersey City. ISBN 0-939773-02-3

Lundry JL (1968) A numerical solution for the minimum induced drag, and the corresponding loading, of non-planar wings, NASA CR-1218

Marchai CA (2000) Aero-hydrodynamics of sailing. Adlard Coles Nautical, London, p. 398, p. 445

McCroskey WJ (1981) The phenomenon of dynamic stall, NASA Tech.Memo No. 81264

Milgram JH (1971) Section data for thin, highly cambered airfoils in incompressible flow, NASA CR-1767

Milgram JH (1998) Fluid mechanics for sailing vessel design. Ann Rev Fluid Mech 30:613–653

Munk MM (1923) The minimum induced drag of aerofoils, NACA Report No. 121

Oossanen Peter van (1981) Method for the calculation of the resistance and side force of sailing yachts, paper presented at Conference on 'calculator and computer aided design for small craft—the way ahead', R.I.N.A., 1981

Prandtl L (1927) On the frictional resistance of air (German). Gottinger Ergebnisse 3:1

Schlichting H, Truckenbrodt E (1979) Aerodynamics of the airplane. McGraw-Hill, New York

Slooff JW (1984) On wings and keels. Int Shipbuild Prog 31(356):94–104

Slooff JW (1985) On wings and keels (II), AIAA 12th Annual Symposium on Sailing ('The Ancient Interface'), Seattle, Wa, 21–22 Sept 1985

The USAF Stability and Control Digital Datcom. AFFDL-TR-79-3032

Theodorsen T (1935) General theory of aerodynamic instability and the mechanism of flutter. NACA Rept 496

Tinoco EN, Gentry AE, Bogataj P, Sevigny EG, Chance B (1993) IACC Appendage Studies, 11th Chesapeake Sailing Yacht Symposium, Annapolis, Md., 1993

Van der Vooren J (2006) Streamwise vorticity in viscous, compressible, steady flow about aircraft. Aerosp Sci Technol 10:288–294

Vladea J (1934) Fuselage and engine nacelle effects on airplane wings, NACA TM No. 736

Von Karman Th (1921) On laminar and turbulent friction (German). Z angew Math Mech 1:223

Whitcomb RT (1976) A design approach and selected windtunnel results at high subsonic speeds for wing-tip mounted winglets, NASA TN D-8260

Winter H (1937) Flow phenomena on plates and airfoils of short span', NACA TM 798

Woodward DS, Lean DE (1992) Where is high-lift today?—A review of past UK research programmes, AGARD CP-415, Oct 1992

Young AD (1989) Boundary layers, AIAA Education Series. ISBN 0-930403-57-6

Chapter 6
Forces Under Water: Hydromechanics

6.1 Functions of Hull and Appendages

Before going into the details of the hydromechanic forces acting on the underwater part of a sailing yacht it is probably useful to summarize the functions of the hull and appendages.

The primary function of the underwater part of the hull is to

- carry the weight of the yacht through displacement of water (buoyancy)

The basic mechanism has already been described in Sect. 5.3 on fluid statics. Other functions are:

- accommodation of ballast (mainly by the keel)
- compensation, partly by the hull but mainly by the keel, of the aerodynamic forces and moments acting on the sails (Sect. 4.1)
- provision of a means for directional control (through the rudder)

For the benefit of boat speed it is required that the functions listed above are performed with the lowest possible hydrodynamic resistance. However, other requirements, e.g. with respect to comfort, seaworthiness or class rules, may enter the design of a sailing yacht in the form of constraints on the shape of the hull and appendages, displacement, position of the centre of gravity and other factors (Larsson and Eliasson 1996).

In the following sections of this chapter we will consider the hydro mechanic forces acting on a sailing yacht in some detail, with reference to the more general discussion on fluid mechanics of the preceding chapter.

© Springer International Publishing Switzerland 2015
J. W. Slooff, *The Aero- and Hydromechanics of Keel Yachts,*
DOI 10.1007/978-3-319-13275-4_6

6.2 Righting Moments and (Static) Stability; Hydrostatics

As already indicated earlier the weight carrying function of the hull and the balancing by the hull of the aerodynamic heeling and pitching moments induced by the sails (see Sect. 4.1) are governed by hydrostatics. With the basics of fluid statics discussed in Sect. 5.3 we can now look at the hydrostatic forces and moments of the hull in some further detail.

Of particular importance for the sail carrying capacity of a yacht is the lateral or transverse hydrostatic righting moment M_{rx} introduced in Sect. 3.2. Figure 6.2.1 illustrates the mechanism and a number of notions involved.

As already indicated in Sect. 3.2 the basic mechanism underlying the righting moment is in the non-alignment of the weight vector $\underline{V}$ and vector $\underline{\Delta}$ representing the hydrostatic or buoyancy force when a yacht is under heel. While the weight vector $\underline{V}$ acts through the center of gravity CG of the yacht, the buoyancy vector $\underline{\Delta}$ acts through the so-called *center of buoyancy* CB. The latter is defined as the (virtual) center of gravity of the volume of water displaced by the hull. It follows from Archimedes' law (Sect. 5.3) that this is the point of application of the buoyancy force. While the position of the center of gravity CG is independent of the attitude of the yacht, the center of buoyancy CB moves outboard with increasing angle of heel. As a consequence, the righting moment M_{rx}, which can be expressed as

$$M_{rx} = \Delta\,\mathrm{gb},\qquad\qquad(6.2.1)$$

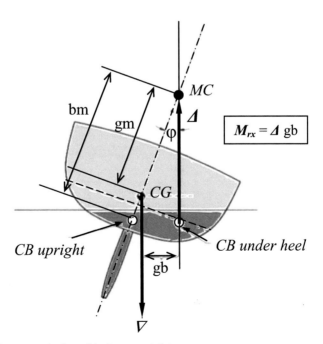

Fig. 6.2.1 *Illustrating the lateral hydrostatic righting moment*

also increases with heel due to the increasing lateral horizontal distance 'gb' between the center of buoyancy and the center of gravity (the righting arm).

Metacentric Height
The point where the buoyancy vector Δ intersects the plane of symmetry is called the (transverse) *metacentre* (MC) and the distance between the metacentre and the center of gravity is called the *metacentric height* (gm). It is readily seen from Fig. 6.2.1 that the expression (6.2.1) for the righting moment can also be written as

$$M_{rx} = \Delta \, \text{gm} \sin \varphi, \qquad (6.2.2)$$

where, in general, gm = gm(φ) is also a function of the angle of heel.[1]

It will be clear from Fig. 6.2.1 that, for a given displacement and heel angle, the position of the metacentre is a function of the cross-sectional shape of the hull and the position of the centre of gravity. The 'gm' of a modern yacht is usually of the order of 0.45 times the beam of the waterline. Hulls with wide and shallow cross-sections have a higher position of the metacentre than hulls which are narrow and deep, as illustrated by Fig. 6.2.2.

Although the position of the metacentre of the hull is mainly determined by the hull-proper (or 'canoe-body'), the effect of the volume and position of the appendages (keel and rudder) is not negligible. This is the case in particular for 'classic' yachts with long, 'integrated' keels.

It can be shown, that the position of the metacentre is almost constant at small angles of heel. This implies that for small angles of heel and a fixed position of the center of gravity the righting moment can be approximated by

Fig. 6.2.2 *Effect of beam and depth on position of metacentre*

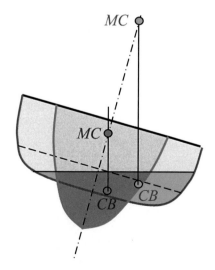

---

[1] In naval architecture the definition of *the* metacentric height is usually limited to angle of heel zero. Here we have adopted to extend the definition to cover all angles of heel and make the metacentric height a function of the heel angle.

$$M_{rx} \approx \Delta \, gm(0) \sin\varphi \quad (\varphi \ll 1 \text{ radian}), \tag{6.2.3}$$

where φ is the angle of heel, as before, and gm(0) is the metacentric height at zero heel.

At larger angles of heel the position of the metacentre may be higher or lower than that at zero heel, also depending on the cross-sectional shape. For semi-circular cross-sections the position is invariant with heel until the water reaches the deck. Hulls that flare out above the (zero heel) waterline have a higher metacentre at large heel angles and for hulls with tumble-home the position of the metacenter at large angles of heel is lower than at small angles (Lewis 1988a) (See Fig. 6.2.3).

The position of the metacentre is also influenced by sinkage (caused, for example, by extra weight, or, as we will see later, boat speed). Sinkage, that is a higher waterline, lowers the metacentre for wide, shallow hulls, in particular with tumble-home. For narrow, deep hulls, in particular with flare, the metacentre is raised when the sinkage increases.

A fundamental relation (Larsson and Eliasson 1996; Lewis 1988a) for the distance 'bm' (see Fig. 6.2.1) between the center of buoyancy CB and metacentre MC at zero heel can be written as

$$bm = \frac{I_T}{\nabla} \tag{6.2.4}$$

The quantity I_T depends on the shape of the waterplane. It can be shown (Larsson and Eliasson 1996; Lewis 1988a) that its value is proportional to the third power of the width of the waterline (i.e. $\div B_{WL}^3$). For a given shape of the waterplane it can be determined by means of numerical integration (Larsson and Eliasson 1996; Lewis 1988a). An approximation formula used by the author is given in Appendix D.

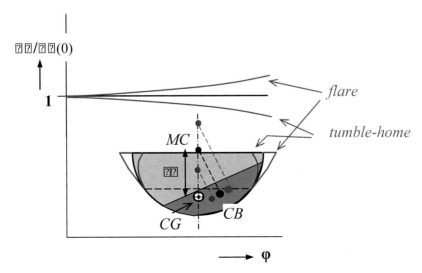

Fig. 6.2.3 *Effect of flare and tumble-home on metacentre at larger angles of heel (schematic)*

In order to be able to determine the metacentric height gm two more quantities are needed: the position of the centre of buoyancy CB and the position of the centre of gravity CG (see Fig. 6.2.1). The position of the centre of buoyancy depends on the shape of the (under-water part) of the cross-sections of the hull, in particular those around amidships. At zero heel it will, for full, almost square cross-sections, be a little smaller than half the distance between the waterplane and the keel line of the hull. This can be expressed as $b0 \leq \frac{1}{2} D_h$, where 'b0' is the distance between the waterplane and the centre of buoyance at zero heel. For deep-V type cross-sections it will be closer to a position at about 1/3 of the distance between the waterplane and the keel line. This can be expressed as $b0 \leq \frac{1}{3} D_h$. The precise position of the centre of buoyancy can be determined through numerical integration techniques (Larsson and Eliasson 1996). An approximation formula is also given in Appendix D.

The position of the centre of gravity depends on the type of yacht and its loading. For cruising yachts the position is often not very far from the water surface. For racing yachts with a large amount of ballast stored in a bulb at the bottom of the keel the centre of gravity is usually in a significantly deeper position.

Stability

The notion of the lateral or transverse, *static stability* of a yacht is directly related to the sign and magnitude of the righting moment. A yacht is said to be statically stable if the sign of the righting moment is such that it tends to bring the yacht back into the zero heel position. An important quantitative measure of stability is the length of the righting arm 'gb' (Fig. 6.2.1), which, as mentioned, is equal to $gm(\varphi) \sin \varphi$. The static stabilitystatic stability characteristics of a yacht are usually summarized in the form of a static stability curve. Such a curve presents the length of the righting arm 'gb' as a function of the angle of heel.

Figure 6.2.4, based on data from Larsson and Eliasson (1996), compares stability curves for two hypothetical 40 ft yachts. One is a modern cruiser-racer with a wide, shallow hull and a fin keel. The other is a more classical yacht with a narrow, deep-V hull and a relatively long keel. The figure illustrates that the initial stability of the wide, shallow hull is substantially larger than that of the narrow hull. However, the stability range, that is the range of heel angles for which the righting moment is positive, is significantly larger for the narrow hull with long keel. Yachts are unstable, or stable upside-down, when the righting moment is negative. Also, the larger the stability range, the better is the resistance against capsizing.

From the point of view of performance (boat speed), it is important to have a large righting moment at angles of heel up to, say, 30°. From the point of view of seaworthiness and safety, however, it is desirable that the range of negative righting moment is as small as possible. Obviously, the stability properties of the classical yacht are better at very large angles of heel (beyond, say, 90°).

Also plotted in Fig. 6.2.4 is the value of the product $gm(0) \sin \varphi$ (dotted lines). The difference between these lines and the actual stability curves can be interpreted as a measure of the 'erosion' of stability as the heel angle increases. Obviously, the yacht with the large beam and short fin keel is more vulnerable to loss of stability at larger heel angles than the more classical yacht.

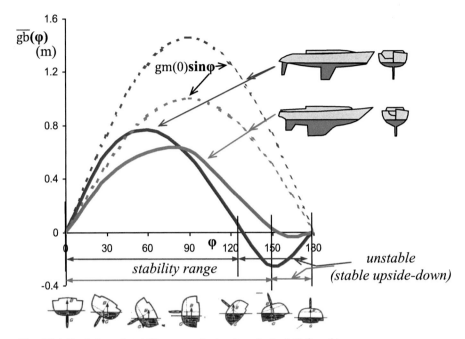

Fig. 6.2.4 *Static lateral stability curves for two hypothetical 40 ft yachts*

As already indicated above, the static stability of a sailing yacht is not only a function of the position of the metacentre but also of the position of the centre of gravity (see Fig. 6.2.1). For conventional yachts, with the centre of gravity at or near the plane of symmetry, the centre of gravity should be as low as possible since this gives a large metacentric height.

The righting arm and hence the static stability, can, however, also be increased by moving the centre of gravity towards the windward side of the yacht by moving ballast. This, of course, is the principle utilized on boats with *water ballast* and on yachts with *canting keels* such as the Open 60 and Volvo 70 class racing yachts. Moving loose 'equipment', including the crew, to the high side of the boat is of course another, but far less effective option.

In the case of water ballast (Fig. 6.2.5a) water is pumped into a ballast tank on the high side of the hull. In the case of a canting keel (Fig. 6.2.5b) a heavy, torpedo-shaped bulb mounted at the end of a narrow strut is swung sideways to windward over an angle of about 40° by a hydraulic mechanism. In both cases the centre of gravity is moved to windward, away from the plane of symmetry of the hull. The result is an increase of the righting arm in such a way that that there is already a positive righting moment at zero heel (see Fig. 6.2.6).

In terms of the expression (6.2.2) for the righting moment the effect of an off-centre position of the centre of gravity can be represented by adding a term $\delta\eta_{CG}\cos\varphi$, where $\delta\eta_{CG}$ is the lateral, off-centre, displacement of the centre of gravity, so that

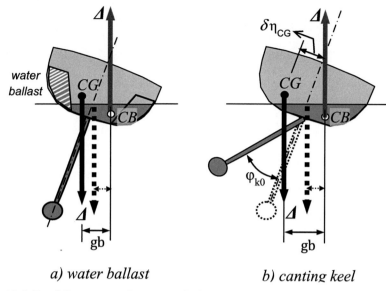

a) water ballast   b) canting keel

Fig. 6.2.5 *Two different ways of improving the lateral static stability by increasing the righting arm: water ballast and a canting keel*

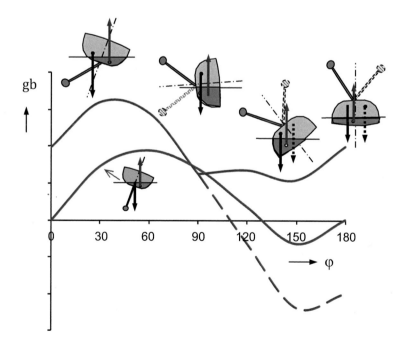

Fig. 6.2.6 *Stability curve of a hypothetical yacht with a canting keel*

$$M_{rx} = \Delta\{gm(\varphi)\,\sin\varphi + \delta\eta_{CG}\cos\varphi\} \qquad (6.2.3)$$

Figure 6.2.6 compares the stability curves of hypothetical yachts of the same size with a canting keel (red curve) and a conventional, non-canting keel (blue curve). The figure illustrates that the increase of the righting arm under normal sailing conditions, that is angles of heel up to 30°, it is quite substantial. With the keel on one side the stability deteriorates rapidly beyond about 90° heel (dotted red line). However, when the keel is swung to the other side when the boat is at 90° heel and beyond, it is possible to have a stability range of 180° (solid red curve). Hence, canting keels offer not only more stability under normal sailing conditions but have also a safety advantage, provided that the canting mechanism is sufficiently robust and keeps working under conditions of extreme heel.

The same applies to a yacht with water ballast, albeit to a much lesser extent, and provided, of course, that the pumping mechanism keeps working under conditions of extreme heel. A further advantage of canting keels is that they can be swung from one side to the other in a matter of a few seconds while it usually takes minutes to pump water ballast from one side to the other.

As already mentioned earlier the static stability of a yacht is important for the sail carrying capacity of a yacht. It is also of some importance for the sea keeping characteristics. However, for the latter, the dynamic stability characteristics, that is the stability in waves and in motion, are even more important. These will be discussed in Sect. 6.8.

Longitudinal Static Stability
The longitudinal static stability of a yacht, as illustrated by Fig. 6.2.7, is subject to the very same mechanisms as the lateral stability. For this reason it will not be discussed in any further detail. As already discussed in Sect. 4.1 it determines the trim-in-pitch angle θ of a yacht.

An important parameter is the longitudinal metacentric height (gm_L in Fig. 6.2.7). Its value is usually of the order of 1.1 times the length of the waterline.

Fig. 6.2.7 *Illustrating longitudinal static stability*

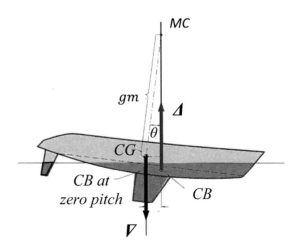

6.3   Hydrodynamic Side Force (Zero Heel)

As already indicated in Sect. 3.2 the hydrodynamic side force acting on a sailing yacht is produced primarily by the keel. A keel, as we have seen in Sect. 5.15, is a lifting surface, attached to a non-lifting body. In the case of a sailing yacht the non-lifting body is the hull. Because yachts have to be able to sail on port as well as on starboard tacks the keel is usually positioned in the plane of symmetry and the shape of the profile section(s) is usually symmetric.

We have also seen in Sect. 5.15 that the lift acting on a lifting surface can be expressed as

$$L = \tfrac{1}{2}\,\rho V^2\, S\, C_L,\qquad\qquad(6.3.1)$$

where, for sufficiently high Reynolds numbers, the lift coefficient C_L can be approximated by

$$C_L = \frac{2\pi E(\alpha - \alpha_0)}{1 + 2E/(Ae)}\qquad\qquad(6.3.2)$$

In Eq. (6.3.2) α and α_0 are, respectively, the actual angle of attack and the angle of attack at zero lift. For foils with symmetrical sections α_0 is usually zero. It is further recalled, see Sect. 5.15, that E is an 'edge factor' that, see Eq. (5.15.14), takes the value zero for vanishing aspect ratio and is equal to $\cos\Lambda$ for very high aspect ratios (Λ is the angle of sweep). A is the geometrical aspect ratio defined as

$$A = b^2/S,\qquad\qquad(6.3.3)$$

where b is the geometrical span and S is the planform area. 'e' is an efficiency factor, equal to b_e/b, that is related to the spanwise distribution of circulation through the effective span b_e. For an elliptic distribution of circulation $e = 1$.

The second term in the denominator of (6.3.2) is associated with the downwash induced by the trailing vorticity. In terms of the induced angle of attack α_i, caused by the trailing vorticity (Sect. 5.15), it can be expressed as

$$\alpha_i \approx C_L/(\pi Ae)\qquad(\alpha\ \text{in radians})\qquad(6.3.4)$$

Obviously, the keel of a sailing yacht can be considered as a lifting surface with a vertical orientation (or an airplane wing standing on its tip), at least when the yacht has zero heel angle. This means that, at least in principle, the expressions for the lift force acting on a lifting surface are equally valid for the side force acting on the keel of a sailing yacht. Hence we can write, for zero heel angle:

$$S = \tfrac{1}{2}\,\rho V_b^2 S_k C_S\qquad\qquad(6.3.5)$$

with

$$C_S = \frac{2\pi E(\lambda - \lambda_0)}{1 + 2E/(A_k e)} \tag{6.3.6}$$

Here we have replaced lift by side force and the angle of attack α by the angle of leeway λ. A_k is the aspect ratio of the keel.

A relatively minor complication is the fact that the keel is attached to a non-lifting body (the hull). We have seen in Sect. 5.17 that this can be dealt with, at least in principle. A bigger problem is, as we shall see below, that the hull is not fully submerged but pierces through the free water surface.

Effect of Free Surface

The question of the effect of the free water surface on the side force and on the induced drag (or drag due to side force) of a sailing yacht has been, and possibly still is, a subject of considerable controversy and confusion[2] (see, for example, the discussion in Ref. Larsson and Eliasson (1996) and those in Refs. Larsson (1990 in Chap. 1) and Milgram (1998 in Chap. 1)). The controversy and confusion is reflected in particular in theoretical models for the rapid estimation of the side force and induced drag of sailing yachts. Such models are usually also underlying the qualitative analysis of experimental results and computational fluid dynamic simulations. Hence, there is a risk that they lead to incorrect conclusions with respect to, for example, the direction into which a keel-hull design should be modified in order to improve its hydrodynamic performance. Fortunately, the controversy and confusion does not, in general, concern the validity of experimental and computational flow simulations as such.

In this author's opinion the problem requires answers to two key questions:

- What is the effect of the free surface on the effective span b_e and the efficiency factor e ($= b_e/b$) in Eqs. (6.3.1)–(6.3.6)
- What is the direct effect of the free surface equal pressure condition (Sect. 5.19) on the side force?

We will try to formulate answers to these questions against the background of the discussion of the properties of the free surface in Sect. 5.19. For the sake of simplicity we will do so first for the idealized case of a simple rectangular lifting fin that protrudes through the water surface (Fig. 6.3.1). At a later stage we will extend this to configurations with a fin attached to a surface piercing hull.

The answer to the first question, that on the effective span, can only be given if we can answer the underlying question of what the configuration is of the trailing vortex system in the presence of the free surface. It is often argued that, at least at low Froude numbers, when the water surface is hardly perturbed, the free surface acts like a reflection plane or a virtual plane of symmetry. This leads to a picture, see Fig. 6.3.1, that is similar to that of Fig. 5.17.1. It implies an effective span b_e equal to about twice the actual span b of the underwater part of the fin. In this picture there is no trailing vorticity in the free surface and the bound vorticity (circulation)

---

[2] Which has not fully by-passed this author!

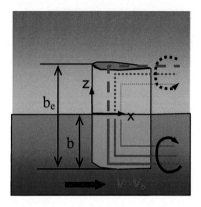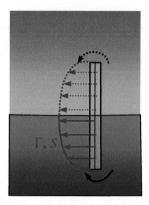

Fig. 6.3.1 *Lifting fin protruding through the free surface; simple, symmetrical reflection plane model*

Γ of the fin runs right up to the free surface at full strength. Although this model is not in conflict with the conditions to be satisfied at the free surface (Sect. 5.19) it poses several problems. The first is that when the value $b_e = 2b$ is substituted into Eqs. (6.3.1)–(6.3.6) it is silently assumed that these expressions, who were derived from wing theory, also hold in the presence of a free water surface. This, however, is not the case. We can see this by realizing that the equal pressure condition on the water surface implies that there can be no lift or side force at all at the waterline (the intersection of the fin with the water surface). So, while the circulation Γ may take some finite, non-zero value at the waterline, the local side force must go to zero (Fig. 6.3.2). As a consequence the total lift or side force must be smaller than in the case without a free surface. Moreover, this must be the case at all Froude numbers, because the equal pressure condition must be satisfied at all Froude numbers.

A second problem of the simple reflection plane model is that it does not take into account that the difference in water level between the two sides of the fin causes some additional, local side force S_h at the waterline (Fig. 6.3.2). This counteracts the loss of side force due to the equal pressure condition. Because the local surface

Fig. 6.3.2 *Lifting fin protruding through the free surface; qualitative model of distribution of side force*

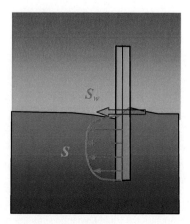

elevation (or depression) is Froude number dependent (Eq. (5.19.14)) the additional side force due to the difference in water level also depends on Froude number.

The third and most practical problem is the general experience (Marchai 2000, pp. 465–469) that the simple reflection plane model leads to an overestimation of the side force and an underestimation of the drag-due-to-lift. Experimental evidence (Gerritsma 1971; Beukelman and Keuning 1975) suggests that, in terms of hydro-dynamic forces, a surface piercing fin behaves as if the effective span is of the order of the actual span (or draught) rather than twice that value.

It has also been argued (Slooff 1984 in Chap. 5) that, at least at high Froude num-bers, the free surface acts as a plane of anti-symmetry. This is found to lead to a spanwise distribution of circulation that goes to zero at the free surface (Fig. 6.3.3). This model has the property that, when inserted into the classical Kutta-Youkowski formula (5.14.1) relating lift and circulation, it always leads to zero lift or side force at the waterline, independent of Froude number. Unfortunately this model implies trailing vorticity in the free surface which, at least at low Froude numbers, when the water surface is (almost) flat, is in conflict with the conclusions of Sect. 5.19. Moreover, the effect on side force of the difference in water level is, again, not taken into account. It is further the author's experience that this anti-symmetric circula-tion model leads to a significant underestimation of the total side force.

To the author's knowledge a simple theoretical model, with a qualitatively proper representation of the free surface effects on circulation and side force, is not avail-able in the literature.[3] Nevertheless, it seems possible to formulate such a model (see Appendix E).

In Appendix E it is shown that, in the presence of a free surface, and at low Froude numbers, the spanwise distribution $S(z)$ of the side force per unit of span on a rectangular fin with zero sweep, is, qualitatively, given by

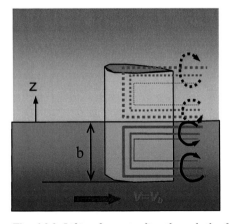
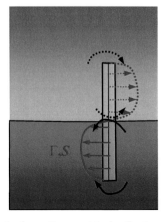

Fig. 6.3.3 *Lifting fin protruding through the free surface; simple, anti-symmetrical reflection plane model*

---

[3] Here, we are, of course, not talking about the basic conservation laws of physics underlying Computational Fluid Dynamic simulations.

$$S(z) = \rho V_b \{\Gamma(z) - \Gamma(0)e^{2\pi A_k \kappa\, z/b}\} \qquad (6.3.7)$$

Here, $\Gamma(z)$ is the spanwise distribution of the circulation, $\Gamma(0)$ the circulation at the waterline ($z=0$) and 'A_k' the geometric aspect ratio of the fin. In Eq. (6.3.7) the first term, $\Gamma(z)$, between parentheses, represents the basic, classical result of lifting surface theory. The term with $\Gamma(0)\, e^{2\pi A_k \kappa\, z/b}$ represents the free surface effect. As explained in more detail in Appendix E, it is inspired by the exponential function (5.19.18) describing the decay, with depth, of the orbital motion of water particles in surface waves (Sect. 5.19). The empirical factor κ (≈ 1.5) in the exponential function has been introduced to account for the fact that a surface piercing body does not generate infinitely long waves normal to the direction of motion like wind driven ocean waves. There is reason to believe that the streamline elevation associated with the finite waves generated by a surface piercing body will decay somewhat faster with depth than in the case of free running ocean waves.

It is also indicated in Appendix E that the local side force S_w at the waterline due to deformation of the free surface (Fig. 6.3.2), is proportional to Fr^4 and, hence, for small Froude numbers, can be safely neglected for many if not most practical purposes.

It is important to note that Eq. (6.3.7) implies that the classical Kutta-Youkowsky law (Eq. (5.14.1)) relating lift (or side force) and circulation is no longer valid in the presence of a free surface. This is also illustrated by Fig. 6.3.4. The figure shows the spanwise distribution of side force $S(z)$ for different fin aspect ratios and an elliptical distribution $\Gamma(z)$ of circulation, i.e.

$$\Gamma(z) = \Gamma(0)\sqrt{\{1 - (z/b)^2\}} \qquad (6.3.8)$$

Fig. 6.3.4 Spanwise distribution of side force of surface piercing fins with an elliptic distribution of circulation

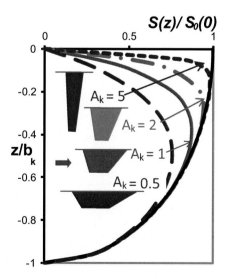

Note that $S(z)$ has been made dimensionless through dividing by the side force $S_0(0)$ at $z=0$ in the absence of free surface effects. The figure illustrates that for small aspect ratios there is a substantial loss of side force due to the pressure relief effect of the free surface. For very high aspect ratios the relative effect of free surface becomes smaller and smaller and the distribution of side force approaches the (elliptic) distribution of the circulation.

It is also indicated in Appendix E that in case of an elliptic distribution of circulation and zero sweep angle the expression for the total side force coefficient can be formulated as:

$$C_S = C_{S_0} \left[1 - \left\{ 2/(\pi^2 \kappa A_k) \right\} (1 - e^{-2\pi \kappa A_k}) \right] \tag{6.3.9}$$

Here, C_{S_0} is the side force without free surface effects. This is given by Eq. (6.3.6) with, however, the aspect ratio 'A_k' replaced by $2A_k$. Note also that the edge factor E and the efficiency factor e in Eq. (6.3.6) should be taken as those for a lifting surface with aspect ratio $A=2A_k$. Obviously, the term

$$\left[1 - \left\{ 2/(\pi^2 \kappa A_k) \right\} (1 - e^{-2\pi \kappa A_k}) \right] \equiv F_{FS} \tag{6.3.10}$$

represents the effect of the free surface. We will call F_{FS} the free surface factor.

Figure 6.3.5 presents the side force coefficient according to Eq. (6.3.9) as a function of the angle of leeway for different aspect ratios ($\lambda 0$ has been assumed zero). The solid lines represent the results with free surface and the dashed lines the classical

Fig. 6.3.5 *Effect of the free surface on the side force of surface piercing fins of different aspect ratios and an elliptic distribution of circulation*

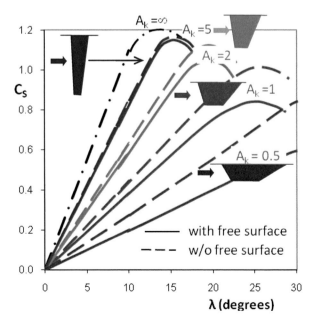

result without free surface effect. It can be noticed that, for small aspect ratios, the lift loss due to the presence of the free surface is substantial.

Appendix E also contains a more refined analysis including the effects of planform sweep and taper on the side force. Figure 6.3.6a and b illustrate the effects of taper on the slope $\dfrac{dC_S}{d\lambda}$ of the side force curve (rate of change of the side force coefficient with angle of leeway), for zero and 45 degrees angle of sweep. The figures illustrate the trend that, for a given aspect ratio, the magnitude of the free surface effect decreases with increasing taper ratio and increasing sweep angle. One rea-

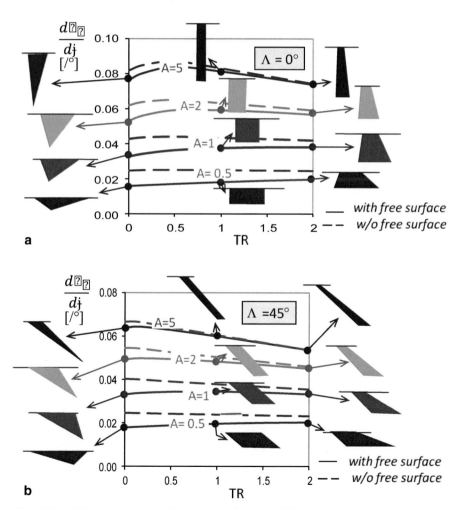

Fig. 6.3.6 *a Lift curve slope of surface piercing fins with different aspect ratios as a function of taper ratio, for zero angle of sweep (Λ = 0°). b Lift curve slope of surface piercing fins with different aspect ratios as a function of taper ratio, for 45° angle of sweep (Λ = 45°)*

son for this is the fact that the span-wise distribution of circulation changes in the sense that the center of the side force shifts in the direction from root to tip with increasing taper ratio and sweep angle (see Figs. 5.15.7 and 5.15.8 in Chap. 5). In terms of Eq. (6.3.7) this means that the circulation $\Gamma(0)$ at the root decreases with increasing taper ratio and increasing sweep angle. As a consequence the loss of lift at the waterline is smaller for high taper ratios and high angles of sweep. Another reason is in the fact that the exponential function in Eq. (6.3.7) takes the form $e^{\pi \kappa A_k (1+TR)z/(b\cos\Lambda)}$ in the case with sweep and taper (see Appendix E). This means that, for $z < 0$, its value decreases for increasing values of TR and Λ.

Figure 6.3.7 shows the lift curve slope as a function of sweep angle for different taper ratios and an aspect ratio $A_k = 1$. Together with Fig. 6.3.6 the figure illustrates that, for small aspect ratios ($A_k < 2$), the most effective fin, that is the fin with the highest lift curve slope, is obtained for large taper ratios ('inverse' taper) and small, negative sweep angles.[4] For an aspect ratio of 5, when the effect of the free surface is smaller, the most effective fin is obtained for a taper ratio of about 0.65 and, about zero sweep (Fig. 6.3.6). For very large aspect ratios, when the free surface effect goes to zero, the optimal taper ratio tends to the classical value of about 0.5.

It is useful to note that when there is no free surface effect, such as in the case of an aircraft wing, the highest lift curve slope for any aspect ratio is always obtained for zero sweep and a taper ratio of about 0.5. It is also obvious that, in all cases, aspect ratio is the most important parameter. For high aspect ratios the second important parameter is angle of sweep. For very low aspect ratios the effects of both sweep and taper become quite small when there is no free surface effect.

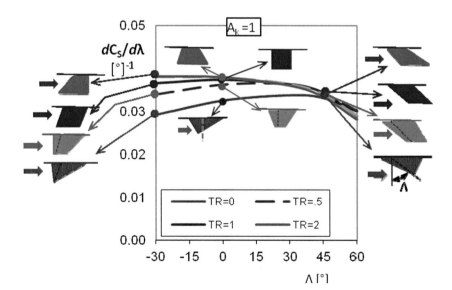

Fig. 6.3.7 *Lift curve slope of surface piercing fins as a function of sweep angle*

---

[4] As already noted in Slooff (1984, 1985 in Chap. 5). It is also the reason why the author proposed a taper ratio of 2 for the basic keel of the 12-Metre yacht Australia II in 1980.

Little is known about the effect of the free surface on maximum lift. There is reason to believe, however, that the free surface causes a loss of maximum lift that is proportional to the loss of the slope of the linear part of the lift curve. The reasoning is that maximum lift is determined by the distribution of the chordwise velocity which also determines the circulation. Hence, the condition of maximum lift is determined by the condition of the maximum circulation. The latter is not, or hardly, affected by the presence of a free surface. However, as we have seen above, the pressure relief caused by the free surface causes a loss of lift which is proportional to the circulation. Recalling further that the effect of aspect ratio on maximum lift is fairly small when there is no free surface (see Fig. 5.15.5), this means that, qualitatively, the lift curves of surface piercing fins behave as indicated in Fig. 6.3.5. With free surface effect there is a significant loss of maximum lift for small aspect ratios.

Effect of the Hull

As already indicated in Sect. 5.17 a hull attached to a fin-keel can be considered as a non-lifting body attached to a lifting surface. It was also seen that the main effect of a non-lifting body like a bulb or hull on the lift or side force generated by a fin is an increase of lift through the mechanism of *'lift carry-over'*. When there is no boundary layer separation in the fin-body junction (which would be the case for a properly designed configuration), the circulation and associated lift at the root of the fin is carried over onto the body. In terms of the expression (6.3.6) for the side force coefficient the main effect is an increase of the *'active span'* b' (Sect. 5.17, Fig. 5.17.5) as well as the *effective span* b_e.

The case of a surface piercing hull is special in the sense that here we have something like a half-body that is subject to the pressure relief effect of the free surface. Assuming that there is no vortex shedding and associated trailing vorticity at the waterline, as in the case of the surface piercing fin, the distribution of circulation must, as before, be that of a symmetrical reflection plane model (Fig. 6.3.8). Due to the mechanism of lift, or rather, circulation carry-over, the circulation at the root of

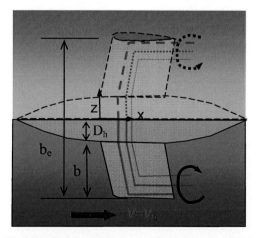
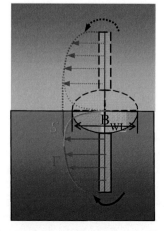

Fig. 6.3.8 *Simple, symmetrical reflection plane model for the circulation and side force of a (fin) keel-hull configuration*

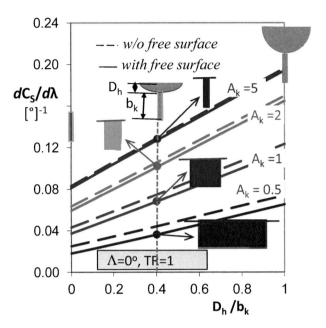

Fig. 6.3.9 *Lift curve slope of fin-hull configurations as a function of the hull depth/fin span ratio*

the fin-keel is, in the absence of boundary layer separation, continued into the hull at constant strength. This means (see Sect. 5.17) that, at least for small hull depth and small hull beam to keel span ratios, the active span b' is about equal to $b_k + D_h$ and the effective span $b_e \approx 2(b_k + D_h)$.

As described in Appendix E, the effect of the free surface on the side force of a fin-hull configuration can be modeled in a similar way as for the surface piercing fin. Figure 6.3.9 illustrates the effect of a hull on the side force for fin-keels of constant chord and zero sweep. Shown is the rate of change per degree leeway of the side force coefficient (or lift curve slope) $\dfrac{dC_S}{d\lambda}$ as a function of the hull depth to keel span ratio D_h/b_k for several values of the aspect ratio of the keel. In considering Fig. 6.3.9 it should be noted that keel areas and aspect ratio are those of the exposed keels. It should also be realized that the keel areas depend on the ratio D_h/b_k. For a given span b_k the area of the high aspect ratio keel is smaller than those of the keels with lower aspect ratio.

The figure shows, that the side force coefficient increases almost linearly with hull depth and that the rate of increase is substantial. Figure 6.3.10 gives an indication of the dependence on taper ratio of the side force on fin-hull configurations of different aspect ratios and zero angle of sweep. Note that the picture is a little different from that (Fig. 6.3.6) for a surface piercing fin without a hull. It can be noticed that for high aspect ratio fins, which do not suffer much from the free surface effect, a low taper ratio is now more effective due to the higher amount of lift carryover from fin to hull (see also Sect. 5.17). For low aspect ratios the most effective fin is still one with a taper ratio of 1 or beyond.

Fig. 6.3.10 *Dependence of the side force of fin-hull configurations on the taper ratio of the fin. (See also Figs. 6.3.6 and 6.3.8)*

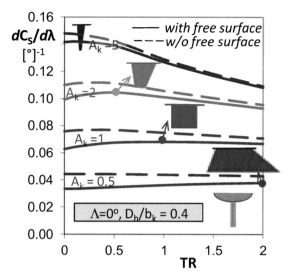

It should be emphasized that the theoretical model underlying Figs. 6.3.9 and 6.3.10 is fairly crude. Nevertheless, it is the author's impression that it is sufficiently accurate to represent the main trends (see also Appendix E). Towing tank tests and/or computational fluid dynamics are, of course, required for more accurate studies.

A final remark to be made is that the assumption that there is no shedding of stream-wise vorticity at the waterline may not be correct for flat bottomed hulls with a tran-som stern. The reason is that in this case stream-normal vorticity will, in general, be present at the location on the waterline where the hull boundary layer detaches from the hull. We have seen in Sect. 5.19 that this may lead to the formation of stream-wise, trailing vorticity when the velocity profile of the boundary layer is 'twisted'. When this happens, some of the circulation carried over from the keel to the hull may be lost at the waterline which means that the water surface will no longer act as a full reflection plane. Unfortunately there does not seem to be any experimental (or computational fluid dynamic) evidence as to whether, or to what extent, this occurs in reality.

Trailing-Edge Flaps or Trimtabs
The foil sections of sailing yacht keels are sometimes equipped with a trailing-edge flap (Fig. 6.3.11), usually called a *trimtab* in the jargon of sailing. The purpose of a trim tab is to reduce the resistance for a given level of side force. The effect of the flap deflection on the lift or side force is similar to that of adding camber to a foil section (Sect. 5.14). In terms of Fig. 6.3.5 it implies that the lift curve is shifted to a higher level of lift. This at the expense of a higher profile drag at zero lift. However, as will be discussed in more detail in Sect. 6.5, the profile drag can be a little smaller for a given non-zero level of the side force.

It has been found (Abbott and Von Doenhoff 1945) that, in two-dimensional flow, the additional lift $\delta C_{L\delta}$ due to the deflection of a plain flap can be expressed as:

Fig. 6.3.11 *Effect of a trailing-edge flap or trimtab on the side force of a keel*

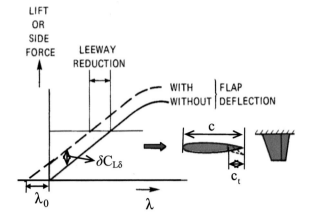

$$\delta C_{L\delta} \approx 2\pi\,\eta_f\,\delta_f \tag{6.3.11}$$

Here, δ_f is the flap deflection angle in radians and η_f is the flap effectiveness. The latter is a function (see Fig. 6.3.12) of the ratio c_t/c between the chord of the flap and the full chord of the foil. For a keel of finite span, the lift curve slope factor 2π in Eq. (6.3.11) should be replaced by the actual lift curve slope in three-dimensional flow described in the preceding paragraphs.

Because of the higher lift level, a keel with a trimtab can, in principle, also have a smaller area than a plain keel. This leads to a further reduction of the viscous resistance through a reduction of the wetted area. The reduction of leeway for a given level of the side force usually also implies a small reduction of the wave-making resistance (See Sect. 6.7).

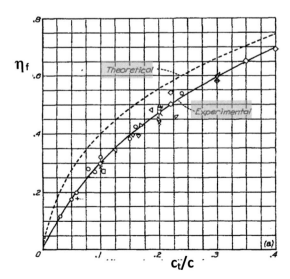

Fig. 6.3.12 *Flap effectiveness of plain flaps (trimtabs)*

Bulbs, Winglets, etc.

The effects of winglets and bulbs on the fluid dynamic characteristics of lifting surfaces have already been discussed in Sects. 5.15 and 5.17, respectively. There we have seen, that the primary objective of winglets is to lower the induced drag. Because winglets do not, in general, increase the lateral area and geometric span of a wing or keel, the effect on lift or side force is usually quite modest and due mainly to the increase of the effective span (see Sect. 5.15).

Bulbs or tip-tanks, mounted at the tip of a lifting surface have also been discussed in Sect. 5.17. There we saw that, depending on the configuration of the bulb, there is usually an increase of lift due to an increase of the active span as well as the effective span.

Because winglets and bulbs are mounted at the tip of a fin-keel, far away from the water surface, the pressure relief effect of the latter is, in general, negligible.

Rudders

The contribution of the rudder(s) to the hydrodynamic side force acting on a sailing yacht is usually much smaller than the side force produced by the keel. The main reason is, that, as we will see later, a rudder produces a lot more induced resistance for the same side force than a keel. This is caused by the fact that the span or draught of a rudder is usually significantly smaller than that of the keel. The consequence is that, under normal sailing conditions, the lift or side force on the rudder should be kept as small as possible. The hydrodynamics governing the forces acting on a rudder are basically the same as those for a keel. There are, however, a few minor differences that should be mentioned. The most important difference is that the rudder usually operates in the downwash (or sidewash) of the keel (Fig. 6.3.13).

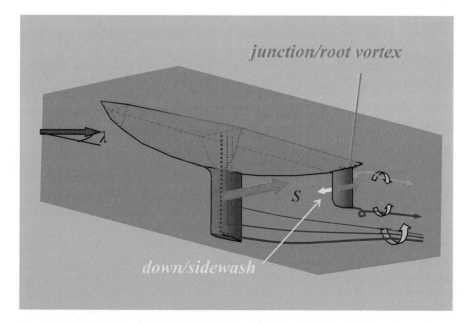

Fig. 6.3.13 *Side force on keel and rudder, trailing vorticity and downwash/side-wash*

This means that the effective angle of attack (Sect. 5.15) of the rudder is smaller than the angle of leeway minus the flow angle induced by the trailing vortex system of the rudder. One can write:

$$(\alpha_e)_{\text{rudder}} = \lambda - \alpha_\varepsilon - (\alpha_i)_{\text{rudder}} + \delta_r \qquad (6.3.12a)$$

or

$$(\alpha_e)_{\text{rudder}} = \lambda\{1 - (\alpha_\varepsilon/\lambda)_{k-r}\} - (\alpha_i)_{\text{rudder}} + \delta_r \qquad (6.3.12b)$$

Here, α_ε is the angle of the downwash (or, rather, side wash) induced by the keel at the position of the rudder and δ_r is the deflection angle of the rudder. Equation (6.3.12) implies that the rate of change due to leeway of the side force on the rudder can be written as

$$\left(\frac{dC_S}{d\lambda}\right)_{\text{rudder, with keel}} = \{1 - (\alpha_\varepsilon/\lambda)_{k-r}\}\left(\frac{dC_S}{d\lambda}\right)_{\text{rudder, no keel}} \qquad (6.3.13)$$

Because $(\alpha_\varepsilon/\lambda)_{k-r} < 1$, as we will see shortly, this expresses that the effective rate of change due to leeway of the side force on the rudder in the presence of the keel is smaller than when the keel would not be present.

It is further noted that the rate of change $\dfrac{dC_S}{d\delta_r}$ of the side force on the rudder due to deflection of the rudder at constant angle of leeway is not effected by the presence of the keel. The reason is that a deflection of the rudder has hardly any effect on the flow about the keel, at least as long as the rudder is sufficiently far downstream of the keel.

Equation (6.3.12) implies also that to obtain zero side force on the rudder it must be deflected. Inserting $(\alpha_e)_{\text{rudder}} = 0$ and $(\alpha_i)_{\text{rudder}} = 0$ in Eq. (6.3.12) gives

$$0 = \lambda\{1 - (\alpha_\varepsilon/\lambda)_{k-r}\} - 0 + \delta_r \qquad (6.3.14)$$

or,

$$\delta_r = -\lambda\{1 - (\alpha_\varepsilon/\lambda)_{k-r}\}, \qquad \text{for } S_r = 0 \qquad (6.3.15)$$

The downwash angle α_ε, induced by the keel, is caused by the trailing vorticity as well as the bound vorticity of the keel. Like the induced angle of attack it is, a function of the lift or side force on the keel, the aspect ratio and the spanwise distribution of circulation (see this section, above, and Sect. 5.15). In addition, it is also a function of the longitudinal distance between the keel and the rudder. When the distribution of circulation of the keel is not elliptical, α_ε also varies in the spanwise direction.

The quantity $\alpha_\varepsilon/\lambda$ (or $\alpha_\varepsilon/\alpha$) is called the downwash parameter. As discussed in more detail in Appendix F, $\alpha_\varepsilon/\lambda$ takes the value 1 at the trailing edge of the keel. In theory, in inviscid flow, the value of α_ε far downstream is equal to twice the induced angle of attack at the location of the keel. In practice, in particular at high angles of attack, the value of α_ε far downstream is a little smaller than twice the induced angle of attack because of the increasing distance in the downstream direction between the plane of the rudder and the trailing vortex sheet. This is probably the reason why the downwash angle at the position of the rudder is sometimes (Gerritsma 1971) taken equal to 1.8 times the induced angle of attack at the position of the keel.

The downwash behind a keel can be estimated (see Appendix F) on the basis of methods and charts for aircraft wings (Silverstein and Katzoff 1939; Etkin and Reid 1996). Figure 6.3.14 gives an impression of the streamwise variation of the downwash behind the root section of keels of different aspect ratios, attached to a hull $(B_{WL}/D_h = 6)$. Note that the streamwise distance $x_r - x_k$ behind the centre of pressure of the keel is expressed in parts of the span (or draft) of keel + hull $(b_k + D_h)$. Note also that the keels have zero sweep and a taper ratio of 1 and that the hull draft to keel span ratio is 0.4.

It can be noticed that the downwash decreases rapidly just aft of the trailing edge of the keel but much more slowly further downstream. Like the induced angle of attack the level of the downwash is mainly determined by the (effective) aspect ratio of keel plus hull.

Figure 6.3.15 illustrates the effect of taper ratio on the downwash parameter for keels of aspect ratio 1 with zero sweep. It can be noted that the downwash behind the root section of a keel decreases with increasing taper ratio. This is the result of changes in the spanwise distribution of circulation. Because the effect of sweep on the spanwise distribution of circulation is similar to that of taper (see Sect. 5.15) the downwash behind the root section of a keel also decreases with increasing angle of sweep.

Fig. 6.3.14 Downwash parameter as a function of the (relative) distance behind the root section of the keel of keel-hull configurations of different aspect ratios

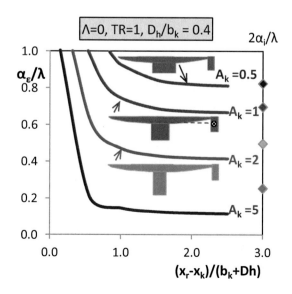

Fig. 6.3.15 Downwash
parameter as a function of
the (relative) distance behind
the keel root of keel-hull
configurations with different
keel taper ratios

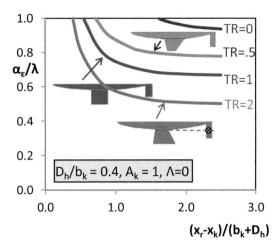

Another difference between rudders and keels is that there is, in general, a small
gap or slot between the rudder and the hull. This means that there is no full lift
carry-over between the rudder and the hull but a loss of lift and an associated root
vortex as in Figs. 5.17.2 and 5.17.3. An additional reason for the occurrence of a
root vortex is the fact that a fillet at the leading edge in the root area (see Fig. 5.17.8)
is usually absent. All this means, of course, that the effective span of a rudder is
even smaller and the induced resistance even higher than might be expected on the
basis of the geometric span. This is even more reason to keep the side force on the
rudder as small as possible. It will also be clear that it is advantageous to seal, if
possible, the gap between the rudder and the hull.

Because the objective of a rudder is to provide manoeuvrability, rudder foil sections are usually chosen to give a high maximum lift coefficient rather than low drag
at moderate values of the lift coefficient. In this way the size of the rudder, for a
given level of manoeuvrability, can be kept small. The latter means, of course, less
frictional resistance.

A final difference between keels and rudders to be mentioned is that, at small angles
of leeway, a (single) rudder usually operates in the wake of the keel. Here, due to the
action of viscosity, the effective flow velocity is somewhat smaller than outside the
wake (see Fig. 5.13.2). As a consequence the lift or side force on the rudder is also
smaller than might be expected on the basis of inviscid flow.

It will be clear that the discussion given above is applicable in particular to 'free-fly-
ing', or 'spade' rudders as utilized nowadays on most sailing yachts. However, most
of what has been said is also applicable to rudders attached to a skeg (Fig. 6.3.16).
A disadvantage of a rudder on a skeg is that the skeg cannot be aligned with the
local flow direction in the downwash field behind the keel. This means that skegged rudders usually require more deflection (and associated additional drag) than a
spade rudder.

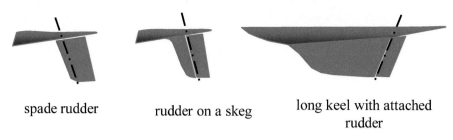

spade rudder   rudder on a skeg   long keel with attached
                                                              rudder

Fig. 6.3.16 Different types of rudder

Rudders attached to a long keel are a different matter. They act more like a flap or trim-tab and are neither very effective nor very efficient as a yawing moment generating device.

The position of the rudder axis is important for the steering force to be applied by the helmsman. We have seen in Sect. 5.14 that the moment experienced by a symmetrical foil section is close to zero with respect to the quarter chord point of the foil. This means that, for a high aspect ratio spade rudder, the steering force is small when the rudder axis is close to the quarter chord line. The best position is a little ahead of the quarter chord line. The rudder is then, properly, just a little 'under-balanced', that is the direction of the force to be applied on the helm corresponds with the direction of the side force on the rudder ('weather helm'). For low aspect ratio foils the position of the centre of pressure is a little closer to the leading edge (see Sect. 5.15). Hence, the rudder axis should then also be positioned a little more forward.

Proper balancing of a skegged rudder or a rudder attached to a long keel is more difficult and requires other, special measures, that are beyond the scope of this book.

6.4   Centre of Lateral Resistance, Hydrodynamic Moments, Vertical Force

As already indicated in Chap. 5 the point of application of the side force is called the centre of pressure. In the terminology of yachting this point is usually called the Centre of Lateral Resistance (CLR). We have also seen in Sect. 3.2, that the distance between the centre of pressure and the and the ξ-η plane determines the hydrodynamic heeling moment (Fig. 3.2.4) and that the hydrodynamic yawing moment is determined by the longitudinal distance between the centre of pressure and the position of the centre of gravity in the (Fig. 3.2.1).

The precise position of the centre of pressure depends on many parameters and can be determined only through computational fluid dynamics or, indirectly, through towing tank tests. It is possible, however, to indicate some general trends on the basis of lifting surface theory and the simple model for the free surface effect introduced in the preceding section.

Vertical Position of the Centre of Lateral Resistance (CLR)

First of all it is clear, that, when the side force on the rudder is small, the size and shape of the keel is the dominant factor for the position of the centre of pressure. For the vertical position of the centre of pressure the draft (or span) of the keel plus hull is the most important quantity. The span-wise load distribution is the next quantity of importance. We have seen already in Sect. 5.15, (Fig. 5.15.9), that, without the presence of a free surface, the centre of pressure of a lifting surface moves outboard with increasing taper ratio and increasing angle of sweep and that this effect is more pronounced for high aspect ratios. Considering Fig. 6.3.4, which shows the span-wise load distribution of surface piercing fins of different aspect ratios with an elliptic distribution of circulation, it is also clear that the relative position of the centre of pressure, in terms of the ratio z/b, goes upwards, that is assumes smaller absolute values of z/b, when the aspect ratio increases. This trend is indicated more clearly in Fig. 6.4.1,[5] which gives the vertical position z_{CP}/b of the centre of pressure as a function of aspect ratio for different values of the taper ratio TR. The figure also indicates that, not surprisingly, the centre of pressure goes downward with increasing taper ratio.

Further qualitative information, in the form of span-wise load distributions for taper ratios 0, 1 and 2 (each for $A_k = 0.5, 1, 2$ and 5), is contained by the small figures on the right hand side of Fig. 6.4.1.

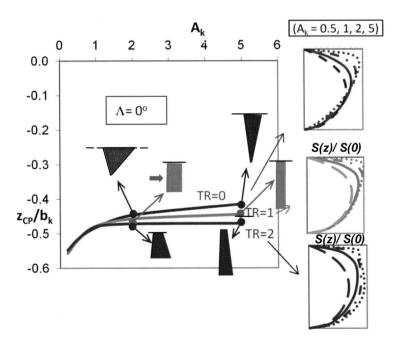

Fig. 6.4.1 *Vertical position of the center of pressure of surface piercing fins as a function of aspect ratio for different taper ratios*

[5] The figure has been constructed on the basis of Eq. (6.3.9) and spanwise distributions of circulation given in Munk (1923 in Chap. 5).

The effect of sweep angle on the position of the centre of pressure is more complex and not explicitly given in Fig. 6.4.1. However, the reader will understand from the discussion in Sect. 5.15 on the effects of plan-form shape on the span-wise distribution of circulation and lift, that the effect of a positive angle of sweep is similar to that of a high taper ratio and that a negative angle of sweep corresponds with a small taper ratio. In fact, it is found (see Fig. 6.4.2) that TR=0 for $\Lambda=0°$ corresponds closely with $\Lambda=-30°$, TR=0.5, that TR=1 for $\Lambda=0°$ corresponds closely with $\Lambda=+30°$, TR=0.5 and that TR=1.3 for $\Lambda=0°$ corresponds closely with $\Lambda=60°$, TR=0.5. Note that the corresponding span loadings are indicated in Fig. 6.4.1 and that plan-forms giving an elliptic distribution of circulation were given in Fig. 5.15.9.

Figure 6.4.3 shows the effect of the presence of a hull on the vertical position of the centre of pressure. Results are shown, in the same format as Fig. 6.4.1, for three different values of the hull depth to fin/keel span ratio and a fin taper ratio of 1. It can be noted that the effect of the hull draft to keel span ratio on the relative vertical position of the centre of pressure is quite small. This is the result of a combination of two counteracting factors: more lift carry-over, but also a higher loss of side force on the hull due to the free surface effect.

What has been described above, for fin-keels attached to a hull, is also applicable to rudders. There are, however, some quantitative differences. One is that the depth of the hull at the position of the rudder is usually much smaller than at the position of the keel. As a consequence there will be less lift carry-over to the hull in the case of the rudder. In addition lift will be lost when, as usual, there is a gap between hull and rudder. Both factors imply that, in terms of Fig. 6.4.3, the vertical position of the centre of pressure for rudders will be close to that corresponding with $D_h/b = 0$.

Fig. 6.4.2 Combinations of sweep and taper giving approximately the same distribution of circulation on a fin-keel

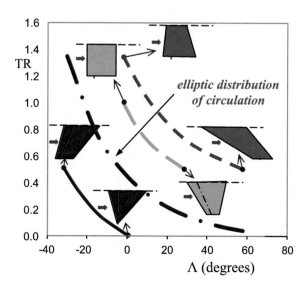

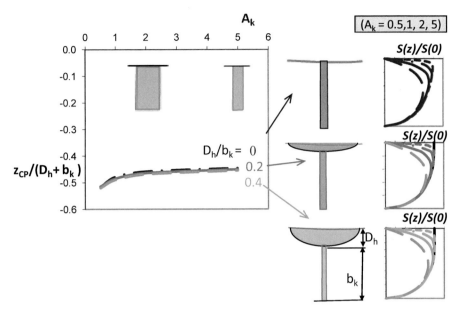

Fig. 6.4.3 *Vertical position of the center of pressure of fin-hull configurations as a function of aspect ratio for different values of hull depth to keel*

Hydrodynamic Heeling Moment

We have seen in Sect. 3.2, that the hydrodynamic heeling moment is the product of the heeling force H_H times the distance (d_{ch}) between the centre of pressure (centre of lateral resistance) and the longitudinal axis (See Eq. (3.2.6) and Fig. 3.2.4). This means that, for a given heeling force, the dependence of the hydrodynamic heeling moment on the parameters defining the keel and hull shape is the same as discussed above for the vertical position of the centre of pressure. In other words, the hydrodynamic heeling moment increases with increasing total draught or keel span, increasing taper ratio and increasing angle of sweep. For a given, fixed keel span, the heeling moment for a given heeling and side force decreases slightly with increasing aspect ratio of the keel.

Longitudinal Position of the Centre of Lateral Resistance; Hydrodynamic Yawing Moment

While the vertical position of the centre of pressure or centre of lateral resistance determines the heeling moment, the longitudinal position is connected to the hydrodynamic yawing moment. The latter can be expressed as

$$M_{Hz} = (M_{Hz})_{hull} + (x_{cg} - x_k)S_k + (x_{cg} - x_r)S_r, \qquad (6.4.1)$$

where the centre of gravity has been taken as the point of reference. As we have seen in Chap. 5, the longitudinal position x_{CP} of the centre of pressure of a plain lifting surface is usually close to the 25 % chord line, except for (very) low aspect ratios when it is closer to the leading edge (Fig. 5.15.19). The vertical position was

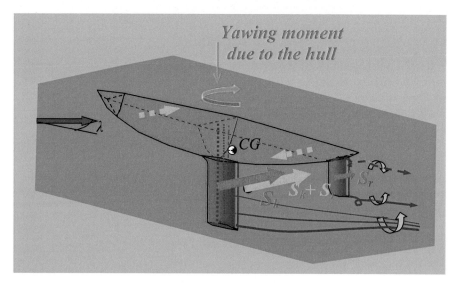

Fig. 6.4.4 *Illustrating the contributions of keel, hull and rudder to the hydrodynamic yawing moment*

found at about 47% of the draught in the preceding paragraph, (Figs. 6.4.1 and 6.4.3). The centre of gravity is usually positioned a little aft of amidships and a little ahead of the trailing edge of the fin-keel (Larsson and Eliasson 1996) (Fig. 6.4.4). This means that the length $x_{cg} - x_k (= > 0)$ of the 'arm' of the yawing moment caused by the keel is of the order of one tenth of the length of the waterline. As a consequence, the contribution of the keel to the hydrodynamic yawing moment is usually positive but not very large.

Of greater importance for the hydrodynamic yawing moment are the contributions of the hull and the rudder. As indicated in Sect. 5.16 a hull type of body at an angle of attack (or leeway) experiences an appreciable moment,[6] as the result of a positive side force on the fore-body and a negative side force of slightly less magnitude on the rear parts. For a fully submerged, ellipsoidal body of revolution the yawing moment was seen to be proportional to the angle of attack and the volume of the body (see Eqs. (5.16.1/2)). This means

$$M_{Hz} \cong \tfrac{1}{2}\rho V^2 2\lambda \ (1 - D/L)\nabla, \tag{6.4.2}$$

where D/L is the diameter/length ratio. The classical definition of a corresponding dimensionless moment coefficient is given by

$$C_{MHz} \equiv M_{Hz}/(\tfrac{1}{2}\rho V_b^2 \nabla) \tag{6.4.3}$$

---

[6] Sometimes called the 'Munk moment', after the German fluid dynamicist Max Munk, who was the first to study the phenomenon.

This gives

$$C_{MHz} \approx 2\lambda \ (1 - D/L) \tag{6.4.4}$$

For a sailing yacht hull with semi-circular underwater cross-sections the yawing moment coefficient would, in a first approximation, also be given by Eq. (6.4.4), with D replaced by $2D_h$. A tentative, more general approximation for sailing yacht hulls is described in Appendix G and given by

$$C_{MHz} \cong 2\lambda \ \{(1 - D_h/L_{WL})f' - D_h/L_{WL}\} \tag{6.4.5}$$

In this expression the factor f' is a function of, primarily, the beam/draft ratio and the displacement/length ratio. Quantities of secondary importance are the block coefficient C_B and the cross-sectional area coefficient C_m.

Figure 6.4.5 illustrates the dependence of the hull yawing moment on the main geometrical parameters. Shown is the rate of change $\dfrac{dC_{MHz}}{d\lambda}$ (per degree) as a function of the draft/beam ratio of the hull for several values of the displacement/length ratio. It is emphasized that the figure is tentative in the sense that it has a theoretical background but has been validated to a limited extent only against experimental or CFD results.

It is further to be noted that the yawing moment of a hull will be influenced by attached appendages. The main effect of a keel on the flow about its hull is to cause 'upwash' in front of and downwash aft of the keel. Qualitatively this implies an increase of the (positive) side force on the front part of the hull and a decrease of the (negative) side force on the rear part. Little is know about the net effect on the yawing moment, but it is likely that the effect is modest.

Under equilibrium conditions the yawing moment of the hull (and that of the sails!) is compensated by the contribution, due to deflection, of the rudder to the hydrodynamic yawing moment. Because of the relatively large distance $(x_{cg} - x_{CPr}, < 0)$ between the rudder and the centre of gravity, a small side force on the rudder already gives a significant, negative, contribution to the hydrodynamic yawing moment.

Fig. 6.4.5 Hull yawing moment coefficient as a function of displacement/ length and beam/draft ratios (trend, zero heel)

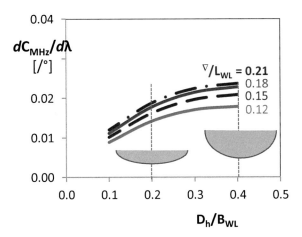

Static, Hydrodynamic, Directional Stability

A sailing yacht, or any ship for that matter, is said to be static-hydrodynamic, directionally stable when, for a fixed position and deflection of the rudder, such that there is hydrodynamic equilibrium, a deviation of course is automatically restored by a 'correcting' yawing moment. At this point, it is useful to rewrite Eq. (6.4.1) in terms of dimensionless coefficients:

$$C_{MHz} = (C_{MHz})_{hull} + C_{Lk}(x_{cg} - x_k)\, S_k/\nabla + C_{Lr}(x_{cg} - x_r)\, S_r/\nabla \quad (6.4.6)$$

In terms of Eq. (6.4.6) static directional stability means that the slope $\dfrac{dC_{MHz}}{d\lambda}$ of the line representing the variation of the yawing moment as a function of the angle of leeway is <0.

We have seen above (Fig. 6.4.5) that the contribution of the hull is destabilizing ($\dfrac{dC_{MHz}}{d\lambda} > 0$). With the centre of gravity at about 55% of the waterline (Larsson and Eliasson 1996) and the centre of pressure of the keel at $\cong 0.45 L_{WL}$ the contribution of the keel is, in general, also destabilizing. This would, of course, be less the case with a more rearward position of the keel.

Because the position of the rudder is usually near the rear end of the waterline, we have $x_{cg} - x_r \cong -0.45 L_{WL}$ and the contribution of the rudder to the directional stability is stabilizing.

As illustrated by Fig. 6.4.6 the variation with leeway of the total hydrodynamic yawing moment is unstable for the configurations considered. The figure shows the hydrodynamic yawing moment as a function of the leeway angle for a hull, a hull

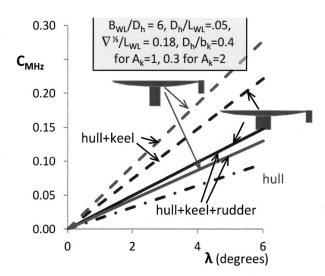

Fig. 6.4.6 *Yawing moment ($x_{ref} = 0.55\,L_{WL}$) as a function of leeway for yachts with keels of different aspect ratio attached to the same hull with $x_k = 0.45\,L_{WL}$ and with the same rudder*

plus keel and a hull plus keel plus rudder configuration. Note that two keels are considered, with different aspect ratios. Both configurations would be less unstable with the keels further aft.

An important aspect of the contribution of the rudder is that it operates in the field of downwash generated by the keel. As discussed in Sect. 6.3, the effective angle of attack of the rudder is reduced significantly by the downwash due to the trailing vortex system of the keel (Eq. (6.3.12)). As a consequence the net response of the rudder to a change in the angle of leeway is significantly smaller, by a factor $\{1 - (\alpha_\varepsilon / \lambda)_{k-r}\}$, see Eq. (6.3.13), than might be expected on the basis of the magnitude of the change in the angle of leeway.

In practice, the rudder experiences an effective change of leeway reduced to a factor of about 0.2–0.8 times the actual change of leeway. The precise value depends primarily on the aspect ratio and the taper ratio of the keel (see Figs. 6.3.14 and 6.3.15). For high aspect ratio keels the reduction is relatively small but it can be quite large for low aspect ratio keels with small taper ratio. This means that the position of the centre of lateral resistance of keel plus rudder is positioned significantly further forward (at about 50 % L_{WL}) than might be expected on the basis of the relative sizes and shapes of the keel and the rudder. It also means that the hydrodynamic directional stability of yachts with a high aspect ratio keel is, in principle, better than for yachts with low aspect ratio keels.

The point is also illustrated by Fig. 6.4.6. The 'average' cruising yacht configuration considered in this figure is directionally less unstable with a keel of aspect ratio 2 than for $A_k = 1$. This in spite of the fact that the direct contribution to the yawing moment of the high aspect ratio keel is larger due to its better lift curve slope. In both cases the yacht is equipped with the same rudder ($A_r = 2$, $TR_r = 1$, $(x_{cg} - x_r)S_r / \nabla \cong -1.3$). The difference in the contribution of the rudder is entirely due to the difference in downwash from the keel.

For a 'classical' yachts equipped with a long keel and attached rudder the characteristics and mechanism governing the hydrodynamic yawing moment and static directional stability are a little different. In particular when keel and hull form a 'blended' configuration as in Fig. 6.4.7, they act more like one (thick) lifting surface of low aspect ratio than as a combination of a hull with separately attached keel and rudder. This implies that there is no 'Munk moment' caused by the hull and that the centre of pressure or centre of lateral resistance is positioned ahead of the 25 % chord-line (see Sect. 5.15, Fig. 5.15.13) at a depth of about 50 % of the draft (see

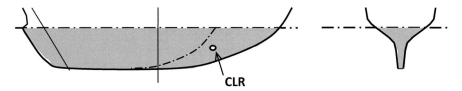

Fig. 6.4.7 *Illustrating the approximate position of the centre of lateral resistance of 'classical' yachts with blended, long keel*

Fig. 6.4.7). This corresponds with, approximately, a point at $\cong 25\%$ L_{WL} from the bow. This is considerably further forward than for yachts with fin keels and separate rudder. It means that the static hydrodynamic directional stability of yachts with long keels is not as good as for yachts with fin keels and separate rudder.

It has also been found (Larsson and Eliasson 1996; Marchai 1983c; Claughton et al. 2012) that the centre of lateral resistance of blended, long keels varies considerably with leeway. This in the sense that the CLR moves aft with increasing angle of leeway. The explanation for this phenomenon is the relatively large effect of side-edge separation of low aspect ratio lifting surfaces (see Sect. 5.15).

We have seen above that, perhaps somewhat surprisingly, sailing yachts are, in general, directionally unstable as far as the hydrodynamics of the hull and appendages are concerned. This means that they are directionally unstable and require continuous rudder action in the 'motoring mode'. Fortunately, the situation is different under sail due to the aerodynamic forces, as we will see in Chap. 8.

It should also be mentioned that the directional stability of a yacht is often more a dynamic than a static problem, in particular when sailing in waves (Marchai 1983c). We will return to the subject in Sect. 6.8 and Chap. 8.

Vertical Force and Hydrodynamic Pitching Moment
In Sect. 3.2, Fig. 3.2.4, the vertical component V_H of the hydrodynamic force was, for the sake of simplicity, related to the total lateral hydrodynamic force $F_{H(lat)}$ according to the expression (see Eq. (3.2.3))

$$V_H = F_{H(lat)} \sin \varphi \approx H_H \sin \varphi, \qquad (6.4.7)$$

where φ is the angle of heel. This would imply that there is no vertical hydrodynamic force when the angle of heel is zero and that the orientation of the vertical force is upward when $\varphi \neq 0$. However, this is not fully correct. Although the vertical component of the total lateral hydrodynamic force, which was seen to be caused mainly by keel and rudder, is zero for $\varphi=0$, there is, even at zero heel, also a (small) vertical force $V_H(0)$ caused by the flow about the hull. Due to the convex curvature of the submerged part of the hull the (average) local flow velocity at the bottom of the hull is higher than the free stream velocity. According to Bernoulli's law (Sect. 4.6) this implies a low pressure at the bottom of the hull which causes a downward directed force ($V_{H(hull)}$), <0, in Fig. 6.4.8). It means also that Eqs. (3.2.3)/ (6.4.7) should be replaced by

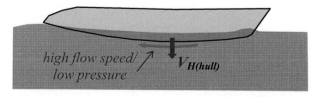

Fig. 6.4.8 *Vertical hydrodynamic force at zero heel, caused by locally high flow velocity at the bottom of the hull*

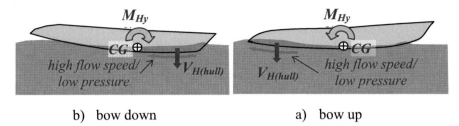

b) bow down   a) bow up

Fig. 6.4.9 *Effect of pitch angle on the hydrodynamic pitching moment (zero heel)*

$$V_H(\varphi) = V_{H(hull)}(\varphi) + F_{H(lat)} \sin \varphi, \qquad (6.4.8)$$

where $V_H(0)$, <0, is the vertical force at zero heel.

The point of application of the vertical hydrodynamic force component $V_{H(hull)}$ depends on the shape of the hull of the yacht but also on the pitch angle. When the yacht acquires a bow-down attitude the point of application of $V_{H(hull)}$ moves towards the bow and there is an bow-down contribution to the hydrodynamic pitching moment (M_{Hy} in Fig. 6.4.9a). When the yacht acquires a bow-up attitude the point of application of $V_{H(hull)}$ moves towards the stern and the contribution of $V_{H(hull)}$ to the hydrodynamic pitching moment is bow-up (Fig. 6.4.9b).

The vertical hydrodynamic force, like the other hydrodynamic force components, scales with the dynamic pressure $\frac{1}{2}\rho V_b^2$ and with the square of the size (a characteristic area) of the yacht. There is also a dependence on Froude number (Sect. 5.19) due to the presence of the free surface and wave making.

When the bottom of the yacht is sufficiently flat and wide, the boat speed is sufficiently high and the yacht is in a bow-up position, high pressure may develop over the middle and forward parts of the bottom of the hull. This causes an upward directed vertical or 'lifting' force (see Fig. 6.4.11) that may cancel or even more than compensate the downward vertical force near the stern. Depending on the shape of the hull, boat speed and initial pitch angle, this may cause the yacht to start (semi-)planing, that is, it is partially lifted out of the water. When this happens, the hydrodynamic resistance (see Sect. 6.5) drops dramatically. Planing does, however,

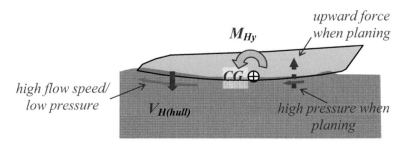

Fig. 6.4.10 *Hull vertical force components under (semi-)planing conditions (zero heel)*

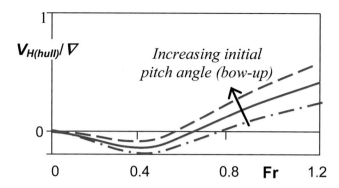

Fig. 6.4.11 *Variation with Froude number (qualitatively) of hydrodynamic vertical force on a hull*

not occur with conventional displacements yachts. It can only happen with 'beamy', flat-bottomed, very light displacement yachts and dinghies.

Figure 6.4.11 illustrates, qualitatively, how the vertical force, as a fraction of the displacement, varies with Froude number (i.e. $V_b/\sqrt{(Lg)}$). Even for very light displacement yachts with a high beam/length ratio planing effects do usually not occur at Froude numbers below about 0.45. It is also noted that Froude numbers above about 0.75 are seldom reached in practice, except, perhaps, by very light displacement yachts under extreme conditions, such as surfing down a wave.

The vertical hydrodynamic force is not the only contributor to the pitching moment. As already indicated in Sect. 3.2, Fig. 3.2.5, there is also a contribution (usually bow down) of the hydrodynamic resistance (to be discussed in the next section). The biggest contribution to the total (hydrodynamic plus aerodynamic) pitching moment of a sailing yacht comes, however, from the sails. The reason is, of course, that the vertical dimensions of the rig and sails are much larger than those of the hull and appendages (see also Fig. 4.1.6).

6.5 Resistance

6.5.1 Decomposition of Drag

We have seen in Sect. 5.2 that, from a 'mechanical' point of view, the fluid dynamic forces acting on an object can be distinguished in pressure forces and friction forces. This applies also to the hydrodynamic drag (or 'resistance' in the terminology of naval architecture) acting on a sailing yacht. It was also indicated in Sect. 5.2 that the drag of thin or slender objects at small angles of attack manifests itself mainly as friction forces but that pressure drag becomes more important for thick bodies and at high angles of attack.

A different division of fluid dynamic drag is obtained when its components are related to the physical mechanisms that are acting in the flow. As already indicated in Sect. 5.11, a part of the fluid dynamic drag of an object can be related to the viscosity of the fluid and the associated boundary layer phenomena. This part is called 'viscous' drag, boundary layer drag or profile drag. On thin or slender objects at small angle of attack this 'viscous' drag consists mainly of friction. However, there is also a pressure part ('form' drag) which becomes more important at high angles of attack and on thick bodies.

In Sect. 5.15 a second kind of drag was identified with the mechanism of induced downwash of three-dimensional lifting surfaces. This 'induced' drag was seen to consist of pressure drag only.

The break-down of drag into the components distinguished above is illustrated by the table below.

TOTAL DRAG	Friction drag	Pressure drag
'Viscous', boundary layer or profile drag		form drag
Induced drag		
Wave-making resistance		

Note that this 'drag matrix' contains a third category of resistance in the form of wave-making resistance. The wave-making resistance of a moving vessel is caused by the presence of the air-water interface. As already mentioned in Sect. 5.19, the Froude number $V_b/\sqrt{(Lg)}$ is the governing parameter for the phenomena associated with the air-water interface. Hence, it is also the governing parameter for the wave-making resistance.

The relative importance of the drag components depends on the configuration of the yacht as well as on the conditions of sailing. Figure 6.5.1 gives a qualitative impression of the relative magnitude of the viscous resistance, induced drag and the wave-making resistance, as a fraction of the total hydrodynamic resistance and as a function of Froude number (or scaled boat speed), for a conventional cruising yacht in upwind sailing conditions. Figure 6.5.2 gives the absolute value (scaled by the total resistance at Fr=0.4) as a function of Froude number for the same type of yacht. The figures are based on data (Keuning and Sonnenberg 1998a) from towing tank tests for a yacht with a 10 m waterline. It can be noted that the 'viscous' resistance dominates at low boat speed but that the wave-making resistance becomes increasingly important at higher boat speeds. The induced drag is usually of the order of 20% of the total resistance. Because of the dependence of the induced drag on keel span/draft (see Sect. 5.15), this would be less for a yacht with a deeper keel but a larger fraction for a yacht with less draft. The induced drag is, of course, also less for sailing conditions with a smaller side force/heeling force, that is at larger

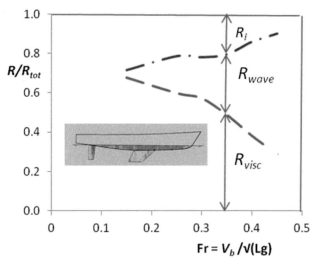

Fig. 6.5.1 *Fractional magnitude of different components of resistance for a cruising yacht in upwind conditions (flat water)*

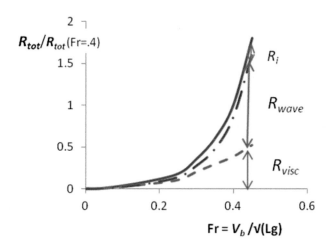

Fig. 6.5.2 *Resistance (scaled) as a function of Froude number (flat water)*

apparent wind angles (see Sect. 4.3). It is zero when the yacht goes dead downwind (apparent wind angle 180°).

It can be found in the literature that experimentalists often use a different way of decomposition of resistance that is based on the observed variation of measured forces with independent variables such as angle of heel, leeway, etc. In such an approach (Keuning and Sonnenberg 1998a) one distinguishes between the resistance at zero heel and 'delta' components due to heel, leeway/side force, etc. The resistance at zero heel ('upright' resistance) is usually further subdivided into a calculated frictional component and the 'residuary' resistance. The latter is defined as the difference between the total resistance at zero heel and the calculated frictional resistance. A decomposition of this kind can be useful for the prediction of resistance on a statistical basis. It does not, however, contribute much to the understanding of the physical mechanisms governing resistance.

In considering Figs. 6.5.1 and 6.5.2 it is important to realize that these are representative for conditions of calm (or flat) water. In a seaway, a yacht picks-up additional resistance due to the encounter with waves. Because of the unsteady, periodic character of the waves and the motion of the yacht, the resistance and its components in a seaway are time dependent. However, it is common practice to consider the hydrodynamic resistance of a yacht in a seaway in a time-averaged sense. The difference between the time-averaged resistance and the resistance under steady conditions in calm water is called the *added resistance* (in waves). As we will see later the added resistance consists mainly of wave-making and induced resistance although the viscous component may not be negligible. The magnitude of the added resistance can be substantial and can easily amount to 20–30 % of the total resistance in calm water.

It will be clear from the discussion given above that each of the viscous, induced and wave-making components of the hydrodynamic resistance is a function of heel, pitch and yaw angle and side force (apart from boat speed). It is further recalled that the condition of equilibrium in yaw usually requires a load on the rudder that causes additional ('trim-in-yaw') resistance.

In the following paragraphs we will consider the components of resistance in some further detail. We will do so first for calm water conditions. The added resistance in waves is discussed in a separate, following paragraph. The effects of heel, pitch and trim-in-yaw are considered in a separate section.

It is further noted that cruising yachts and ocean going racing yacht usually have a propeller that also contributes to the hydrodynamic resistance. The resistance of propellers is considered in Sub-Sect. Propeller Resistance.

6.5.2   Viscous Resistance

It follows from the discussion in Sect. 5.11 that the viscous resistance or profile drag of an object can be expressed as

$$R_v = \tfrac{1}{2}\,\rho V_b^{2} S C_{D_v}, \tag{6.5.1}$$

where S is a suitably chosen reference area and C_{D_v} is the viscous or profile drag coefficient. The latter is given by

$$C_{D_v} = (1+k)\,C_{F_0} \tag{6.5.2}$$

If S is chosen to be the wetted area S_{wet}, C_{F_0} is the friction drag coefficient of (one side of) a flat plate at zero angle of attack. 'k' is a 'form factor' depending on the shape and lift of the object.

As described in Sect. 5.11, C_{F_0} is a function of Reynolds number. It also depends strongly on the state (laminar or turbulent) of the boundary layer and on the roughness of the surface (see Fig. 5.11.3). While there is no difference of opinion on the validity of the friction drag formula for laminar flow on a flat plate (Eq. (5.11.8)), several alternative empirical formulae for turbulent flow can be found in the literature. In the naval architects community it is common practice to use the so-called 'ITTC[7] 1957 model-ship correlation line' (Lewis 1988 in Chap. 5). The latter reads

$$C_{F_0} = 0.075 / \left(\log_{10} Re - 2 \right)^2 \qquad (6.5.3)$$

Fortunately, the differences between results of this formula and those given in Sect. 5.11 are generally smaller than a few percent. Except, perhaps, at low Reynolds numbers ($<5*10^5$), where the differences may be as large as 10% for $Re = 1*10^5$ (see Appendix H).

It is convenient and common practice in sailing yacht hydrodynamics to subdivide the viscous resistance into the contributions of hull, keel and/or centre-boards/dagger-boards, keel-bulb (if applicable) and rudder. For conventional sailing yacht configurations the viscous resistance of the hull is about 60–70% of the total viscous resistance. About 20–30% is due to the keel and about 5–15% is due to the rudder.

Hulls and Bulbs
The viscous resistance of hulls and bulbs can be expressed as

$$R_v = \tfrac{1}{2}\, \rho V_b^2 S_{wet} C_{D_v}, \qquad (6.5.4)$$

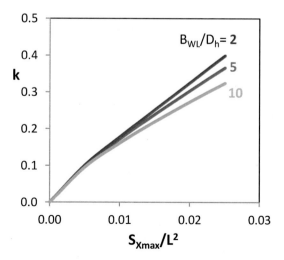

Fig. 6.5.3 *Form factor of sailing yacht hulls as a function of cross-sectional area/ length ratio*

---

[7] International Towing Tank Conference.

with C_{D_v} given by Eq. (6.5.2). Figure 6.5.3 gives the form factor of sailing yacht hulls as a function of the cross-sectional-area/length ratio for several values of the beam/draft ratio B_{WL} / D. The curves are based on a form factor formulation given in Appendix H. As described in more detail in Appendix H, the cross-sectional-area/length ratio is the most important quantity determining the form factor. The beam/draft ratio is another factor, of secondary importance.

As argued also in Appendix H, the viscous resistance of yachts with a wide, transom type stern and flat underbody may be a little smaller than suggested by Fig. 6.5.3, because there is less convergence of the boundary layer flow towards the stern.

The form factor of keel bulbs (or, in general, bodies of revolution) is similar to that given by Fig. 6.5.3 for /B_{WL}/D_h = 2. However, the values of the cross-sectional area along the horizontal axis must be multiplied with a factor 2. The reason is that, due to the presence of the free water surface, the effective cross-sectional area of a hull is twice the geometric cross-sectional area of the underwater part. For keel bulbs this does not apply. The Reynolds number is, of course, to be based on the length of the bulb.

As already discussed in Sect. 5.11 the friction drag is low when there is a significant amount of laminar flow (Fig. 5.11.3). This holds, proportionally, also for the total viscous resistance. In this context it is appropriate to recall from Sect. 5.11 that partial laminar flow is possible for Reynolds numbers up to 5–10 millions,[8] or even higher values, when there is a favourable pressure gradient on the front part of the hull. A prerequisite for laminar flow is, as mentioned, that the surface of the underwater body is sufficiently smooth. The question 'how smooth?' will be addressed later in this section.

Appendages
The viscous resistance of appendages like keels, rudders, centre boards and dagger boards is usually also expressed as in Eq. (6.5.1). However, it is common practice, as in the aeronautical community, to choose the projected area S_{proj} as the reference area S.

Hence, for keels and rudders one writes

$$R_v = \frac{1}{2}\, \rho V_b^{\,2} S_{proj} C_{D_v},\qquad\qquad (6.5.5)$$

with

$$C_{D_v} = (1+k)\, 2\, C_{F_0}\qquad\qquad (6.5.6)$$

The factor 2 in the expression for the viscous drag coefficient is a consequence of the choice of the projected or planform area as the reference area: the equivalent flat plate is now two-sided. The Reynolds number determining the flat plate friction coefficient C_{F_0} is based on the (average) chord length of the lifting surface.

As discussed in some detail in Appendix H, the form factor 'k' for lifting surfaces depends on the thickness/chord ratio t/c in the first place. Other factors of

---

[8] As indicated by Fig. 5.9.1 this corresponds with a boat speed of 2–3 kts for a medium size (L_{WL}=10 m) yacht.

importance are the angle of sweep, the aspect ratio and the lift coefficient. The taper ratio and variation of thickness/chord ratio along the span is usually accounted for by taking the (weighted) average of root and tip values of the chord length and other parameters. Figure 6.5.4 illustrates the variation of 'k' with the thickness/chord ratio for different aspect ratios.

The dependence on the angle of sweep is illustrated by Fig. 6.5.5 and the dependence on lift by Fig. 6.5.6. The figures show that the effects of aspect ratio, sweep angle and lift on the viscous resistance of lifting surfaces are substantial.

As already discussed in Sect. 5.17, camber, as usually applied in dagger boards, and flap deflection (or 'trimtab'), as sometimes applied on keels, are also known to increase the viscous resistance, at least at zero lift. As discussed in Appendix H, the direct effect of camber on the minimum viscous resistance is usually quite small.

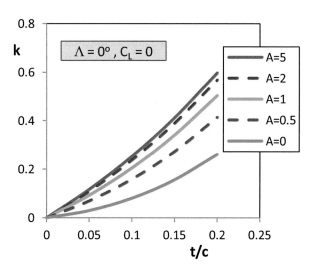

Fig. 6.5.4 *Dependence of the form factor of keels and rudders on aspect ratio*

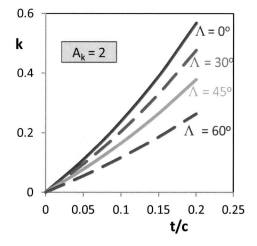

Fig. 6.5.5 *Dependence of the form factor of keels and rudders on angle of sweep*

Fig. 6.5.6 Dependence of form factor of keels and rudders on lift coefficient

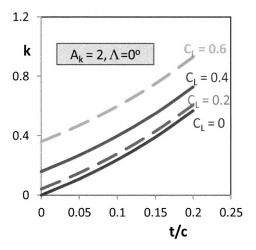

However, the indirect effect on the viscous resistance for a given lift coefficient is significant, as shown in Fig. 6.5.7. The figure illustrates that it pays to apply camber for non-zero lift coefficients, with the optimum amount of camber depending on the level of the lift coefficient. Because the effect of a trailing-edge flap or trimtab is similar to that of camber, it also pays to apply a trimtab on a keel, at least in principle. However, the drag penalty of a trimtab, for a given shift of the lift coefficient at which the viscous drag has its minimum value, is much larger than in the case of camber (see Appendix H). As a consequence, the tab deflection must be kept small and the profit of a lower viscous resistance for a given lift or side force is quite modest. This is illustrated by Fig. 6.5.8, which shows estimated drag polars, based on the analysis given in Appendix H, for 10% thick foils fitted with a 25% chord trimtab.

Fig. 6.5.7 Drag polars of two-dimensional 10% thick foils for different camber/chord ratios

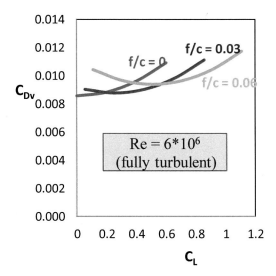

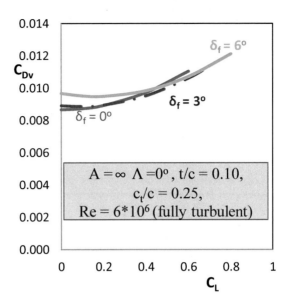

Fig. 6.5.8 *Estimated drag polars of two-dimensional, 10% thick foils with a 25% plain flap*

Because of the smaller streamwise dimensions and the associated lower Reynolds numbers, the possibility of the occurrence op partial laminar flow on appendages like keels, centre-boards, dagger boards and rudders, is larger than for hulls. As shown in Fig. 5.9.1, Reynolds numbers ($< \approx 10^6$) permitting partial laminar flow on keels and rudders can, even for fairly large yachts, occur at all boat speeds below about 10 kts. A prequisite is, again, that the surface is sufficiently smooth.

Additional viscous resistance is caused by the mutual fluid dynamic interference between a hull and its appendages. Even in the case of a properly shaped fillet at the hull-appendage junction, with fully attached flow (Sect. 5.17), there is additional viscous resistance caused by higher supervelocities induced by the hull on the appendages and vice versa. As discussed in Appendix H this additional viscous resistance can be a significant fraction, upto, say, 10% of the total viscous resistance.

Unfortunately there are no simple methods available in the literature for estimating the precise magnitude of the interference effects on the frictional resistance. The precise magnitude can probably be assessed only through Computational Fluid Dynamics. Towing tank tests cannot discriminate between these and other components of resistance.

Effects of Surface Roughness
As already indicated in Sect. 5.11, the effect of surface roughness on the viscous resistance is very important. Surface roughness can affect the viscous resistance in two different ways. The first is that it can cause early transition from laminar to turbulent boundary layer flow. As illustrated by Fig. 5.11.3 there is a substantial increase of the friction drag of a body when the position of transition is moved forward. Secondly, surface roughness can also increase the friction drag of a boundary layer that is already fully turbulent.

It has been found (Young 1989 in Chap. 5) that in a laminar boundary layer there is a certain, critical height, k^*_1, of surface roughness or an excrescence, below which there is no effect on transition. As the size of the excrescence is increased above this value, transition moves forward in the wake of the excrescence. A second critical height, $k^*_2 \approx 2 k^*_1$, is met when the point of transition has reached the position of the excrescence. The boundary layer then exhibits a wedge of turbulent flow with its apex at the excrescence and an apex semi-angle of about 10° (Fig. 6.5.9). In the case of uniform, distributed roughness there will be turbulent flow everywhere behind a line that connects points where the critical roughness height is exceeded.

It appears further that the critical roughness height depends on the local thickness of the boundary layer. This in the sense that the critical roughness height decreases with decreasing local thickness of the boundary layer. It means, that the critical roughness height decreases with increasing Reynolds number and is smaller near the front than at the rear of a body. There is also a strong dependence of the critical roughness height on the shape of the excrescence (Abbott and Von Doenhoff 1949; Young 1989 in Chap. 5). The critical height of protuberances with a relatively large lateral dimension can be smaller by a factor 5 or more than that of protuberances that are slender in the direction normal to the surface.

An empirical expression, due to Van Driest and Blumer (1960) for the critical height of spherical roughness elements distributed in a narrow band can, for incompressible flow, be written as

$$k^*_2/L = 42.6 \, \mathrm{Re}^{-3/4} (U_e/V_b)^{-3/4} (x/L)^{1/4} \qquad (6.5.7)$$

This expression can be considered to give a rough indication of the critical roughness in general. Figure 6.5.10 presents the critical height according to Eq. (6.5.7) as a function of the streamwise position x/L (or x/c), for different values of the Reynolds number, for $U_e/V_b = 1$. The latter is the case for a flat plate at zero angle of attack. In the case of a supervelocity on the body ($U_e/V_b > 1$), the effective Reynolds number is proportionally higher and for a negative supervelocity ($U_e/V_b < 1$) the Reynolds number is proportionally lower.

Also indicated in the figure is the approximate position of transition (x_{tr}) when the surface is smooth. Because x_{tr} depends not only on the Reynolds number but

Fig. 6.5.9 Turbulence wedges behind isolated excrescences

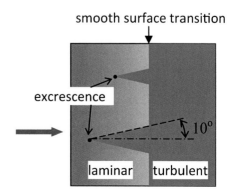

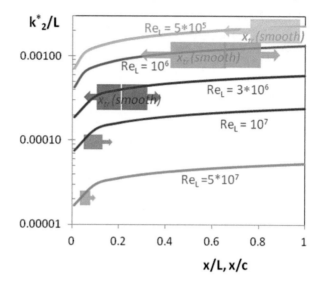

Fig. 6.5.10 *Critical roughness height for boundary layer transition as a function of streamwise position for different Reynolds numbers*

also on other quantities, the local pressure gradient in particular (see Sect. 5.11), its precise location cannot be given for an arbitrary body. As indicated by the arrows, transition occurs more forward when the pressure increases locally in the streamwise direction (decelerating flow). It occurs further downstream when the flow is accelerating.

It is useful to consider the implications of Fig. 6.5.10 in terms of absolute dimensions for the hull and appendages of a sailing yacht. For a hull with a waterline length of 10 m, sailing at a boat speed of about 4 kts (which corresponds with a Reynolds number of about 10^7) the figure implies that there is probably always immediate transition at the bow when the local typical roughness height is larger than about $0.0001 * 10$ m, that is about 1 mm. Because of the uncertainty involved with the shape of the roughness, this could also be as low as 0.2 mm. When the surface is appreciably smoother than that, (say a factor 2, because $k^*_1 \approx 0.5\, k^*_2$), there is a chance of some laminar flow just aft of the bow. For bigger hulls and/or higher boat speeds significant laminar flow is out of the question anyhow.

For a smaller hull and/or lower boat speeds the flow is a little more forgiving. For a waterline length of 5 m and a boat speed of only 2 kts, corresponding with a Reynolds number of about $3*10^6$) the critical roughness height near the bow is about $0.0002*5$ m, that is also about 1 mm. We now may expect laminar flow with an extent of the order of 20 % of the length of the waterline when the roughness height near the bow is smaller than about 0.1–0.5 mm.

For a keel with a chord length of 1 m and a boat speed of 4 kts, corresponding with a Reynolds number of about 10^6, Fig. 6.5.10 implies that the critical roughness height near the leading edge is about $0.0004*1$ m, that is, typically, about 0.4 mm. However, it could also be a factor 5 smaller, that is about 0.08 mm, depending on the specific shape and density of the roughness. So, for a keel with a chord length of 1 m and a boat speed of 4 kts, one may expect laminar flow of significant extent when the roughness height near the leading edge is smaller than about 0.04–0.2 mm. As indicated by Fig. 6.5.10, the critical roughness height increases rapidly aft of the leading edge.

As mentioned before, surface roughness can also increase the skin friction in a boundary layer that is already turbulent. Here also there is a critical or admissible height, k^+, below which there is no effect of roughness on the development of the boundary layer (Young 1989 in Chap. 5). This height corresponds roughly with the height of the viscous, laminar sub-layer mentioned in Sect. 5.11. k^+ can be estimated through the expression (Young 1989 in Chap. 5).

$$k^+/L \approx 5\ \mathrm{Re}^{-1}\sqrt{(C_f/2)},\qquad\qquad (6.5.8)$$

where C_f is the local skin friction coefficient, given by

$$C_f = .0592\ \mathrm{Re}_x^{-1/5}\qquad\qquad (6.5.9)$$

When this admissible height k^+ is exceeded, the skin friction and viscous resistance of a body increases to above the value for a smooth wall.

Figure 6.5.11 presents the admissible roughness height of a turbulent boundary layer on a logarithmic scale, as a function of the streamwise position, for different Reynolds numbers. The curves are representative for surfaces that are densely and uniformly covered with sand grains of a certain size. Like in the case of the critical roughness height for transition, the admissible height of other types of distributed roughness (Young 1989 in Chap. 5), with a different density or texture, can differ by as much as a factor 5. In terms of absolute dimensions, Fig. 6.5.11 implies, that a hull with a waterline length of 10 m, sailing at a boat speed of about 4 kts

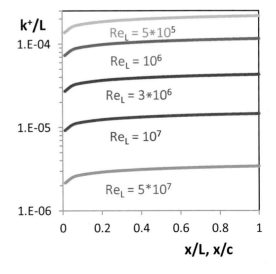

Fig. 6.5.11 *Admissible roughness height for turbulent boundary layers as a function of streamwise position, for different Reynolds numbers*

(which corresponds with a Reynolds number of about 10^7), will not experience any additional resistance due to roughness when the height of the equivalent sand roughness in the bow area is smaller than about 0.00001*10 m=0.1 mm.

For longer hulls and/or higher flow velocities/boat speeds this number will be correspondingly smaller. For example, for a hull of 20 m, sailing at a boat speed of 10 kts the number will be as small as about 0.03 mm. For a keel with a chord length of 1 m and a boat speed of 4 kts, that is $Re \approx 10^6$, the admissible equivalent sand roughness near the leading edge is about 0.00007 * 1 m=0.07 mm.

If these numbers are compared with those for the critical roughness height for transition, it is clear that when a surface is smooth enough to avoid turbulent drag due to roughness it is, in general, also smooth enough to avoid forced transition from laminar to turbulent boundary layer flow.

When, for a given roughness height, the Reynolds number (velocity) is increased, the admissible roughness height is first exceeded near the bow or the leading edge of an appendage. At this point the friction drag coefficient C_{D_F} begins to increase to values above those of a smooth wall. When the Reynolds number become so high that the admissible height is also exceeded at the rear of the body, C_{D_F} becomes independent of the Reynolds number. This is illustrated qualitatively by Fig. 6.5.12 (see also Fig. 5.11.3). When C_{D_F} has become constant, one speaks of 'fully developed roughness flow'. The flat plate friction drag coefficient is then given by

$$C_{F_0} \text{ (rough)} \approx 0.032(k_s/L)^{1/5}, \qquad (6.5.10)$$

where k_s/L is the so-called equivalent sand roughness, that is the roughness of comparable size sand grains.

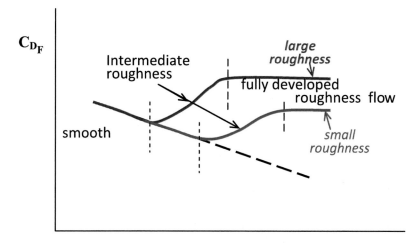

Fig. 6.5.12 *Variation (qualitatively) of friction drag with Reynolds number for a rough surface*

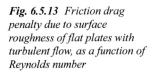

Fig. 6.5.13 *Friction drag penalty due to surface roughness of flat plates with turbulent flow, as a function of Reynolds number*

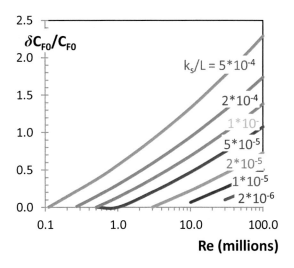

Figure 6.5.13, based on data from Prandtl and Schlichting (1934), gives the drag increase due to roughness of flat plates with turbulent flow as a function of Reynolds number (logarithmic scale) for different values of the relative equivalent sand roughness k_s/L. The figure illustrates that the drag penalty of surface roughness can be huge. For a yacht with a waterline length of 10 m, sailing at 6 kts, which corresponds with a Reynolds number of about 20 million, the friction drag of the hull is about 30% higher than the 'smooth' value when the height of the surface (equivalent sand) roughness is about 0.00002*10 m=0.2 mm. For a 20 m hull sailing at 10 kts, or $Re \approx 70$ million, the same roughness would cause a friction drag penalty of about 60%. For a keel with a chord length of 2 m, operating at 10 kts (or $Re \approx 7$ million) this would also be about 60% (for $k_s/L=0.2/2000=0.0001$). It should, again, be born in mind that the roughness figures are indicative only. As mentioned before, the effect of roughness of a given height depends also on the density and texture and can vary as much as a factor 5.

The practical implications of the numbers given above become clear if we compare them with numbers for the roughness of various types of surfaces and surface finish. The roughness height of smooth surfaces covered with anti-fouling paint is known (Larsson and Eliasson 1996; Candries et al. 2001) to be of the order of 0.04–0.1 mm, depending on the type of paint and the brush or paint-roller. Recalling that the admissible roughness height for the hull of a 20 m yacht at 10 kts is about 0.03 mm, this implies that large yachts may incur a significant drag penalty at high boat speeds if the underwater part is painted without special care. After fine-sanding and polishing, the roughness height of anti-fouling paint can, according to Marchai (2000, p. 272), be as low as 0.02 mm. This is probably small enough to avoid a drag penalty at most if not all conditions. A prerequisite is, of course, that the basic surface, underneath the paint, is sufficiently smooth.

The effect of marine fouling on frictional resistance can be quite dramatic. Barnacles, several millimeters in height, that is with a value of k/L of the order of

0.0001– 0.0003, can triple the friction drag when densely packed. The effect of a layer of algae or weed is probably a little less dramatic, but, with a roughness height of about 0.5 mm (Candries et al. 2001), can still imply doubling of the frictional resistance.

It should, finally, be mentioned that, in principle, there is also a dependence of the viscous resistance on Froude number. The reason is that, due to wave forming, the wetted area of the hull varies with boat speed. In addition, there may also be a (small) effect of the wave form on the velocity distribution.

  To the author's knowledge, methods for estimating these effects do not exist. The only possible way for calculating the effect of Froude number on the viscous resistances is probably through Computational Fluid Dynamics.

6.5.3   Induced Resistance

As discussed in Sect. 5.15 the induced drag of lifting surfaces, like the appendages of a sailing yacht, is a consequence of the downwash generated by the trailing vortices of the lifting surface. For unbounded flows, that is without the presence of a free water surface, the induced drag was seen to be proportional to the lift or side force squared and inversely proportional to the square of the span of the lifting surface (see Eq. (5.15.3)). In a dimensionless form, the induced drag coefficient C_{D_i} can be expressed as, Eq. (5.15.5):

$$C_{D_i} = C_L{}^2/(\pi A e) \qquad (6.5.11)$$

Here, C_L is the lift coefficient, A is the aspect ratio and e is the so-called efficiency factor. The latter was seen to be a function of the spanwise distribution of lift, which, in turn, is a function of the taper ratio, angle of sweep and the aspect ratio (see Fig. 5.15.11).

  As we have seen in Sect. 5.15, the induced drag of a lifting surface can also be written as

$$C_{D_i} = C_L \alpha_i, \qquad (6.5.12)$$

where α_i is the induced angle of attack due to the downwash induced by the trailing vortices. It follows from Eqs. (6.5.11) and (6.5.12) that α_i can be expressed as

$$\alpha_i = C_L/(\pi A e) \qquad (6.5.13)$$

It follows from the discussion in Sect. 6.3 and Appendix E, that for a surface piercing fin at zero heel, the downwash is the same as for a fin of twice the span, with a plane of symmetry at the water surface. 'α_i' can then be written as

$$\alpha_i \approx C_{S_0}/(2\pi A_k e) \qquad (6.5.14)$$

In Eq. (6.5.14), C_{S_0} is the side force coefficient of the 'double model' fin without free surface effect, as given by Eq. (6.3.2). 'A_k' is the aspect ratio of the submerged part of the fin and e is the efficiency factor of the fin plus its mirror image (see Sect. 5.15, Fig. 5.15.11 and Appendix B).

With the free surface effect taken into account, the induced resistance of a surface piercing fin can be written as

$$C_{D_i} = C_S \alpha_i, \tag{6.5.15}$$

with α_i given by (6.5.14). C_S is the side force coefficient in the presence of the free surface, as described in Sect. 6.3 and Appendix E. That is, for an elliptic distribution of circulation, it can be expressed as Eq. (6.3.9).

Figure 6.5.14 gives the induced resistance (coefficient) as a function of the side force (coefficient) squared for surface piercing fins of different aspect ratios with an elliptic distribution of circulation.

The figure illustrates that the presence of the free surface causes a higher induced resistance, in particular for low aspect ratio fins. For an aspect ratio as low as 0.5, a value representative for shallow draft cruising yachts, the increase of the induced resistance due to the free surface is as much as 60%. The effect of the free surface is, practically, negligible (<5%) for aspect ratios of 5 and higher.

We have seen in Sect. 5.15 that, without a free water surface, the induced drag of a lifting surface of given aspect ratio has a minimum when the spanwise distribution of circulation is elliptic. It was also indicated that an elliptic distribution of circulation can be obtained for specific combinations of sweep angle and taper ratio (see Fig. 5.15.9).

Fig. 6.5.14 *Induced resistance of surface piercing fins with an elliptic distribution of circulation, as a function of the side force squared, for different aspect ratios*

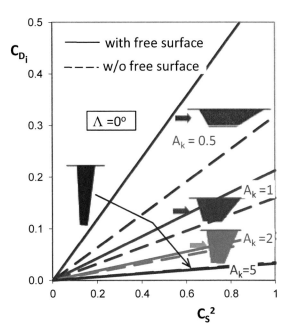

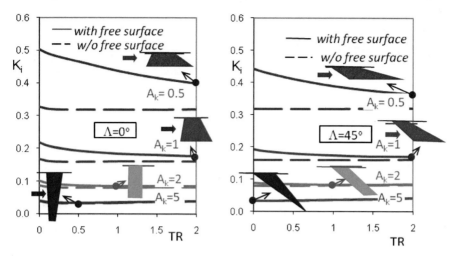

Fig. 6.5.15 *Induced drag factor of surface piercing fins of different aspect ratio as a function of taper ratio*

It is interesting to consider the effect of taper and sweep angle on the induced resistance. For this purpose it is useful to rewrite Eqs. (6.5.15) and (6.5.14) as

$$C_{D_i} = K_i C_S^2 \qquad (6.5.16)$$

The factor K_i is known as the induced drag factor and, for a surface piercing fin, can be written as

$$K_i = 1/(F_{FS} 2\pi A_k e), \qquad (6.5.17)$$

where F_{FS} (<1) is the free surface factor introduced in Sect. 6.3.

Figure 6.5.15 presents the induced drag factor as a function of taper ratio for different aspect ratios and two angles of sweep. It can be noticed that, for small aspect ratios (say $A_k < 2$), the induced resistance is the lowest for high taper ratios (TR > 2) and that sweep appears to have a favourable effect, at least for small aspect ratios and small taper ratios. This is also shown by Fig. 6.5.16, which gives the induced drag factor as a function of the angle of sweep for different taper ratios and aspect ratio 1.

The variation of the induced drag with sweep and taper is the combined result of two effects. One is the well known dependence of the induced drag on the spanwise distribution of circulation (Sect. 5.15). The other, which is dominant for low aspect ratios, is the decrease of the loss of side force due to the free surface effect with increasing taper ratio and angle of sweep.

For high aspect ratios, when the free surface effect becomes negligible, the minimum induced drag is, again, realized for TR ≈ 0.5 for Λ = 0° and TR ≈ 0.1 for Λ = 45° (See Fig. 6.5.15). The reader may recall from Sect. 5.15 that these are values corresponding with an elliptic distribution of circulation (Fig. 5.15.9).

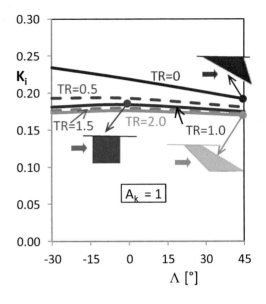

Fig. 6.5.16 *Induced drag factor of surfacing piercing fins as a function of sweep angle, for different taper ratios*

Effect of the Hull

The relation (6.5.15) holds also in the presence of a hull. Therefore, the effect of the latter on the induced drag can be estimated by incorporating the effect of the hull in the expressions for the side force coefficient C_S and the induced angle of attack α_i (see Sect. 6.3 and Appendices C and E).

Figure 6.5.17 presents the induced drag factor as a function of the (relative) hull depth D_h/b for fin keels with a constant chord, zero sweep and aspect ratios

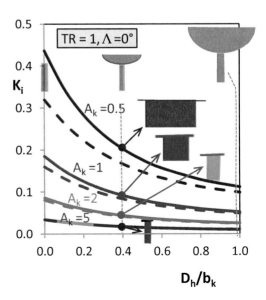

Fig. 6.5.17 *Induced drag factor for fin-hull configurations as a function of hull depth/fin span ratio*

ranging from 0.5 to 5. The figure illustrates that the effect of the hull is large, in particular for low aspect ratio fins.

The dependence on taper ratio of the induced drag factor of a fin keel with zero sweep, attached to a hull, (see Fig. 6.5.18), is similar to that of a surface piercing fin without a hull (Fig. 6.5.15). It appears that, for low aspect ratios, a large taper ratio is also favourable in the presence of a hull. The main difference is a lower level of the induced drag, in particular for low aspect ratios, due to the increased effective span of the fin in the presence of the hull.

Figure 6.5.19 illustrates the dependence on the angle of sweep for fin keels with an aspect ratio of 1 and taper ratios between 0 and 2, attached to a hull with a hull depth/keel span ratio of 0.4. It can be noticed that the effect of sweep is almost negligible, except for very small taper ratios.

Bulbs, Winglets, etc.

The effect of bulbs and winglets on the induced drag of lifting surfaces has already been described in Sects. 5.17 and 5.15, respectively. Their main effect was seen to be an increase of the effective span. This implies, of course, a decrease of the induced drag. It applies also to a bulb or winglets attached to the fin keel of a sailing yacht. Because the position of a bulb or winglets is usually relatively far from the water surface, the effect of the latter on the hydrodynamic functioning of bulbs and winglets is, in general, negligible. This means that the effect of a bulb or winglets on the induced resistance of a fin keel can be represented by substituting the appropriate value (see Sects. 5.15 and 5.17) for the efficiency factor e in Eq. (6.5.17).

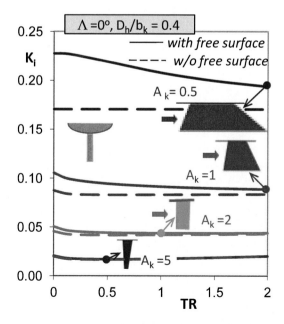

Fig. 6.5.18 *Induced drag factor of fin keels of different aspect ratio, attached to a hull, as a function of taper ratio*

Fig. 6.5.19 *Induced drag factor of fin keels of different aspect ratio, attached to a hull, as a function of the angle of sweep*

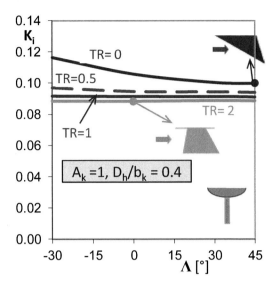

Rudders
Because rudders are lifting surfaces also, they also produce induced drag. The mechanism is basically the same as that of keels. This means that the effects of the free surface and shape parameters like aspect ratio, sweep and taper are, qualitatively, the same as for keels. The main difference is that the rudder is positioned in the downwash field produced by the keel (see Sect. 6.3). The effect of this on the induced resistance of the rudder can, approximately, be taken into account by rewriting Eq. (6.5.15) as

$$(C_{D_i})_{rudder} = (C_S)_{rudder}\{(\alpha_i)_{rudder} + (\alpha_\varepsilon)_{keel}\} \qquad (6.5.18)$$

Here, α_ε represents the (average) downwash induced by the keel at the position of the rudder. An impression of its possible magnitude can be obtained from Fig. 6.3.14. Because of the downwash induced by the keel, the induced drag of a rudder in the proximity of the keel is usually much larger than would be the case if the rudder were outside the influence of the keel. In addition, the downwash induced by the rudder itself is usually larger than that produced by a keel for the same side force. There are two reasons for the this. As already mentioned in Sect. 6.3, one is the smaller span of the rudder. The other is the (narrow) gap between the top of the rudder and the hull, which reduces the effective span.

On the other hand, there is also a compensating effect of the rudder on the induced drag of the keel. This is caused by the fact that the rudder, when producing a side force, causes some upwash at the position of the keel which reduces the induced drag experienced by the keel. The effect can be expressed as

$$(C_{D_i})_{keel} = (C_S)_{keel}\{(\alpha_i)_{keel} + (\alpha_\varepsilon)_{rudder}\}, \qquad (6.5.19)$$

where $(\alpha_\varepsilon)_{rudder}$ (<0) represents the upwash induced by the rudder at the position of the keel. Because the keel is not in close proximity with the trailing vorticity of the rudder, $(\alpha_\varepsilon)_{rudder}$ is, in general, quite small. The net effect of the interaction between keel and rudder is, in general, a larger induced resistance for a given total side force for the keel-rudder combination when the rudder produces part of the side force. In other words, it is advisable, form the point of view of hydrodynamic resistance, to keep the side force on the rudder as small as possible.

Summary

Summarizing the discussion on induced resistance we can now conclude the following:

- As for all lifting surfaces, the span or, for a given area, the aspect ratio of a keel is the most important parameter; the induced resistance is proportional to the square of the lift or side force and inversely proportional to the square of the span
- The effect of the hull is to increase the effective span through the mechanism of lift carry-over
- Due to the pressure relief effect of the free water surface there is a significant difference between sailing yacht keels and, for example, aircraft wings in the dependence of induced resistance on taper and angle of sweep. For aircraft wings the minimum induced drag for a given lift is obtained for all combinations of sweep and taper that lead to an elliptical spanwise distribution of circulation. This is also the case for high aspect ratio fin keels, for which the free surface effects are small. For keels with a low aspect ratio, the loss of side force (not circulation!) near the water surface is an important factor for the induced drag at a given level of side force: the larger this loss of side force, the higher the induced drag for a given side force.
- The loss of side force due to the free surface effect decreases with increasing taper ratio and sweep angle. Keels with 'inverse' taper and/or high sweep angles have, therefore, less induced drag for a given side force when the aspect ratio is small.
- The rudder of a sailing yacht also causes induced resistance. The mechanism is basically the same as for keels. There is a difference in the sense that the rudder operates in the field of downwash induced by the keel and the keel experiences some upwash induced by the rudder when the latter produces a side force. Because of the smaller span of the rudder it is advisable to keep the side force on the rudder as small as possible. The latter is also a matter of sail trim, as we will see in Chaps. 7 and 8.

6.5.4 Wave Making Resistance

Wave Making

Every yachtsman is, of course, familiar with the phenomenon of wave making. A sailing yacht, or, for that matter, any vessel moving through or near a water surface, is known to generate waves. These waves are known to exhibit a certain pattern. As already discussed in Sect. 5.19, they are not to be confused with the wind driven waves that form a seaway.

Fig. 6.5.20 *Basic mode of deformation of the water surface around a moving vessel*

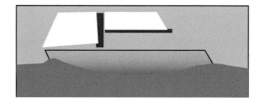

As discussed in some detail in Sect. 5.19, a basic mechanism underlying the generation of surface waves is the deformation of the water surface caused by the velocity and pressure perturbations that are generated by the moving vessel. Basically, this surface deformation is of the form sketched in Fig. 6.5.20:

- The water level rises where the local fluid velocity is low, that is at the bow and at the stern
- The water level sinks where the local fluid velocity is high, that is amidships

The main parameter governing the magnitude of the deformation of the water surface was seen to be the Froude number Fr, defined by (Sect. 5.19)

$$\mathrm{Fr} = V_b/\sqrt{(Lg)}, \tag{6.5.20}$$

where V_b is the speed of the vessel (boat speed), L is a characteristic length such as the length of the waterline and g is the gravitational acceleration. More precisely, the magnitude of the surface deformation was seen to be proportional with Fr^2 (see Eq. (5.19.14)).

Because the surface deformations travel with the same speed as the boat, they give the appearance of travelling waves. That, however, is not the full story. As we have seen in Sect. 5.19, surface waves prefer to travel at a propagation speed c_w that is a function of the wavelength λ_w:

$$c_w = \sqrt{(g\lambda_w/2\pi)} \tag{6.5.21}$$

The other way around is also true: If a surface disturbance travels at a speed V_b it tends to generate waves with a wave length λ_w equal to

$$\lambda_w = 2\pi V_b^2/g \tag{6.5.22}$$

Equations (6.5.21) and (6.5.22) represent the situation that the direction of motion of the wave front is perpendicular to the wave front itself. When (the normal to) the wave front is at an angle ψ with the direction of propagation (See Fig. 6.5.21) Eq. (6.5.22) takes the form

$$\lambda_w = 2\pi V_b^2\cos^2\psi_w/g \tag{6.5.23}$$

This can also be expressed in terms of the Froude number:

Fig. 6.5.21 *Relation between wave length, wave angle and boat speed*

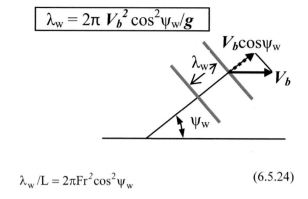

$$\lambda_w = 2\pi \, V_b^2 \cos^2 \psi_w / g$$

$$\lambda_w / L = 2\pi \mathrm{Fr}^2 \cos^2 \psi_w \qquad (6.5.24)$$

While Eqs. (6.5.23/24) says something about the relation between wave length and wave angle for a given boat speed it does not say anything about the wave pattern. Some information about the pattern of waves generated by a ship can be obtained by considering the ship from a large distance, as a moving point. This was first accomplished by Lord Kelvin (Kelvin 1904). It leads to a wave pattern of the type shown in Fig. 6.5.22. It is characterized by a series of divergent waves radiating out from the point that represents the position of the ship and a system of transverse waves travelling behind the point. Perhaps the most striking aspect is that the whole pattern is contained within two straight lines that start from the point representing the position of the ship and make an angle of about 19.5° with the line of motion of the ship. Moreover, this angle turns out to be independent of boat speed (See Appendix I for a (partial) explanation).

As illustrated by Fig. 6.5.23 the real wave pattern about a ship as seen from a large distance is very similar. Near the ship, the pattern is a little more complicated and depends also on details of the ship's geometry (Lewis 1988b). A schematic diagram is given in Fig. 6.5.24. Two systems of diverging and transverse waves are indicated. One originating from the bow and one from the stern. Additional, similar wave systems may be formed at locations where there are rapid changes in the geometry of the hull. The diverging waves are roughly contained by envelopes of the Kelvin-type described above.

Fig. 6.5.22 *Kelvin ship-wave pattern*

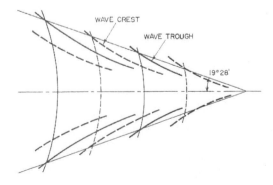

Fig. 6.5.23 *Kelvin type ship-wave pattern in reality. (Source: http://www.wikipedia. com)*

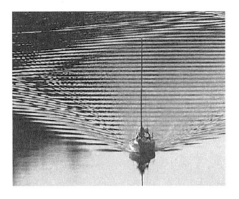

Fig. 6.5.24 *Schematic of near-field wave system of a ship*

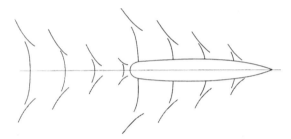

The distance between successive wave crests roughly follows the relations (6.5.22), (6.5.23) and, hence, increases with boat speed. A particular situation arises when the wavelength of the transverse waves is equal to the length L_{WL} of the waterline of the ship. In this situation the crests and troughs of the basic mode of deformation of the water surface (Fig. 6.5.20) almost coincide with the crests and troughs of the bow wave system. As a consequence, they strengthen each other. In addition, the bow and stern systems of waves get into some state of resonance in which they strengthen each other further.

It is easily understood that this happens when λ_w in Eq. (6.5.22) is equal to L_{WL}, that is at a boat speed V_b^* given by[9]

$$V_b^* = \sqrt{(gL_{WL}/2\pi)} \approx 1.25 \sqrt{L_{WL}}   \quad [m/s] \qquad (6.5.25)$$

Note that when L_{WL} in Eq. (6.5.25) is expressed in meters and the gravitational constant 'g' in m/s², V_b^* is in m/s. For the boat speed to be expressed in knots and L_{WL} in meters, the constant 1.25 in Eq. (6.5.25) should be replaced by about 2.43, or about 1.34 when L_{WL} is given in feet. As an example, Eq. (6.5.25) gives a critical boat speed of about 7.3 kts for a boat with a length of waterline of 9 m.

The critical boat speed can also be expressed in terms of Froude number. This gives (see Eq. (6.5.24)):

---

[9] This boat speed (V_b^*) is sometimes called the critical boat speed or the 'hull speed'.

$$\text{Fr}^* = \sqrt{(1/2\pi)} \approx 0.40 \qquad (6.5.26)$$

Some, but less severe, interference between the basic surface deformation mode and the bow and stern wave systems takes place also at lower boat speeds, when the wave length is a full fraction (such as $\frac{1}{2}, \frac{1}{3}, \frac{1}{4}$, etc.) of the length of the waterline. This corresponds with Froude numbers of about 0.28, 0.23 and 0.20, respectively.

When the wavelength is such that the bow wave system creates a trough where the stern wave system creates a crest, the two wave systems will attenuate each other. This is the case when the ratio of waterline length to the wave length is an uneven multiple of $\frac{1}{2}$ (such as 3/2, 5/2 etc.). The latter corresponds with Froude numbers of 0.33 and 0.25, respectively.

The interference between the bow and stern waves and the basic mode of surface deformation is illustrated by Fig. 6.5.25. Shown, schematically, is the longitudinal distribution of crests and troughs of the transverse waves for different Froude numbers.

It should be noted that the wave height does not represent the proper scaling with respect to boat length but follows the scaling with Froude number given by Eq. (5.19.14).

It should be mentioned that, in practice, the characteristic Froude numbers mentioned above are a little higher than those indicated. The reason is that most sailing

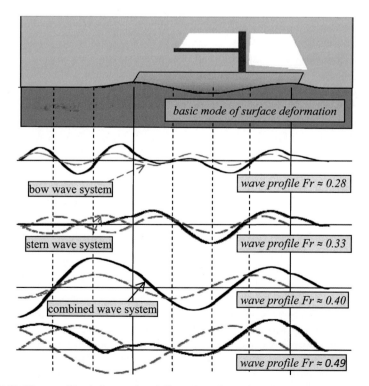

Fig. 6.5.25 *Wave profiles (schematic) at different Froude numbers (wave height not to scale)*

yacht hulls have overhangs at the stern as well as the bow. This causes the effective length of the waterline when sailing to be a little larger than when just floating.

An important effect of the wave forming is that it causes (additional) changes in the vertical position ('sinkage') and the pitch attitude of a yacht with varying boat speed. We have seen already in Sect. 6.4 that sinkage develops with increasing boat speed due to the downward directed vertical hydrodynamic force acting on the bottom of the hull (see Fig. 6.4.7). Additional sinkage is caused by the fact that, for most Froude numbers, the wave pattern exhibits a trough about amidships (see Fig. 6.5.25). Because this is close to where, for most yachts, the longitudinal distribution of displacement has its maximum, the yacht sinks deeper into the water ('digs a hole for itself').

For Froude numbers beyond 0.4 a large difference develops between the water level at the bow and that at the stern (see also Fig 6.5.25). This causes the yacht to acquire a significant pitch angle (bow-up) at high Froude numbers. This, as already mentioned in Sect. 6.4, has a significant effect on the hydrodynamic vertical force and pitching moment (see Figs. 6.4.8 and 6.4.10). It is the main cause of the increase of the vertical hydrodynamic force with Froude number for Froude numbers above 0.4 shown by Fig. 6.4.11.

As discussed in some detail in the next paragraph, sinkage and pitch attitude are also important for the hydrodynamic resistance, the wave-making resistance in particular.

Resistance

The most important effect of wave-making is that the deformation of the water surface causes additional pressure drag called wave-making resistance. This wave-making resistance can be related to the energy that is required to sustain the system of waves generated by the yacht.

With the preceding paragraph in mind, it is not surprising that the Froude number is an important parameter for the wave-making resistance. Other quantities of importance are related to the geometry and the attitude of the yacht. In principle any hydrodynamic disturbance caused by the flow of water about a surface vehicle causes wave-making resistance. It will also be clear, that a hydrodynamic disturbance close to the water surface will generate more wave-making resistance than a disturbance that is situated far away from the surface (deeply submerged submarines do not have wave-making resistance!).

The displacement volume of the hull of a yacht is usually the most important geometrical quantity for the wave-making resistance. This is understandable because the displacement volume of the hull is relatively large and surface piercing. The importance of parts that are farther away from the water surface and/or smaller, such as appendices, is far less because they cause much less disturbance of the water surface. However, it has been found (Beukelman and Keuning 1975; Keuning and Sonnenberg 1998a) that the displacement volume of (large) keels is not always negligible, in particular under heel when a keel gets closer to the water surface.

Because the side force and circulation acting on a keel, rudder and hull also cause perturbations of the flow and the free surface, they are another source of wave-making resistance. Surprisingly, this kind of wave-making resistance seems to be completely disregarded in the world of naval architecture.

We have seen in Sect. 5.19 that gravity forces, like the wave-making resistance, scale like

$$F_g \div \rho g$$

per unit of volume. This means that the total wave-making resistance $R_{w\nabla}$ due to the displacement volume of a moving vessel scales like

$$R_{w\nabla} \div \rho g \nabla, \tag{6.5.28}$$

where ∇ is the displacement volume. Equation (6.5.28) can also be written as

$$R_{w\nabla} \div V, \tag{6.5.29}$$

where V is the displacement weight.

Equation (6.5.29) suggests that a suitable definition of a dimensionless coefficient for the wave-making resistance due to volume displacement is

$$C_{Rw\nabla} \equiv R_{w\nabla} / V \tag{6.5.30}$$

An alternative definition is given in Appendix J.

Because of the wave-interference phenomena described above, the wave-making resistance coefficient $C_{Rw\nabla}$ is a function of Froude number and the geometrical properties of the hull. It has been found (Larsson and Eliasson 1996; Keuning and Sonnenberg 1998a) that among the geometry parameters that describe the shape of the hull (in the upright position) the displacement/length ratio ∇/L_{WL}^3 or $\nabla^{1/3}/L_{WL}$ is the most important for the wave-making resistance coefficient. Of secondary importance are quantities like B_{WL}/D_h, the prismatic coefficient C_P and the longitudinal position of the centre of buoyancy LCB.

Fig. 6.5.26 *Effect of displacement/length ratio on the wave-making resistance due to displacement*

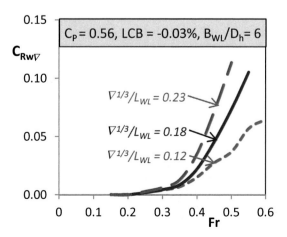

Fig. 6.5.27 *Effect of beam/*
draft ratio on the wave-making
resistance due to displacement

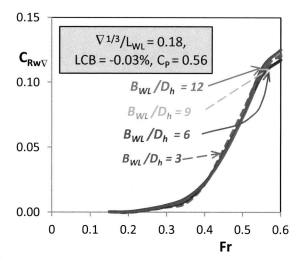

Fig. 6.5.28 *Effect of*
prismatic coefficient on the
wave-making resistance due
to displacement

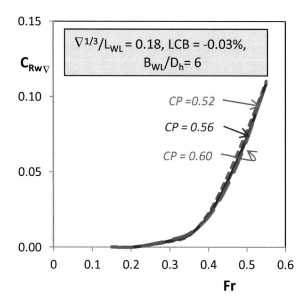

Figures 6.5.26, 6.5.27, 6.5.28 and 6.5.29 provide an impression of the dependence of the wave-making resistance on the displacement/length ratio, the beam/draft ratio, the prismatic coefficient and the longitudinal position of the centre of buoyancy. The figures have been constructed according to the empirical formulae for the 'residual' resistance given in Keuning and Sonnenberg (1998a). As mentioned earlier the residual resistance is defined as the difference between the total resistance as measured in a towing tank and the estimated frictional resistance at zero heel and leeway. In the method of Keuning and Sonnenberg (1998a)

Fig. 6.5.29 *Effect of position of the longitudinal centre of buoyancy on the wave-making resistance due to displacement*

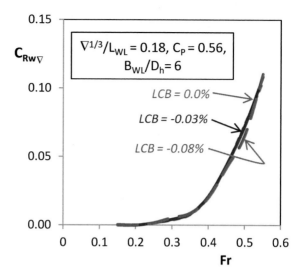

the frictional resistance is estimated on the basis of Eq. (6.5.3) without the use of a form factor. However, the Reynolds number is based on 70% rather than 100% of the length of the waterline. This implies a friction drag coefficient that is some 10% higher. Effectively, this corresponds with a viscous resistance with a form factor of about 0.1 for a Reynolds number based on 100% of the length of the waterline. In any case, it should be borne in mind that the residuary resistance is, to some extent, contaminated with viscous pressure drag and not completely identical with the wave-making resistance.

This, however, is not important for the present purpose of illustrating the dependence of the wave-making resistance due to displacement on the main geometrical properties of the hull of a sailing yacht.

Figures 6.5.26, 6.5.27, 6.5.28 and 6.5.29 illustrate that the displacement/length ratio is, by far, the most important of the geometrical parameters of the hull for the wave-making resistance. While the variation of the wave-making resistance over the range of practical values of the displacement/length ratio considered is more than 50%, the variation with the prismatic coefficient and the position of the longitudinal centre of buoyance is of the order of 10–15%.

It is further important to note that the effects of sinkage and variation of pitch angle due to wave forming are included in Figs. 6.5.26, 6.5.27, 6.5.28 and 6.5.29. This in the sense that the results are representative for a situation in which there is no external pitching moment and the yacht is free to adopt the appropriate sinkage and pitch attitude. In Fig. 6.5.28 this is reflected in the fact that the wave-making resistance at high Froude numbers is relatively low for high values of the prismatic coefficient. This is caused primarily by reduced sinkage due to fuller bow and stern sections. In Fig. 6.5.29 it is reflected in the lower wave-making resistance at high Froude numbers for the most rearward position (−0.08%) of the longitudinal centre of buoyancy.

There is, in principle, also a (weak) dependence of the wave-making resistance due to displacement volume on the angle of yaw or leeway. This is due to the fact that the perturbation velocity caused by the volume of the hull increases slightly with leeway. As discussed in more detail in Appendix J, the effect is proportional to the square of the angle of leeway. This means that, in general, it is quite small and negligible, except, perhaps, in case of large angles of leeway and a high displacement/length ratio.

The effect of leeway on the wave-making resistance due to displacement volume is not to be confused with wave-making resistance due to side force. The latter is due to the perturbation of the water surface that is caused by the bound vorticity (circulation) and the trailing vorticity of the keel and rudder.

As also discussed in Appendix J, the wave-making resistance due to side force R_{wS} can be expressed as

$$R_{wS} = \tfrac{1}{2}\,\rho V_b^2 S_k C_{RwS},$$ (6.5.31)

with

$$C_{RwS} = K_w C_S^2,$$ (6.5.32)

Note that the coefficient C_{RwS} is, like the induced resistance coefficient (Eq. (6.5.16)), proportional to the square of the side force coefficient C_S. It is, through the factor K_w, also proportional to the size of the keel, in particular the ratio b_k/L_{WL}, and inversely proportional to the slope $\dfrac{dC_S}{d\lambda}$ of the lift (or side force) curve and the keel aspect ratio (see Appendix J for further details). Because the lift curve slope also decreases with decreasing aspect ratio (see Sect. 6.3) the wave-making resistance due to side force varies more rapidly with the aspect ratio of the keel than the induced resistance.

Little is known about the absolute magnitude of the wave-making resistance due to side force and its variation with Froude number. Figure 6.5.30 gives a (tentative) comparison of the relative magnitudes of the wave-making resistance due to side force and the induced resistance. The analysis in Appendix J suggests that at a Froude number of 0.35 it is of the order of 10–35% of the induced resistance. The latter is of the order of 20% of the total resistance for an 'average' cruising yacht in upwind conditions. This would mean that, for an 'average' cruising yacht with a keel aspect ratio of about 1, the wave-making resistance due to side force is of the order of 2–7% of the total resistance. For yachts with a very low aspect ratio keel this would be significantly more. For (racing) yachts with a high aspect ratio keel, the wave-making resistance due to side force is probably negligible for most practical purposes.

Other Effects of Wave Making
In principle, wave-making has effects on all the hydrodynamic forces acting on a sailing yacht. However, we have seen already in Sect. 6.3, that the effect of Froude number (that is wave making) on the rate of change $\dfrac{dC_S}{d\lambda}$ of the side force coeffi-

Fig. 6.5.30 *Estimated relative importance of wave-making resistance due to side force and induced resistance as a function of keel aspect ratio*

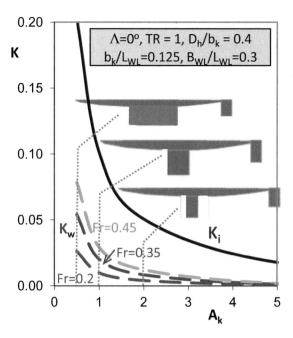

cient with angle of leeway is negligible for most practical purposes, at least for low

Froude numbers. To the author's knowledge that is also the case for the hydrodynamic heeling moment. The effect on the frictional (or 'viscous') resistance can, it seems (Keuning and Sonnenberg 1998a), also be disregarded.

There is, however, a measurable effect of Froude number on the resistance-due-to-side-force (Keuning and Sonnenberg 1998a). It is the author's impression that the major part of this is due to the wave-making resistance due to side force discussed above. Possibly, there is also a (small) effect on the induced resistance due to the fact that the free surface no longer acts as a flat mirror surface for the bound and trailing vorticity (Sect. 6.3) but, rather, as a 'warped mirror surface' that reduces the effective span of a keel plus hull.

Another significant effect of wave making is that on stability. As will, for example, be discussed in more detail in Sect. 6.8, a wave crest amidships in beam seas reduces the effective beam at the waterline of a yacht. Because beam, as we have seen in Sect. 6.2, is the most important parameter for the static lateral stability of a yacht this implies a reduced stability.

6.5.5 Added Resistance in Waves

Introduction
Every yachtsman will have experienced that a sailing yacht, or any type of ship for that matter, can suffer from a significant, or even dramatic, loss of boat speed in a seaway. The phenomenon is of a dynamic nature, depending on the periodic motion

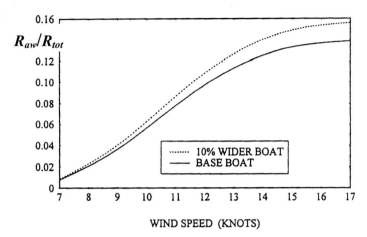

Fig. 6.5.31 *Added resistance in waves as a function of wind speed for an IACC type of yacht. (Milgram 1998 in Chap. 1)*

of the ship and the frequency of encounter with the waves. It is, of course, also a function of the wave height.

The loss of boat speed in waves is a consequence of an apparent increase of resistance when a yacht sails in waves. This apparent, time-averaged, additional resistance is known as 'added resistance in waves'. As already mentioned in Sect. Decomposition of Drag, the added resistance of a yacht sailing in waves can be as much as 15–30% of the total resistance in calm water.

As an example Fig. 6.5.31 presents the added resistance in waves as a function of wind speed for IACC type of yachts sailing to windward (Milgram 1998 in Chap. 1).

The added resistance is given as a fraction of the total resistance. The wave conditions underlying the data in the figure are representative of fully developed, long-crested (uni-directional) seas.

Basic Mechanisms

The mechanisms underlying the phenomenon of added resistance can be described as follows. As a consequence of the encounter with waves, a yacht responds with periodic motions in all 6 degrees of freedom (See Fig. 3.1.7). This dynamic behaviour is known as the sea-keeping characteristics of a yacht. We will address these in some further detail in Sect. 6.8.

For the resistance, the motions in pitch and heave are the most important (Fig. 6.5.32). The dominant effect is that of the pitching motion (Keuning 1998; Lewis 1989). However, the effect of rolling may not always be negligible, in particular under heel and in oblique seas.

As a consequence of the orbital motion of water in the waves (Sect. 5.19, Fig. 5.19.2) and the motions of the yacht, the under-water part of the hull and its appendages experience periodic variations in:

- the speed and direction of the flow of water relative to the (component of the) yacht (Fig. 6.5.33)

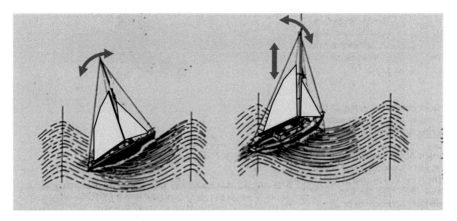

Fig. 6.5.32 *Yacht motion in a seaway*

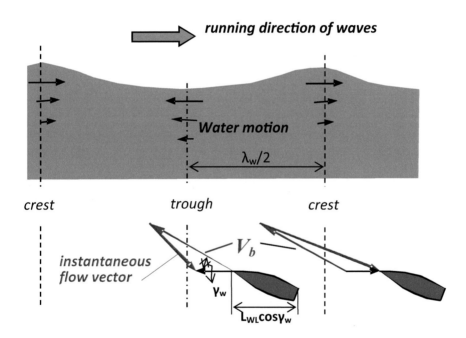

Fig. 6.5.33 *Variation of instantaneous flow vector in waves*

- the component of the gravity force in the direction of motion, which can be positive or negative (Fig. 6.5.34)
- displacement volume and wetted surface

We have seen in the preceding sections that the hydrodynamic forces on a vehicle are proportional to the square, or even a higher power, of the flow velocity. This implies that if the instantaneous boat speed varies between $V_b + \Delta V_b$ and $V_b - \Delta V_b$,

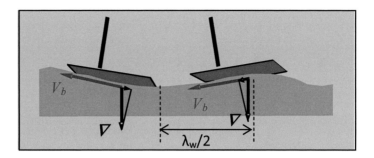

Fig. 6.5.34 *Variation in waves of the component of the gravity force in the direction of sailing*

with V_b as the time-averaged value, the hydrodynamic forces vary between $c_*(V_b + \Delta V_b)^2$ and $c_*(V_b - \Delta V_b)^2$, where 'c' is a constant that depends on the shape, size and attitude of the yacht. This means that the time-averaged value of the hydrodynamic forces is approximately proportional to $V_b^2 + \Delta V_b^2$ rather than just V_b^2. (The average of squares is larger than the square of the average). This is a common mechanism for the increase in waves of all the hydrodynamic forces, resistance as well as side force.

The variation in side force under heel and in oblique seas also causes an additional increase in the time-averaged induced resistance because of its quadratic dependence on lift or side force. Moreover, the side force varies also due to the periodic variation of the angle of attack of keel and rudder that is caused by rolling, swaying, yawing and the orbital motion of the water particles in waves.

The most important contribution to the added resistance of a yacht in waves is caused by additional wave making. Due to the heaving and pitching motion of the yacht, triggered by the waves, the hull generates, periodically, additional waves causing additional wave-making resistance. The mechanism is very similar to that of a body that plunges up and down in the water and radiates energy away from itself through the waves that are generated.

In oblique seas and under heel there may also be added induced resistance due to orbital and rolling motion induced, time-varying loads on the keel. Unfortunately there is hardly any information on this aspect of added resistance in the literature. Marchai (2000, p. 658) mentions, however, that it was found in towing tank tests on models of a 5.5-m class yacht that *'the resistance increase associated with rolling of quite large amplitude was in the order of about 2% only'*. This suggests that the added induced resistance due to rolling motion in waves is probably not very large.

For wavelengths that are small compared to the dimensions of the yacht, the diffraction of waves by the hull may also give a (small) contribution to the added resistance.

In the following we will consider yacht motions and the associated added resistance in waves in some detail. Rolling motions will be addressed in further detail in the context of dynamic stability and sea keeping (Sect. 6.8).

Yacht Motions The motions of a yacht in a seaway are the result of the excitation forces due to the running surface waves, the buoyancy forces, the inertia forces and damping. It is often instructive to describe the motions of a ship in waves in terms of the dynamic mass-spring-damper system introduced in Sect. 5.18 (see Figs. 5.18.3 and 5.18.4). The equilibrium of forces in the heaving motion in waves can, for example, be formulated as

$$(m + m')\ddot{Z} + b\dot{Z} + kZ = F_0 \sin \omega_e t, \tag{6.5.33}$$

where Z is the vertical displacement. The right hand side of the equation represents the external forces due to the encounter with waves. F_0 is the amplitude of the external force component in the vertical direction and ω_e the angular frequency of encounter with the waves. The role of the restoring (spring) force kZ is played by the hydrostatic (buoyancy) forces. The hydrodynamic damping forces are represented by the term $b\dot{Z}$. The in-phase hydrodynamic inertia forces (Sect. 5.18) are contained by the added mass term $m'\ddot{Z}$.

  Damping causes the amplitude of the 'natural' motion to decrease with time (Sect. 5.18, Fig. 5.18.2). For sailing yachts the damping is such that the 'natural' oscillations of heaving and pitching motions disappear within only a few cycles. Because the hydrodynamic forces increase with boat speed (squared), the damping of the motions of a yacht also increases with boat speed.

  Resonance phenomena can occur when the frequency of the encounter of the yacht with waves (which causes the external, excitation forces) is about equal to the natural frequency of a 'mode of motion' such as the pitching motion. This can result in violent ship motions and a correspondingly high added resistance.

The heaving and pitching motions of a ship are, in general, coupled. That is, a heaving motion induces pitching and vice versa. For ships with a large asymmetry between fore and aft, such as most sailing yachts, the coupling is stronger than for ships which are almost symmetric fore and aft. For keel yachts under heel and in oblique waves there is also a coupling between the heaving and rolling motions and between the pitching and rolling motions.

Excitation Forces

The hydrodynamic excitation forces and moments caused by waves acting on a yacht are primarily caused by the orbital motion of the water particles in the waves (Sect. 5.19), relative to the position, speed and heading of the yacht (Lewis 1989). For the heaving motion the magnitude of the excitation force is proportional to the wave height. For the pitching and rolling motions the magnitude of the excitation moments is proportional to the effective maximum wave slope. For the pitching motion the effective maximum wave slope is given by $\theta_w \cos \gamma_w$, where $\theta_w = 2\pi(\zeta_w / \lambda_w)$, λ_w is the wavelength and ζ_w is the wave amplitude (half wave height). γ_w, the wave incidence, is the angle between the course of the ship and the direction from which the waves are running.[10]

---

[10] In the literature of naval architecture it is common practice to define the wave direction by the symbol μ ($= 180° - \gamma_w$). For head waves we have $\gamma_w = 0°$, $\mu = 180°$ and for stern waves $\gamma_w = 180°$, $\mu = 0°$. In most, but not all conditions γ_w will be equal to the true wind angle γ.

For the rolling motion the effective maximum wave slope is given by $\theta_w \sin\gamma_w$.

The ratio between the length (of the waterline) of the boat and the wavelength is another factor of importance for the magnitude of the external excitation forces. It is easily understood from Figs. 6.5.33 and 6.5.34 that a ship will be most easily excited into oscillatory motion by waves when the (projected) length of the ship is about equal to or slightly larger than half the wavelength. For the pitching motion this means that this mode of motion is most easily triggered when $L_{WL}\cos\gamma_w \cong \frac{1}{2}\lambda_w$. For the rolling motion the corresponding condition is $B_{WL}\sin\gamma_w \cong \frac{1}{2}\lambda_w$. On the other hand there will be hardly any excitation in pitch when $\lambda_w \ll L_{WL}\cos\gamma_w$ and hardly any excitation in roll when $\lambda_w \ll B_{WL}\sin\gamma_w$. Similar conditions apply to the heaving motion (See Appendix L.1 for further details).

Figures 6.5.35, 6.5.36 and 6.5.37 provide a qualitative impression of the dependence of the excitation forces and moments on wave length and wave incidence. The excitation levels are given in terms of 'wave excitation factors'. As discussed in some more detail in Appendix L.1 these represent the excitation level for a given

Fig. 6.5.35 *Wave excitation factor for pitching motion as a function of wave incidence and wavelength (qualitatively, zero heel, constant wave height)*

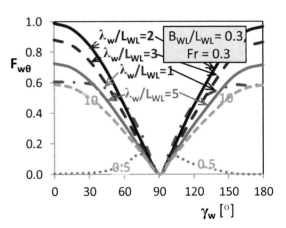

Fig. 6.5.36 *Wave excitation factor for heaving motion as a function of wave incidence and wavelength (qualitatively, zero heel, constant wave height)*

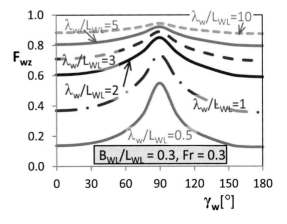

Fig. 6.5.37 *Wave excitation factor for rolling motion as a function of wave incidence and wavelength (qualitatively, zero heel, constant wave height)*

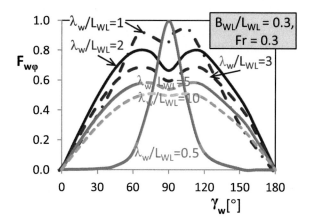

wave height normalized by the maximum excitation for that wave height. The excitation factors are shown as a function of the wave incidence γ_w for different values of the wavelength/boat-length ratio. It is emphasized that only qualitative significance is to be attached to these figures. The precise dependence of the wave excitation forces on wave length and incidence probably depends strongly on the specific configuration of a yacht. The amount of fore/aft asymmetry and the appendage configuration, for example, are believed to be of particular importance, but have not been taken into account in the construction of these figures (see Appendix L.1).

For the pitching motion (Fig. 6.5.35) it is important to notice is that the wave excitation attains a maximum for head and stern waves with a wave length corresponding with $\lambda_w/L_{WL} \cong 2$. There is, at least for hulls with fore/aft symmetry, no excitation in pitch for waves that are coming in from abeam ($\gamma_w \cong 90°$). It can further be noted that there is a small asymmetry in the curves with respect to the condition of beam seas ($\gamma_w = 90°$). This is caused by the effect of boat speed, which, however, appears to be small (Lewis 1989).

For the heaving motion (Fig. 6.5.36) the maximum excitation is seen to occur for $\gamma_w \cong 90°$, at least for relatively short wavelengths. For long waves the effect of wave incidence is quite small.

The corresponding figure for the rolling motion is given in Fig. 6.5.37. Here it can be noticed that the excitation is large for $\lambda_w/L_{WL} \cong 1$ and a fairly wide range of angles around $\gamma_w \cong 90°$. An exception is formed by relatively short waves ($\lambda_w/L_{WL} \cong 0.5$), when there is a rather sharp peak at $\gamma_w \cong 90°$.

'Spring' (Buoyancy) Forces, Damping Forces and Added Mass

In terms of the mass-spring-damper system, the spring constant k for the heaving motion of a sailing yacht (see Eq. (6.5.33)) is proportional to the water plane area S_{WP}.

The damping forces are associated with the out-going waves that are generated by the hull of the yacht through the plunging motion. The damping coefficient is proportional to the product of the energy contained by the out-going waves due to the heaving motion and boat speed (Lewis 1989). For a yacht in a significant seaway, the magnitude of these gravitational damping forces is of the same order as the inertia forces and much larger than that of the frictional forces. The latter represent the only mechanism for the damping in the case of a fully submerged non-lifting body.

The added mass type hydrodynamic forces acting on a semi-submerged body in oscillatory motion are also much larger than those of a fully submerged object. This is also caused by the outgoing waves which imply larger inertia forces in the fluid surrounding the body. For a sailing yacht in heaving motion the added mass can be of the order of 1.8 times the real mass (Lewis 1989).

Because gravitational free surface effects are dominated by inertia effects at high frequency, the damping goes to zero at very high frequencies. It also goes to zero for very low frequencies when there is hardly any generation of additional waves. The added mass approaches a constant value (about equal to that of a fully submerged body with the same length and volume) for very high frequencies.

In case of the pitching motion or the rolling motion one has to consider a 'rotational version' of the dynamic system of Sect. 5.18. For the pitching motion this can be written as

$$(I_{yy} + I_{yy'})\ddot{\theta} + b_\theta \dot{\theta} + k_\theta \theta = M_0 \sin \omega_e t \qquad (6.5.34)$$

Here, the spring constant k_θ is proportional to the product of the displacement weight with the longitudinal metacentric height. The damping forces $b_\theta \dot{\theta}$ are associated with the outgoing waves generated by the pitching motion. It has been found (Keuning 1998; Lewis 1989) that the pitch damping increases with increasing displacement/length ratio and increasing beam/draft ratio of the hull. The quantity $I_{yy'}$ represents the moment of inertia around the y-axis of the added mass. It has been found (Keuning 1998) that for sailing yachts $I_{yy'}$ is of the order of 0.7 times the moment of inertia I_{yy} of the real mass.

In case of the rolling motion the spring constant is proportional to the product of the displacement weight with the lateral or transverse metacentric height. The hydrodynamic damping of the rolling motion is caused primarily by the side force and associated rolling moment due to the keel. For hulls with a near-semicircular cross-section the contribution to the hydrodynamic damping of the rolling motion is almost negligible (Marchai 1983a). This is also the case for the added mass effects of the hull.

For all modes of motion of a sailing yacht the situation is a little more complex than as modeled by the simple mass-spring-damper system of Sect. 5.18. The main difference is that, in reality, the different modes of motion are coupled and that the damping coefficients and added terms are also a function of the frequency, boat speed and amplitude.

The pitching and rolling motions of a sailing yacht are also influenced by the aerodynamic forces acting on the sails (Skinner 1982). In upwind conditions the aerodynamic damping of the pitching motion is caused by the increase of the aerodynamic drag of the sails due to the pitching oscillation (see also Sect. 7.10). In case of the rolling motion the variation of the aerodynamic heeling force due to rolling, in addition to the hydrodynamic side force on the keel, causes a large amount of damping, at least in upwind conditions of sailing. The effect of the sails on the heaving motion is negligible, except, possibly, under large angles of heel.

Natural Frequencies

The natural frequency (see Sect. 5.18) of the heaving motion of a yacht can be understood as follows: When the yacht, in still water, is suddenly pushed downward by some vertical, instantaneous force acting through the centre of gravity, it will start popping up and down with a certain 'natural' frequency. This is called the natural frequency (f_{0z}) in water of the heaving motion. Similarly, a yacht will begin pitching up en down in a certain frequency when it is suddenly excited by a moment of force around the pitching axis. This is called the natural frequency ($f_{0\theta}$) of the pitching motion. Needless to say that the rolling motion also has its natural frequency ($f_{0\varphi}$).

Figure 6.5.38 gives an impression of the order of magnitude of the frequency of the natural heaving motion as determined according to formulae given in Lewis (1989) and reproduced in Appendix K. Shown is the frequency in cycles/second as a function of the length of the waterline for several values of the length/displacement ratio. The natural frequencies of the motions of a ship are determined primarily by the mass properties and the dimensions, in particular the 'spring constant k (Sect. 5.18). For the heaving motion the most important quantities are the displacement or mass and the waterplane area (See Appendix K). The figure illustrates that the frequency of the natural heaving motion of a ship decreases with increasing size and mass.

Figure 6.5.39 illustrates the effect of the beam/length ratio: the frequency increases with increasing beam/length ratio.

The most important quantities determining the frequency of the natural pitching motion are the length of the ship and the mass moment of inertia around the pitching axis (See also Appendix K). Another quantity of importance is the longitudinal metacentric height (gm_L).

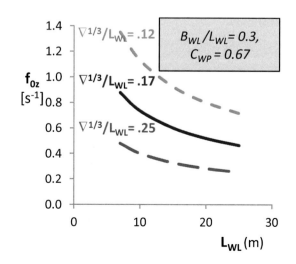

Fig. 6.5.38 *Frequency of the natural heaving motion of sailing yachts (estimated) as a function of length and length/displacement ratio*

Fig. 6.5.39 Frequency of
the natural heaving motion
of sailing yachts (estimated)
as a function of length and
beam/length ratio

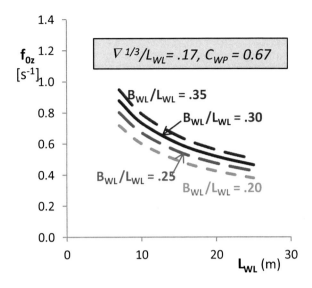

Fig. 6.5.40 Frequency of the
natural pitching motion of
sailing yachts (estimated) as a
function of length and longitudinal gyradius

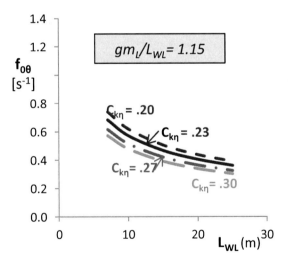

Figure 6.5.40 gives the frequency of the natural pitching motion as a function of the boat length for several values of the, so-called, longitudinal radius of gyration or gyradius (the coefficient C_k is the ratio between the gyradius and the length of the waterline). As discussed in more detail in Sect. 2.5 and Appendix K, the radius of gyration of an object is the distance from the axis of rotation at which the concentrated mass of the object has to be positioned in order to have the same moment of inertia as the actual object. Note that the figure applies to a gm_L/L_{WL} ratio of 1.15, which seems to be a representative value for most sailing yachts (Larsson and Eliasson 1996).

Frequency of Encounter
The frequency f_e at which a yacht meets (regular) waves is a function of boat speed, the wavelength and the direction of the waves. It is easily verified that

$$f_e = (c_w + V_b \cos\gamma_w)/\lambda_w \qquad (6.5.35)$$

In this expression c_w is the propagation velocity of the waves, and, as before, V_b is boat speed, λ_w is the wavelength and γ_w is the wave direction angle. Recalling from Sect. 5.19 that the wave propagation speed is a function of the wavelength:

$$c_w = \sqrt{(g\lambda_w/2\pi)}, \qquad (6.5.36)$$

Equation (6.5.35) can also be written as

$$f_e = \sqrt{\frac{g}{2\pi\lambda_w}} + (V_b/\lambda_w)\cos\gamma_w \qquad (6.5.37)$$

or

$$f_e = \sqrt{\frac{g}{L_{wL}}} \left[\frac{1}{\sqrt{\dfrac{2\pi\lambda_w}{L_{wL}}}} + \frac{Fr}{\lambda_w/L_{wL}}\cos\gamma_w \right] \qquad (6.5.38)$$

Figure 6.5.41 presents the frequency of encounter as a function of boat speed for several values of the wave length and a wave direction angle $\gamma_w = 45°$. Note that this wave direction angle is roughly representative for cruising yachts beating upwind.

The variation with the wave direction angle for a wave length of 25 m is shown by Fig. 6.5.42. It can be noticed that the frequency of encounter increases linearly with boat speed and decreases with increasing wavelength and wave direction angle.

Fig. 6.5.41 *Frequency of encounter with regular waves as a function of boat speed and wave length*

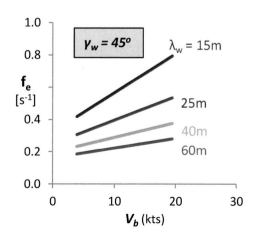

Fig. 6.5.42 *Frequency of encounter with regular waves as a function of boat speed and wave length*

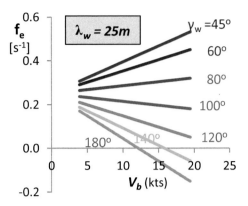

Note that the frequency of encounter becomes negative for following and quartering waves and (very) high boat speeds. In this situation the yacht is overtaking the waves.

It is also worth noting that the frequency range in Fig. 6.5.40 corresponds roughly with that of the natural frequency of the heaving motion of heavy to moderately heavy sailing yachts (see Fig. 6.5.38). This is also true for the pitching motion (Fig. 6.5.40).

Hydrodynamic Unsteadiness
As discussed in Sect. 5.18 the so-called reduced frequency k_f, defined by

$$k_f = 2\pi fL/(2V) = \pi f \sqrt{(L/g)}/Fr$$

is a measure of the unsteadiness of the flow about an oscillating body. With the frequency range known it is possible to estimate the reduced frequencies k_f of the pitching and heaving motions. For $f \cong 0.5$ (typical natural frequency of the pitching motion of a sailing yacht) and a Froude number of 0.35 we get $k_f \cong 2.5$ for a waterline length of 10 m. This means that the flow about the hull of a sailing yacht in heaving and pitching oscillations will, in general, be of a strong unsteady nature. This implies, see Sect. 5.18, that there is a significant phase lag or time lag, through damping, between the periodic variation of the flow about the hull and the motion. At the same time there will be a substantial 'added mass' effect.

For modern yachts, with a characteristic keel chord length of, say, 1/5th to 1/8th of the length of the waterline, we have $k_f \cong 0.7$ for the keel. This means that the flow about the keel of a yacht in pitching motion will also exhibit a considerable time lag, but also a reduction of the amplitude of the hydrodynamic forces (Sect. 5.18). The latter at least as long as the flow about the keel is attached. For rudders, with their smaller longitudinal dimensions, the unsteady flow effects will be somewhat smaller.

For classical yachts, with long keels with a chord length of the order of the length of the waterline, the flow about the keel in a seaway will exhibit a larger time lag. However, the associated reduction of the amplitude of the hydrodynamic forces may not be necessarily larger than for modern yachts with fin keels because of the effect of aspect ratio on the lift deficiency factor (See Sect. 5.18, Eq. (5.18.8)).

Synchronisms

As already mentioned, the ratios between the frequency of encounter (f_e) of a yacht with waves and the 'natural' frequency of a mode of motion is very important for the response of the yacht to its encounter with the waves. This follows also from the discussion in Sect. 5.18. There we have seen that the oscillatory motion of an object can become catastrophic when the natural frequency of the motion and the frequency of the excitation forces is about the same. For the motion of ships in waves this means that the heaving, pitching and/or rolling motion can become quite violent when the frequency of encounter with waves coincides with the natural frequency of the motion. This is called a condition of synchronism or resonance. The conditions (boat speed, wave length, wave direction angle) of resonance can be determined by equating the frequency of encounter f_e with the natural frequencies (f_{0z}, $f_{0\theta}$, $f_{0\varphi}$, respectively).

For the heaving motion of sailing yachts this is, in terms of dimensionless quantities, illustrated by Fig. 6.5.43. The figure presents the Froude number Fr_s (dimensionless boat speed) of synchronistic encounter as a function of the relative wavelength for several values of the length/displacement ratio and 'average sailing yacht' values of the beam/length ratio and waterplane coefficient. Note that the wave direction angle (45°) is representative for close-hauled sailing to windward.

Bearing in mind, (see the paragraph on *excitation forces,* Fig. 6.5.36, above) that heaving motions are most effectively excited for values of $\lambda_w / L_{WL} > \cong 1.5$, Fig. 6.5.43 indicates that, at Froude numbers (0.3–0.4) representative for sailing to windward in high wind speed, heavy yachts ($\nabla^{1/3}L_{WL} > \cong$ 0.2 to 0.25) are more sensitive for resonant heaving motion[11] than light yachts. This, because for the heavy yachts resonance occurs at relatively large wave lengths. The latter is potentially also more dangerous because of the associated larger wave heights.

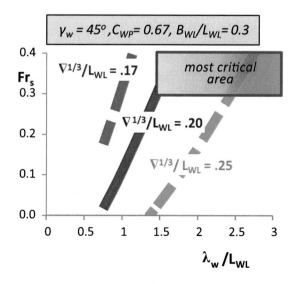

Fig. 6.5.43 *Froude number for resonance conditions of the heaving motion as a function of wavelength and length/displacement ratio*

---

[11] An impression of the effect of beam/length ratio can be obtained from Fig. 6.5.39.

Fig. 6.5.44 *Froude number for conditions of resonance for the pitching motion of sailing yachts as a function of wavelength and gyradius*

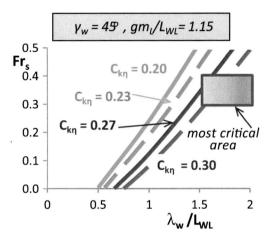

Figure 6.5.44 presents similar information for the pitching motion. It appears that, for a Froude number of about 0.35, resonance in pitch may occur at relatively short wavelengths ($\lambda_w/L_{WL} \cong 1$ to 1.5) for yachts with a relatively small gyradius ($C_k \cong 0.20$–0.23). For yachts with a relatively large gyradius ($C_k \cong 0.27$ to 0.3) this happens at somewhat larger wavelengths ($\lambda_w/L_{WL} \cong 1.5$ to 2).

We have also seen in Fig. 6.5.35 that yachts are most easily excited in pitching oscillation for wave lengths corresponding with $\lambda_w/L_{WL} \cong 2$. This means that yachts with a large gyradius are most sensitive for pitching oscillations in upwind conditions.

Particularly violent motion may be expected when the natural frequencies of the heaving and pitching motions coincide. From the preceding figures and discussion it follows that this will be the case for moderately heavy yachts ($\nabla^{1/3}/L_{WL} \cong 0.2$) with an average gyradius ($C_{k\eta} \cong 0.25$).

Figure 6.5.45 illustrates the influence of the wave direction angle on the Froude number for resonance of the pitching motion for yachts with an average radius

Fig. 6.5.45 *Froude number for conditions of resonance for the pitching motion of sailing yachts as a function of wavelength and wave direction angle*

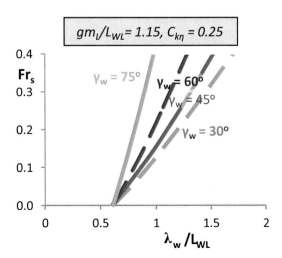

of gyration ($C_{kn} \cong 0.25$). For large wave direction angles the resonance conditions
are seen to shift to values of λ_w / L_{WL} well below 1, which do not hurt as much as
values $> \cong 1$. Besides, the wave excitation for pitching goes to zero for $\gamma_w \to 90°$
(Fig. 6.5.35). This means that resonance conditions become less important for large
wave direction angles. This is also the case for the heaving motion, albeit probably
somewhat less so because of a smaller effect of wave incidence on the wave excitation factor for the heaving motion (Fig. 6.5.35).

A particular form of synchronism may occur in following seas ($\gamma_w \cong 180°$) when the
boat speed is equal to the wave propagation speed. In terms of Eq. (6.5.35) this is
the case when

$$f_e = (c_w + V_b \cos\gamma_w)/\lambda_w = 0 \qquad\qquad (6.5.39)$$

This is the condition known as *surfing*. As illustrated by the right hand side of
Fig. 6.5.34 the gravity force then has a component in the direction of sailing for a
prolonged period of time. The result is an exceptionally high boat speed, usually
well beyond the critical boat speed (or hull speed) that corresponds with $Fr \cong 0.4$.

Figure 6.5.46 shows the Froude number for surfing conditions as a function of wave
length. The figure illustrates that the boat speed (Froude number) required for prolonged surfing increases with the wavelength.
    It should be noted that surfing is not likely to occur when the wavelength is sig-
nificantly smaller than about 1 boat length ($\lambda_w / L_{WL} < 1$). The reason is, as may be
apparent from Fig. 6.5.34, that a yacht cannot take full advantage of the wave slope
for small wavelengths.

Added Resistance
It will be clear from the discussions in the preceding paragraphs, that the motion and
the associated added resistance of a sailing yacht in waves is determined partly by
the characteristics of the sea state (spectrum of wavelengths, wave height and slope

Fig. 6.5.46 *Froude number for surfing conditions as a function of wave length and wave direction*

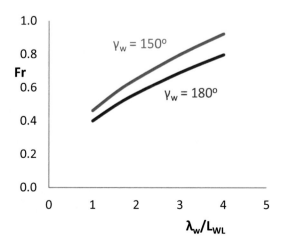

and the direction of wave propagation relative to the direction of sailing) and partly by the properties (dimensions and mass) of the yacht.

It is generally assumed (Keuning 1998) that the added resistance of a sailing yacht is mainly caused by the heaving and pitching motions. However, there is no doubt that there can also be an effect of rolling motion, in particular under heel and in oblique seas. Unfortunately it seems that, as mentioned earlier, there is hardly any quantitative information on the effect of rolling on added resistance. The main reason for this is the fact that it is not possible to simulate the behaviour of sailing yachts in oblique seas in a towing tank. However, there is, as already mentioned, some indication (Marchai 2000, p. 658) that the effect of rolling is small.

The most important quantities for the heaving and pitching motions of a yacht were seen to be the (relative) wavelength λ_w/L_{WL}, wave height and wave direction (γ_w), as well as the displacement/length ratio and the longitudinal moment of inertia (or the gyradius) of the yacht. Factors of secondary importance are the beam/length ratio and the prismatic coefficient.

Since the added resistance in waves (R_{aw}) appears to consist mainly of additional, periodic wave-making resistance, caused by the displacement volume of the hull, it scales, in principle, also like (see Sect. Wave Making Resistance)

$$R_{aw} \div \rho g \nabla \qquad (6.5.40)$$

When elaborated in more detail (see Appendix L.2) it can be found that a more specific scaling for the added resistance in regular waves is

$$R_{aw} \div \rho g \nabla \ (\zeta_w/T_e)^2/(L_{WL}C_P), \qquad (6.5.41)$$

where ζ_w is the wave amplitude, T_e the period of encounter and C_P the prismatic coefficient of the hull. However, it appears (Keuning and Sonnenberg 1998a) that, surprisingly, a commonly adopted form of scaling is

$$R_{aw} \div \rho g \, h_w^2 \, L_{WL} \qquad (6.5.42)$$

This leads to a non-dimensional coefficient for the added resistance defined by

$$C_{Raw} = R_{aw}/(\rho g \, h_w^2 \, L_{WL}) \qquad (6.5.43)$$

It is to be expected that C_{Raw} as defined by Eq. (6.5.43) will be a function of the quantities determining the natural heaving and pitching motions and boat speed or Froude number. The latter in particular because boat speed determines the frequency of encounter. It is further to be expected that the added resistance in waves will be relatively high when the natural heave and pitching motions exhibit a large amount of damping.

As an example, Fig. 6.5.47 presents the characteristics of the heaving and pitching motion and the added resistance of two different yacht models. Shown are the results of towing tank tests (Gerritsma and Keuning 1987) in regular waves (head seas,

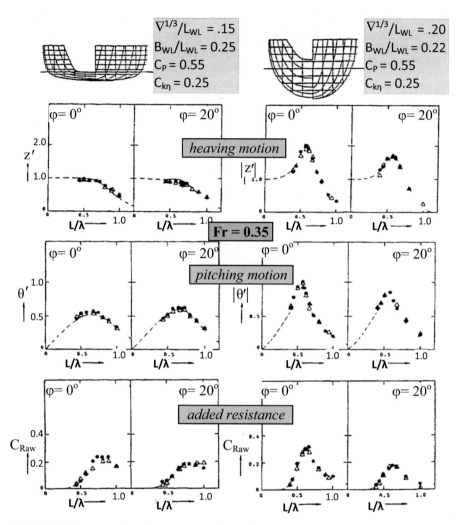

Fig. 6.5.47 *Heaving and pitching motion and added resistance in regular waves (head seas) for two different yacht models. (Gerritsma and Keuning 1987)*

$\gamma_w = 0°$),[12] at a Froude number of 0.3. The results are given in terms of the dimensionless amplitudes[13] $|z'|$ and $|\theta'|$ of the heaving and pitching motions and the added resistance coefficient as a function of the boat length/wave length ratio L_{WL}/λ_w.

---

[12] In a towing tank it is usually not possible to test at wave direction angles other than 0°. However, the results are believed to be representative for non-zero wave direction angles also if the ratio LWL/l_w is interpreted as $LWL\cos\gamma_w/l_w$ and the effective wave slope is chosen as $2\pi(\zeta_w/l_w)\cos g_w$ (Gerritsma and Beukelman 1972).

[13] Defined by $|z'| = |z|/\zeta_w$ and $|\theta'| = |\theta|/(2\pi \zeta_w/L_{WL})$, where $|z|$ and $|\theta|$ are the amplitudes of the heaving and pitching motion, respectively.

The main difference between the two models is that one model represents a fairly heavy displacement yacht ($\nabla^{1/3}/L_{WL} = 0.20$) and the other a light displacement yacht ($\nabla^{1/3}/L_{WL} \cong 0.15$). It can be noticed that the peak amplitude of both the heaving and pitching motion as well as the added resistance is larger for the heavy yacht. It can also be noticed that for the light displacement yacht the added resistance has its maximum at $L_{WL}/\lambda_w \cong 0.8$. For the heavy yacht this occurs at $L_{WL}/\lambda_w \cong 0.65$. According to the formulae for the resonance conditions, given earlier in this section, resonance in heave at a Froude number of 0.3 is to be expected at $L_{WL}/\lambda_w \cong 1.25$ for the light displacement yacht and at $L_{WL}/\lambda_w \cong 0.6$ for the heavy yacht. The corresponding number for the pitching motion of both yachts is $L_{WL}/\lambda_w \cong 0.7$. The fact that the resonance conditions for pitch and heave almost coincide for the heavy yacht explains the larger peak amplitudes of the latter.

The numbers also explain why the maximum added resistance occurs at a somewhat higher value of the boat length/wave length ratio L_{WL}/λ_w for the light displacement yacht. They also illustrate that the effect of the pitching motion is dominant.

In a real seaway the added resistance depends, of course, on the contents of the wave spectrum (Sect. 5.19). It has been found (Lewis 1989) that the response and the associated added resistance of a ship in a seaway containing waves of different lengths and heights can be determined as the weighted sum of the response in the regular waves over all the wavelengths contained by the spectrum of the seaway. It has also been found (Keuning and Sonnenberg 1998a) that the added resistance depends primarily on the significant wave height ($h_{w\,\frac{1}{3}}$) and the average wave period (T_{wave}) of the spectrum. As in regular waves, the most important properties of the yacht were found to be the displacement/length ratio and the longitudinal radius of gyration. Quantities like the beam/length ratio and the prismatic coefficient are of less importance, at least for longer wave length. The time-averaged added resistance in a realistic seaway is usually expressed like Eqs. (6.5.42) and (6.5.43), but with the wave height h_w replaced by the significant wave height $h_{w\,\frac{1}{3}}$. I.e.:

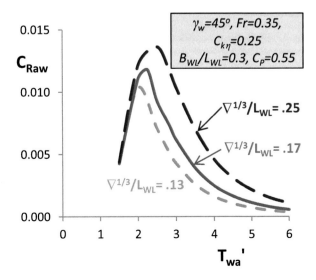

Fig. 6.5.48 *Variation of the average added resistance (coefficient) in upwind conditions in average ocean waves as a function of the normalized average wave period, for different length/ displacement ratios*

Fig. 6.5.49 *Variation of the average added resistance (coefficient) in upwind conditions in average ocean waves as a function of the normalized average wave period, for different values of the gyradius*

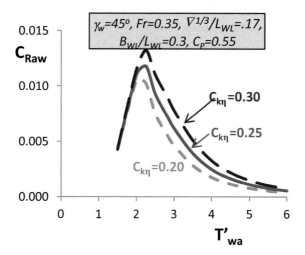

$$R_{aw} = \rho g\, h_{w_{1/3}}^{2}\, L_{WL} C_{Raw} \qquad (6.5.44)$$

Figure 6.5.48 gives an impression of the dependence of the coefficient C_{Raw} of the time-averaged added resistance on the dimensionless average wave period $T'_{wa} = T_{wa}\sqrt{(g/L_{WL})}$. (Note that $T'_{wa} \cong T_{wa}$ for a boat length of about 10 m). The figure has been constructed on the basis of the data contained by Keuning and Sonnenberg (1998a). It is representative for upwind sailing in average ocean waves described by the so-called Brettschneider wave spectrum (Sect. 5.19) but does not contain any effect of rolling or added induced resistance. The figure illustrates that the added resistance increases with increasing displacement/length ratio. It is also clear that the peak in the added resistance shifts towards longer wave periods when the displacement/length ratio increases. The increase of the added resistance and the shift to longer wave periods is also the case for increasing values of the gyradius (Fig. 6.5.49).

The latter two effects are understandable when it is realized that larger displacement/length ratios and larger gyradii lead to longer natural periods of the heaving and pitching motions. It should also be noted that the added resistance increases with boat length because of the scaling (6.5.44) and shifts also to longer wavelengths because $T_{wa} = T'_{wa}\sqrt{(L_{WL}/g)}$.

Figure 6.5.50 gives an impression of the variation of the added resistance as a function of the wave direction angle and Froude number. The dependence on the wave direction angle can be seen to be quite strong, with the added resistance going to zero in beam seas ($\gamma_w = 90°$). It is somewhat questionable whether the latter will also be the case when a yacht experiences significant rolling motion.

The effect of Froude number is seen to be modest.

There is little information on the effect of side force and heel angle on the added resistance. Reportedly (Gerritsma and Keuning 1987), the effect of heel is not very large and the effect of side force (angle of leeway) hardly measurable.

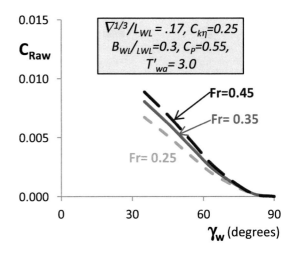

Fig. 6.5.50 *Variation of the average added resistance (coefficient) as a function of wave direction angle, for different Froude numbers*

Effect of Wave Spectrum

Because of the strong dependence on wave height and wavelength or wave period, the added resistance depends also, quite strongly, on the wave spectrum of a sea state. In this respect, the ratio λ_w/L_{WL} between wave length and boat length is of prime importance. In the preceding paragraphs we have seen that the added resistance in upwind conditions usually attains a maximum for $\lambda_w/L_{WL} \cong 1.5$. This means that yachts are particularly sensitive for sea states with an average wave period corresponding with $\lambda_w/L_{WL} \cong 1.5$ and a relatively large wave height at that wavelength. In terms of average wave period this means $T'_{wa} \cong 2.5$ or

$$T_{wa} = T'_{wa}\sqrt{(L_{WL}/g)} \cong 2.5\sqrt{(L_{WL}/g)}, \text{ (seconds)} \qquad (6.5.45)$$

For a yacht with a length of about 10 m this means an average wavelength of about 10–15 m or an average wave period of about 2.5 s. For a 20 m yacht the corresponding figures are about 25 m and about 3.5 s.

In terms of wave spectra (see Sect. 5.19) these are very low values. This is illustrated by Fig. 6.5.51, reproduced from Keuning (2006). The figure compares the wave spectra of different types of sailing environments (ocean, North Sea, estuary) for a wind speed of 5 Bft. Note that the energy density $S(\omega_w)$ is given as a function of wave length/boat length ratio for a boat length of 40 ft rather than the average wave period or frequency. Note also that the estuary type of environment has the highest energy density (relative wave height) at small wavelengths. Also shown as a function of the wavelength is the coefficient C_{Raw} of the added resistance in regular waves as calculated for a 40 ft yacht (IMS-40 design).

Comparing the added resistance curve with the wave spectra it is immediately clear from the amount of overlap between the resistance and wave spectra curves that the yacht will experience the highest added resistance in the estuary environment. This is confirmed by Fig. 6.5.52, also from Keuning (2006), which gives the

Fig. 6.5.51 *Comparison of wave spectra for different sailing environments with the added resistance properties of a 40 ft yacht. (Keuning 2006)*

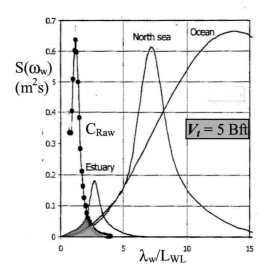

calculated added resistance as a function of the true wind angle for a wind speed of 5 Bft in the three sailing environments considered.

Wings in Waves

A particular form of added resistance in waves (positive or negative!) may occur when a sailing yacht has a keel (and/or a rudder) with winglets. When sailing upwind, the combined result of the orbital motion and the motion of the yacht is that the wings experience upwash on the rising side of a wave and downwash on the falling side (Fig. 6.5.53). This causes an unsteady lift force L with an average positive component in the direction of sailing. The net result is an unsteady thrust force that is to be added to (or, rather, subtracted from) the induced resistance. When sailing with the waves, the relative motions are, unfortunately, such, that an additional resistance is experienced.

The phenomenon just described is basically the same as that described in Sect. 5.18 for a lifting surface in heaving motion. For winged keels it was first noted by the author (Slooff 1985 in Chap. 5) and considered in more detail by Milgram

Fig. 6.5.52 *Average added resistance in waves of a 40 ft yacht in different sailing environments. (Keuning 2006)*

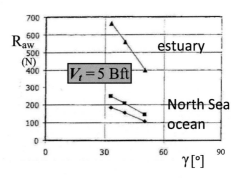

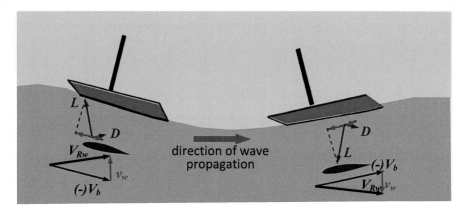

Fig. 6.5.53 *Forces in waves on the winglets of a winged keel*

(1998 in Chap. 1). The latter has calculated that the dynamic thrust of winglets in waves can be as much as 10–20% of the viscous resistance of the winglets.

6.5.6 Propeller Resistance

A final component of hydrodynamic resistance to be addressed is the resistance of a sailing yacht's propeller when sailing and the engine is not running. Depending on the size and type of the propeller, its resistance can be considerable.

The resistance of a propeller consist of friction drag as well as pressure drag. Most of the pressure drag is form drag due to the displacement thickness (see Sect. 5.11) of the wake of the propeller, in particular when the propeller is not free to rotate. Because the propeller blades will usually carry some lift, there will also be some induced drag.

The resistance of a propeller is, like lift and other drag components, usually expressed like

$$R_{prop} = \tfrac{1}{2}\rho V_b^2 S_{prop} C_{Dprop} \qquad (6.5.46)$$

Here, S_{prop} is the frontal area of the propeller and C_{Dprop} its (dimensionless) drag coefficient. The frontal area of a propeller can be expressed as

$$S_{prop} = A_R (\pi/4) D_{prop}^2, \qquad (6.5.47)$$

where D_{prop} is the diameter of the propeller and A_R is the ratio between the blade area and the disc area of the propeller. For two-bladed propellers A_R is of the order of 0.3. For three-bladed props A_R is of the order 0.45. However, the diameter of three-bladed propellers is usually a little smaller, such, that the effective frontal area is only some 5–10% larger.

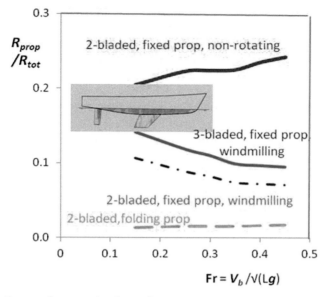

Fig. 6.5.54 *Fractional magnitude of propeller resistance for a cruising yacht ($L_{WL} = 10$ m) in upwind conditions (flat water)*

According to Keuning (1985), the drag coefficient C_{Dprop} is about 1.2 for a fixed 3-bladed, non-rotating propeller. For a 2-bladed propeller this is about 1.0. For fixed-bladed propellers that are free to rotate, C_{Dprop} is about 0.5 and decreases with boat speed. The rate of decrease with boat speed of a wind milling propeller is about the same as the decrease with Reynolds number of the friction drag coefficient of a flat plate (Fig. 5.11.3).

For a two-bladed folding propeller C_{Dprop} is about 0.05 and also decreases with boat speed. The resistance of a feathering propeller is even smaller; about half that of a folding propeller.

Figure 6.5.54 gives an impression of the relative importance of the propeller resistance for a yacht with a waterline length of about 10 m with an ordinary size propeller, that is, a diameter of about 5 % of the length of the waterline. For a 2-bladed, locked (non-rotating) propeller the resistance is huge, about 22 % of the total resistance. For fixed-bladed propellers that are free to rotate, the resistance is much smaller, about 10 % of the total resistance. For a folding propeller this is about 2 % of the total resistance.

6.6   Hydrodynamic Efficiency

We have seen in Chap. 4 (see also Sect. 3.2), that the hydrodynamic drag angle ε_H, defined by

$$\varepsilon_H \equiv \arctan(R/S) \qquad (6.6.1)$$

is a very important parameter for the performance of sailing yachts. This is the case in particular when sailing to windward. Maximizing the velocity made good to windward was seen to require minimizing of the hydrodynamic (as well as the aerodynamic) drag angle. The latter obviously requires minimizing the resistance/ side force ratio (or maximizing the side force/resistance ratio).

For the purpose of fully understanding the performance mechanisms of sailing it is useful to consider the dependence of ε_H on the conditions of sailing and the geometrical properties of a yacht. Figure 6.6.1 serves part of this purpose by comparing, qualitatively, resistance polars for high and low boat speed conditions. The figure is representative for a conventional cruising yacht with a keel aspect ratio of about 1.

In the preceding sections we have seen that, at low boat speeds, the basic hydrodynamic resistance of a sailing yacht, that is the resistance at zero side force, con-

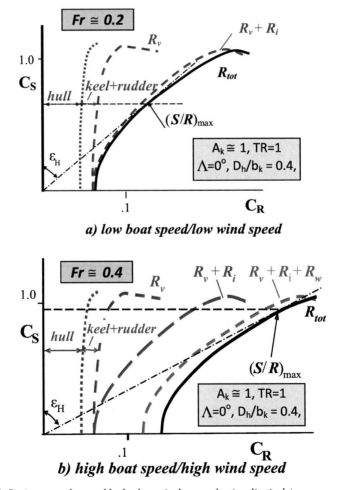

Fig. 6.6.1 Resistance polars and hydrodynamic drag angles (qualitatively)

sists mainly of viscous or frictional resistance (R_v in Fig. 6.6.1a). At high boat speed the wave-making resistance (R_w) becomes the dominant factor and there is, usually, also some added resistance due to (wind driven) waves. As a consequence, the total resistance (R_{tot}) polar is shifted to higher values of the resistance at high boat speeds (Fig. 6.6.1b). This implies an increase of the minimum hydrodynamic drag angle from about 10° to about 17° when the Froude number is increased from about 0.2 to about 0.4. Note also that the minimum value of ε_H at high boat speeds occurs at a higher value of the side force coefficient than at low boat speeds.

The draft of the keel, or, for a given keel area, the aspect ratio A_k, is the most important geometrical property for the minimum hydrodynamic drag angle. Figure 6.6.2 compares resistance polars for low boat speed conditions (Fr≈0.2) for conventional cruising yachts with the same hull and the same keel area but with different keel span and aspect ratios. The figure illustrates the strong dependence of the minimum drag angle on the keel aspect ratio. For a keel aspect ratio of 2 the minimum drag angle is about 8°. For A_k=1 the minimum value of ε_H is about 10°. For A_k=0.5 it is about 14°, i.e. almost twice the value for A_k=2.

It can also be noted that the side force coefficient at which the drag angle has its minimum increases with aspect ratio. Since

$$C_S = S/(½\rho V_b^2 S_k),\qquad(6.6.2)$$

this means, that for a given level of side force (sail dimensions), the area of high aspect ratio keels can be smaller than the area of low aspect ratio keels.

It will further be clear, that any increase of the resistance at zero side force, due to, for example, surface roughness, a blocked, non-folding propeller, or added resistance in waves, will inevitably lead to a larger minimum hydrodynamic drag angle. A larger minimum drag angle is also incurred when a significant part of the side

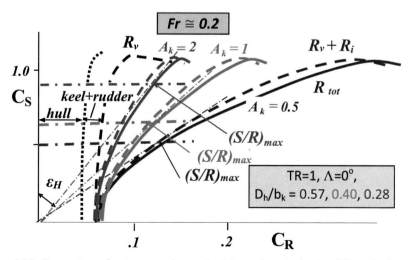

Fig. 6.6.2 *Comparison of resistance polars and minimum drag angles for different keel aspect ratios (conventional cruising yacht, keel area constant)*

force is carried by the rudder. This is mainly caused by the fact that the span or draft of the rudder is usually smaller than that of the keel. The smaller span implies a larger induced drag factor, as discussed in (Sub-)Sect. Induced Resistance.

6.7 Effects of Heel, Yaw Balance and Trim-In-Pitch

The attitude of a sailing yacht, as reflected in the angles of heel and trim-in-pitch, can have a significant or even a large effect on the hydrodynamic forces and moments. Of these, the angle of heel is, by far, the most important. This is primarily so because the angle of heel a yacht can adopt can be quite large; upto, say, 30°.

The (maximum) angle of trim-in-pitch is, in general, much smaller and its effect on the hydrodynamic forces and moments is correspondingly smaller

6.7.1 Angles in Different Coordinate Systems

Prior to discussing the effects of heel and trim-in-pitch on the hydrodynamic forces and moments in some detail it is useful to consider the relations between governing quantities like angle of leeway, angle of attack, force components etc. and the angles of heel and pitch. For this purpose we look first at the different coordinate systems, introduced in Sect. 3.1 (Fig. 3.1.6) and recall that

- The *angle of leeway* λ is the angle in the horizontal plane between the x-axis and the projection in the horizontal plane of the ship's ξ-axis, due to rotation about the ζ-axis or the z-axis
- The *angle of heel* φ is the angle between the vertical (z-) axis and the projection of the top (ζ-) axis of the boat in the z-η plane, due to rotation about the ξ-axis or the x-axis
- The *pitch angle* θ is the angle between the longitudinal (ξ-) axis of the ship and the horizontal plane, due to rotation about the η-axis or the y-axis

We also recall that the lateral hydrodynamic forces acting on a sailing yacht are governed essentially by the angle of attack α of the keel, that is the angle between the direction of the undisturbed flow (i.e. the direction of sailing) and the plane of the keel. For zero heel we have $\alpha = \lambda$. For non-zero angles of heel and pitch it can be shown, using the theory of rotation matrices (http://en.wikipedia.org/wiki/Rotation_matrix in Chap. 3), that this should be replaced by

$$\tan\alpha = (\tan\lambda\,\cos\varphi - \sin\theta\,\sin\varphi)/\cos\theta \qquad (6.7.1)$$

For small pitch angles θ this can be approximated by

$$\tan\alpha \approx \tan\lambda\cos\varphi \qquad (6.7.2)$$

When the angle of leeway is also small this can be approximated by

$$\alpha \approx \lambda \cos \varphi, \tag{6.7.3}$$

Equations (6.7.1)–(6.7.3) imply that the angle of attack of the keel of a sailing yacht is, effectively, reduced under pitch and heel.

6.7.2   *Effects of Heel on Vertical Force and Side Force*

We recall from Sect. 6.4 that the total vertical hydrodynamic force V_H can be expressed as

$$V_H(\varphi) = V_{H(hull)}(\varphi) + F_{H(lat)} \sin \varphi, \tag{6.7.4}$$

Here, $V_{H(hull)}(\varphi)$, <0, is the vertical force acting on the hull. The term $F_{H(lat)} \sin \varphi$ represents the contribution of the lateral force due to the lift on the keel and lift carry-over on the hull (see Fig. 6.7.1). As already discussed in Sect. 6.4, the vertical force on the hull is caused by supervelocity due to the longitudinal, convex curvature of the hull. It increases with increasing displacement/length ratio ∇/L_{WL}^3 or cross-sectional-area/length ratio S_{Xmax}/L_{WL}^2.
   $V_{H(hull)}(\varphi)$ also increases with heel, at least for hulls with a large beam/draft ratio. The reason is that the longitudinal, convex curvature, at the deepest point of the hull under heel, increases with the angle of heel for hulls with a large beam/draft ratio.

Fig. 6.7.1 *Hydrodynamic forces under heel in the lateral plane*

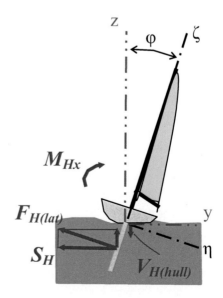

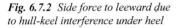

Fig. 6.7.2 Side force to leeward due
to hull-keel interference under heel

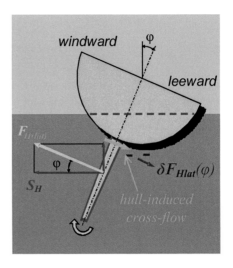

At the same time, the associated area with relatively high supervelocity, or low
pressure(suction), moves to the leeside of the (under-water part of the) hull (see
the area with '−' signs in Fig. 6.7.2). With it, the point of application of $V_{H(hull)}(\varphi)$,
(Fig. 6.7.1), also moves to leeward with increasing heel angle.

Because of the asymmetric position of the keel relative to the hull under heel, the
high-velocity/low-pressure area on the hull induces also low pressure at the leeward
(pressure) side of the top of the keel. At the same time, the upper part of the keel will
experience some cross-flow due to the asymmetric flow about the hull (Fig. 6.7.2).
The combined effect of this is that, at zero leeway, but under heel, the upper part of
the keel experiences a lateral force component $\delta F_{Hlat}(\varphi)$ induced by the hull that
'points the wrong way'.

From the discussion just given it follows that the total lateral force component
normal to the plane of the keel can be written as

$$F_{H(tot)}(\varphi) = F_{H(lat)}(\lambda, \varphi) + \delta F_{Hlat}(\varphi) \tag{6.7.5}$$

The lateral force component $\delta F_{Hlat}(\varphi)$, < 0, induced by the hull at zero leeway, is
caused primarily by the asymmetry of the under-water part of the hull under heel.
$F_{H(lat)}(\lambda, \varphi)$ is the lateral force produced by the keel under heel as a result of the
angle of attack of the keel.

Because the vertical force $V_{H(hull)}(\varphi)$ on the hull under heel has a component normal to the plane of symmetry of the yacht, it also contributes to the heeling force.
The heeling force can, therefore, now be written as

$$H_H(\varphi) = F_{H(tot)}(\varphi) - V_{H(hull)}(\varphi)\sin\varphi, \tag{6.7.6}$$

where $V_{H(hull)}(\varphi) < 0$. There is, of course, a corresponding contribution to the heeling moment through Eq. (3.2.5). Note that Eq. (6.7.6) reduces to Eq. (3.2.2) for
vanishing angle of heel φ.

For small angles of pitch and leeway the total horizontal component S_H of the total lateral force can be expressed as

$$S_H(\varphi) \approx \left\{ F_{H(lat)}(\lambda, \varphi) + \delta F_{H(lat)}(\varphi) \right\} \cos\varphi, \tag{6.7.7}$$

Dividing by the dynamic pressure $\frac{1}{2}\rho V_b^2$ and the keel area S_k this can be written as

$$C_S(\varphi) \approx \{C_L(\lambda, \varphi) + \delta C_L(\varphi)\} \cos\varphi \tag{6.7.8}$$

Here, $C_L(\lambda, \varphi) \equiv F_{H(lat)}(\lambda, \varphi) / (\frac{1}{2}\rho V_b^2 S_k)$ is the lift coefficient of the keel (plus hull) under heel due to leeway and $\delta C_L(\varphi) \equiv \delta F_{H(lat)}(\varphi) / (\frac{1}{2}\rho V_b^2 S_k)$ is the lift coefficient under heel at zero leeway.

The lift coefficient $C_L(\lambda, \varphi)$ of the keel under heel due to leeway can also be expressed as

$$C_L(\lambda, \varphi) = \frac{dC_L}{d\lambda}(\varphi)\lambda \approx \frac{dC_L}{d\alpha}(\varphi)\cos(\varphi)\lambda \tag{6.7.9}$$

The lift slope $\dfrac{dC_L}{d\alpha}(\varphi)$ of the keel plus hull under heel is slightly smaller than at zero heel. There are two reasons for this. One is that the (average) distance between the keel and the tip of its mirror image with respect to the water surface becomes smaller with increasing angle of heel. This causes a reduction of the effective span (see Fig. E.4 in Appendix E) and the effective aspect ratio of the keel and, hence, a reduction of the lift curve slope. The other reason is that the smaller distance between the keel under heel and the water surface causes an increase of the loss of lift/side force due to the presence of the water surface.

It follows from Eqs. (6.7.8) and (6.7.9) that

$$\frac{dC_S}{d\lambda}(\varphi) = \frac{dC_L}{d\alpha}(\varphi)\cos^2(\varphi) \tag{6.7.10}$$

Equation (6.7.10) expresses that the side force acting on a keel for given angle of leeway is roughly proportional to the square of the cosine of the angle of heel. Figure 6.7.3 illustrates that the loss of side force due to heel for a given angle of leeway can be considerable. The figure has been constructed on the basis of the theoretical-empirical model described in Appendix E and the results of the towing tank tests contained by Refs. Gerritsma et al. (1977) and Gerritsma and Keuning (1987).

The yacht configuration considered in Fig. 6.7.3 is representative for an 'average cruising yacht configuration' ($\nabla^{1/3}/L_{WL} = 0.18$, $B/D_h = 6$, $D_h/b_k = 0.4$, $A_k = 1$, $TR = 1$, $\Lambda = 0°$). Note that at $30°$ angle of heel the angle of leeway for zero side force is about $1.5°$ and the loss of side force is as much as 30–50% at moderate angles of leeway.

Figure 6.7.4 presents the variation of the rate of change with leeway $\dfrac{dC_S}{d\lambda}(\varphi)$ of the side force, divided by the rate of change at zero heel, as a function of the heel angle

Fig. 6.7.3 *Effect of heel angle on side force for an 'average' cruising yacht configuration*

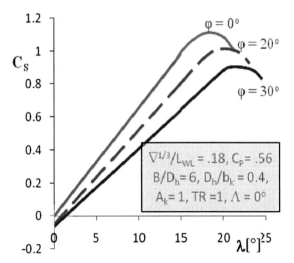

Fig. 6.7.4 *Variation of relative side force slope with angle of heel for keels*

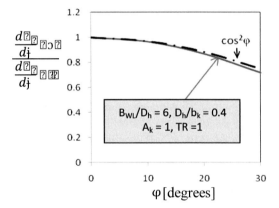

for the same 'average cruising yacht configuration' as in Fig. 6.7.3. Also shown is the factor $\cos^2\varphi$ as it appears in Eq. (6.7.10). From the comparison it is clear that $\cos^2\varphi$ is the dominating factor in the effects of heel on the side force. The effect of the reduced effective span and the increased free surface effect are small. It is the author's impression that this is the case for most practical combinations of hull and keel geometries. (The absolute value of the rate of change of the side force with angle of leeway is, of course, strongly dependent on the hull and keel geometries, see Sect. 6.3).

Figure 6.7.5 gives an impression of the effect of hull and keel geometry on the side force (or, rather, lift force) at zero leeway due to heel. The figure is also based on the theoretical-empirical model described in Appendix E. Shown is the effect of the three most important parameters: the beam/draft ratio of the hull, the taper ratio and the aspect ratio of the keel. The effect of sweep angle of the keel was found to be insignificant. This is, within reasonable bounds, also the case for the section coefficient C_m.

Fig. 6.7.5 *Lift coefficient of keel plus hull due to heel at zero leeway (keel area constant)*

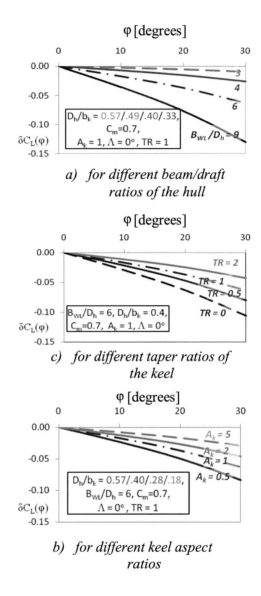

a) *for different beam/draft ratios of the hull*

c) *for different taper ratios of the keel*

b) *for different keel aspect ratios*

It can be noticed that a large beam/draft ratio of the hull, a small taper ratio and a small aspect ratio of the keel are unfavourable.

The effects of heel on the side force of rudders are quite similar to those of keels. That is, for given angles of leeway and rudder deflection, the lift is reduced by, roughly, a factor $\cos\varphi$ and the side force by, roughly, a factor $\cos^2\varphi$.

There is, in all probability, also a side force on the rudder when the hull is under heel at zero leeway. However, there is no information available about the magnitude.

There is also an effect of heel on the downwash induced by the keel at the position of the rudder (see Sect. 6.3). This in the sense that the expression (6.14) for the effective angle of attack of a rudder at zero heel now takes the form

$$(\alpha_e)_r = \lambda \, \cos\varphi \, \{1 - (\alpha_\varepsilon/\alpha)_{k-r}\} - (\alpha_i)_r + \delta_r \qquad (6.7.11)$$

Recall from Sect. 6.3 that $(\alpha_\varepsilon/\alpha)_{k-r}$ is the downwash parameter, that is the downwash induced by the keel at the position of the rudder per degree angle of attack of the keel. Due to the fact that heel reduces, slightly, the effective span of a keel or rudder, there is also a small effect of heel on the downwash parameter and the induced angle of attack of the rudder itself. The effect is of the same magnitude as the difference between the two lines in Fig. 6.7.4 and negligible for most practical purposes. A second, more important effect of heel on the downwash is through the side force of the keel due to heel at zero leeway. Under heel, the lift on the keel, for a given angle of attack, is a little smaller, (by an amount $\delta C_L(\varphi)$, <0), than at zero heel. Because the downwash is proportional to the lift, the downwash parameter $(\alpha_\varepsilon/\alpha)_{k-r}$ has, under heel, to be multiplied by a factor $\{C_L(\lambda, \varphi) + \delta C_L(\varphi)\} / C_L(\lambda, \varphi)$.

In this way the reduction of the downwash at the position of the rudder due to the loss of lift of the keel is taken into account.

From the discussion given above, it follows that for the side force on the rudder under heel there holds

$$\left\{\frac{dC_S}{d\lambda}(\varphi)\right\}_r \approx \cos^2\varphi \, f_{\varepsilon\varphi} \left\{\frac{dC_S}{d\lambda}(0)\right\}_r \qquad (6.7.12a)$$

and

$$\left\{\frac{dC_S}{d\delta}(\varphi)\right\}_r \approx \cos\varphi \left\{\frac{dC_S}{d\delta}(0)\right\}_r \qquad (6.7.12b)$$

The factor

$$f_{\varepsilon\varphi} = [1 - (\alpha_\varepsilon/\alpha)_{k-r}\{1 + \delta C_L(\varphi)/C_L(\lambda, \varphi)\}]/[1 - (\alpha_\varepsilon/\alpha)_{k-r}] \qquad (6.7.13)$$

represents the effect of the downwash from the keel, including the effect of heel. Because $\delta C_L(\varphi)$ is, in general, <0, $f_{\varepsilon\varphi}$ is usually >1. Its value increases with decreasing lift due to leeway of the keel and increasing value of the downwash parameter $(\alpha_\varepsilon/\alpha)_{k-r}$. This means that the effect of heel on the lift of the rudder is compensated, at least partly, by a reduction of the downwash from the keel. This is illustrated by Fig. 6.7.6. The figure presents the rate of change with heel of the side force on the rudder in parts of the rate of change at zero heel. Note that this is the same as the factor $\cos^2\varphi \, f_{\varepsilon\varphi}$ in Eq. (6.7.12a). Note also that the effect of heel is even overcompensated by a reduction of the downwash from the keel for hulls with a large beam/

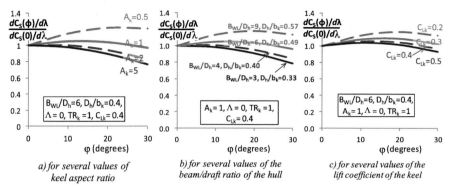

*a) for several values of
keel aspect ratio*

*b) for several values of the
beam/draft ratio of the hull*

*c) for several values of the
lift coefficient of the keel*

Fig. 6.7.6 *Effect of heel on the side force produced by the rudder for different keel-hull configurations and different lift levels of the keel*

draft ratio (large absolute value of $\delta C_{Lk}(\varphi)$), low aspect ratio keels (large value of $(\alpha_\varepsilon/\alpha)_{k-r}$) and small angles of leeway (small $C_L(\lambda,\varphi)$).

Yachts with a large beam/draft ratio of the hull, in particular near the stern, may experience still another phenomenon under heel, with fairly dramatic consequences for the rudder effectiveness. As illustrated by Fig. 6.7.7a, a rudder positioned in the plane of symmetry of a hull may be lifted partially out of the water at large angles of heel. This is the case in particular when a large heel angle is accompanied by a substantial bow-down angle of the trim-in-pitch. This, of course, is the case when sailing upwind in high wind speeds.

   When the top of the rudder is lifted out of the water there is a large reduction of the rudder effectiveness. This is caused by a reduction of the effective span, an increase of the loss of lift due to the free surface effect and a reduction of the effective area of the rudder. For this reason yachts with broad stern sections are sometimes equipped with twin rudders, as in Fig. 6.7.7b. Although the rudder on the weather side loses, practically, all of its effectiveness, the leeside rudder will be fully submerged, and, when positioned at an angle of about 20°, will not lose any effectiveness under heel. At small angles of heel both rudders will be operable but each with slightly less effectiveness than a single rudder. For this reason, the total

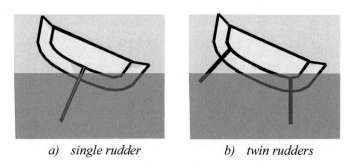

a)   single rudder   *b)   twin rudders*

Fig. 6.7.7 *Effect of heel on the submersion of rudders*

surface area of the rudders can be substantially less than twice (say, between 1.1 and 1.5) that of a single rudder. There is, of course, a penalty to be paid in the form of a higher frictional resistance at small angles of heel when both rudders are fully submerged, that is at low wind/boat speeds. On the other hand there may also be a small advantage in the sense that twin rudders experience a little less downwash from the keel because they are somewhat farther away from the plane of the keel.

6.7.3 Effects of Heel on Hydrodynamic Yawing Moment, Static Directional Stability and Yaw Balance

We have seen in Sect. 6.4 that the hydrodynamic yawing moment can be expressed as

$$M_{Hz} = (M_{Hz})_{\text{hull}} + (x_{\text{ref}} - x_{\text{CPk}})S_k + (x_{\text{ref}} - x_{\text{CPr}})S_r \qquad (6.7.14)$$

Or, in dimensionless form

$$C_{MHz} = (C_{MHz})_{\text{hull}} + C_{Lk}(x_{\text{ref}} - x_{\text{CPk}})\, S_k/\nabla + C_{Lk}(x_{\text{ref}} - x_{\text{CPr}})\, S_r/\nabla, \qquad (6.7.15)$$

where we have distinguished contributions due to the hull, keel and rudder. x_{ref} is the point of reference for the yawing moment. It is useful to note here that if x_{ref} is chosen to be the longitudinal position of the aerodynamic centre of the sails there is no contribution of the sails to the change in yawing moment due to a deviation from the course of sailing.

In the preceding sub-section we have seen that for a given angle of leeway and rudder deflection both S_k and S_r vary with heel like $\approx \cos^2\varphi$ or $\cos^2\varphi\, f_{\varepsilon\varphi}$, respectively. The total side force on keel and rudder can be written as

$$S_k(\varphi) \approx \delta L_k(\varphi)\, \cos\varphi + \frac{dS_k}{d\lambda}(0)\lambda\, \cos^2(\varphi) \qquad (6.7.16a)$$

and

$$S_r(\varphi) \approx \delta L_r(\varphi)\, \cos\varphi + \frac{dS_r}{d\lambda}(0)\lambda\cos^2(\varphi)\, f_{\varepsilon\varphi} \qquad (6.7.16b)$$

Because $\delta L_k(\varphi)$ is usually not very large, the contribution of the keel to the total yawing moment varies with heel like $\cos^2\varphi$ for large angles of leeway λ but like $\cos\varphi$ for $\lambda \to 0$. For the rudder it is a little bit more complicated. Little is known about the magnitude of $\delta L_r(\varphi)$, but it is reasonable to assume that this is also fairly small. It is further recalled from the preceding subsection that the factor $\cos^2(\varphi)f_{\varepsilon\varphi}$ depends on the aspect ratio of the keel and the beam/draft ratio of the hull. For an 'average' cruising yacht it was seen to be of the order of one (see Fig. 6.7.6).

The effect of heel on the yawing moment of the hull $(M_{Hz})_{\text{hull}}$ can be split in two parts:

1. The yawing moment due to heel $\{\delta M_{Hz}(\varphi)\}_{\text{hull}}$ at zero leeway

2. The effect of heel on the rate of change $\left\{\dfrac{dM_{Hz}}{d\lambda}(\varphi)\right\}_{\text{hull}}$ with leeway

The yawing moment due to heel at zero leeway depends on the asymmetry that the under-water part of a hull may adopt under heel. For a hull with circle segment cross-sections the under-water part remains symmetric, so that there is no effect of heel on $(M_{Hz})_{\text{hull}}$. However this is no longer the case for a hull with cross-sections that deviate from the shape of a circle segment, in particular for $B_{WL}/D_h > 2$. In this situation one may expect a non-zero yawing moment under heel for $\lambda=0$. The mechanism of this is basically the same as that of the non-zero lift force $\delta C_L(\varphi)$ under heel for $\lambda=0$, described in the preceding sub-section (see also the last paragraph of Appendix G).

Figure 6.7.8 gives an impression of the variation of the zero leeway yawing moment coefficient $\delta C_{MHz}(\varphi)$ as a function of the angle of heel for a range of values of the two most important geometry parameters B_{WL}/D_h and D_h/L_{WL}. The figure is based on the (tentative) analytical-empirical model described in Appendix F. It can be noticed that the yawing moment due to heel at zero leeway becomes larger for increasing values of the beam/draft and draft/length ratios of the hull.

Figure 6.7.9 gives an impression of the dependence on the angle of heel of the rate of change $\left\{\dfrac{dM_{Hz}}{d\lambda}(\varphi)\right\}_{\text{hull}}$ with leeway of the yawing moment of the hull of a sailing yacht. The figure is also based on the formulation given in Appendix G.

Shown is $\left\{\dfrac{dM_{Hz}}{d\lambda}(\varphi)\right\}_{\text{hull}}$ as a function of the angle of heel for several values of the

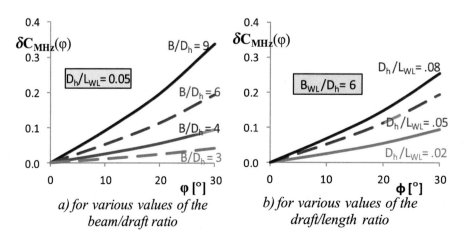

a) for various values of the
beam/draft ratio

b) for various values of the
draft/length ratio

Fig. 6.7.8 *Yawing moment due to heel of sailing yacht hulls at zero leeway*

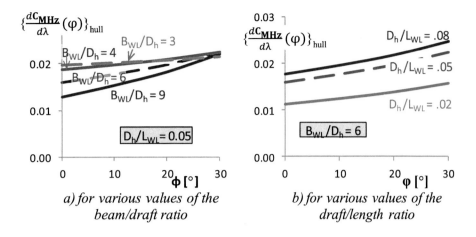

a) for various values of the
beam/draft ratio

b) for various values of the
draft/length ratio

Fig. 6.7.9 *The rate of change (per degree) with leeway of the yawing moment of a sailing yacht
hull as a function of the angle of heel*

beam/draft ratio B_{WL}/D_h and the draft/length ratio D_h/L_{WL}. The rate of change
with leeway of the hydrodynamic yawing moment is seen to increase with the angle
of heel, in particular for hulls with a large beam/draft ratio. The main reason for this
is the increase with heel of the actual draft of hulls with a large beam/draft ratio.

Figure 6.7.10 gives an impression of the effect of heel on the total hydrodynamic
yawing moment of an 'average' cruising yacht configuration (moment reference
point at $x_{cg} \cong 0.55\ L_{WL}$). Shown is the yawing moment coefficient as a function of
the angle of leeway for zero heel and a heel angle of $20°$ for an 'average' hull with
two different keels (see also Fig. 6.4.6). Both keels have a sweep angle of $0°$ and a

Fig. 6.7.10 *Effect of heel on
the total hydrodynamic yawing
moment of an 'average' cruising
yacht with different keels (no
rudder deflection)*

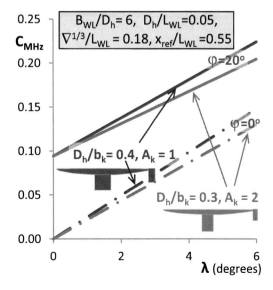

taper ratio of 1. One has an aspect ratio of 1 and for the other $A_k = 2$. The net effect of heel on the slope of the lines and, thus, on the static directional stability is small. This is due to the fact that the effect of heel on the rate of change with leeway of the yawing moment of the hull (Fig. 6.7.9) and the effect of heel on the contribution of the rudder (Fig. 6.7.6) are relatively small.

Yaw Balance

As already mentioned before, the hydrodynamic yawing moment of hull and keel and the aerodynamic yawing moment of the sails must, in equilibrium conditions, be balanced by the yawing moment due to deflection of the rudder. For the purpose of considering the effect of heel on this 'yaw balance', we will assume that the longitudinal position of the centre of effort of the sails is at 36% of the length of the waterline from the bow.[14] This means that the aerodynamic yawing moment of the sails is zero with respect to a point at $x = 0.36 \, L_{WL}$.[15] It is then convenient to consider the hydrodynamic yawing moments with respect to this point because in equilibrium conditions the total hydrodynamic yawing moment with respect to this point must be zero as well. We will do so for several different hypothetical yacht configurations. The base configuration ($B_{WL}/D_h = 6$, $A_k = 1$) is representative for an 'average' cruising yacht and is characterized by the data given in Fig. 6.7.11.

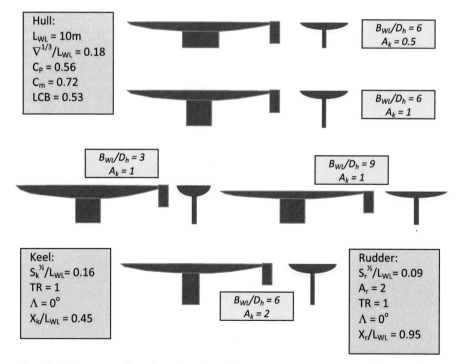

Fig. 6.7.11 *Overview of hypothetical yacht configurations considered for yaw balance analysis*

---

[14] According to Larsson and Eliasson (1996) this is a representative value for upwind sailing.

[15] At least for small angles of heel. For non-zero heel angles there is an additional aerodynamic couple to weather formed by the aerodynamic thrust and hydrodynamic resistance vectors (see Fig. 4.1.1).

The other configurations were obtained by varying the most important quantities for the effects of heel, i.e. the beam/draft ratio of the hull ($B_{WL}/D_h = 3, 6$ and 9) and the keel span or aspect ratio ($A_k = 0.5, 1.0$ and 2.0). In all cases side force and resistance were determined for conditions (Fr=0.35) representing upwind sailing.

Figure 6.7.12 gives an impression of how the load (lift coefficient, C_{LR}) on the rudder required for directional balance, as calculated, depends on the total side force coefficient C_S and the angle of heel φ for the hypothetical yacht configurations described above. Note that these are force 'polar' type of graphs of the kind introduced in Sect. 4.3, Fig. 4.3.3. It can be noticed that, in general, the load on the rudder required for balance in yaw, in terms of the lift coefficient C_{Lr}, increases with increasing side force and angle of heel. This is the case in particular for yachts with a large beam/draft ratio and a shallow, low aspect ratio keel. It can also be noticed that, for zero heel, balance in yaw requires hardly any force on the rudder, in particular for yachts with a large beam/draft ratio and a deep, high aspect ratio keel. Needless to say that the high rudder loads required for balance in yaw at conditions with a large side force and a large angle of heel, imply a large deflection of the rudder (weather helm), in particular for yachts with a high beam/draft ratio and a shallow, low aspect ratio keel. This, of course, is in agreement with every yachtsman's experience.

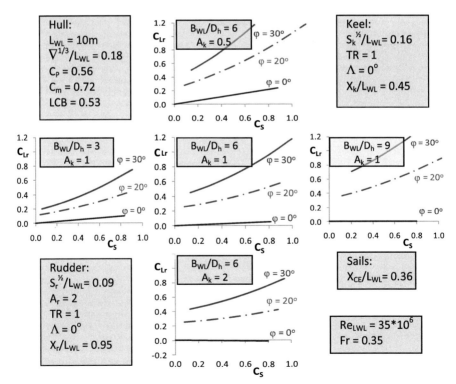

Fig. 6.7.12 *The load (lift coefficient) of the rudder, required for balance in yaw, as a function of the total side force and the angle of heel for several hypothetical yacht configurations. (See also Fig. 6.7.11)*

6.7.4   Effects of Heel on Resistance

Heel also affects the various components of hydrodynamic resistance distinguished in Sect. 6.5.

Considering the *viscous resistance* we recall from Sub-Sect. Viscous Resistance that, for a given Reynolds number, the viscous resistance of a yacht's *hull* is determined by the wetted area and the supervelocities due to the volume and shape of the hull. It appears (Keuning and Sonnenberg 1998a) that, depending on the shape of the hull, the effect of heel on the wetted area can be significant. The most important parameter has been found to be the beam/draft ratio.

Figure 6.7.13 illustrates the effect of heel on the wetted area. The figure is based on an empirical formula given in Keuning and Sonnenberg (1998a). Shown, for several values of the beam/draft ratio, is the wetted area under heel divided by the wetted area at zero heel as a function of the angle of heel. It can be noticed that there is a significant reduction (up to 30%) of the wetted area under heel for yachts with a large beam/draft ratio. This may be recognized by dinghy sailors. Many have experienced that heel can be beneficial for boat speed under conditions of low wind speed, when the frictional (viscous) resistance matters most.

Little is known about the effect of heel on the super-velocities due to volume of a sailing yacht hull. It follows from Sect. Viscous Resistance, that, in principle, this could be modeled by introducing the effects of heel in the form factor 'k' (see Eq. (6.5.2) and Fig. 6.5.3). One might expect that heel has hardly any effect on the form factor when the hull has a semi-circular cross-section, i.e. for $B_{WL}/D_h \cong 2$. For high values of B_{WL}/D_h one would expect 'k' to increase slightly with the angle of heel. This, because the actual, effective beam/draft ratio would decrease with increasing angle of heel for 'beamy' hulls. Figure 6.5.3 suggests, however, that the effect would hardly be significant.

The effect of heel on the *viscous resistance of appendages* is, in all probability, totally negligible in most conditions of sailing. The reason is that there is no effect

Fig. 6.7.13 *Effect of heel on the wetted area of sailing yacht hulls*

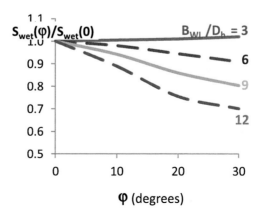

of heel on the wetted area of fully submerged bodies and that the super-velocities due to thickness, and for a given lift, of keels and rudders are not influenced by heel to any significant extent. The only possible exception to the latter is when the lift on a keel or rudder is strongly affected by heel. We have seen in the preceding sub-sections that this may happen for hulls with a large beam/draft ratio. In that case the keel and/or rudder will experience some additional viscous resistance due to lift.

For certain yacht configurations and under certain conditions of sailing the wetted area of a keel or rudder may not be independent of the angle of heel. This may be the case for keels under wide-bodied yachts at very large angles of heel and at high boat speeds when wave-making generates a deep wave trough near amidships. Under such conditions the root part of the keel may start piercing through the surface with a corresponding reduction of the actual wetted area. Another example is formed by twin rudders, as already mentioned in a preceding sub-section.

Induced Resistance
As described in Sect. 6.5, the induced resistance of a sailing yacht depends, for a given lift, primarily on the effective aspect ratio of the keel and the pressure relief effect of the free surface. For zero angle of heel the induced resistance coefficient can be written as (see Eq. (6.5.16))

$$C_{D_i} = K_i C_L^2, \qquad (6.7.17)$$

with the factor K_i given by (see Appendix E, Eq. (E.64))

$$K_i = (1/F_{FS}) \, K_i(0), \qquad (6.7.18)$$

where

$$K_i(0) = \frac{\{1+(4/\pi)F_R \, D_h/b_k\}}{\left\{1+\dfrac{D_h}{b_k}\right\}\pi A_e} \qquad (6.7.19)$$

and

$$A_e = 2A_k(e_k + D_h/b_k) \qquad (6.7.20)$$

is the effective aspect ratio. F_R ($\cong 1$) is a factor that depends on the spanwise distribution of circulation.

Recall also that in Eqs. (6.7.17–6.7.20) C_L is the lift coefficient of keel + hull and that the factor F_{FS} (<1) models the effect of the free surface. 'e' is the efficiency factor introduced in Sect. 5.15. It is a function of the spanwise distribution of circulation.

We have seen earlier (Sub-Sect. 6.7.2) that, for zero heel, lift and side force are the same but that they differ by a factor $\cos\varphi$ for non-zero values of heel.

This means, that Eq. (6.7.17) can also be written as

$$C_{D_i} = K_i C_S^2 / \cos^2\varphi \qquad (6.7.21)$$

As described in some detail in Appendix E, the induced resistance under heel can be approximated by the expression

$$C_{D_i} = K_i \{C_L(\varphi) - 0.75\delta C_L(\varphi)\}^2 \qquad (6.7.22)$$

or

$$C_{D_i} = K_i \{1 - 0.75\delta C_L(\varphi)/C_L(\varphi)\}^2 C_L^{\,2}(\varphi) \qquad (6.7.23)$$

This can also be written as

$$C_{D_i} = K_i(\varphi)C_L^{\,2}(\varphi) \qquad (6.7.24)$$

where

$$K_i(\varphi) = \{1 - 0.75\delta C_L(\varphi)/C_L(\varphi)\}^2 K_i(0) \qquad (6.7.25)$$

and $K_i(0)$ is given by (6.7.19).

As described above, in Sub-Sect. 6.7.2, and in Appendix E, the effects of heel on the effective aspect ratio of the keel and the free surface effect are probably negligible for most if not all practical purposes. In terms of Eq. (6.7.24) and (6.7.18/19/10) this means that the variations of $F_{SF}(\varphi)$ and e_k with ϕ are small.

It is important to note that, under heel, the induced drag is no longer a linear function of the square of the lift. This is illustrated by Fig. 6.7.14. The figure gives the induced drag coefficient as a function of the lift-coefficient-squared for a light displacement yacht configuration. The configuration is characterized by $B_{WL}/L_{WL} = 0.25$, $B_{WL}/D_h = 9.6$, $D_h/b_k = 0.24$, an effective aspect ratio $A_e = 1.6$ (keel + hull) and a lift coefficient due to heel at zero leeway of $\delta C_L(\varphi = 30°) = -0.12$ (for $\varphi = 30°$).

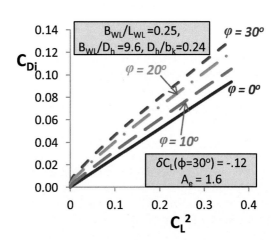

Fig. 6.7.14 *Example of effect of heel on the non-linear dependence of the induced resistance on lift*

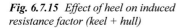

Fig. 6.7.15 *Effect of heel on induced resistance factor (keel + hull)*

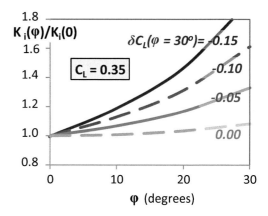

The effect of the lift $\delta C_L(\varphi)$ due to heel at zero leeway on the induced resistance is quite substantial. This is illustrated by Fig. 6.7.15. The figure shows the ratio of the induced drag factor $K_i(\phi)$ and its value $K_i(0)$ at zero heel as a function of the heel angle for several values of the lift due to heel at zero leeway. Note that the values of the latter are given for 30 degrees of heel and that the dependence of $\delta C_L(\varphi)$ on the geometry of hull and keel was given in Fig. 6.7.5. Note also that the total lift coefficient $C_L = 0.35$, which is a representative value for upwind sailing. Figure 6.7.15 illustrates that the induced resistance of an 'average' cruising yacht at 20 degrees of heel, for which $\delta C_L(\varphi = 30°)$ is of the order of -0.05, is about 20 % more than at zero heel. For narrow yachts with a high aspect ratio keel, for which $\delta C_L(\varphi = 30°)$ is close to zero, the effect of heel will be smaller. It will be larger for yachts with a large beam/draft ratio and a low aspect ratio keel for which $\delta C_L(\varphi = 30°)$ may be as large as -0.10 to -0.15 (see Fig. 6.7.5).

It should be realized that the discussion just given applies to the induced resistance of keel plus hull only. As we have seen already in Sub-Sect. 6.5.3, there is, in general, also a significant contribution by the rudder. The latter is, of course, also a function of the angle of heel. Although the mechanisms involved are basically the same as for a keel plus hull, there are a number of differences of importance. The first, as already discussed in Sub-Sect. 6.5.3 Induced Resistance, is that the rudder operates in the downwash field of the keel. This was to seen to vary with the angle of heel (Sub-Sect. 6.7.2). Little or nothing is known about the side force on the rudder under heel at zero leeway and the effect of this on the induced resistance of the rudder. The most important effect of heel on the induced resistance of the rudder is through the high load on the rudder that may be required for balance in yaw (see Fig. 6.7.12).

As an example Fig. 6.7.16 gives the induced drag of the rudder, as a fraction of the total induced drag, as a function of the total side force and angle of heel for the 'average' hypothetical yacht configuration considered in Fig. 6.7.11. While the relative contribution of the rudder to the total induced resistance is small for small angles of heel it becomes larger than 50 % for angles of heel above, say 20°! This would be smaller for yachts with a small beam/draft ratio of the hull and/or a high

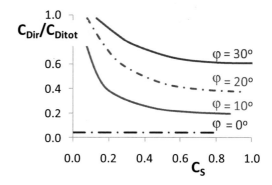

Fig. 6.7.16 Induced resistance of the rudder, as a fraction of the total induced resistance, of an 'average' cruising yacht configuration that is balanced in yaw

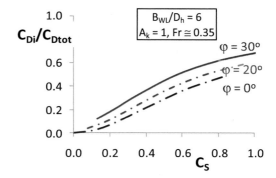

Fig. 6.7.17 Induced resistance of keel plus rudder, as a fraction of the total resistance, for an 'average' cruising yacht configuration, sailing upwind, that is balanced in yaw

aspect ratio keel. It would even be larger for yachts with a larger beam/draft ratio and/or a keel of lower aspect ratio. The main reason for the large fractional contribution of the rudder is the yawing moment of the hull at zero leeway due to heel.

The contribution of the induced resistance of keel plus rudder, as a fraction of the total resistance. Also increases with heel significantly. This is illustrated by Fig. 6.7.17 for the same 'average' cruising yacht. The figure shows the contribution of the induced resistance of keel plus rudder to the total resistance as a function of the side force coefficient for several angles of heel. It can be noticed that for a side force coefficient representative for upwind sailing ($C_s \cong 0.3$) the contribution of the induced resistance at 30° of heel is twice as large as at $\varphi = 0°$.

Wave-Making Resistance

It appears from the literature, that the variation with heel of the wave-making resistance due to volume is usually less than 20% of the wave-making resistance at zero heel. As discussed in some detail in Appendix J, the variation with heel of the wave-making resistance due to volume depends primarily on the (maximum) section coefficient C_m, the beam/draft ratio B_{WL}/D_h and the longitudinal position LCB of the centre of buoyancy.

Figure 6.7.18 shows the ratio $R_{WV}(\varphi)/R_{WV}(0)$ of the wave-making resistance under heel divided by the wave-making resistance at zero heel as a function of the angle of heel, for several values of the section coefficient (a), the beam/draft

Fig. 6.7.18 Varia-
tion of wave-making
resistance with angle of
heel. a Dependence on
the section coefficient.
b Dependence on beam/
draft ratio. c Depen-
dence on the longitudinal
position of the center of
buoyancy

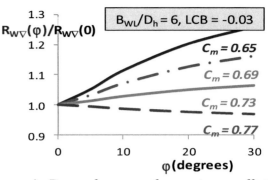

a) Dependence on the section coefficient

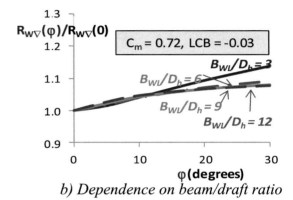

b) Dependence on beam/draft ratio

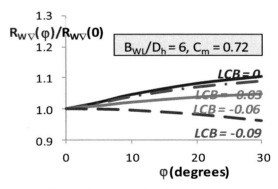

c) Dependence on the longitudinal position of
the center of buoyancy

ratio (b) and the position of the longitudinal centre of buoyancy (c). The curves
are based on a tentative empirical-analytical model described in Appendix J. The
figures reflect that for narrow hulls with 'lean' sections (small C_m) the centre of
buoyancy comes closer to the water surface when the hulls heels over. The opposite
is the case for 'beamy' hulls with full sections (large C_m). As discussed in Sect. 6.5

the wave-making resistance of a yacht usually increases if the centre of buoyancy comes closer to the water surface. As a consequence the wave-making resistance of a narrow hull with 'lean' sections increases more rapidly with heel than that of a 'beamy' hull with full sections.

It is also noted that a particular situation arises when the under-water part of the cross-section of the hull has the shape of a circle segment. This corresponds with a certain value, C_m^*, of the section coefficient C_m, that is a function of the beam/draft ratio B_{WL}/D_h (see Fig. E.5). It will be clear that in this situation the flow about the hull and, hence, the wave-making resistance, becomes independent of heel. In general, the wave-making resistance will, as reflected in Fig. 6.7.18a), increase with heel for 'lean' sections with $C_m < C_m^*$ but will decrease for 'full' sections.

The effect of the beam/draft ratio B_{WL}/D_h on the variation with heel of the wave-making resistance is seen to be relatively small (Fig. 6.7.18b). However, there appears to be a significant effect of the position of the longitudinal centre of buoyancy (Fig. 6.7.18c). This in the sense that when the position of the LCB is further aft (more negative values of LCB) the wave-making resistance increases less rapidly with heel.

Very little is known about the effect of heel on the wave-making resistance due to side force. It can be argued, however, that for a given level of lift on the keel of a yacht, the wave-making resistance due to side force will increase with heel like $1/\cos2\varphi$ like the wave-making resistance due to volume of yachts with a narrow hull. This because the centre of lateral resistance (or centre of pressure) of a keel yacht, like the centre of buoyancy of a narrow hull, gets closer to the water surface when the yacht heels.

Added Resistance in Waves
The effect of heel on the added resistance in waves has already been touched upon in Sect. 6.5, Fig. 6.5.46. This figure gives an impression of the response in waves of the heaving and pitching motions and the added resistance of two different types of yacht at zero and 20 degrees of heel. The figure reflects the impression (Skinner 1982; Keuning and Sonnenberg 1998a) that the effect of heel on the added resistance in waves of a sailing yacht is, in general, small.

An exception is formed, apparently, by yachts with a small beam/draft ratio B_{WL}/D_h. Results of towing tank tests (Skinner 1982) indicate that for such yachts the added resistance in waves tends to decrease with heel (see the right-hand side of Fig. 6.5.47). A satisfactory explanation for this behavior is, unfortunately, not available in the literature. Because added resistance in waves consists mainly of wave-making resistance (see Sect. 6.5), and wave-making resistance was seen to increase with heel for yachts with lean sections and small beam/draft ratios, one would expect the added resistance in waves of such yachts to increase also with heel.

The mechanism behind the observed, opposite behaviour is, as yet, unknown.

It is worth noting in this context that keel yachts under heel in a seaway may be subject to the same mechanism as that described in the paragraph on the '*Propulsion potential of heaving motion*' of lifting surfaces in Sect. 5.18 and the paragraph on '*Wings in waves*' in Sect. Added Resistance in Waves. In the latter paragraph we saw that (approximately) horizontal wings, such as those of a winged keel, may produce a net forward thrust when moving upwind in a seaway. There is no doubt that the same may happen with an ordinary plane keel under heel. Nothing is know about the possible magnitude of this effect, but there is little doubt that it increases with increasing effective span or aspect ratio of the keel (apart from increasing angle of heel). If the effect is significant it might explain (part of) the net reduction with heel of the added resistance of the yacht with the deep hull in Fig. 6.5.47. This because the effective span of the configuration with the deep hull is almost twice that of the configuration with the high beam/draft ratio. (Both configurations are fitted with the same keel of aspect ratio 0.65, taper ratio 0.63 and 45° sweep).

It is further important to realize that in the experiments underlying Fig. 6.5.47 the models were subject to heaving and pitching oscillations only and were not in rolling oscillation. A yacht sailing under heel in oblique seas will also be subject to rolling oscillations. Although this is not likely to cause 'added' wave-making resistance it will cause time-varying loads on the keel and some associated 'added' induced resistance. This in particular in beam seas when the roll excitation by waves is most effective and the aerodynamic damping of the rolling motion due to the sails is smaller than in upwind conditions. To the author's knowledge this is not accounted for in available methods for the estimation of the added resistance of sailing yachts in waves.

6.7.5   Effects of Heel on Hydrodynamic Efficiency

Having established the effects of heel on side force and resistance we are now in a position to determine the effect of heel on the hydrodynamic efficiency and the corresponding hydrodynamic drag angle. This has been done for the hypothetical yacht configurations described earlier (see also Fig. 6.7.11). In all cases side force and resistance were determined for conditions (Fr = 0.35) representing upwind sailing with balance in yaw, assuming the longitudinal position of the centre of effort of the sails to be at 40 % of the length of the waterline. The results, in terms of hydrodynamic resistance polars are summarized in Fig. 6.7.19.

The first that strikes (again) is the enormous effect of the keel aspect ratio on the hydrodynamic efficiency. Or, rather, the effect of keel span for a given keel area (see the vertical column of polars). As already discussed in Sect. 6.6 this is almost entirely due to the induced resistance. We have seen in the preceding section that this is even more the case with balance in yaw through deflection of the rudder, in particular for hulls with a large beam/draft ratio.

Secondly, it is also clear from the horizontal row of polars that, for a given keel size and shape, narrow hulls lead to a higher hydrodynamic efficiency than beamy hulls. This picture is, however, partly misleading in the sense that the total draft of hull plus keel is larger for the narrow hull $(B_{WL}/D_h = 3)$ than for the beamy hull $(B_{WL}/D_h = 9)$. When the keel aspect ratio is adapted so that the total draft of hull plus keel is equal to that of the base configuration (2.08 m), the difference with the base configuration is found to be much smaller.

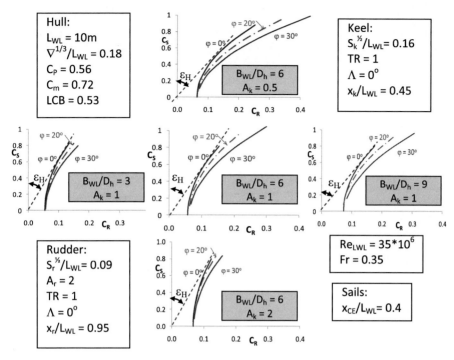

Fig. 6.7.19 *Dependence on the angle of heel of hydrodynamic resistance polars of several hypothetical yacht configurations that are balanced in yaw. (See also Fig. 6.7.11)*

Figure 6.7.20 presents the minimum hydrodynamic drag angle ε_{Hmin} as a function of the angle of heel of the various configurations. In can be noticed that the minimum drag angle increases with heel appreciably, in particular for configurations with a shallow draft keel (small A_k) and a large beam/draft ratio of the hull. The main reason is the increase under heel of the induced resistance of keel and rudder.

Fig. 6.7.20 *Minimum hydrodynamic drag angle as a function of the angle of heel*

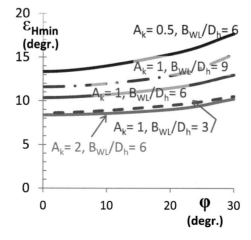

The increase at $20°$ ((degree symbol between thin space)) of heel is about 10% for the base configuration ($A_k = 1$, $B_{WL}/D_h = 6$), about 7% for the configuration with deep keel ($A_k = 2$, $B_{WL}/D_h = 6$) and as much as 13% for the configuration with shallow keel ($A_k = 0.5$, $B_{WL}/D_h = 6$).

We have seen in Chap. 4 that the hydrodynamic drag angle is an important quantity when sailing upwind. Hence, the implication of the figures given above is that 'beamy' yachts with a shallow draft keel must be sailed upwind at smaller angles of heel than narrow yachts with a deep keel.

6.7.6 Effects of Trim-In-Pitch

Under influence of the propulsive force of the sails a yacht experiences a bow-down pitching moment that causes a change in the pitch attitude of the yacht (trim-in-pitch). As already touched upon in Sects. 4.1 and 6.2 the pitch angle θ that a yacht adopts depends on the magnitude of the aerodynamic pitching moment caused by the sails, the hydrodynamic pitching moment caused by the hull plus appendages and the longitudinal metacentric height (gm_L, see Fig. 6.2.7). We have seen in Sect. 4.1 that the total, aerodynamic plus hydrodynamic, pitching moment M_y can be expressed as (see Eq. (4.1.10))

$$M_y = M_{Ay} + M_{Hy} = Td_z - V_A d_x, \tag{6.7.25}$$

where T is the propulsive force and V_A the vertical aerodynamic force. The quantities d_z and d_x are the vertical and longitudinal distances, respectively, between the center of pressure (centre of effort, CE) of the sails and the centre of pressure (centre of lateral resistance, CLR) of the underwater body (see Fig. 4.1.6). Because, in general, $T \gg V_A$ and $d_z \gg d_x$, and we have $T = R$ in equilibrium conditions of sailing, Eq. (6.7.25) can be approximated as

$$M_y \approx Rd_z \tag{6.7.26}$$

We have also seen (Eq. (4.1.9)) that the longitudinal equilibrium in pitch can be written as

$$M_y = M_{ry} \tag{6.7.27}$$

It follows from Sect. 6.2 that for the righting moment there holds

$$M_{ry} \approx -\nabla gm_L \sin \theta \tag{6.7.28}$$

This means that the trim-in-pitch angle is approximately given by

$$\theta \approx -Rd_z/(\nabla gm_L), \quad (\theta \ll 1 \text{ radian}) \tag{6.7.29}$$

It follows from Eq. (6.7.29) that the trim-in-pitch angle θ increases (bow down) with increasing hydrodynamic resistance, i.e. with increasing boat speed. Under upwind conditions and moderate wind speeds θ is usually of the order of (minus) 1°, or less. This means that there is hardly any effect on the hydrodynamic characteristics.

At high wind speeds, in particular when reaching or running, the bow-down trim-in-pitch angle may be as large as 3–4°. As discussed in some detail in Keuning and Sonnenberg (1998a) this can lead to some additional 'residuary resistance' (mostly wave-making resistance). It appears that, at high Froude numbers (Fr > 0.5), the magnitude of this can be of the order of 5–20 % of the total wave-making resistance due to volume. This means that it is not negligible in some practical situations of sailing. It also appears (Keuning and Sonnenberg 1998a) that this 'additional residuary resistance' due to trim-in-pitch increases with increasing beam/draft ratio of the hull.

It is important to note that the trimming moment can be influenced by the crew. Not only by the trimming of the sails, but also, and in particular for light displacement yachts and dinghies, by moving equipment and/or crew members for or aft.

6.8 Dynamic Stability and Sea-Keeping

The notion of dynamic stability in general was already introduced in Sect. 5.18. In the context of the periodic motions of a yacht in a seaway, it was also encountered in the Sub-Sect. on added resistance in waves in Sect. 6.5. More specifically the dynamic stability of a sailing yacht is usually associated with ship motions and the implications thereof for comfort, safety and controllability ('sea-keeping').

Heaving and pitching motions were already discussed in Sub-Sect. Added Resistance in Waves in the context of added resistance in waves. Here, we will address their importance in relation to seaworthiness. We will also consider rolling motions and dynamic directional stability, i.e. periodic motions around the yawing axis. The latter is important for controllability.

Effect of Waves on Righting Moment
In addition to the dynamic stability aspects of periodic ship motions, there are also dynamic effects, caused by waves, on the static hydrodynamic stability. One example is the effect of the waves generated by the yacht itself on the righting moment. For a yacht running at hull speed (Fr ≅ 0.4) in flat water, the water level around amidships is appreciably lower than at zero boat speed (see Fig. 6.5.25). As a consequence the effective beam and the resulting righting moment may be reduced significantly, in particular for hulls with a shallow draft.

Similar effects can occur in a seaway. In beam and oblique seas the (lateral) righting moment will, in general be smaller on a wave crest than in still water, while the opposite is the case in a wave trough. The effect, illustrated by Fig. 6.8.1, is most pronounced when the wave length is of the same order as the lateral dimensions of

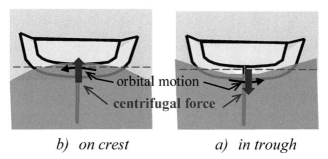

b) on crest   *a) in trough*

Fig. 6.8.1 *Effect of wave shape on effective beam*

the yacht. An additional factor is that the orbital motion of the water particles (centrifugal forces) near a wave crest causes a reduction of the apparent weight density of the water particles (see Sect. 5.19). This causes an additional loss of (dynamic) righting moment when a yacht is at a wave crest (Marchai 1983b).

It has been found (Marchai 1983b) that in extreme conditions the loss of lateral stability can be as much as 40 %. Needless to say that such a reduction of stability increases the risk of capsizing, in particular when the yacht has a small lateral mass moment of inertia.

It should also be mentioned that in head seas and following seas there can be a similar reduction of the longitudinal static stability.

Amplitudes of and 'g-forces' Due to Pitching and Rolling Motions
As mentioned earlier, periodic heaving and pitching motions are important for the added resistance in waves. They also play a role in comfort and safety. This through the resulting accelerations ('*g-forces*') that crew members may be subject to. In this respect pitching motions are more important than heaving motions because of the high accelerations that may occur at the bow and the stern.

As already mentioned in Sect. 6.5 a sailing yacht pitching in waves behaves somewhat like a (rotating) mass-spring-damper system (see also Sect. 5.18). In this analogy the buoyancy forces assume the role of the spring forces and the damping forces are due primarily to the energy contained by the out-going waves generated by the pitching motion. The in-phase hydrodynamic inertia forces (which act as an 'added mass' moment if inertia) are also due to a large extent to the generation of waves due to pitching. Under heel, additional damping forces can be generated by keel and rudder. The external driving force is caused by the wind-driven running surface waves.

An estimate of the (relative) levels of the amplitude of the pitching motion of sailing yachts can be obtained from the theory of mass-spring-damper systems. For this purpose one has to determine the natural frequency of the pitching oscillation of the yacht and the frequency of encounter with waves (Sub-Sect. Added Resistance in Waves, Appendix K). Information about the pitch damping and added mass moment of inertia of ships and sailing yachts can be found in the literature (Keuning 1998; Lewis 1989; Skinner 1982) (see also Sub-Sect. Added Resistance in Waves).

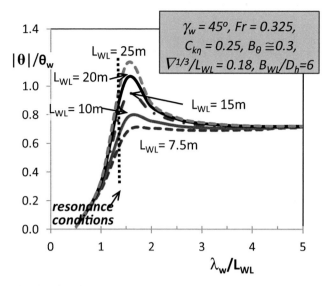

Fig. 6.8.2 *Amplitude of pitching motion in waves as a function of wavelength and boat length (qualitatively, normalized by wave slope)*

The results of such an estimate, for 'average' cruising yachts, in upwind conditions of sailing, are presented in Fig. 6.8.2. What Fig. 6.8.2 illustrates, is that, not surprisingly, the amplitude of the pitching motion in waves is relatively large when the frequency of encounter with waves is approximately equal to the natural frequency of the pitching motion of the yacht. The latter is the case for $\lambda_w/L_{WL} \cong 1.4$ (resonance conditions). It can further be noticed that larger yachts experience larger amplitudes. This is caused by the fact that, in a relative sense, the damping is smaller for the larger yachts. The main reason behind this is that the damping is smaller for lower frequencies (Lewis 1989).

Figure 6.8.3 presents the amplitude of the local accelerations at the bow or stern as a function of the frequency of encounter for the same conditions as in Fig. 6.8.2. Also indicated in this figure is the frequency at which humans are most sensitive for motion sickness (seasickness) (Golding et al. 2001; O'Hanlon and McCauley 1973). This is seen to be in the range 0.17–0.20 periods per second, which is a factor 1.5–3 lower than the frequency at which the acceleration level has its maximum. It means that the risk of seasickness due to the pitching motion on board of yachts sailing in waves is probably small, in particular for small yachts.

It is useful to note that Figs. 6.8.2 and 6.8.3 are applicable for 'average' cruising yachts. One should expect that the maximum amplitude of the pitching oscillation and the associated acceleration levels will be somewhat higher for yachts with less damping. This is the case for yachts with a smaller displacement/length ratio and/or a smaller beam/draft ratio of the hull (Keuning 1998; Lewis 1989). The effect of the gyradius seems to be less pronounced (Keuning 1998).

However, it is to be expected that a larger gyradius will move the resonance condition to lower frequencies (see Fig. 6.5.40).

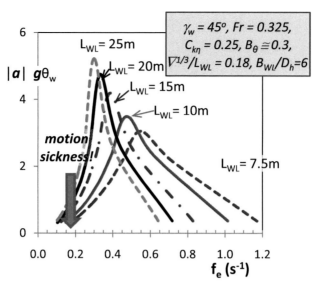

Fig. 6.8.3 *Amplitude of vertical acceleration due to pitching motion in waves as a function of frequency of encounter and boat length (qualitatively, normalized by the wave slope)*

When a similar analysis is performed for the rolling motion it is found that there are several differences. The first is the shorter 'arm' of the rolling moments; the beam of a sailing yacht being a factor 3–5 smaller than the length. The second is a lower natural frequency of the rolling motion (about a factor 2 lower, as we will see below). Because the acceleration is proportional to the frequency squared this implies a reduction of the acceleration level by another factor 4 or so. A third, counteracting factor is less hydrodynamic damping, in particular at low boat speeds, and a smaller added mass moment of inertia.

As already mentioned in Sect. 6.5 the hydrodynamic damping for the rolling motion is embodied primarily by the periodically varying, counter-acting side force on the keel. Because the side force varies proportionally with the square of the fluid velocity experienced by the keel, the hydrodynamic damping is quite effective for moderate and high boat speeds but small for low boat speeds and low roll frequencies.

The natural frequency of the rolling motion depends on boat length, beam/draft ratio and lateral radius of gyration (see Appendix K). The dependence is similar to the dependence of the pitching motion shown in Fig. 6.5.40. However, as illustrated by Figs. 6.8.4 and 6.8.5 the frequency level is about a factor 2 lower. The frequency is also seen to decrease with increasing length and lateral gyradius $C_{k\xi}$ and decreasing beam/length ratio.

The conditions for resonance in the rolling motion are somewhat different from those of the pitching motions. This is caused by the factor 2 difference in the natural frequency. Besides, rolling motions are most effectively triggered by beam seas (see Sect. 6.5, Fig. 6.5.37).

Fig. 6.8.4 Frequency of the natural roll-ing motion of sailing yachts as a function of length and lateral gyradius (estimated trend)

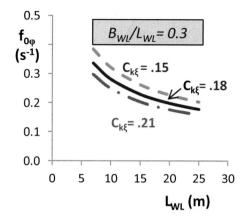

Fig. 6.8.5 Frequency of the natural rolling motion of sailing yachts as a function of length and beam/length ratio (estimated trend)

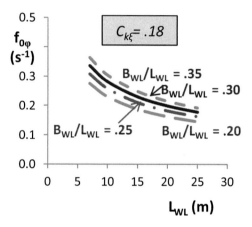

Conditions for resonance in the rolling motion are illustrated by Fig. 6.8.6. Recalling that the hydrodynamic damping is small only for low boat speeds, the figure indicates that significant amplitudes of the rolling motion are to be expected at all wave direction angles but only for wave lengths of the order of twice the boat length $(\lambda_w/L_{WL} \cong 2)$. In broad reaching and downwind conditions of sailing $(\gamma_w > 120°)$ resonance can also happen for shorter wavelengths. Low boat speed (Froude number) resonance conditions are particularly risky because of the small hydrodynamic damping under such conditions.

We have also seen in Fig. 6.5.37 that the excitation into rolling motion is largest for $60° < \gamma_w < 120°$ and $\lambda_w/L_{WL} \cong 1$. Summarizing, this means that quartering seas $(\gamma_w > \cong 120°)$ and relatively short wavelengths in combination with low boat speeds are the most critical for rolling motions.

It is useful to note that under such conditions the reduced frequency $k_f = \omega L/(2V_b)$ of the unsteady flow about the keel of a sailing yacht is typically of the order of 0.5. This means that time lag phenomena involving reduced or possibly even negative damping (Sect. 5.18) are probably important.

Fig. 6.8.6 *Trend of resonance conditions for the rolling motion as a function of wavelength and wave direction*

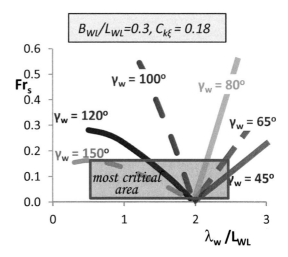

Note also that Fig. 6.8.6 applies to boats with a beam/length ratio of 0.3 and a roll gyradius of 0.18. For 'beamier' yachts and yachts with a smaller roll gyradius the critical wavelengths are somewhat shorter. The opposite is the case for yachts with a smaller beam/length ratio and/or a larger value of the gyradius.

Figure 6.8.7 gives an impression of the amplitude of the rolling motion in quartering seas ($\gamma_w = 120°$) as a function of wavelength and boat speed (Froude number). The figure is believed to be indicative for an 'average' cruising yacht configuration as shown in Fig. 6.7.11. It can be noticed that the amplitude is quite high for low Froude number (Fr = 0.05 and 0.1), but decreases rapidly with boat speed. Figure 6.8.7 is based on a mass-spring-damper model as described in Sect. 5.18. The hydrodynamic damping has been estimated by means of an extended 'lifting-line'-type

Fig. 6.8.7 *Amplitude of rolling motion in waves of an 'average' cruising yacht, as a function of wavelength and boat speed (Froude number) (qualitatively, normalized by wave slope)*

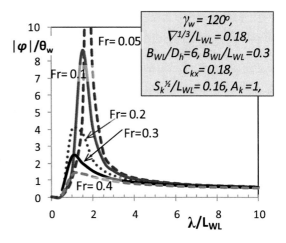

of method (Katz and Plotkin 1991 in Chap. 5; Speer 2010), developed by the author, for rotating motion about the x-axis of keel-hull configurations. The method is semi-empirical in the sense that it utilizes experimental data for the keel section characteristics. Results were found to agree also reasonable well with experimental data for zero boat speed conditions (Masyama et al. 2008; Klaka et al. 2001).

Figure 6.8.7 illustrates that boat speed is an important parameter for the hydro-dynamic damping of the rolling motion. Keel area and draught are also factors of importance. This because the lateral hydrodynamic forces acting on a sailing yacht are proportional to the area of the keel and the hydrodynamic rolling moment proportional to the draught. Keel area is particularly important in conditions of extreme rolling at low boat speeds when the flow about the appendages may no longer be attached. Under such conditions the appendages are less effective and may even create negative damping (see the discussion on *'dynamic stall'* in Sect. 5.18). In this respect high aspect ratio keels are more vulnerable because they have smaller stall angles than low aspect ratio keels (see Fig. 6.3.5).

Figure 6.8.8 presents a similar picture for different wave direction angles at a constant, low Froude number (Fr=0.1). It can be noticed that the risk of large roll amplitudes is relatively high in beam seas.

An impression of the g-forces due to rolling at the maximum beam position of a yacht is given in Fig. 6.8.9. The conditions of sailing are representative for a wave direction angle of 120° and a low Froude number (0.1). Note that the acceleration is normalized by the maximum wave slope and given as a function of the frequency of encounter with the waves. It can be noticed that the acceleration levels are comparable with those due to pitching oscillations (Fig. 6.8.3). Note, however, that the Froude number (boat speed) is quite low in the case of Figs. 6.8.8, 6.8.9. It follow

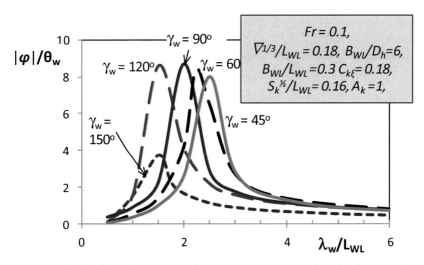

Fig. 6.8.8 Amplitude of the rolling motion of an 'average' cruising yacht at low boat speed, as a function of wave length and wave direction (qualitatively, normalized by wave slope)

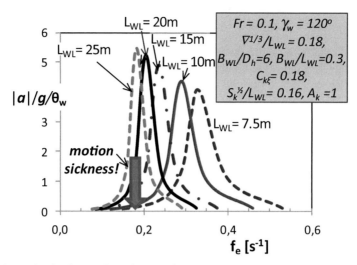

Fig. 6.8.9 *Amplitude of vertical acceleration due to rolling motion in waves as a function of frequency of encounter and boat length (qualitatively, normalized by wave slope)*

from Fig. 6.8.7 that the acceleration levels would be a factor of about 4 smaller for normal Froude numbers (0.3–0.4) of sailing.

Nevertheless, there may, at least at low boat speeds, be a risk of low comfort for large yachts. This because for a length of 20–25 m the resonance frequency is about the same as the frequency ($\cong 0.2$) at which humans are most susceptible for motion sickness (O'Hanlon and McCauley 1973).

It is further important to mention that for most conditions of sailing the hydrodynamic damping of the rolling motion is augmented by aerodynamic damping from the sails. In conditions with small sheeting angle (upwind) and high apparent wind speeds the aerodynamic damping is even much larger than the hydrodynamic damping, as we will see later (Sects. 8.5 and 8.6). This because of the much larger span (height) of the sails, and the correspondingly larger rolling moment, as compared to the span (draught) of the keel.

A more critical contribution of the sails, that can even lead to a dynamic instability in the rolling motion, may occur for large sheeting angles and true wind angles of around 180°. As already touched upon in Sect. 5.18 this is due to the fact that under such conditions the aerodynamic lift forces can enhance the rolling motion. In terms of the preceding discussion this can be considered as a situation with negative damping. We will return to this subject in Chaps. 7 and 8.

Dynamic Directional Stability

The dynamic stability of motions around the third, yawing axis of a yacht is important for the directional controllability. This, of course, also in particular in waves, which cause a periodically varying excitation in yaw.

A relatively simple way to address dynamic stability in yaw is to consider this in terms of a mass-spring-damper system (Sect. 5.18). Without external, wave-induced forces, the equilibrium of moments in the yawing motion can then be written as

$$(I_{zz} + I'_{zz})\ddot{\psi} + b_\psi \dot{\psi} + k_\psi \psi = 0 \tag{6.8.1}$$

Here, ψ is the instantaneous angle of yaw with respect to the time-averaged course. In the steady or fully time-averaged situation ψ is equal to the angle of leeway λ. The angular velocity and angular acceleration of the yawing motion are represented by $\dot{\psi}$ and $\ddot{\psi}$, respectively. I_{zz} is the mass moment of inertia around the yawing axis and I'_{zz} the moment of inertia of the added mass (Sect. 5.18). The rate of change $\dfrac{dM_{Hz}}{d\psi}$ of the yawing moment with the angular velocity of rotation plays the role of the damping coefficient b_ψ. The torsional spring constant k_ψ is equal to minus the rate of change $\dfrac{dM_{Hz}}{d\psi}$ of the restoring yawing moment with the angle of yaw. In the quasi-steady situation one would have

$$k_\psi = -\frac{dM_{Hz}}{d\psi} \approx -\frac{dM_{Hz}}{d\lambda} \tag{6.8.2}$$

The theory of damped harmonic motions (http://en.wikipedia.org/wiki/Harmonic_oscillator in Chap. 5) teaches that if, after an initial disturbance, the amplitude of the motion is to decrease with time (which defines a stable system), an essential condition to be satisfied is $k_\psi > 0$. Depending on the magnitude of the damping, the motion will be periodic or non-periodic. In the latter case the system is said to be overdamped.

The condition $k_\psi > 0$ implies, see Eq. (6.8.2), that

$$\frac{dM_{Hz}}{d\lambda} < 0 \tag{6.8.3}$$

In other words, a ship or yacht can only be dynamically stable in yaw when it has static directional stability. We have seen in Sect. 6.4 that, without sails, many, if not most yachts, do not have static, hydrodynamic directional stability with the rudder fixed (this applies also to ships in general). Hence, they also lack dynamic directional stability, at least when sailing under engine power. This means that course keeping requires continuous rudder action. As we will see in Chap. 8 it is another matter under sail.

As discussed in more detail in the literature (Keuning 1998; Lewis 1989), the yawing motion of a surface ship is usually coupled to other modes of motion, such as sway and roll in particular. When the coupling between sway and yaw is taken into account the condition for dynamic directional stability takes the form

$$-\frac{dS}{d\psi}\frac{dM_{Hz}}{d\dot{\psi}} + \left(\frac{dS}{d\dot{\psi}} - V\right)\frac{dM_{Hz}}{d\psi} > 0 \tag{6.8.4a}$$

or

$$\frac{\dfrac{dM_{Hz}}{d\psi}}{\dfrac{dS}{d\psi}} - \frac{\dfrac{dM_{Hz}}{d\dot{\psi}}}{\dfrac{dS}{d\dot{\psi}} - V} > 0 \qquad\qquad (6.8.4\text{b})$$

The latter form can be interpreted as that the centre of lateral resistance due to sway (or leeway) should be positioned further aft than the centre of lateral resistance due to rotation in yaw.

Note that in the quasi-steady situation, when $\dfrac{dM_{Hz}}{d\dot{\psi}}$ and $\dfrac{dS}{d\dot{\psi}}$ are zero, Eq. (6.8.4) reduces to (6.8.3) and Eq. (6.8.4b) implies that the position of the centre of lateral resistance due to leeway should be aft of the centre of gravity.

It depends on the configuration of a yacht whether the condition (6.8.4) for dynamic directional stability is satisfied or not. For conventional surface ships the directional stability is known (Lewis 1989) to depend on both the shape of the hull and the shape, size and position of appendages. The directional stability appears to decrease with increasing displacement/length ratio and increasing beam/draft ratio of the hull and increasing block coefficient. Fins are known (Lewis 1989) to improve the directional stability when their size and aspect ratio is increased. This positive effect is most pronounced for aft-positioned fins.

For sailing yachts an additional, unknown, aspect of the dynamic directional stability is the effect of the downwash from the keel on the forces on the rudder in unsteady or periodically varying flow conditions. In (quasi-)steady flow the downwash from the keel was seen to reduce the effectiveness of the rudder, in particular for low aspect ratio keels (Sect. 6.4). For low reduced frequencies this mechanism is probably quite similar, except, probably, for a phase shift between the time-varying load on the keel and the now also time-varying downwash from the keel at the position of the rudder. There is, therefore, no reason to believe that the dynamic aspects of the directional stability of a sailing yacht are essentially different from those of other types of surface ships (at least with the sails down). In other words, the dynamic directional stability properties of a sailing yacht will benefit from small displacement/length and small beam/draft ratios and big, aft-positioned keels and rudders with high aspect ratios.

While directional stability is good for course keeping it is also known to impair manoeuvrability. The latter in the sense that the turning radius of a yacht with good directional stability will, in general, be larger than that of yacht with marginal or negative directional stability. This implies that, as often, a compromise has to be made by the yacht designer. However, there is reason to believe that this compromise can be less severe when the movable (rudder) fraction of the total appendage area is large (Lewis 1989).

Dynamic Directional Stability in Waves

While the dynamic directional stability of a sailing yacht in calm water is already far from simple, it is an even more complex matter in waves. In regular waves, the equilibrium of hydrodynamic yawing moments can be expressed as

$$(I_{zz} + I'_{zz})\ddot{\psi} + b_\psi\,\dot{\psi} + k_\psi\,\psi = M_0\sin\omega_e t \qquad (6.8.5)$$

Here, M_0 represents the amplitude of the external yawing moment caused by the waves and ω_e is the angular frequency of encounter. 't' is the time. The first $(\ddot{\psi})-$ term represents the moment due to inertia, the second $(\dot{\psi})-$ term the hydrodynamic damping and the third $(\psi)-$ term the restoring moment. In the case of a yacht that is directionally unstable ($k_\psi < 0$), the implication of Eq. (6.8.5) is that the yaw angle ψ diverges in time. Active rudder deflection is needed to control the situation. When a yacht is directionally stable ($k_\psi > 0$) the yawing motion becomes periodic with an amplitude that depends on the external yawing moment M_0 induced by the waves and the hydrodynamic damping.

The basic mechanism is similar to that of pitching and rolling oscillations. This means that for $k_\psi > 0$ the yawing motion also has a natural (circular) frequency ω_0 given by

$$\omega_0 = \sqrt{\frac{k_\psi}{I_{zz} + I'_{zz}}} \qquad (6.8.6)$$

It means also that some form of resonance, with severe yawing, may occur when $\omega_e \cong \omega_0$. However, since k_ψ is usually quite small (if not <0), the natural frequency is quite low and resonance will occur only for very low frequencies of encounter (following or overtaking waves of long wavelength), if at all.

Like in the case of the pitching and rolling motions, the excitation in yaw depends on the wave slope, the direction of the waves relative to the orientation of the yacht and the ratio between wavelength and boat length. The influence of boat speed (Froude number) appears to be small (Lewis 1989).

Figure 6.8.10 gives a qualitative impression of the relative magnitude of the external yawing moment caused by incident waves. It is given in terms of an excitation factor ($F_{w\psi}$) as a function of the wave incidence and the wavelength. As discussed in some detail in Appendix L.1, $F_{w\psi}$ can be considered as a dimensionless amplitude of the external yawing moment due to the waves, normalized by the wave height. For reasons of symmetry the excitation is zero for $\psi_w = 0$ and $180°$, at least for zero heel. For a ship with longitudinal (fore/aft) symmetry the excitation must also be zero for $\gamma_w = 90°$. Because most sailing yachts do not have fore/aft symmetry, the minimum excitation will, in general, not be zero, in particular through the presence of the rudder. Also, it may occur at a wave incidence angle that differs somewhat from $90°$. The maximum excitation is seen to occur for $\psi_w \cong 45°$ and $\psi_w \cong 135°$ for wavelengths around $\lambda_w/L_{WL} \cong 2$. Qualitatively, this is in agreement with the theoretical and experimental results given in Thomas et al. (2006).

Fig. 6.8.10 *Wave excitation factor for yawing motion as a function of wave incidence and wavelength (qualitatively, zero heel, constant wave height)*

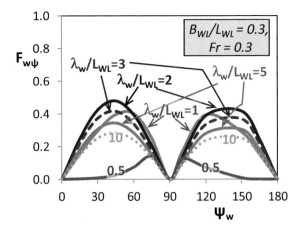

However, there is more to it. As already discussed before, an essential role in the mechanism of motion excitation by waves is played by the orbital motion of the water particles in waves (Sect. 5.19, see also Appendix L.1). For the excitation of the yawing motion this is illustrated by Fig. 6.8.11. When sailing in following or overtaking waves, the orbital motion of the water particles is such, that with the stern near a wave crest and the bow near a trough, an external yawing moment is

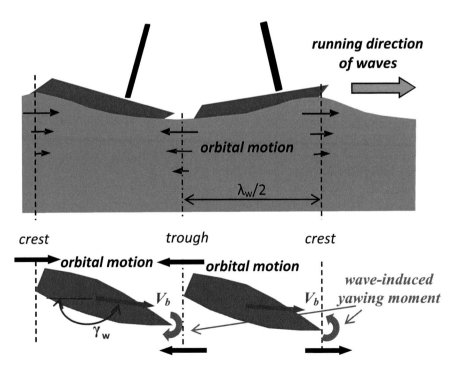

Fig. 6.8.11 *Illustrating the effect of the orbital motion of water particles on the wave-induced yawing moment*

caused by the wave that attenuates the yawing motion and, hence, is destabilizing (Lewis 1989; Thomas et al. 2006). With the stern in a trough and the bow on a crest this moment is of opposite sign and thus stabilizing. This mechanism is most pronounced when $L_{WL} \mid \cos \gamma_w \mid \cong \frac{1}{2} \lambda_w$. It is easily verified that the opposite is the case when sailing into the waves. It has been found (Lewis 1989; Thomas et al. 2006) that the magnitude of the destabilizing, wave-induced yawing moment in the downwind, stern-on-crest/bow-in-trough situation is, in general, so large, that it overrides any directional-stability-in-calm-water property that a ship or yacht may have. This implies that there is large risk of broaching (turning broadside to the waves), in particular for low frequencies of encounter ω_e. The reason is that, for low frequencies of encounter, a ship or yacht may be exposed to the destabilizing wave-induced yawing moment for a relatively long period of time. In addition there may, as already mentioned above, be a risk of resonance for small ω_e, due to the low natural frequency (if any) of the yawing motion. A large and effective rudder and heavy rudder action is required to control a yacht in such situations. The problem is even more severe when the yawing and rolling motions are strongly coupled and the aerodynamic moment induced by the sails is also destabilizing. We will discuss this in some further detail in Chaps. 7 and 8.

It is further clear, that the mechanisms just described will be most pronounced when a yacht has a hull with relatively large amounts of lateral area near the bow and/ or stern. Appendage configurations with separate, aft-positioned rudders are also vulnerable in this respect. The reason is that they are very effective in producing a local side force due to local cross flow at a large distance from the centre of gravity and an associated large yawing momentv (which, of course, is good for maneuverability). For long keels with attached rudder this is not the case. The resulting centre of side force caused by a wave-induced cross-flow at the trailing edge of a long keel is positioned much further forward and the associated yawing moment is much smaller (see also Sects. 5.15 and 6.4).

The mass moment of inertia about the yawing axis is another quantity of importance. Heavy yachts with a large moment of inertia about the yawing axis will respond slower to a yawing moment. Consequently they will have acquired less yaw than a light displacement yacht by the time the yawing moment due to the orbital motion of the waves has changed sign.

Altogether, the discussions just given suggest that classical, heavy yachts with long keel and attached rudder will have better dynamic directional stability properties in following and overtaking seas than modern light displacement yachts with separate keels and rudders.[16] Recalling (see the preceding paragraphs) that the opposite is the case for the calm water conditions, one may conclude that, here also, a yacht designer will have to adopt some form of compromise.

---

[16] This conclusion is similar to that of Marchai (1983c). However, the additional argument of augmented directional stability due to unsteady flow effects (time lag and reduced amplitude of forces) on long keels with attached rudders is not very convincing because of the low frequencies of encounter (of the order of 0.1 cycles/s or smaller), the higher boat speeds in following and overtaking seas and the smaller lift deficiency for low aspect ratios (Sect. 5.18).

References

Abbott IH, Von Doenhoff AE, Stivers LS (1945) Summary of airfoil data, NACA Report No. 824

Ba M, Coirier J, Guilbaud M (1990) Theoretical and experimental study of the hydrodynamic lift-ing flow around yawed surface-piercing bodies. 2nd international symposium on performance enhancements for marine applications, Newport

Beukelman W, Keuning JA (1975) The influence of fin keel sweep-back on the performance of sailing yachts. 4th HISWA symposium

Candries M, Anderson CD, Atlar M (2001) Foul release systems and drag. Proceedings of the PCE conference, Antwerp, pp 273–286

Claughton A, Pemberton R, Prince M (2012) Hull-Sailplan balance, "lead" for the twenty-first century. 22nd international HISWA symposium, Amsterdam, 12 and 13 Nov 2012

ESDU Aerodynamic Series. www.esdu.com

Etkin B, Reid LD (1996) Dynamics of flight: stability and control. Wiley, New York

Gerritsma J (1971) Course-keeping qualities and motions in waves of a sailing yacht. 3rd AIAA symposium on the aerodynamics and hydrodynamics of sailing

Gerritsma J, Beukelman W (1972) Analysis of the resistance increase in waves of a fast cargo ship. International shipbuilding progress

Gerritsma J, Keuning JA (1984) Experimental analyses of five keel-hull combinations, Report no. 637, Ship Hydromechanics Laboratory, Delft University of Technology

Gerritsma J, Keuning JA (1985) Further experiments with keel-hull combinations. Delft Univer-sity of Technology, Ship Hydromechanics Laboratory, Rept. Nr 699-P

Gerritsma J, Keuning JA (1987) Speed loss in waves, Report 772-P, Ship Hydromechanics Labora-tory, Delft University of Technology, Dec 1987

Gerritsma J, Moeyes G, Onnink R (1977) Test results of a systematic yacht hull series. 5th HISWA symposium yacht architecture

Golding JF, Mueller AG, Gresty MA (2001) A motion sickness maximum around the 0.2 Hz fre-quency range of horizontal translational oscillation. Aviat Sp Environ Med 72(3):188–192. ISSN 0095-6562

Holtrop J, Mennen GGJ (1978) A statistical power prediction method. ISP 25

Holtrop J, Mennen GGJ (1982) An approximate power prediction method. ISP 29

Kelvin L (1904) Deep-sea ship-waves. Proceedings of the royal society of Edinburgh

Keuning JA (1985) Resistance of sailing yacht propellers. Delft University of Technology, Ship Hydromechanics Laboratory, Report nr. 656

Keuning JA (1998) Dynamic behaviour of sailing yachts in waves, Chap 6 of Claughton et al. (1999 in Chap. 5)

Keuning JA (2006) An approximate method for the added resistance in waves of a sailing yacht, Report 1481-P, Ship Hydromechanics Laboratory, Delft University of Technology

Keuning JA, Sonnenberg UB (1998a) Approximation of the hydrodynamic forces on a sailing yacht based on the delft systematic yacht hull series. Proceedings of the 15th HISWA interna-tional symposium on yacht design and yacht construction, Amsterdam, Nov 1998

Keuning JA, Sonnenberg UB (1998b) Developments in the velocity prediction based on the Delft systematic yacht hull series, Report no. 1132-P, Ship Hydromechanics Laboratory, Delft Uni-versity of Technology, March 1998

Keuning JA, Vermeulen KJ (2003) The yaw balance of sailing yachts upright and heeled, Report no. 1363-P, Ship Hydromechanics Laboratory, Delft University of Technology

Klaka K, Krokstad J, Renilson MR (2001) Prediction of yacht roll motion at zero forward speed. 14th Australasian fluid mechanics conference Adelaide university, Adelaide, 10–14 Dec 2001

Kroo I (2007) Form factor. http://adg.stanford.edu/aa241/drag

Larsson L, Eliasson RE (1996) Principles of yacht design. Adlard Coles Nautical, London. ISBN 0-7136-3855-9

Letcher JS Jr (1965) Balance of helm and static directional stability of yachts sailing close-hauled. JRAeS 69(652)

Lewis EV (ed) (1988a) Principles of naval architecture, vol I—stability and strength. SNAME, Jersey City. ISBN 0-939773-00-7

Lewis EV (ed) (1988b) Principles of naval architecture, vol II—resistance, propulsion and vibration. SNAME, Jersey City. ISBN 0-939773-01-5

Lewis EV (ed) (1989) Principles of naval architecture, vol III—motions in waves and controllability. SNAME, Jersey City. ISBN 0-939773-02-3

Marchai T (1983a) Seaworthy yachts, Practical Boat Owner, No. 199, July 1983

Marchai T (1983b) Stability in motion, Practical Boat Owner, No. 201, Sept 1983

Marchai T (1983c) Directional stability, Practical Boat Owner, No. 202, Oct 1983

Marchai CA (2000) Aero-hydrodynamics of sailing. Adlard Coles Nautical, London, pp 465–469, 272, 658

Masyama Y, Fukasawa T, Onishi T (2008) Dynamic stability and possibility of capsizing of small light sailing cruiser due to wind. International conference (RINA) on innovation in high performance sailing yachts, Lorient

McWilkinson DG, Blackerby WT, Paterson JH (1974) Correlation of full-scale drag predictions with flight measurements on the C-141 aircraft-phase II, wind tunnel test, analysis and prediction techniques. Vol 1-drag predictions, wind tunnel data analysis and correlation. NASA CR-2333, Feb 1974

O'Hanlon JF, McCauley ME (1973) Motion sickness incidence as a function of the frequency and acceleration of vertical sinusoidal motion. Canyon Research Group Inc., Technical report. 1 July 1972–30 June 1973

Prandtl L, Schlichting H (1934) Das Widerstandgesetz rauher Platten, Werft-Reederei-Hafen, 1–4

Skinner GT (1982) Sailing vessel dynamics: investigations into aero-hydrodynamic coupling. Master's thesis, MIT, Cambridge

Silverstein A, Katzoff S (1939) Design charts for predicting downwash angles and wake characteristics of plain and flapped wings. NACA Report No. 648

Speer T (2010) 'Minimum induced drag of sail rigs and hydrofoils. http://www.tspeer.com. Accessed 2011

Thomas G, Harris D, d'Amancourt Y, Larkins I (2006) The performance and controllability of yachts sailing downwind in waves. 2nd high performance yacht design conference, Auckland, 14–16 Feb 2006

Torenbeek E (1982) Synthesis of airplane design. Kluwer, Netherlands

Van Driest ER, Blumer CB (1960) Effect of roughness on transition in supersonic flow. AGARD Rept. 255

Van Oossanen P (1981) Method for the calculation of the resistance and side force of sailing yachts. Paper presented at conference on 'calculator and computer aided design for small craft—the way ahead,' R.I.N.A.

Vroman R (2003) Principles of ship performance, EN200 courses, Chap 7, resistance and powering of ships. http://www.usna.edu. Accessed 2007

Whicker LF, Fehlner LF (1958) Free stream characteristics of a family of low aspect ratio control surfaces for application to ship design. DTRC report 933

Chapter 7
Forces Above the Water Surface: Aerodynamics

7.1   Functions of Hull, Rig and Sails

As in the case of the description of the hydrodynamic forces and moments acting on the under-water part of a sailing yacht, in the preceding chapter, it may be useful to consider the functions of the 'above-the-water' parts of a sailing yacht before entering the description of the aerodynamic forces and moments acting on these parts.

The most important function of the 'above-the-water-surface' parts of a sailing yacht is to provide the force that propels the yacht forward. The sails in particular are, of course, meant for this purpose.

We have seen in Chap. 4 that the sails should, for a given wind speed and direction of sailing, provide (most of) the driving force T while keeping the side force S_A as small as possible (Fig. 7.1.1). We have also seen that this requires:

- In upwind conditions (small apparent wind angle β):

  - A high lift/drag ratio L/D and a high lift L at low wind speeds
  - The highest possible L/D but a moderate lift L at high wind speeds

- In half wind (reaching) conditions ($\beta \cong 90°$): the highest possible maximum lift (L_{max}) but a moderate drag D
- In downwind conditions ($\beta \cong 180°$): the largest possible drag D but a small lift force L

A very important practical aspect of the sails is that they have to be reefable and storable in heavy weather conditions and in the harbor. This requires that they are made of thin, foldable materials. Without this requirement (and with enough money to spend!) aircraft-wing type sails would be a serious option, as demonstrated by BMW-Oracle in the 2010 America's Cup race.

© Springer International Publishing Switzerland 2015
J. W. Slooff, *The Aero- and Hydromechanics of Keel Yachts,*
DOI 10.1007/978-3-319-13275-4_7

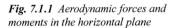
Fig. 7.1.1 *Aerodynamic forces and moments in the horizontal plane*

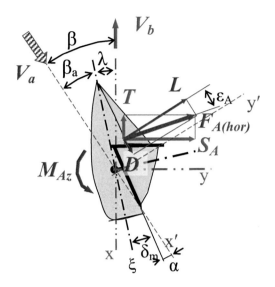

The function of the rigging (mast, spreaders, stays, shrouds, booms, etc.) is, in the first place, nothing else but to carry the sails and to transfer the forces acting on the sails to the hull. However, it is clear that the rigging should perform this job preferably without impairing the aerodynamic functioning of the sails. Unfortunately it is far from clear what this requires in terms of aerodynamic shaping of the various parts of a sailing rig. This is due to the fact that the aerodynamic requirements for different points of sailing can be completely opposite, as summarized above. It means, for example, that mast and rigging should preferably not decrease the lift and/or increase the drag of the sails in upwind and reaching conditions. However, they should preferably produce as much drag as possible when sailing downwind.

Even if we would know what kind of shapes are favourable for different points of sailing, the requirements would almost certainly be conflicting for any object with a fixed shape. Given the fact that sailing yachts spend most of their time with sailing upwind, shaping for low drag in such conditions is probably a reasonable choice that is adopted by most yacht designers.

The main functions of the above-the-water part of the hull are:

- Provision of space in cabin and cockpit for crew and equipment (partly together with the under-water part of the hull)
- Provision of a support base for mast and rigging

As in the case of the rigging these functions are to be performed preferably without impairing the aerodynamic functioning of the sails. Again, this leads to different aerodynamic requirements for different points of sailing which we will not repeat here (see the preceding paragraph).

For a hull there are two additional aspects that have important consequences for the aerodynamics. The first is that, in general, the dimensions of the hull of a sailing yacht are of the same order as the dimensions of the sails. This means that the effect of the hull on the aerodynamics is relatively large. In general significantly larger than the effect of the rigging. Secondly, the possibilities for aerodynamic shaping of the hull are severely limited by all sorts of practical, constructional or formal (class rules, safety) constraints.

Nevertheless, and in spite of all limitations, the hull of a sailing yacht does not have negative effects on the aerodynamics only. We will return to this in Sect 7.6.

7.2   Wind and Wind Gradient

Sailing is an activity that can be done only when there is a wind blowing and the wind experienced by an observer or object on the surface of the earth is nothing else but the flow of air in the earth's atmosphere. Like in the case of the flow about any arbitrary body the flow of atmospheric air about the earth creates a boundary layer (Sect 5.11) on its surface. In other words, sailing is an activity that takes place in the boundary layer of the flow within the earth's atmosphere (Fig. 7.2.1).

We have seen in Sect 5.11 that the flow in the boundary layer of an object is of a turbulent nature when the Reynolds number is sufficiently large. Because of the enormous scale of the atmospheric boundary layer, the flow within it (i.e. the wind) is always turbulent. This means that the flow velocity (wind speed and wind direction) varies continuously in space and time. The range of scales, both in space and time, of these fluctuations is also enormous. Spatially it varies from microscopic to hundreds or even thousands of meters. In time it varies from microseconds to minutes.

Figure 7.2.1 depicts the time-averaged velocity profile of the atmospheric boundary layer. The velocity profile is given in terms of the dimensionless quantity

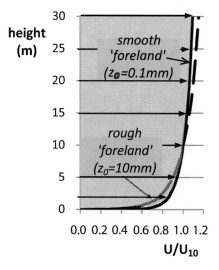

Fig. 7.2.1 *The atmospheric (turbulent) boundary layer, where sailing takes place*

U/U_{10}, where U is the (horizontal component of) the wind velocity and U_{10} is the wind velocity at a height of 10 m. It has been found (Simiu and Scanlan 1986) that this can be expressed as

$$U/U_{10} = \{(1/k_{turb}) \ln(z/z_0)\}, \qquad (7.2.1)$$

where ln is the natural logarithm function, z is the height above the surface and z_0 is the so-called roughness length. 'k_{turb}' is a universal constant equal to about 0.4. For a smooth sea, the roughness length z_0 is of the order of 0.1 mm. Behind wooded farmland it is of the order of 10 mm. A common value adopted in the International Measurement System (IMS) is 1 mm.

The figure illustrates that near the surface, the time-averaged wind speed increases rapidly with height. The enormous scale is reflected in the height above the surface at which the wind velocity hardly increases any further with increasing height. In boundary layer terminology this is the thickness of the boundary layer. Depending on the wind speed and the roughness of the landscape or 'seascape' the thickness of the atmospheric boundary layer can be hundreds of metres. This means that sailing takes place in the lower portions of the atmospheric boundary layer.

The precise form and extent of the time-averaged velocity profile depends on the wind speed, the turbulence intensity, the vertical temperature gradient of the atmosphere (atmospheric stability) and the roughness of the 'foreland'. The latter is the land or sea area upwind of the geographical position that is considered. As indicated in Fig. 7.2.1, the time averaged velocity increases less rapidly with height when the 'foreland' is rough (hills, tall buildings) than when the 'foreland' is smooth (low, flat lands or sea).

Because the wind velocity varies with the height above the water surface, the true wind speed V_t experienced by a sailing yacht should be defined in relation to the height above the water surface. A logical way to do so would be to couple the true wind speed to the height of the masthead, which is usually the position of the wind velocity sensor (anemometer) of a sailing yacht. This, however, has the obvious drawback that the definition would be based on a different height for every type and size of yacht. For this reason the true wind speed is usually referenced to a height of 10 m above the water surface. In other words, $V_{tref} = U_{10}$ in terms of Fig. 7.2.1.

In principle, the wind velocity profile can vary in terms of both speed and direction. Since very little is known about the latter it is common practice to assume that the true wind direction does not vary along the length of the mast.

We have seen in Sect. 3.3 (Fig. 3.1.1) that the wind experienced by the sails of a sailing yacht (the apparent wind V_a) is the vector sum of the true wind V_t and the reciprocal—V_b of the boat speed. With $V_t = V_t(z)$ varying in the vertical direction, (z), this means that the apparent wind V_a also varies along the length of the mast. This now applies to the wind speed as well as the wind direction (Fig. 7.2.2). As a consequence the apparent wind profile acquires a certain amount of 'twist'. That is, the apparent wind angle β, which depends on the ratio V_b/V_t and the true wind angle γ, varies with height.

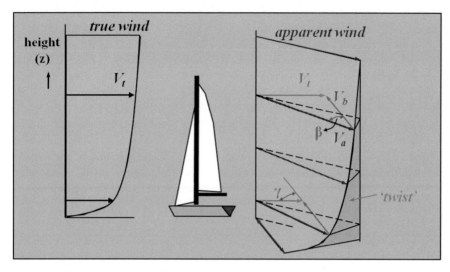

Fig. 7.2.2 *Illustrating the twist of the apparent wind profile*

The twist of the apparent wind profile can be considered for different true wind speeds and different apparent wind angles. It is then found (Fig. 7.2.3a that the twist decreases when the true wind speed increases. The twist of the apparent wind profile increases with increasing true and apparent wind angles (Fig. 7.2.3b).

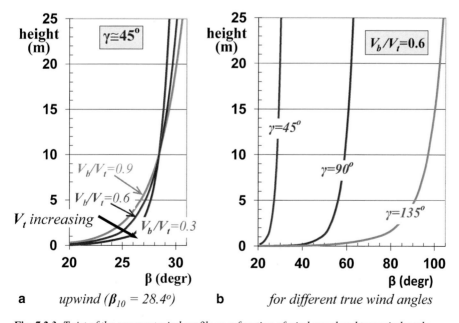

Fig. 7.2.3 *Twist of the apparent wind profile as a function of wind speed and true wind angle*

Figure 7.2.3a gives, for close-hauled conditions, the variation with height of the apparent wind angle for several values of the ratio V_b/V_t between boat speed and the reference true wind speed (V_{t10}). Recall, when considering this figure, that the ratio V_b/V_t decreases with increasing true wind speed (see Sects. 3.3 and 4.1). The figure illustrates that, in close-hauled conditions, the twist of the apparent wind profile between 1 and 15 m above the water surface varies between, say, 2.5° at high wind speeds (corresponding with $V_b/V_t = 0.3$) to roughly 6° at low wind speeds ($V_b/V_t = 0.9$). For large true wind angles γ this can easily amount to 25° (See Fig. 7.2.3b).

Figure 7.2.4 presents the variation with height of the magnitude of the apparent wind speed for the same conditions as in Fig. 7.2.3. For upwind conditions Fig 7.2.4a, the variation between 1 and 15 m above the water surface of the magnitude of the apparent wind speed is seen to be of the order of 20% for low wind speeds ($V_b/V_t = 0.9$) to about 30% for high wind speeds ($V_b/V_t = 0.3$). For large apparent wind angles (Fig. 7.2.4b, it can be more than 40% ($\gamma = 135°$). These figures illustrate clearly that the non-uniformity of the flow of air to which the sails of a yacht are exposed is quite substantial. The effect of this non-uniformity on the aerodynamic forces is, not surprisingly, significant. This will be addressed in some detail in a later section.

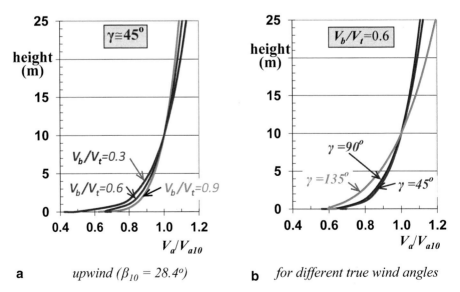

a   upwind ($\beta_{10} = 28.4°$)   b   for different true wind angles

Fig. 7.2.4 *Variation with height of the magnitude of the apparent wind speed as a function of boat speed and true wind angle*

7.3   Apparent Wind Angle, Angle of Attack and Sheeting Angle

We have seen in Sect. 3.4, Chaps. 4 and 5, that the aerodynamic forces acting on an object like a sail of given shape and dimensions are determined primarily by the magnitude of the apparent wind speed and the angle of attack.

In Sect. 3.4 the angle of attack α was seen to be determined by the apparent heading angle β_a and the sheeting angle δ (see also Fig. 7.1.1):

$$\beta_a = \alpha + \delta \qquad (7.3.1)$$

Because the apparent heading angle β_a is coupled to the apparent wind angle β through the angle of leeway λ

$$\beta_a = \beta - \lambda, \qquad (7.3.2)$$

Equation (7.3.1) can also be expressed as

$$\beta = \alpha + \delta + \lambda \qquad (7.3.3a)$$

or

$$\alpha = \beta - \delta - \lambda \qquad (7.3.3b)$$

Equation (7.3.3) implies that, because the apparent wind angle β varies with height, the angle of attack α also varies with height. Besides, the sheeting angle δ of horizontal sections of a sail also varies (increases) with height under the influence of the aerodynamic loads. While the reference value of the apparent wind angle β is usually taken as its value at a height of 10 m (β_{10}), the sheeting angle of the foot (base) or the boom of the sail (see Fig. 2.3) is commonly taken as the reference for the sheeting angle.

As will be discussed in some more detail in Sect. 7.12, the geometrical twist of a sail depends in particular on the aerodynamic normal force acting on the sail plus the stresses in the sail cloth caused by the tension in the sheet and the boom vang or the kicking strap and position of the traveller (if applicable). There is not much information available as to how the geometrical twist of a sail can vary along the length of the mast. However, it seems (Marchai 2000; Fossati et al. 2008) that, for a properly trimmed sail, this variation is in general (mildly) non-linear. The latter in the sense that, going upwards from the boom, the local sheeting angle first increases at an almost constant rate but the rate of increase becomes less towards the top of the sail.

Figure 7.3.1 gives an example of how the geometrical twist of a sail can vary along the length of the mast. The figure has been derived from data presented in Ref. Fossati et al. (2008) for an IMS class sail configuration.

*Fig. 7.3.1 Example of the variation along
the mast of the geometrical twist of a sail
(dimensionless quantities)*

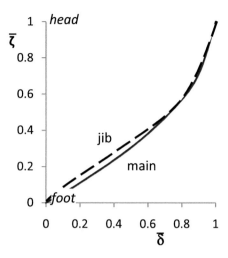

Shown is the relative twist angle $\bar{\delta}$ versus the dimensionless distance $\bar{\zeta}$ along the length of the mast, where

$$\bar{\delta} = (\delta - \delta_0)/(\delta_h - \delta_0) \tag{7.3.4}$$

and

$$\bar{\zeta} = (\zeta - \zeta_0)/(\zeta_h - \zeta_0) \tag{7.3.5}$$

Here, δ_0 and δ_h are the sheeting angles at the foot and at the top (head) of the sail, respectively, and ζ_0, ζ_h the corresponding coordinates in the direction of the mast.

It is emphasized that Fig. 7.3.1 is meant to give an indication only for the case of a reasonably well trimmed sail. Apart from the fact that it is based on data from a wind tunnel model it is to be expected that the variation depends also on the plan-form, cut and material of the sail. Nevertheless, the example can be used to get an impression of the vertical variation of the angle of attack that a sail can be exposed to. Figure 7.3.2 serves this purpose. Shown is the vertical variation of the angle of attack α ('aerodynamic twist') due to the twist of the apparent wind profile and a geometrical twist of the kind indicated in Fig. 7.3.1. Note that the foot of the sail is assumed to be positioned at 2 m above the water surface and the head (top) of the sail at 17 m. The figure on the left (a) represents a condition of upwind sailing ($\gamma = 45°$), that on the right (b) a condition of broad reaching ($\gamma = 135°$). In both cases a boat speed to true wind speed ratio V_b/V_t of 0.6 has been assumed. Also in both cases the total amount of geometrical twist $\delta_h - \delta_0$ has been chosen such that the angle of attack is approximately constant in the vertical direction. For the upwind condition this means $\delta_h - \delta_0 \cong 3°$ and for the broad reaching condition $\delta_h - \delta_0 \cong 16°$. The reference sheeting angle δ_0 of the foot of the sail is about 7° for the upwind case and about 60° for the broad reaching condition. Note that is corresponds roughly with the range of optimum sheeting angles indicated in Fig. 4.4.4.

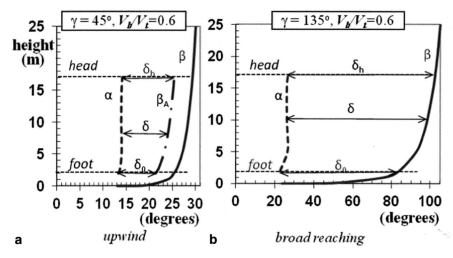

Fig. 7.3.2 *Example of vertical variation of angle of attack due to the twist of the apparent wind profile and the geometrical twist of a sail*

What Fig. 7.3.2 attempts to illustrate is that it is possible to, approximately, compensate the twist of the apparent wind profile by geometrical twist of the sail.[1] Whether the resulting almost constant angle of attack is always the optimum is another matter. We will come back to this in some of the sections to follow.

7.4 Single Sails

7.4.1 Introduction

In terms of the general fluid dynamic considerations of Chap. 5 the sails of a sailing yacht are thin lifting surfaces. Unlike the lifting surfaces below the water surface (appendages), sails are made of thin, deformable and foldable material.

Conventional sails are made of some woven fabric. Woven fabrics have, by themselves, a number of drawbacks: they are

- porous
- 'rough'
- subject, in principle, to permanent deformation under continuous, repetitive strain

Porosity has been found (Marchai 2000, pp. 510–516) to reduce the lift force on a sail, in particular under upwind conditions at high wind speeds. The loss of lift and associated driving force can be as much as 10 %. The porosity of a woven fabric

---

[1] As indicated by Fig. 7.3.2b), the very lower part of a sail is, perhaps, an exception.

can be reduced or even eliminated by applying some kind of filler or coating. This, however, has only a minor effect on the roughness of the surface. Preferably, the latter should be avoided also, because, as discussed in Sect. 5.11, it leads to additional friction drag (see also (Sub-)Sect. Viscous Resistance) and a loss of maximum lift.

The sensitivity for permanent deformation depends, of course, on the material of the fibres, but also on the pattern of weaving and the direction in which the fibres are applied. The latter can be chosen such that the fibres are optimally conditioned to absorb the loads and the resulting stresses.

Modern (but expensive!) sails are made of laminated materials. A laminate can, for example, consist of two skin films of Mylar or a similar material with a matrix of polyester, Kevlar or carbon fibres in between. The advantages of such materials are obvious: zero porosity, very low roughness and a high resistance to plastic deformation. There is, as always, also a negative aspect (apart from the price): a risk of delamination under repetitive (fatigue) loads.

We have seen in Chap. 4 that the aerodynamic characteristics required of a sail are different for different points of sailing. Upwind sailing was found to require a high lift/drag ratio at a high level of lift, reaching requires the maximum amount of lift but with a moderate amount of drag and downwind sailing requires the maximum amount of drag but with a small amount of lift. While each of these conditions requires its own angle of attack and corresponding sheeting angle, they all need sails with a substantial amount of sectional camber (see also Sect. 5.14), albeit that, as we will see shortly, one condition requires more camber than the other.

7.4.2   *Flow and Force Characteristics of Sail Sections*

The fluid dynamic behaviour of thin, highly cambered, sharp edged sail section is distinctly different from the behaviour of the thick, round nosed sections that are commonly used in aircraft wings and sailing yacht keels. In terms of flow properties at small to moderate angles of attack, the most characteristic feature of sail sections is the presence of a laminar separation bubble at the sharp leading edge (Fig. 7.4.1). Because of the sharp edge the laminar separation bubble is always present, except for one particular angle of attack when the attachment point of the flow is exactly at the leading edge. This is case 'B' of Fig. 7.4.1. The particular angle of attack is called the 'ideal' angle of attack.

For angles of attack below the ideal angle of attack the attachment point of the flow moves to the upper surface and the leading edge separation bubble is formed on the lower surface. This the case 'C' in Fig. 7.4.1.

For angles of attack larger than the 'ideal' angle of attack the attachment point moves to the lower side of the section (pressure or windward side) and a separation bubble appears on the upper surface (suction or leeward side, case 'A'). The length of the separation bubble increases with increasing angle of attack. Beyond a certain angle of attack, which depends on the amount of camber and the Reynolds number, the reattached, and now turbulent, boundary layer will also separate when

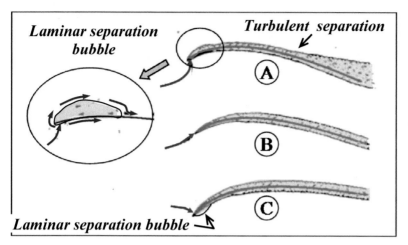

Fig. 7.4.1 *Typical flow characteristics of sail sections at small to moderate angles of attack*

it approaches the trailing edge. We will see later that for (very) low Reynolds numbers and/or (very) large camber, the turbulent boundary layer may already exhibit separation at the ideal angle of attack.

When the angle of attack is increased further, the turbulent separation moves forward and is overtaken by the separation bubble at the leading edge. There is then separated flow over all of the upper surface (Fig. 7.4.2).

Lift and Drag
The flow properties just described cause the aerodynamic force and moment characteristics of sail sections to be quite different from those of the thick, round nosed sections that are applied in keels. For the lift force at small to moderate angles of attack this is illustrated by Fig. 7.4.3. Shown is the lift coefficient as a function of the angle of attack for a sail section (NACA 65 type mean line (Abbott and Von Doenhoff 1949)) with a camber/chord ratio of 12 % and for a keel section (NACA 63-012) with a thickness/chord ratio of 12 %. The curves are based on data from Milgram (1971) and Abbott and Von Doenhoff (1949) and are representative for fully turbulent flow at typical sail and keel Reynolds numbers, respectively.

What strikes first is the difference in lift level. For small angles of attack the lift of the sail section is about just as high as the maximum lift of the keel section.

Fig. 7.4.2 *Flow pattern about sail section at high angle of attack*

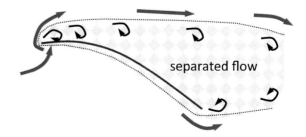

separated flow

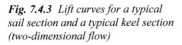

*Fig. 7.4.3 Lift curves for a typical
sail section and a typical keel section
(two-dimensional flow)*

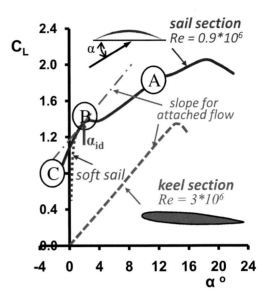

Secondly, where the lift slope of the keel section is constant (at about 2π per radian or 0.1 per degree) until close to maximum lift, the behaviour of the lift curve of the sail section is quite non-linear. For very small angles of attack the lift curve slope is appreciably larger than that of the keel section (point 'C' in Figs. 7.4.1 and 7.4.3). For angles of attack above, say, 4°, the slope is much smaller. The transition takes place where the slope is about equal to that of the keel section (attached flow). This is the point '**B**' that is also indicated in Fig. 7.4.1, which corresponds with the 'ideal' angle of attack α_{id} at which there is attached flow at the leading edge. With a leading edge separation bubble on the upper surface ($\alpha > \alpha_{id}$, point '**A**') the lift curve slope is much smaller. This is caused by a loss of suction on the front part of the upper surface due to the leading edge bubble and a boundary layer (aft of the bubble) that separates before it reaches the trailing edge. With the leading edge separation bubble on the lower surface ($\alpha < \alpha_{id}$, point '**C**') the lift curve slope is larger.

It should be noted that the lift curve shown is representative for a 'hard' sail foil. A 'soft' sail will start losing its shape for $\alpha < \alpha_{id}$. This causes the lift curve to behave more like the dotted line in Fig. 7.4.3 for angles of attack below α_{id}.

Figure 7.4.4 compares the drag characteristics of the same typical sail and keel sections. It can be noticed that the minimum drag is much higher for the sail section but is obtained for a much higher value of the lift. As a result the minimum drag angles (ε_{min}), or maximum lift/drag ratios of the two sections are about the same.

It can also be noticed that the sail section with the sharp leading edge is much more sensitive for variation of the lift coefficient or angle of attack than the thick, round-nosed keel section. As a consequence the lift/drag ratio of the sail section decreases more rapidly from its maximum value (or minimum drag angle) for a small deviation of the angle of attack than the keel section. This can also be seen in Fig. 7.4.5, which shows the lift/drag ratio as a function of the angle of attack.

Fig. 7.4.4 *Drag polars for a typical sail section and a typical keel section (two-dimensional flow)*

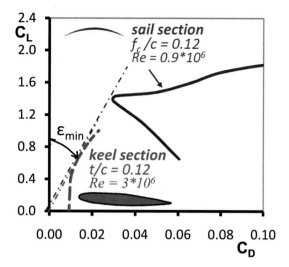

Fig. 7.4.5 *Lift/drag ratio as a function of angle of attack for a typical sail section and a typical keel section (two-dimensional flow)*

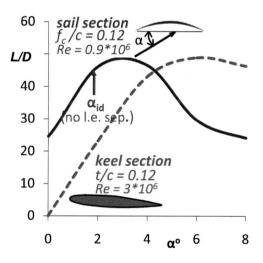

Figure 7.4.5 also illustrates that the maximum lift/drag ratio or minimum drag angle is obtained for an angle of attack that is slightly higher than the 'ideal' angle of attack α_{id}. The latter is, as mentioned above, the angle of attack for which the attachment point in the flow is located precisely at the sharp leading edge and for which there is no leading edge separation bubble.

It is, perhaps, interesting to note that much higher lift/drag ratios and, hence, much smaller minimum drag angles, have been obtained for special, so-called 'high lift' airfoils of the type first developed by Liebeck (1973). This is illustrated by Fig. 7.4.6 which compares drag polars for a 12% thick high lift section with 7% camber and a thin sail section with 12% camber. The Reynolds number is $3*10^6$ and the flow is assumed to be fully turbulent on both surfaces. The curves are based

Fig. 7.4.6 *Drag polars for a typical sail section and a high lift section (two-dimensional flow)*

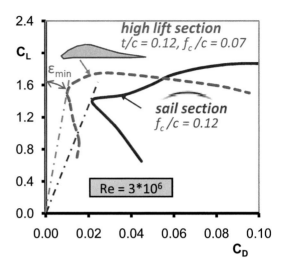

on data presented in Marchai (2000, p. 240, 444, pp. 302–325) and Olejniczak and Lyrintzis (1994) with the drag data adapted for differences in Reynolds number and boundary layer condition.

It can be noted that the minimum drag angle of the high-lift section is about half that of the sail section. Moreover, the minimum drag angle occurs at a slightly higher lift coefficient. The main reason for this is that the airfoil has been designed so as to avoid boundary layer separation up to close to the maximum lift coefficient.

Needless to say that such section characteristics would be extremely interesting for application in (wing) sails. Not in the least because it can be estimated, using the empirical relations from Appendix H, that for a thinner section with the same upper surface, implying more camber, the lift and lift/drag ratio might even be up to 40 % higher. However, and unfortunately, it does not seem to be possible to realize such shapes with foldable materials. Inverting the shape when tacking from port to starboard and the other way around is another problem.

Moment

The behaviour of the moment coefficient of a thin, sharp-edged sail section is also quit different from that of round-nosed foils. This is illustrated by Fig. 7.4.7. The figure shows the coefficient $C_{M^{1/4}}$ of the moment with respect to the quarter chord point as a function of the angle of attack for a sail section and for a symmetrical keel section. While the $C_{M^{1/4}}$ of the keel section is practically constant and zero, except for close to maximum lift, that of the sail section is highly negative ('nose down' or 'to weather') and varying in a strongly non-linear way. The negative level is caused by the large amount of camber (see also Sect. 5.14). The variation with angle of attack is caused by the boundary layer separation at the leading edge and further downstream towards the trailing edge. Fully attached flow occurs only at the 'ideal' angle of attack α_{id} (provided, as we will see shortly, that the Reynolds number is sufficiently high).

An implication of the variation of the moment with angle of attack is that the contribution of the sails of a sailing yacht to the yawing moment can vary significantly

Fig. 7.4.7 *Variation of moment coefficient with angle of attack (two-dimensional flow)*

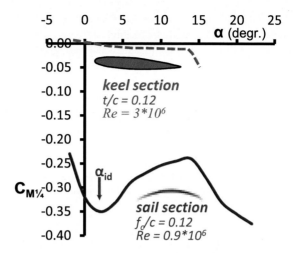

with varying apparent wind angle and apparent heading. This means that rudder action or sail trim is required to maintain or restore balance in yaw. Starting out from a balanced situation at $\alpha=\alpha_{id}$, the curve in Fig. 7.4.7 indicates that the sails develop a less negative moment, that is a yawing moment to lee, when the angle of attack changes, irrespective of the sign of the change in angle of attack. A more negative moment (yawing moment to weather) develops when the angle of attack is increased beyond about 15°, that is when boundary layer separation on the lee side of the sail moves forward progressively.

Effects of Type and Amount of Camber
In sail maker's and sailing practice the shape of a section of a sail is, in general, characterized by four quantities (see Fig. 7.4.8):

- the maximum camber (or 'draft') f_c, in parts of the chord length c. That is f_c/c, usually expressed as a percentage
- the position x_f of maximum camber, in parts x_f/c of the chord length, measured from the leading edge
- the entry angle ι_0, that is the slope at the leading edge, in degrees
- the exit angle ι_1, that is the slope at the trailing edge, in degrees

When considering similar shapes, that is shapes which can be described mathematically by the same type of formula (such as a circular arc or a parabolic arc), the entry and exit angles follow from the values of f_c and x_f. In such case the last two parameters suffice to describe the profile shape.

Fig. 7.4.8 *Quantities defining the shape of a sail section profile*

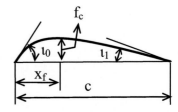

It will be clear from Fig. 7.4.8 that increasing the maximum camber f_c will, in general, lead to an increase of both the entry angle and the exit angle. Moving the position of maximum camber forward (reducing x_f), will increase the entry angle ι_0 and reduce the exit angle ι_1. It is the other way around when the position of maximum camber is moved further aft.

The two most important aerodynamic characteristics of a sail, maximum lift and minimum drag angle (or maximum lift/drag ratio), are influenced by camber in different ways.

For the level of lift at a given angle of attack the exit angle is the most important. Increasing the exit angle implies, for attached flow and a given angle of attack, that more circulation and hence more lift will have to be generated in order to satisfy the Kutta condition of smooth flow detachment at the trailing edge (Sect. 5.14). There are, however, limits to the maximum amount to which the exit angle can be increased. Increasing the exit angle implies an increase of aft camber. The latter leads to larger pressure gradients on the aft part of the suction surface towards the trailing edge. Depending on the Reynolds number this may lead to a more forward position of the separation point of the turbulent boundary layer and a corresponding loss of lift. Increasing the lift level for a given angle of attack will, in general, also lead to an increase of the maximum lift. Unless, again, the Reynolds number is too low to prevent early boundary layer separation.

Increasing the entry angle means that the leading edge region of the sail section is inclined more into the direction of the incoming flow. This leads to an increase of the 'ideal' angle of attack α_{id} and an associated increased level of lift at α_{id}. The provision for the latter is, again, that the Reynolds number is high enough to prevent early boundary layer separation on the aft part of the suction surface towards the trailing edge.

Figure 7.4.9 gives an impression of the dependence of the lift behaviour of a sail section on the amount of camber. Shown are curves of the lift coefficient C_L as a

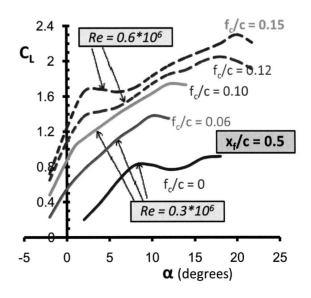

Fig. 7.4.9 *Dependence of the lift coefficient C_L, as a function of the angle of attack α, on the amount of camber f_c of a sail section (two-dimensional flow)*

function of the angle of attack α for foil sections with different amounts of camber, all with the position of maximum draft at 50% of the chord. The figure is based on data from Milgram (1971) and Marchai (2000, p. 240, 444, pp. 302–325). Note that the Reynolds number for the sections with the smaller amount of camber (0, 6 and 10%) are a little lower than those for the sections with 12 and 15% camber. This difference is believed to be small enough to be qualitatively insignificant.

The figure illustrates that the level of lift increases strongly with the amount of camber. At small angles of attack, close to the ideal angle of attack α_{id}, the lift is, roughly, proportional with the amount of camber.

The effect of the amount of camber on the drag characteristics at small to moderate angles of attack is shown in Fig. 7.4.10. The figure presents drag polars (lift coefficient C_L versus drag coefficient C_D) for different amounts of camber. The conditions are the same as those of Fig. 7.4.9.

It can be noticed that, in agreement with the description in Appendix H, the minimum drag increases with increasing camber/chord ratio f_c/c. It can also be seen that the smallest minimum drag angle ε_{min} is obtained for a camber/chord ratio f_c/c equal to about 0.10.[2] For larger values of f_c/c the minimum drag increases rapidly, in particular at low Reynolds numbers. For smaller values of f_c/c the benefit of the increased lift level is not big enough.

It is useful to note at this point that Fig. 7.4.10 implies that sails with a moderate amount of camber, say $f_c/c \cong 9\%$, are a good choice for sailing close-hauled. This because, as discussed in detail in Chap. 4, a small aerodynamic drag angle ε_A is

Fig. 7.4.10 Drag polars (lift coefficient C_L versus drag coefficient C_D) of sail sections with varying amounts of camber (two-dimensional flow)

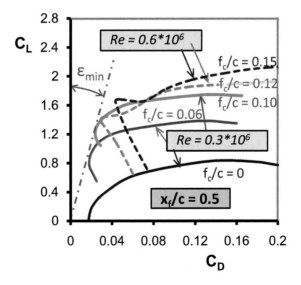

---

[2] A value of 0.08 with the maximum camber at about 33% of the chord was found in a computational fluid dynamic optimisation study (Chapin et al. 2008).

favourable for sailing upwind. However, at high wind speeds it may, for a fixed sail area, be advantageous to realize the minimum drag angle at a lower lift coefficient in order to avoid excessive heel and excessive hydrodynamic resistance due to side force. This means that, for a given size of the sails, high wind speeds require less camber.

Figure 7.4.9 implies that a large amount of camber is favourable for reaching conditions, which require a high maximum lift coefficient.

With respect to the moment (coefficient $C_{M\frac{1}{4}}$) it can be remarked that the larger the amount of camber, the more negative the moment coefficient $C_{M\frac{1}{4}}$ will be. Although there is very little information about the moment of sail sections available in the literature, it is to be expected on theoretical grounds (Marchai 2000, p. 240, 444, pp. 302–325) that the level of the moment will be approximately proportional to the camber/chord ratio f_c/c. This means, for example, that the level of the moment coefficient of a section with 6% camber will be about half that of the section with 12% camber depicted in Fig. 7.4.7. That of a section with 15% camber will be correspondingly (25%) more negative. In terms of yachting the latter means a 25% larger yawing moment to weather.

An indication of the dependence of lift on the position x_f of maximum camber is given by Fig. 7.4.11, for a fixed value (0.1) of the camber/chord ratio f_c/c. The figure should be considered as indicative. In the absence of systematic data in the literature it has been constructed by the author on the basis of more incidental data available in Milgram (1971); Marchai (2000, p. 240, 444, pp. 302–325); Abbott and Von Doenhoff (1949) and Goovaerts (2000).

As already mentioned earlier, an increase of the amount of camber of a sail section implies an increase of both the entry angle and the exit angle. A more forward location of the position x_f of maximum camber was seen to imply an increase of the entry angle and a decrease of the exit angle (For a more rearward location of x_f it is, of course, the other way around). As mentioned earlier, and as shown in Fig. 7.4.11, this means that a more forward position x_f of maximum camber leads to a larger

Fig. 7.4.11 *Estimated effect of chordwise position x_f/c of maximum camber on lift (coefficient C_L versus angle of attack α, two-dimensional flow)*

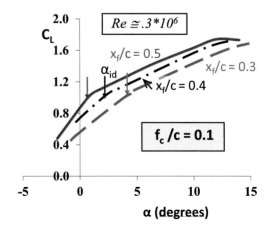

Fig. 7.4.12 *Estimated effect of the chordwise position x_f/c of maximum camber on the drag polar (C_L versus C_D) of sail sections (two-dimensional flow)*

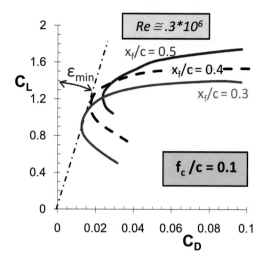

value of the ideal angle of attack α_{id} but a lower lift level for the same angle of attack. Note that the kink in the C_L-α curves corresponds roughly with the ideal angle of attack α_{id}.

The effect of variation of the position of maximum camber on the drag polar is shown in Fig. 7.4.12. It can be noticed that when the position of maximum camber is moved forward the minimum drag becomes lower and occurs at a lower lift coefficient. For sections with 10% camber ($f_c/c = 0.10$), the drag angle seems to attain a minimum when the position of maximum camber is at about 30–40% of the chord behind the leading edge. Results of computational fluid dynamic optimisation studies (Chapin et al. 2008) suggest that with the position of maximum camber at $x/c = 0.30$ the minimum drag angle attains its lowest value for $f_c/c \cong 0.08$. For the performance of a sailing yacht this means (see Chap. 4) that a moderate amount of camber (about 8–10%), with the position of maximum camber at about 30–40% of the chord, is favourable for close-hauled conditions. This in particular at high wind speeds, when the sails have to be set for moderate lift to avoid excessive heel.

Figure 7.4.12 implies that a more rearward position of maximum camber is favourable for reaching conditions when the sails have to be set for maximum lift.

It is further to be expected, on theoretical grounds (Abbott and Von Doenhoff 1949), that the moment of a thin section will be less negative when the position of maximum camber is moved forward. Theoretical data (Abbott and Von Doenhoff 1949) suggest that the reduction is of the order of 30% (less negative) when the position of maximum camber is moved from 50 to 30% of the chord. In terms of sailing this means a smaller yawing moment to weather when the position of maximum camber is moved forward.

Lift and Drag at Very High Angles of Attack
Unlike the keel, the sail(s) of a sailing yacht are required to operate over a very large range of angles of attack. In the case of a dead-run up to about 90° (see Fig. 4.3.9, Sect. 4.3). Hence it is of interest to consider the lift and drag of sail sections over the full range of angles of attack, that is from 0 to 90°.

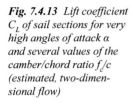

Fig. 7.4.13 Lift coefficient C_L of sail sections for very high angles of attack α and several values of the camber/chord ratio f_c/c (estimated, two-dimensional flow)

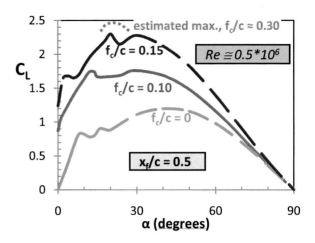

Figure 7.4.13 presents the lift coefficient of thin sail sections for angles of attack up to 90°. The lift coefficient C_L is shown as a function of the angle of attack for three different values of the camber/chord ratio f_c/c, all with the position of maximum camber at half chord ($x_f/c = 0.5$). The curves have been constructed on a basis of measured data (Milgram 1971; Marchaj 2000, p. 240, 444, pp. 302–325; Abbott and Von Doenhoff 1949; Goovaerts 2000; Bethwaite 2010, pp. 209–215) available to the author. Also shown is the estimated absolute maximum for sections with the maximum draft at half chord, which, according to numerical simulations (Chapin et al. 2005, 2008) is obtained for a camber/chord ratio of about 0.3. The maximum lift may even be a little higher when the position of maximum draft is further backward (Collie et al. 2004).

It can be noticed that the lift decreases gradually when the angle of attack is increased beyond that of maximum lift, with a more rapid drop towards zero when the angle of attack approaches 90°.

Note, again, that the effect of camber on the general level of lift is large; the lift level is more than doubled when the camber/chord ratio f_c/c is increased from zero (flat plate) to 0.15.

Figure 7.4.14 presents similar information for the drag coefficient C_D. The increase of C_D with the angle of attack α is most pronounced for angles of attack α around those of maximum lift ($α \cong 20°$). Beyond that, the drag tapers off to a maximum value close to 2 at $α = 90°$. Note that the effect of camber/chord ratio on drag is much smaller than that on lift (compare Figs. 7.4.13 and 7.4.14).

Effects of Reynolds Number (Scale Effects)
The sensitivity to variations in Reynolds number of sharp-edged sail sections differs from that of round-nosed foils with thickness. For sharp-edged sail sections two different mechanisms can be distinguished. First of all the flow separation at the sharp leading edge is always present (except when $α = α_{id}$) and, hence, independent of the Reynolds number. The length of the separation bubble at the leading edge is determined primarily by the shape of the camber line and the circulation around the section (see also Sect. 5.14).

Fig. 7.4.14 The drag coefficient C_D of thin sail sections
upto very high angles of
attack α and several values
of the camber/chord ratio f_c/c
(estimated, two-dimensional
flow)

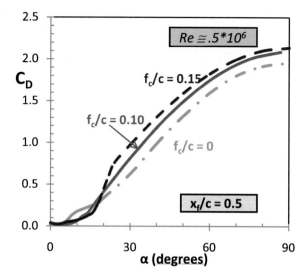

The separation of the turbulent boundary layer further downstream is dependent of Reynolds number. For a given amount of camber it will be present below
a certain, critical, Reynolds number but will disappear when the Reynolds number
is increased beyond this critical value. Turbulent boundary layer separation, when
present, has an indirect effect on the length of the leading edge separation bubble
through a reduction of the circulation.

Because the pressure gradients that the boundary layer is subject to increase with
camber and angle of attack, the critical Reynolds number for turbulent separation
just in front of the trailing edge increases with camber and angle of attack. For
large amounts of camber and low Reynolds numbers there can already be turbulent
boundary layer separation for $\alpha = \alpha_{id}$. This is illustrated by Fig. 7.4.15, which gives

Fig. 7.4.15 Critical Reynolds number for trailing
edge separation of thin sail
sections at the ideal angle
of attack as a function of
camber/chord ratio fc/c
(estimated, two-dimensional
flow)

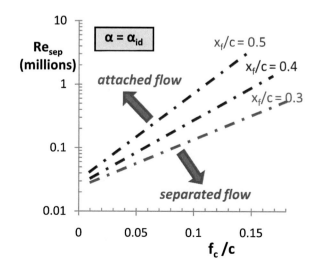

an indication of the critical Reynolds number Re_{sep} at the ideal angle of attack as a function of the camber/chord ratio f_c/c. Note that the Reynolds number is plotted on a logarithmic scale for three values of the position x_f/c of maximum camber. The critical Reynolds number is seen to increase with increasing camber/chord ratio f_c/c and with increasing value of the position x_f/c of maximum camber. This means that sections with a forward position of maximum camber are less vulnerable to trailing edge separation than sections with the position of maximum camber further aft.

We have seen earlier (Sect. 5.14) that boundary layer separation causes a strong increase of profile drag. For sail sections this means that the profile drag at the ideal angle of attack will increase rapidly when the Reynolds number drops below the relevant line in Fig. 7.4.15. Because a large profile drag implies a large drag angle, this contains an important consequence for the practice of sailing close-hauled, when a small drag angle is of prime importance (Chap. 4): At low Reynolds numbers, that is for small yachts and/or low apparent wind speeds (see Sect. 5.9), the sails should be set with a small amount of camber and/or with a forward position of maximum camber when sailing close-hauled.

As an example let us consider a medium-size yacht sailing, close-hauled, with an apparent wind speed of 5 kts. Figure 5.9.2 tells us that the Reynolds number of the sail(s) is then of the order of 0.5×10^6. Figure 7.4.14 teaches that for such a Reynolds number, with the position of maximum camber at mid-chord, the camber of the sail(s) should be smaller than about 9% in order to avoid boundary layer separation. With the position of maximum camber at 40 or 30% the maximum allowable amount of camber is about 12 or 18%, respectively.

For angles of attack below or above the ideal value α_{id}, when there is a separation bubble at the leading edge, the variation of the aerodynamic forces with Reynolds number is quite modest as long as the turbulent boundary layer aft of the separation bubble is attached. The lift for a given angle of attack will then hardly change with Reynolds number and the profile drag will be roughly proportional to $Re^{-1/5}$, like the friction drag of a flat plate. Under these conditions the lift/drag ratio will also be roughly proportional to $Re^{-1/5}$.

7.4.3   Effects of a Mast or Head-Foil on Sail Section Characteristics

Sails have to be attached to something rigid in order to keep them aloft. In the case of a single sail or a mainsail the 'something rigid' is a mast. In the case of a jib or genoa type head sail this is, nowadays usually, a small forestay profile or head-foil.

The effects of circular mast sections of different size have been investigated in wind tunnels by Milgram (1978) and Wilkinson (1984) and computationally by Chapin et al (2005). Chapin et al. (2004) also mentions wind tunnel experiments by Haddad and Lepine (2003). Milgram (1978) also investigated elliptical mast sections. Mast sections of different forms and shapes were studied by Gentry (1976). Some results of wind tunnel tests with mast sections of different shapes are reported by Marchai (2000, pp. 328–338), as well as results of wind tunnel tests on 3D sails with mast (Marchai 1964). Bethwaite (2010, pp. 209–215) describes results of tests with circular and wing-like mast sections.

The general picture emerging from these studies is that the presence of a mast attached to the luff of a sail causes a reduction of lift and an increase of the drag of the sail section. In detail, this trend depends on the relative dimensions of the mast and sail sections and, of course, on the apparent wind angle and sheeting angle. The ratio D/c between the (effective) diameter 'D' of the mast section and the chord length 'c' of the sail appears to be the most important parameter.

A logical starting point for analyzing the flow about a mast section is to consider the flow about a circular cylinder. This type of flow is known to be strongly dependent on Reynolds number (Prandtl and Tietjens 1934b in Chap. 1). For Reynolds numbers ($Re_D = VD/v$) between 2000 and 3000 the flow is unsteady, producing the Von Kármán vortex trail introduced in Sect. 5.18. For a medium size sailing yacht the Reynolds number based on the chord length of the sail was seen to be of the order of 10^6 (Fig. 5.9.2). This means that for a mast section it is of the order of $(D/c)*10^6$, where D is the diameter or thickness of the mast and 'c' the chord length of the sail. Because for most yachts D/c is of the order of 0.05–0.1, the mast section Reynolds number Re_D is of the order of $1*10^5$. This means that the flow will, in general, be steady, except at very low wind speeds and/or for very small mast diameters.

While the lift of a circular cylinder in steady flow is zero because of the symmetry of the flow field, the drag depends strongly on Reynolds number. Figure 7.4.16 gives the drag coefficient C_D of a circular cylinder, based on the diameter of the cylinder, as a function of Reynolds number for the range of Reynolds numbers that is applicable to sailing yachts. For $Re_D < 1.5*10^5$ the drag coefficient is practically constant with a value of about 1.2. For $Re_D > 5*10^5$ it is also constant, but with a value of about 0.3. In between there is a transition region.

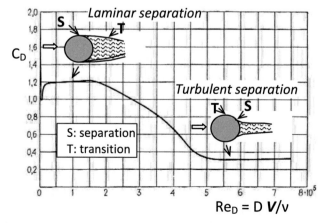

Fig. 7.4.16 *Drag coefficient of a circular cylinder as a function of Reynolds number. (Prandtl and Tietjens 1934b in Chap. 1)*

It appears (Prandtl and Tietjens 1934b in Chap. 1) that for $Re_D > 1.5*10^5$ the flow is characterized by laminar separation of the boundary layer while for $Re_D > 5*10^5$ the boundary layer is turbulent at separation. Because the turbulent boundary layer separates further downstream (Sect. 5.12), the wake and 'dead water' region behind the cylinder is considerably thinner than in the laminar case. As a result there is less form drag (pressure drag).

With a sail attached, the wake of the cylinder is cut in two parts, each forming a separation bubble (Fig. 7.4.17). On a smooth mast of (near-)circular cross-section the boundary layer at the separation point will, in general, be laminar. It may be turbulent on a mast with roughness or when the shape of the mast section and the pressure distribution is such that there is boundary layer transition ahead of the separation point. The position of transition is of great importance for the drag and the characteristics of the wake/separation bubbles.

When the sail is attached at a point opposite to the stagnation point on the mast, as is the case when the apparent heading is zero, the mast will not develop lift. This is no longer the case when the mast is at angle of attack. The asymmetry introduced by the presence of the sail then causes the mast to produce lift. This is reinforced through upwash induced by the sail when the latter carries lift, but reduced by wider sheeting of the sail. In this situation the part of the wake (separation bubble) of the mast on the lee side of the sail will be larger than the part on the windward side.

When the mast carries lift it generates downwash at the position of the sail. This means that the lift of the sail is reduced by the lift on the mast. A further reduction of the lift of the sail is caused by increased asymmetry between the boundary layers on the suction and pressure sides of the sail. The latter is caused by the fact that, at angle of attack and under lift, the separation bubble on the lee side (or suction side) originating from the wake of the mast, will, in general, be larger than the bubble on the pressure side.

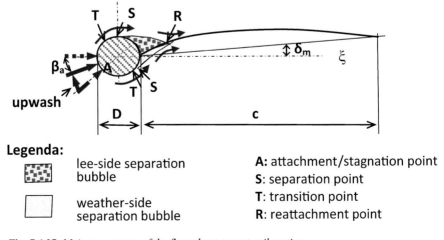

Legenda:

lee-side separation bubble

weather-side separation bubble

A: attachment/stagnation point
S: separation point
T: transition point
R: reattachment point

Fig. 7.4.17 *Main parameters of the flow about a mast-sail section*

The interference between mast and sail is also fairly complex in terms of drag, although due only to viscous effects. Several different mechanisms can be distinguished. In the first place there is the basic drag (mostly form drag) of the mast cylinder (Fig. 7.4.16). The mast will incur additional form drag when it produces lift. There is, however, more to it. Under conditions with lift, the point of transition from laminar to turbulent flow in the boundary layer will, in general, be more forward on the suction side of the mast. This means that on the suction side of the mast the boundary layer may now be turbulent at separation, with an associated reduction of form drag (see Fig. 7.4.16).

For the sail the drag mechanisms are equally complex. Because the sail is in the wake of the mast, it experiences, effectively, a lower onset velocity. This tends to reduce the drag (as well as the lift) of the sail. The lower lift level of the sail also tends to reduce the drag. However, the increased asymmetry, due to the mast, between the boundary layers on the suction and pressure sides of the sail causes an increase of the form drag.

Figure 7.4.18, adapted from Wilkinson (1984) presents a generalized picture of the flow field and the pressure distribution around a sail-mast combination under

Fig. 7.4.18 *Generalized picture of the flow field about a mast-sail combination. (After Wilkinson (1984))*

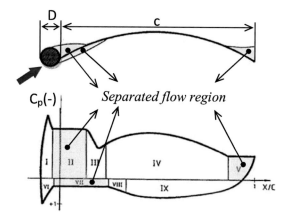

REGION	DESCRIPTION
I	Upper Mast Attached Flow Region
II	Upper Separation Bubble
III	Upper Reattachment Region
IV	Upper Aerofoil Attached Flow Region
V	Trailing Edge Separation Region
VI	Lower Mast Attached Flow Region
VII	Lower Separation Bubble
VIII	Lower Reattachment Region
IX	Lower Aerofoil Attached Flow Region

close-hauled conditions. The most important features are the separation bubbles formed at the mast on both sides of the sail. The main difference between the flow about a sail section with a sharp leading-edge and a sail section with a mast in front is in the absence, in the case with mast, of an 'ideal' angle of attack at which there is no leading-edge separation. With a mast present there are always separation bubbles on both sides of the mast, irrespective of the angle of attack. The angle of attack has only influence on the size (length and thickness) of the separation bubbles. With increasing angle of attack the size of the bubble on the suction or lee side increases, while the bubble on the pressure or weather side becomes smaller. The (relative) size of both bubbles increases also with increasing ratio between the diameter D of the mast and the chord length 'c' of the sail section.

The ever-present separation bubbles at the mast have two consequences for the flow about the sail. The first is that, as already mentioned above, they reduce the lift and increase the form drag of the sail. The second is that they make the flow somewhat less sensitive for changes in angle of attack through the absence of an 'ideal' angle of attack at which there is no leading edge separation. A consequence of the latter is that the lift curve of a sail section attached to a mast exhibits less of the 'kink' that, as described above, is characteristic for sail sections with a sharp leading edge.

This is illustrated by Fig. 7.4.19, which presents lift curves for a sail section with 12 % camber and different size (D/c) circular mast sections, as measured in a wind tunnel (Chapin et al. 2005).[3] The figure shows that for small angles of attack there is a substantial loss of lift due to the presence of the mast. This loss of lift increases with increasing diameter (D/c) of the mast. However, and perhaps somewhat sur-

Fig. 7.4.19 *Effect of masts with circular cross section on the lift of a thin sail section. (From wind tunnel tests (Chapin et al. 2005), 2D flow)*

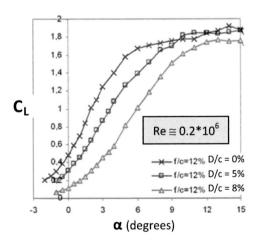

---

[3] Note that the lift curve for mast diameter zero (D/c=0) differs significantly from that given in Fig. 7.9.3 for the same camber/chord ratio ($f_c/c = 0.12$). This illustrates that tests in different wind tunnels do not always tell exactly the same story! Differences in the effects of the presence (or absence!) of the walls of the test section of the tunnel or the condition of the surface of the model are usually the reason. However, the trend observed in one facility is usually trustworthy.

Fig. 7.4.20 *Effect of a mast of circular cross section on the drag polar of a sail section. (Chapin et al. 2005)*

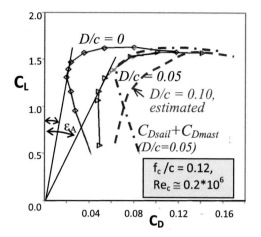

prisingly, the loss of lift is much smaller for angles of attack near maximum lift ($\alpha \cong 15°$).

Figure 7.4.20 shows drag polars (C_L versus C_D) for a sail section of 12 % camber with and without a mast with a diameter/chord ratio of 0.05. The figure reflects results of numerical, viscous flow simulations given in Chapin et al. (2005). Note that the minimum drag is about doubled by the presence of the mast and the minimum drag angle ε_A more than doubled. This in spite of the fact that the diameter/chord ratio is relatively small (0.05). A more common value for the average along the length of the mast for cruising yachts is possibly close to 0.1, for which the dotted line provides an estimated drag polar. Much larger values, up to 0.3, with even much larger drag penalties are incurred near the mast head where the local chord of the sail is small.

It is useful to notice in Fig. 7.4.20 that for moderate lift coefficients, the drag of the sail-mast combination is considerably smaller than the sum of the individual drag coefficients of sail and mast in isolation. As already mentioned above, this is probably due to a change from laminar to turbulent separation of the boundary layer on the suction side of the mast.

It has further been found (Chapin et al. 2004) that the optimal camber of the sail (for maximum lift/drag ratio or minimum drag angle) seems to increase with increasing diameter/chord ratio D/c. For sail sections with the maximum draft at 50 % of the chord and Reynolds numbers of the order of $0.2*10^6$, the camber/chord ratio f_c/c for maximum *L/D* for a sail without a mast was found to be about 9 %. With a mast-diameter/sail-chord ratio D/c of 0.05 this was found to be about 13 %. The reason is probably that the separation bubble on the leeside of the mast is a little shorter for higher camber/chord ratios. In case of the latter the free shear layer separating the bubble from the outer flow intersects the sail a little more forward. This is in agreement with the finding (Bethwaite 2010) that flattening a sail behind a mast gives a longer separation bubble.

Shape of the Mast

It appears from the limited amount of information available in the open literature that the shape of a mast section is at least as important as its size. As already indicated above there is reason to believe that the size of the separation bubble on the lee-side of the mast is the governing factor in determining the effects of the mast on the aerodynamics of the sail. This means that the point of separation of the boundary layer on the lee-side of the mast is an important parameter. As described in Sect. 5.12 the point of separation of a boundary layer depends in general on two factors: the pressure distribution and the state of the boundary layer (laminar or turbulent). Any boundary layer has the tendency to detach from the surface when it is exposed to an adverse pressure gradient over a sufficiently long distance. In addition thin, high Reynolds number, turbulent boundary layers are more resistant to separation than laminar boundary layers.

Figure 7.4.21 gives schematic pictures of the shape and size of the separation bubbles on different mast-sail combinations in upwind conditions. Figures (a) and (b) depict mast sections of circular and elliptic shape with laminar separation of the boundary layer on the lee-side (suction side) of the mast. The point 'A' (attachment point or stagnation point) indicates where stagnation streamline hits the surface of the mast and the air particles come to a standstill. This is where the pressure has its maximum value (stagnation pressure, see Sects. 5.6 and 5.14). From there, the air accelerates rapidly to attain its maximum velocity (or minimum pressure) immediately ahead of the point labeled 'S'. For the circular mast section the maximum velocity is about twice the free stream velocity and the point 'S' is located at 90° around the circumference of the circle from the attachment point. When the boundary layer at the point of maximum velocity is laminar, which, in general, will be the case, the boundary layer will separate almost immediately after it has passed the point of maximum velocity and will form a separation bubble downstream of the point 'S'. The streamline that separates the bubble from the main stream is almost a straight line and tangent to the mast section circle at the point 'S'. It reattaches at the point 'R' where it intersects the curve that represents the sail. Transition from laminar to turbulent flow in the shear layer takes place at the point 'T', just ahead of the reattachment point. On the lower surface (or weather-side) the process is similar, but reattachment takes place further downstream because of the curved shape of the sail.

The process is also similar when the cross-section of the mast is elliptic, as in the (b) sketch of Fig. 7.4.21. However, because the point of maximum velocity and the associated separation point is a little more forward on the lee-side and a little further backward on the weather-side, the separation bubble on the lee-side will be a little longer and that on the weather side will be shorter than in the case of the circular section. Whether this causes the elliptic shape to be better or worse than the circular shape for the same thickness or width 'D' of the mast is difficult to tell. There is reason to believe (Milgram 1978; Gentry 1976; Bethwaite 2010, pp. 209–215) that any difference in efficiency depends intricately on details of the flow such as the Reynolds number, the precise position of the stagnation point (which depends on the angle of attack), free stream turbulence and surface roughness. It will further be clear that an elliptic mast will loose its advantage, if any, for large apparent wind

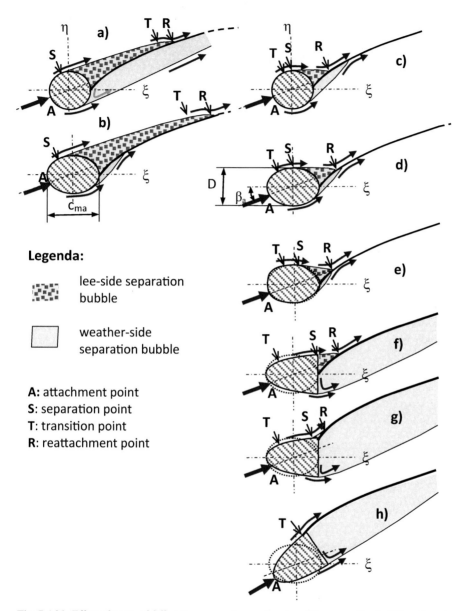

Fig. 7.4.21 *Effect of masts of different cross-section on the size of the separation bubbles of mast-sail combinations under close-hauled conditions (schematic and qualitative)*

angles β_a, when the section dimension in the direction normal to the apparent wind vector is larger than that of a mast with circular cross-section (we will return to this aspect shortly).

Apart from this, it should be mentioned that a mast with an elliptic section with the same width as a circular section will have somewhat larger strength and stiffness. This can of course be used to reduce the lateral dimension of the elliptic section relative to that of the circular shape, which is advantageous for the aerodynamic efficiency (L/D).

The sketches (c) and (d) of Fig. 7.4.21 give a qualitative indication of what happens when the boundary layer on the mast can be forced into transition from the laminar to the turbulent mode before it reaches the point of minimum pressure. Because a turbulent boundary layer is more resistant to separation due to an adverse pressure gradient, the position of the separation point 'S' is further downstream than in the case of laminar separation. As a result, the leeside separation bubbles are much shorter than in the laminar case. This has a significant, positive effect on the aerodynamic efficiency. On the weather side the effect is probably somewhat smaller for the elliptical section because the point of maximum velocity/minimum pressure is already further downstream.

Forcing the boundary layer into transition before it reaches the point of maximum velocity/minimum pressure may, however, not be an easy task. The reason is that laminar boundary layers are quite stable against disturbances in strongly accelerating flows (Sect. 5.11) such as occur between the stagnation point and the point of minimum pressure. Gentry (1976) has found that this problem can be overcome by subtle shaping of the cross-section of the mast. By introducing a mild 'knuckle' in the forward half of what is basically an elliptic section a short plateau of almost constant pressure can be created between the knuckle and the maximum width of the section. This creates excellent conditions for tripping the boundary layer into transition by introducing a small disturbance in the form of a thin tripping wire or distributed roughness. As indicated in the (e) sketch of Fig. 7.4.21 this also leads to small separation bubbles on both sides of the mast.

Results of wind tunnel tests with mast sections of a parabolic shape are reported by Marchai (2000, pp. 328–338). As indicated in the (f) sketch of Fig. 7.4.21 the separation points 'S' on a mast with a parabolic section will be positioned at the sharp trailing edges that close the parabola. That is, provided that there is no laminar separation ahead of the trailing edge. The latter is probably the case when the angle of attack is sufficiently small and there is boundary layer transition at some point between the leading edge and the trailing edge.[4] As indicated in sketch (f) such a configuration leads also to a small separation bubble on the leeside of the mast but not so on the weather side. That the leeside, where the boundary layer on the sail is heavily loaded, is more important than the weather side is confirmed by the fact that Marchai (2000, pp. 328–338) mentions an improvement in aerodynamic efficiency (L/D) in close-hauled conditions of about 20 % of the parabolic configuration (f) over an elliptic configuration with the same overall dimensions. Even better results (improvement of about 40 %) were obtained when the sail was allowed to slide to the leeward side of the blunt trailing edge of the mast (configuration (g) of Fig. 7.4.21).[5]

---

[4] Marchai (2000, pp. 328–338) mentions an angle of attack of 10°, corresponding with an apparent wind angle of 25°.

[5] Similar effects have been found when the sail was allowed to slide to the lee-side of a mast with circular cross-section (Marchai 2000, pp. 328–338).

It is to be expected that for large apparent wind angles laminar separation at the leading edge is, in all probability, also unavoidable for parabolic sections. This means that any advantage over an elliptic or circular section is lost under such conditions. As already indicated above, the dimension in the direction normal to the chord of the sail of elliptic and parabolic mast sections is then larger than that of a mast of the same thickness with a circular cross-section. This will result in a very large separation bubble on the lee-side (Fig. 7.4.22).

It is useful, in this context, to introduce the notion "effective mast diameter" D_{me}. This can be defined as (see Fig. 7.4.22):

$$D_{me} = \sqrt{\left\{ (D \cos \delta)^2 + (2c_{ma} \sin \delta)^2 \right\}} \qquad (7.4.1)$$

For small sheeting angles δ (small apparent wind angles) this reduces to

$$D_{me} \approx D, \quad (\delta \to 0), \qquad (7.4.1a)$$

while

$$D_{me} \approx 2c_{ma} \quad \text{for } \delta \to 90^{\circ} \qquad (7.4.1b)$$

For a mast with a circular cross-section we have $D = c_{ma}$. However, it can be argued that the factor 2 in Eq. (7.4.1) should be replaced by 1 if the sail is permitted to slide to the lee-side of the mast and that Eq. (7.4.1a) holds for all δ if the mast is permitted to rotate.

Configuration (h) of Fig. 7.4.21 indicates what will happen when the parabolic mast of configuration (g) is permitted to rotate. Separation on the leeside can then probably be avoided completely and the bubble on the weather side can be expected to be

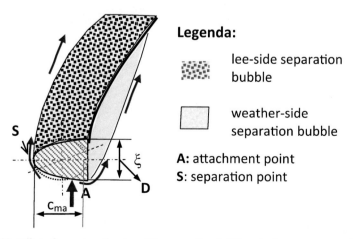

Legenda:

lee-side separation bubble

weather-side separation bubble

A: attachment point
S: separation point

Fig. 7.4.22 *Effect of a mast on the size of the separation bubbles at a large apparent wind angle (schematic and qualitative)*

significantly smaller than in the case that the mast is fixed. Hence, the aerodynamic efficiency will be better than that of configuration (g). While this will already be the case in close-hauled conditions, it will be clear that the advantage of a rotating mast is particularly large under conditions with large apparent wind angles. In terms of effective mast diameter D_{me} (see Eq. (7.4.1)) one can say that for a rotating mast D_{me} is more or less independent of the sheeting angle and the apparent wind angle β_a and takes the value corresponding with Eq. (7.4.1a).

Not surprisingly, it has also been found (Chapin et al. 2004) that the effects of a rotating mast with a truncated airfoil-like section (or 'wingmast' (Bethwaite 2010, pp. 209–215)), as in Fig. 7.4.23, on the aerodynamics of the attached sail are almost negligible. When expressed in terms of effective mast diameter the latter is probably of the order of only 1%. The research by Bethwaite (2010, pp. 209–215) on wingmast section shapes that avoid laminar separation should be mentioned in this context.

Figures 7.4.24 and 7.4.25 summarize the effect of a (fixed) mast on the main aerodynamic characteristics of sail sections with cam ber/chord ratios between 3 and 121/2%. The figures are based on the limited amount of experimental data for masts with circular cross-section (markers) that is available to the author. Since the Reynolds numbers in these experiments are relatively low, the flow about the mast is, in all probability, of the type with laminar separation.

Fig. 7.4.23 *Rotating mast with wing-like section: minimal aerodynamic losses? (Photo reproduced from Ref. Chapin et al. (2004))*

Fig. 7.4.24 *Trend of reduction (relative) of the maximum lift of a sail section due to the presence of a mast, as a function of the (effective) diameter/chord ratio*

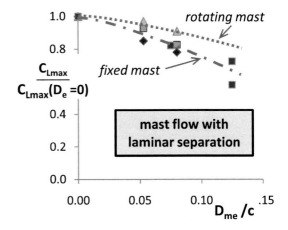

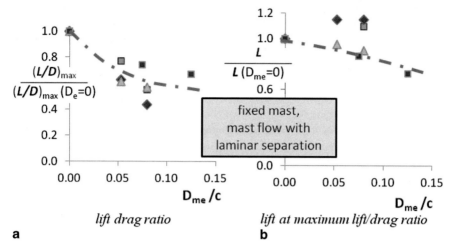

Fig. 7.4.25 *Trend of the dependence on the (effective) mast diameter of the maximum lift/drag ratio and the corresponding lift of a sail section attached to a mast*

As indicated in the figures there is an appreciable amount of scatter in the data. A clear trend in the dependence on camber/chord ratio of the sail section could not be found.

The effect on maximum lift is given in Fig. 7.4.24 in terms of the ratio between C_{Lmax} for the actual value of D_{me}/c and C_{Lmax} for $D_{me}/c = 0$ (no mast). The figure shows that maximum lift is reduced by almost 30 % for $D_{me}/c = 0.10$ and even more so for higher values of D_{me}/c. Also shown is an estimate, based on Refs. Chapin et al. (2004), Paton and Morvan (2007), for the case that the mast is permitted to rotate.

Figure 7.4.25 summarizes the effect of a fixed mast on the maximum lift drag ratio *L/D* and the corresponding level of lift *L*, again, and in both cases, in dimensionless form, that is normalized by the values without mast (D/c=0).

As shown in Fig. 7.4.24a, the maximum lift drag ratio is reduced by about 40 % by the presence of a mast for a mast-diameter/sail-chord ratio of 0.10. The effect on the lift level at which the maximum lift/drag ratio is attained (Fig. 7.4.25b) appears to be of the same order as the loss of maximum lift.

The figures just discussed are probably also applicable to masts of non-circular cross-section if the diameter/chord ratio D/c is interpreted as the effective ratio D_{me}/c as defined by Eq. (7.4.1).

As already mentioned, the Figs. 7.4.24 and 7.4.25 are based on experimental data for low Reynolds numbers with, in all probability, laminar flow about the mast. This means that the reduction of maximum lift and lift/drag ratio by the presence of the mast is probably appreciably smaller when the flow about the mast is of the turbulent type with a much thinner wake (Fig. 7.4.16). The latter may be the case for high Reynolds numbers or when the boundary layer on the mast is forced into transition by some form of roughness.

It is finally worth mentioning that the presence of a mast has also a significant effect on the position of the centre of pressure of a sail section. Test data (Milgram 1978) suggest that the centre of pressure moves forward when a sail section is attached to a mast. At moderate lift coefficients and Reynolds numbers of about $1*10^6$ the forward shift appears to be of the order of 10% of the chord for a diameter/chord ratio of $D/c = 0.10$.

Head-Foils

The configuration (h) of Fig. 7.4.21 is, qualitatively, very similar to that of a jib or genoa attached to a rotating, profiled forestay or head-foil. The main difference is in the size. While the dimension of a mast section is of the order of 10% of the sail chord, that of a head-foil is of the order of 1%. As a consequence the performance of a jib or genoa attached to a head-foil is not necessarily worse than that of a hanked-on sail. Depending mainly on the thickness/chord ratio of the foil, the performance of a sail with head-foil may even be better than that of the hanked-on sail. Marchai (2000, pp. 604–606) mentions improvements in lift/drag ratio of up to 10% for properly shaped head-foils with a thickness/chord ratio of about 25%.

A remark to be made here is that the examples given by Marchai seem to concern head-foils with a single, centred groove for attaching the sail (See Fig. 7.4.26a). In

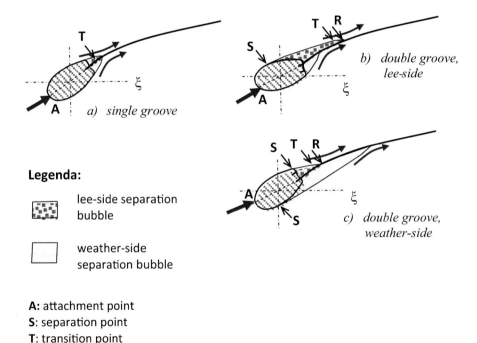

Fig. 7.4.26 *Flow patterns about single- and double-grooved head-foil-jib/genoa combinations (schematic and qualitative)*

such a case the longitudinal stress in the sail tends to turn the head-foil properly in line with the front part of the sail section. This leads to (very) small separation bubbles on both sides of the head-foil.

Some head-foils have double grooves. The advantage is that a new jib or genoa can be hoisted while the other one is still up, so that the time spent with a loss of thrust while changing sails is as short as possible. In the experience of this author a disadvantage of the double groove system is that the longitudinal forces in the sail tend to turn the head-foil out of line with the front part of the sail section. With the sail in the lee groove the head-foil acquires an incidence that is too large in relation with the front part of the sail (See the Fig. 7.4.26b). The result is a forward shift of the minimum pressure and the (laminar) separation point on the lee side of the foil, a longer separation bubble and more drag/less lift. With the sail in the weather groove the head-foil tends to turn the other way resulting in an angle of attack that is too small (See the Fig. 7.4.26c). Although this leads to a larger separation bubble on the weather side, the bubble on the lee-side is shorter than with the sail in the lee groove.

Since the effect of the lee side bubble dominates, this configuration is not as bad as the one with the sail in the lee groove. In the author's experience the effect is noticeable at low and moderate wind speeds.

A similar phenomenon that most yachtsman are familiar with is the poor performance of head sails when they are reefed by furling, in particular when the rolled-up portion of the sail is on the leeside. As illustrated by Fig. 7.4.27 (a) this is caused by

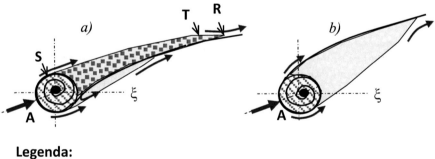

Legenda:

▨ lee-side separation bubble

☐ weather-side separation bubble

A: attachment point

S: separation point

T: transition point

R: reattachment point

Fig. 7.4.27 *Flow separation bubbles behind jib/genoa furling rolls (schematic and qualitative)*

the fact that a large separation bubble develops on the leeside behind the furling roll. When the furling roll is on the weather side (b) figure of Fig. 7.4.27), the separation bubble on the leeside is probably much smaller or even absent. The bubble on the weather side is larger but less harmful than a bubble on the leeside.

The author is not aware of any quantitative information on these phenomena. It can be expected, however, that the effects are about the same as for a mast with the same diameter/chord ratio.

7.4.4   '3D' Effects: Effective Aspect Ratio

As already discussed in Sect. 5.15, the most important parameter of lifting surfaces, like sails, in three dimensions is the aspect ratio 'A' defined as

$$A = b^2/S, \tag{7.4.2}$$

where 'b' is the span of the lifting surface and 'S' the projected area (see Fig. 5.15.3). In sailing yacht terminology the aspect ratio of a single sail with mast, such as the mainsail in Fig. 2.4.1, would be defined as

$$A_M = P^2/S_M, \tag{7.4.3}$$

where 'P' is the height of the sail and 'S_M' the sail area. The main effect of the finite span of a lifting surface is, as discussed in Sect. 5.15, to reduce the effective angle of attack through the downwash induced by the trailing vorticity ('tip' vortices) shed at the trailing edge (Figs. 5.15.2 and 5.15.3).

Although the downwash induced by the trailing vorticity, for a given lift, is mainly determined by the span or aspect ratio, it is also influenced by the spanwise distribution of lift or circulation. For a given angle of attack, the spanwise lift distribution is also determined by other factors such as the shape of the planform, the proximity of other bodies and the spanwise variations of geometrical twist and camber. In the case of the sail of a sailing yacht there are additional factors in the form of the gradient and twist of the apparent wind profile (Sect. 7.2).

As already mentioned in Sects. 5.15 and 5.17 the effect on the aerodynamics of a lifting surface of most of the other factors mentioned above can be conveniently, though approximately, modelled by introducing the notions effective span (b_e) and effective aspect ratio (A_e):

$$b_e = eb \tag{7.4.4}$$

$$A_e = Ae \tag{7.4.5}$$

The factor e in these definitions is the so-called aerodynamic efficiency factor. For a single, isolated, plane lifting surface with an elliptical lift distribution, 'e' was seen to take the value 1. For a non-elliptical lift distribution there holds $e < 1$. The presence of other bodies such as winglets, reflection planes (like the water surface) can be modelled by adopting values of $e > 1$ (Sect. 5.17). At this stage we will not bother about the dependence of 'e' on the configuration of a yacht and its sails. This will be addressed in a later sub-section.

Lift and Drag at Low to Moderate Angles of Attack
We have seen in Sect. 5.15 that, assuming that the aerodynamic efficiency factor 'e' is known, the lift coefficient $C_L(\alpha)$ of a plane lifting surface with attached flow at a (small) angle of attack 'α' can be approximated by the following formula:

$$C_L(\alpha) = \frac{dC_L}{d\alpha}\,\alpha = 2\pi E\alpha/\{1 + 2\pi E/(\pi A_e)\} \qquad (7.4.6)$$

Here, $\dfrac{dC_L}{d\alpha}$ is the lift curve slope per radian, equal to $2\pi E/(1 + 2E/A_e)$. The lift curve slope depends on the effective aspect ratio A_e and an 'edge factor' 'E' which depends mainly on the geometric aspect ratio and the angle of sweep (Eq. (5.15.14)).

It follows from Abbott and Von Doenhoff (1949) that for lifting surfaces without twist, with a constant, cambered foil section all over the span, Eq. (7.4.6) can be generalized to read

$$C_L(\alpha) \cong \frac{dC_L}{d\alpha}\,(\alpha - \alpha_0) + \delta C_L(A, \alpha) \qquad (7.4.7)$$

where α_0 is the angle of attack for zero lift. The term $\delta C_L(A, \alpha)$ is a (small) correction for low aspect ratios ($A \lesssim {\sim}2$) that has been described in Sect. 5.15. It models the effect on lift of side edge (tip) separation. For Bermuda type sails, with their asymmetrical, triangular planform (Sect. 2.3), only the foot of the sail will contribute to $\delta C_L(A, \alpha)$. For low aspect ratio ($A \cong 1$), square rigged sails the magnitude of the term $\delta C_L(A, \alpha)$ at high angles of attack can be substantial.

As an alternative to Eq. (7.4.7), the lift can be related to the lift coefficient C_{L2D} of the basic foil section:

$$C_L(\alpha) \cong C_{L2D}(\alpha_{2D}) + \delta C_L(A, \alpha) \qquad (7.4.8)$$

The angle of attack α is given by

$$\alpha = \alpha_e + \alpha_i \qquad (7.4.9)$$

with the effective angle of attack α_e given by (Abbott and Von Doenhoff 1949)

$$\alpha_e = (\alpha_{2D} - \alpha_0)/E + \alpha_0 \qquad (7.4.10)$$

and the induced angle of attack as

$$\alpha_i = C_L / (\pi A_e) \tag{7.4.11}$$

The drag can be approximated as

$$C_D(\alpha) \cong C_{D2D}(\alpha_{2D}) + C_L^2 / (\pi A_e) + \delta C_L(A, \alpha) \tan\alpha \tag{7.4.12}$$

with the second term representing the induced drag and the third term representing the effect on drag of the side edge separation (Sect. 5.15, Appendix C.2).

Figures 7.4.28 and 7.4.29 illustrate the trends of the effect of aspect ratio on the lift and drag characteristics of a single, Bermuda-type sail. The figures have been constructed on the basis of the formulae given above[6] and section data, based on Ref. Milgram (1978), for a sail with 10% camber and the maximum draft at 50% of the chord. The sail has been assumed to have zero aerodynamic twist and to be attached to a mast with an effective diameter/chord ratio D_{me}/c of 8%.

It can be noticed that the angle of attack for a given lift increases rapidly for small aspect ratios. For an effective aspect ratio of 3 the angle of attack is about twice that of the 2D section (or $A_e = \infty$). The effect of aspect ratio on the maximum lift is small, but large on the angle of attack for maximum lift. This behaviour is, of course, similar to that of wing-type lifting surfaces of different aspect ratio, discussed in Sect. 5.15.

The effect of aspect ratio on the lift-drag polar of a sail (Fig. 7.4.29) is also similar to that of an aircraft-type wing (Fig. 5.15.6). That is, due to the induced drag, the drag increase with lift is much larger for low aspect ratios than for high aspect ratios. Note also, however, that there is a quantitative difference in the sense that the minimum drag coefficient of a sail(+ mast) is much higher than that of an aircraft-type wing but occurs at a higher lift coefficient. As a consequence there is also a substantial difference between the minimum drag angles. For a sail with an effective aspect ratio of 3, attached to a mast, the minimum drag angle is about 12° while it is of the order of only 4° for an aircraft wing.

Also indicated in Fig. 7.4.29 is the locus of the points where the drag angle attains its minimum. This shows that the lift coefficient for the minimum drag angle decreases with decreasing aspect ratio. For a yachtsman this means that, in close-hauled conditions, when a small aerodynamic drag angle is most important (Sect. 4.5), a low aspect ratio sail should be set for a lower lift coefficient than a sail of high aspect ratio with the same section. This applies in particular to yachts with a high sensitivity of the hydrodynamic resistance to side force and/or 'tender' yachts and/or for high wind speeds. It also suggests that, in such conditions, there may be a point in applying less camber in low to moderate aspect ratio sails. This is confirmed by Fig. 7.4.30, which shows lift-drag polars for a sail (plus mast) with an effective aspect ratio of 3 and different amounts of camber. The figure shows that the minimum drag angle is obtained for a camber/chord ratio f_c/c of about 6%. This

---

[6] In the author's experience results for sails based on these formulae agree reasonably well with wind tunnel data given by Marchai (1964, 2000d, pp. 328–338, 2000, pp. 604–606).

Fig. 7.4.28 *Illustrating the effect of aspect ratio on the lift of a sail + mast (zero aerodynamic twist)*

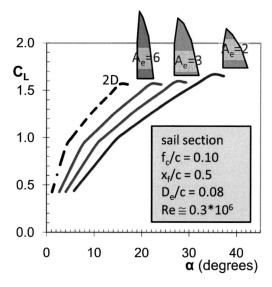

Fig. 7.4.29 *Effect of aspect ratio on the lift-drag polar of a sail + mast (zero aerodynamic twist)*

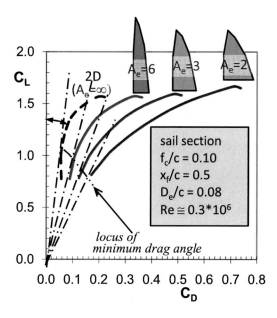

is considerably less than the optimum upwind camber for two-dimensional flow (or very high aspect ratios). As mentioned earlier and shown in Fig. 7.4.30, the latter is of the order of 10 %.

It should be emphasized, that Fig. 7.4.29 does not imply that a high aspect ratio leads automatically to the most efficient rig for upwind conditions. The reason is, apart from the heeling moment, that tall, high aspect ratio rigs require, in general, thicker masts to cope with the larger bending moments. It is easily verified that this

Fig. 7.4.30 *Effect of camber on the minimum drag angle of a sail-mast configuration of finite aspect ratio (zero aerodynamic twist)*

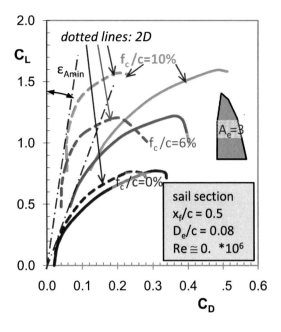

means that, for the same sail area, the mast-diameter/sail-chord ratio D/c should be proportional to the aspect ratio of the rig. As an example, when the mast-diameter/sail-chord ratio D/c of an aspect ratio 3 rig is 0.08, about twice that value will be required for an aspect ratio 6 rig. The consequences of this for the aerodynamic efficiency are quite large. This is illustrated by Fig. 7.4.31 which shows drag polars

Fig. 7.4.31 *Effect of aspect ratio on the lift-drag polar of sail + mast configurations of comparable strength/stiffness (zero aerodynamic twist)*

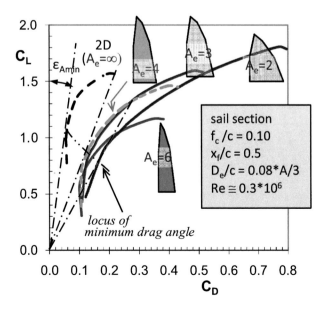

for different aspect ratios and associated, proportional values of D/c. The effect on maximum lift in particular is dramatic for the A=6 rig and the minimum drag angle is now even a fraction worse than for A=3. It appears that, of the cases considered, the best aerodynamic efficiency (minimum drag angle) is obtained for an aspect ratio of about 4.

The situation will, of course, be less dramatic when the mast is thinner as a result of the application of better materials (e.g. carbon fibre) or a better structural design. It is also much less dramatic in the case of rotating, streamlined masts.

Heel is another factor that comes into the picture in a real situation. Taller rigs imply more heel, with associated loss of aero- and hydrodynamic performance. This means that, for a single sail with mast, the optimum aspect ratio will probably be less than the value of 4 mentioned above. Another important aspect is that the presence of a foresail will reduce the negative effect of the mast on the total aerodynamic performance. We will return to this subject in Sect. 8.3.

Because of the dependence on aspect ratio of the effective angle of attack of a sail, the sheeting angle required for optimal propulsion also depends on aspect ratio (recall that $\delta = \beta_a - \alpha$). This is illustrated by Fig. 7.4.32 which shows the sheeting angle δ as a function of the apparent wind angle β for the same sail configurations as in Figs. 7.4.28 and 7.4.29. The figure shows that higher aspect ratios require larger sheeting angles, in particular when the objective is to maximize the driving force.

Effect of Aspect Ratio on Lift and Drag at Very High Angles of Attack
At high and very high angles of attack, when the flow about a lifting surface like a sail is massively separated, the effect of aspect ratio on lift and drag is more complex. As long as the foil produces some lift and circulation there will still be downwash and some associated induced drag. However the dominating mechanism

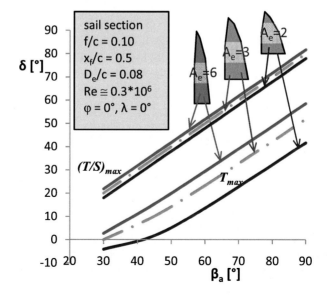

Fig. 7.4.32 *Effect of aspect ratio on the sheeting angles required for the maximum propulsive efficiency (T/S)$_{max}$ and the maximum driving force T$_{max}$ respectively (upwind conditions, zero aerodynamic twist)*

sail section
f/c = 0.10
x$_f$/c = 0.5
D$_e$/c = 0.08
Re $\cong$ 0.3*10^6
$\varphi = 0°$, $\lambda = 0°$

δ [°]

$(T/S)_{max}$

T_{max}

A_e=2

A_e=3

A_e=6

β_a [°]

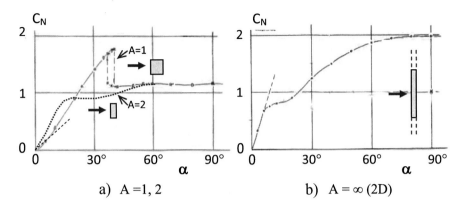

Fig. 7.4.33 *Normal force coefficient of rectangular flat plates as a function of angle of attack. (Winter 1937 in Chap. 5; Hoerner 1965)*

is now reduced lift and increased pressure drag due to the displacement effects of the areas with separated flow. Under such conditions, with flow separation at all edges, the effects of Reynolds number are totally negligible and the tangential force is practically zero. This means (see Eq. (5.2.1)), that the lift is proportional to cos α and the (pressure) drag proportional to sinα. Moreover, it has been found (Hoerner 1965) that the normal force is almost constant for very high angles of attack. As shown in Fig. 7.4.33, the normal force coefficient C_N of square, flat plates (A=1) takes the value (Hoerner 1965) $C_N \approx 1.17$ for $\alpha > \cong 40°$. For an aspect ratio of 2 this is $C_N \approx 1.20$ for $\alpha > \cong 55°$ (Winter 1937 in Chap. 5). The corresponding values in two-dimensional flow (A=∞) are $C_N \approx 1.96$ for $\alpha > \cong 75°$. Note that there is a significant difference in behaviour between plates of aspect ratio 2 (and larger) and plates of aspect ratio 1 (and smaller). This is caused by the side edge vortex burst phenomenon of low aspect ratio lifting surfaces, described in Sect. 5.15.

An implication of the behaviour just described is that, beyond a certain, large angle of attack, that depends on aspect ratio, the lift and drag coefficients of a thin lifting surface can be approximated by

$$C_L \approx C_{N90} \cos\alpha \qquad\qquad (7.4.13)$$

and

$$C_D \approx C_{N90} \sin\alpha, \qquad\qquad (7.4.14)$$

respectively, where C_{N90} is the normal force coefficient at 90° angle of attack. The latter is a function of aspect ratio, as already implied by Fig. 7.4.33 and as shown more specifically in Fig. 7.4.34. The latter figure shows that the normal force coefficient of flat plates at 90° is practically constant at $C_{N90} \cong 1.2$ for values of $1/A > \cong 0.2$, that is for $A < \cong 5$. In other words: C_{N90} is practically constant for all

Fig. 7.4.34 *Dependence on aspect ratio of the normal force of flat and cambered plates at 90° angle of attack*

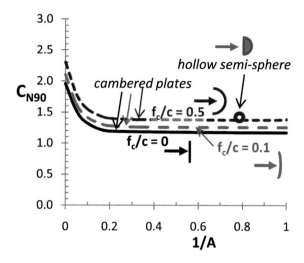

aspect ratios that are applicable to sails and much smaller than in two-dimensional flow. The reason for this is that for square plates (A=1) it is 'easier' for the surrounding fluid to enforce a higher, closer to atmospheric, pressure on the backside of the plate (backpressure) than for elongated plates (A≪1, A≫1).

Figure 7.4.34 also shows the normal force at 90° for cambered plates. The largest normal force for an angle of attack of 90° is obtained for a hollow, semi-cylinder (f/c=0.5). Also shown is the value of C_{N90} for a hollow, semi-sphere, which is the 3D body with the highest resistance (Hoerner 1965).

It is perhaps interesting to note here, that the formulae 7.13/7.14 seem to form the basis of a sailing theory underlying the famous square-rigged tea clippers of the nineteenth century (Hoerner 1965). Needless to say that such a theory could work only because these ships sailed almost always under conditions with large apparent wind angles involving angles of attack of 45° and beyond.

It will be clear from the discussion given above that the effects of aspect ratio on the lift and drag of a sail at very high angles of attack α are different from those at small and moderate α. This is the case for drag in particular since the latter is most sensitive to the level of the backpressure at angles of attack close to 90°.

Figure 7.4.35 gives an impression of the variation of the lift coefficient over the whole range of angles of attack up to α=90° for Bermuda-type sails of different aspect ratio, again with a mast-diameter/sail-chord ratio D/c of 0.08.

The curves for the different aspect ratios have been constructed by matching the results for small and moderate angles of attack (Fig. 7.4.28) with results of the formula (7.4.13) for very high angles of attack.[7] Note that the effect of aspect ratio, while very significant at small and moderate angles of attack, is almost negligible

---

[7] Application of this procedure gives, in the author's experience, a reasonable agreement with experimental results for rigs considered by Marchai (2000 in Chap. 4) and classical wind tunnel data for cambered plates (without mast) by Eiffel (1910).

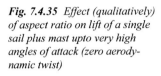

Fig. 7.4.35 Effect (qualitatively) of aspect ratio on lift of a single sail plus mast upto very high angles of attack (zero aerodynamic twist)

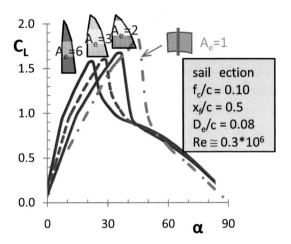

for angles of attack beyond, say, 50°. This is caused by the fact that, with flow separation at all edges, the aerodynamic forces are dominated by the back pressure of the sail. As already mentioned above, the latter is almost independent of aspect ratio for very large angles of attack.

For comparative purposes Fig. 7.4.35 also shows the lift curve for a square rigged sail of aspect ratio 1 with the mast positioned at the centre of the pressure side of the sail. Because the sail has been assumed to have camber in the horizontal plane the yards should be cambered also, like a modern **Dynarig** (see Sect. 7.13). The figure shows that the square rigged sail is superior to the Bermuda rigs in the range of angles of attack between, approximately, 40 and 50°.

It should also be mentioned that Fig. 7.4.35 is representative, qualitatively, for situations with zero aerodynamic twist. This means that the angle between the apparent wind vector and the local chord of the sail is constant along the length of the mast (Sect. 7.2). When the sail does have aerodynamic twist in the sense that the angle of attack at the top of the sail is smaller than at the foot, the lift will be somewhat smaller and the peak in Fig. 7.4.35 will be less pronounced.

Figure 7.4.36 presents the drag (coefficient) as a function of the angle of attack for the same sail configurations as in Fig. 7.4.35. Shown is the coefficient of the total drag C_{Dtot} as well as the induced drag C_{Di}. It can be noted that for angles of attack approaching 90° there is very little difference between the sail configurations considered. Recall that this is also the case for lift (Fig. 7.4.35) and that this is caused by the fact that under these conditions there is total flow separation from all edges.

It can also be noted that the Bermuda type rigs exhibit a local maximum in drag for an angle of attack somewhere between, say, 25 and 40°, the precise angle of attack depending on aspect ratio. This is due to the induced drag, which, as we have seen before, attains its maximum at maximum lift (compare with Fig. 7.4.35).

The local maximum of the drag is much more pronounced for the square-rigged sail configuration. This is caused mainly by the induced drag, which is much higher as a consequence of the higher maximum lift and the smaller aspect ratio (cf. the second term of Eq. (7.4.12)). As shown by Fig. 7.4.36, the local drag maximum for

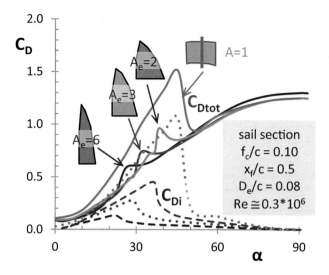

Fig. 7.4.36 *Effect (qualitatively) of aspect ratio on drag of a single sail plus mast for very high angles of attack (zero aerodynamic twist)*

$\alpha \cong 45°$ is even higher than the drag at $\alpha \cong 90°$. This has, potentially, an interesting consequence for the optimum sheeting angle when sailing downwind, as we will see shortly.

Lift-drag polars for the full range of angles of attack for the sail configurations considered above are compared in Fig. 7.4.37. We have seen already that a high aspect ratio is, in general, favourable for a high aerodynamic efficiency and effectiveness

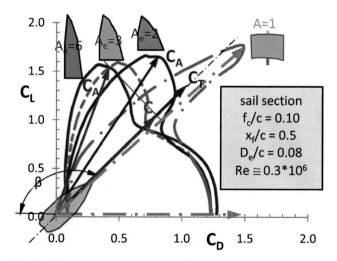

Fig. 7.4.37 *Effect (qualitatively) of aspect ratio on the lift-drag polar of a single sail plus mast (zero aerodynamic twist)*

of a sail at small apparent wind angles. In such conditions the sails operate at small to moderate angles of attack.

Figure 7.4.37 illustrates that this is not the case for large apparent wind angles when the sails operate at high or very high angles of attack. Shown are, in addition to the polars, the total aerodynamic force vectors (coefficients C_A) for an angle of attack close to maximum lift for an aspect ratio 6 rig and an aspect ratio 2 rig. Also shown are the corresponding driving force components (coefficients C_T) for the case of an apparent wind angle β of about 130°.[8] It is clear that for this condition the driving force of the aspect ratio 2 rig is substantially larger than that of the aspect ratio 6 rig. Moreover, the opposite is the case for the heeling force, that is the component of the total aerodynamic force perpendicular to C_T.

It can be verified through Fig. 7.4.37 that the driving force potential of the aspect ratio 2 rig is better than that of the aspect ratio 6 rig for all apparent wind angles in the range $\cong 80° < \beta < \cong 140°$, although the associated heeling force in the range $\cong 80° < \beta < \cong 110°$ is smaller for the high aspect ratio rig. The high aspect ratio rig is obviously superior (in terms of driving force potential) for $\beta < \cong 80°$, and, but only marginally so, for $\beta > \cong 140°$.

It is, finally, interesting to note that the aspect ratio 1 square rig has the highest driving force potential for all apparent wind angles above, approximately, 120°, up to and including $\beta = 180°$. This is due to the high values of the maximum lift and the maximum drag. The latter in particular leads to a situation in which the maximum driving force in the dead run is obtained for an angle of attack of 45° (see Fig. 7.4.37), that is, for a sheeting angle of 135° or 45°. This, however, is accompanied by a large heeling force (lift), so that it may not be a very good option for high wind speeds.

7.4.5 Factors Determining the Effective Aspect Ratio

As discussed in some detail in Sect. 5.15, the effective aspect ratio A_e of a lifting surface like a sail is defined by (see Eq. 5.15.20):

$$A_e = A \, b_e/b = Ae, \tag{7.4.15}$$

where A is the geometrical aspect ratio, b the geometrical span and b_e the effective span. The latter is usually defined as the span of a fictitious lifting surface with an elliptic span loading that, for the same lift, has the same (average) downwash as the actual lifting surface. This means that $b_e = b$ or $e = 1$ when the actual lifting surface has an elliptical span loading and that $b_e < b$ or $e < 1$ for all other (planar) situations.

It is also recalled from Sect. 5.15 that when the span loading of a planar, isolated lifting surface is elliptic, the downwash and the associated induced drag adopt a minimum value. This would suggest that for upwind conditions of sailing, when a

---

[8] Recall that this kind of presentation was introduced in Sect. 4.3, Fig. 4.3.3.

low drag level is important, a sail should preferably have an elliptic span loading. For a planar, isolated lifting surface without sweep in uniform flow this would require an elliptical planform or a trapezoidal one with a taper ratio of about 0.45 (Fig. 5.15.11).

However, there is more to it. We have seen already in Sect. 5.15 that the span-wise lift distribution that gives a minimum induced drag is no longer elliptic when the lifting surface is non-planar, as, for example, in the case of a wing or keel with winglets. This is also the case when the lifting surface is in the proximity of a wall, as in Fig. 5.17.2. Depending on the distance between the root of the lifting surface and the wall, the effective span/geometrical span ratio can then be greater than 1, as was shown in Fig. 5.17.3. A sail that operates near the water surface experiences a similar phenomenon. In addition there is the effect of the hull, the effects of the wind gradient and the twist in the apparent wind profile and the twist of the sail. In the following paragraphs we will discuss each of these factors in some detail.

Effect of the Proximity of the Water Surface
We have seen in the preceding chapters that for a lifting surface, like the fin-keel of a sailing yacht, the water surface acts like a reflection plane. That is, the water surface acts as if there is a mirror image of the lifting surface on the other side of the water surface (Sect. 6.3). This also the case for a lifting surface on the air side of the 'ancient interface', i.e. a sail.

There is, however, also a difference. In the case of a surface piercing fin or keel the difference in density between air and water makes that the water surface is ex-perienced as a 'soft' surface by the fin or keel. This causes a pressure relief effect on the water side of the free surface. For a sail this is not the case; the water surface is now experienced as a solid surface and there is no pressure relief effect.

Figure 7.4.38 illustrates, qualitatively, what the mechanism is that determines the downwash due to trailing vorticity and the effective span of a sail near the

Fig. 7.4.38 *Illustrating the effect of the presence of the water sur-face on the downwash experienced by a sail*

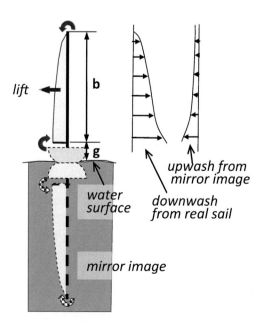

water surface. Ignoring, for the time being, the presence of the hull, the figure attempts to indicate that, due to the reflection plane effect of the water surface, a sail experiences less downwash than when there is no water surface. This reduction of downwash can be represented by the upwash induced by the trailing vorticity of the (virtual) mirror image of the sail. This upwash is caused mainly by the mirror image of the 'tip vortex' of the boom of the sail. The top of the mirror-image sail causes some downwash, but this is much smaller due to the greater distance between the top of the mirror-image sail and the real sail.

The mechanism just described is, of course, the same as that described in Sect. 5.17 for the interference between a lifting surface and a plane wall. As already illustrated by Fig. 5.17.3 and, again, by Fig. 7.4.38, an important consequence is that there is an increase of the effective span that depends, primarily, on the ratio g/b between the width of the gap and the span of the lifting surface. The effective aspect ratio is doubled when the gap width is zero but drops rapidly to ≈ 1 with increasing gap width. For the sail of a sailing yacht the gap width is usually between, say 1 and 20 % of the span of the sail. Because of the presence of the hull the effective gap width will, in general, be smaller (as we will see in a later paragraph). This means that the increase of the effective span due to the proximity of the water surface can be significant.

An associated phenomenon, shown again, in Fig. 7.4.39 (see also Fig. 5.17.4), is that there is also a change in the spanwise distribution of lift. When the gap width is zero the spanwise loading is the same as that of one half of a lifting surface with twice the geometrical aspect ratio (in this case 6 instead of 3).When the width of the gap is very large as compared to the span of the lifting surface ($g/b \rightarrow \infty$) the lift distribution of the sail and the effective span approach those of an isolated lifting surface of the same planform. For a rectangular planform (TR = 1) this will be symmetrical but for a triangular, Bermuda type rig (TR = 0) this will be non-symmetric and qualitatively as indicated in the figure for $g/b \rightarrow \infty$. Note that the loading (lift) goes to zero at the foot of the sail for any gap width different from zero.

For reference purposes the figure also shows elliptical span loadings. As discussed before, this is the spanwise distribution of lift (or load, or circulation) for which the induced drag adopts a minimum value in the case of an isolated planar lifting surface of given span or aspect ratio. When the gap width is zero ($g/b = 0$ in Fig. 7.4.39) the loading that leads to the minimum induced drag has the form of the semi-ellipse shown in the left side of Fig. 7.4.39. When the gap width is much larger than the height of the mast ($g/b = \infty$) this optimal loading takes the form of the full ellipse shown on the right.

An interesting question for the upwind performance of a sailing yacht is: "what is the spanwise distribution of lift or circulation that leads to the minimum induced drag for a sail (lifting surface) in the proximity of the water surface (a plane wall)?" The question has been studied in, amongst others, Sparenberg and Wiersma (1976); Wiersma (1977) and Sugimoto (1992). The answer depends, not surprisingly, on the relative width of the gap. Through the heeling moment, it also depends, indirectly, on wind speed.

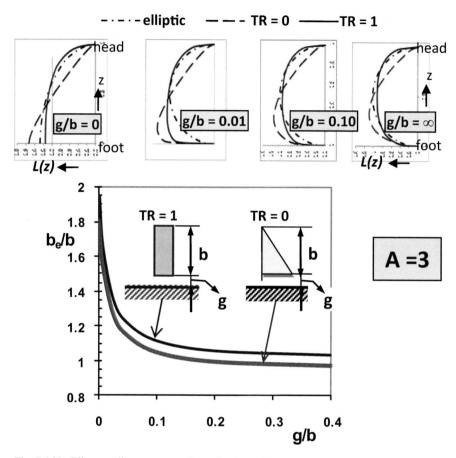

Fig. 7.4.39 *Effect on effective span and span loading of the gap between the foot of a sail and the water surface*

Figure 7.4.40 presents lift distributions, all for the same total lift, for minimum induced drag for several values of the gap width. The curves in the left ('(a)') figure present the case that there is no constraint on heeling moment. This is representative for sailing conditions with (very) low wind speed and stiff yachts. The figure illustrates that with a small gap, the lift is shifted towards the foot of the sail as compared to a full elliptic ($g/b = \infty$) distribution. In the figure (b) on the right the lift distributions are subject to a constraint on the heeling moment (reduction of about 10%, for the same total lift) that is representative for sailing conditions at higher wind speeds and tender yachts. As compared to the situation without constraint there is an even larger shift of the loading towards the foot, in particular when the gap is narrow.[9] For $g/b = 0$ the lift distribution is seen to adopt a bell-type shape of the sort also found for aircraft wings with optimal spanwise loading and a constraint

---

[9] The value $g/b = 0.07$ is representative for the effective gap width of an 'average' sailing yacht.

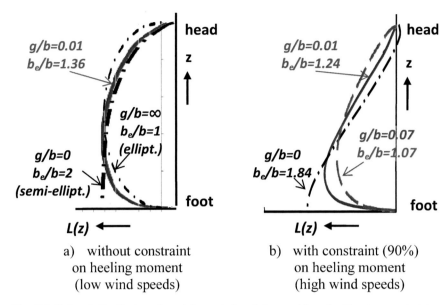

a)   without constraint
     on heeling moment
     (low wind speeds)

b)   with constraint (90%)
     on heeling moment
     (high wind speeds)

Fig. 7.4.40 *Load distributions for minimum induced drag, with and without a constraint on heeling moment*

on root bending moment (Jones 1950). It has been shown that in such, constrained conditions the downwash varies linearly along the span, rather than being constant as in the case of an elliptic lift distribution. For the sail (Wiersma 1977) the lift is seen to be slightly negative near the head. For a flexible, deformable surface like the sail of a sailing yacht this is, of course, not a feasible situation.

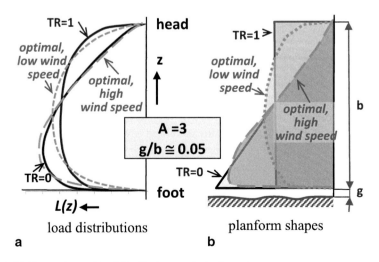

load distributions

planform shapes

a   b

Fig. 7.4.41 *Comparison of load distributions and planform shapes*

In Fig. 7.4.41a the optimal lift distributions of Fig. 7.4.40 are compared with the lift distributions given in Fig. 7.4.39 for rectangular (TR = 1) and triangular (TR = 0) planforms. It can be noticed that, for a relative small gap (g/b = 0.05), the optimal distribution for high wind speeds, resembles that for the triangular planform. This means that the effective span of a rig with the optimal, constrained load distribution is only slightly better than that of a rig with triangular planform.

The corresponding estimated sail planform shapes are compared in the right, (b), figure. The figure illustrates that the triangular, Bermuda type rig (TR = 0) is not quite so bad as is sometimes stated by those having the elliptical distribution in mind, at least for conditions of upwind sailing in high wind speeds. Triangular planforms do, however, have other drawbacks. They are, for example, prone to early stall near the tip. This is due to the spanwise variation of downwash as well as the low local Reynolds numbers near the tip.

For low wind speeds, when the heeling moment hardly matters, in particular for stiff yachts, the optimal planform is one with a shorter boom and a wider chord at the head of the sail.

It will further be clear from Fig. 7.4.40 that for a smaller gap the optimal planform will be closer to a semi-elliptical one for low wind speeds and closer to a bell shape for high wind speeds. For a (much) wider gap the optimal planform will be a little closer to a full elliptic shape.[10]

Effects of the Presence of the Hull
The air-exposed part of the hull of a sailing yacht plays a significant but rather complex role in the total aerodynamic forces. The basic reason for this is that the dimensions of the topsides and the freeboard are usually far from negligible in comparison to those of the sails. As will be discussed in some more detail in Sect. 7.6, the aerodynamic forces acting on the hull are substantial by itself. In addition there is a mutual interaction between the airflow about the sails and the airflow about the hull. Besides, the character and strength of this interaction vary strongly with the apparent wind angle.

Unfortunately, experimental data on the effects of the presence of the hull on the aerodynamics of the sails is almost non-existent. It is possible, however, to give some analytical, qualitative considerations.

The effective span of the rig of a sailing yacht (i.e. the downwash and its spanwise distribution) is influenced by the presence of the hull through several different mechanisms. The most conspicuous one is of a geometrical nature. With a hull between the foot of the sail and the water surface the effective width of the gap between the sail foot and the water surface is smaller than the real distance between the two (Fig. 7.4.42). The reason is that in addition to the water surface the deck of the hull also acts as a solid 'plane of reflection' for the downwash (Fig. 7.4.38). When the sheeting angle is small, the deck of the hull is closer to the foot of the sail than the water surface. The effective gap width (g_{we}) will then have a value somewhere between that of the distance g_w from the foot of the sail to the water

---

[10] As discussed in Sect. 5.15 a more practical alternative for the elliptic planform is a straight-tapered (symmetric) one with a taper ratio of about 0.45.

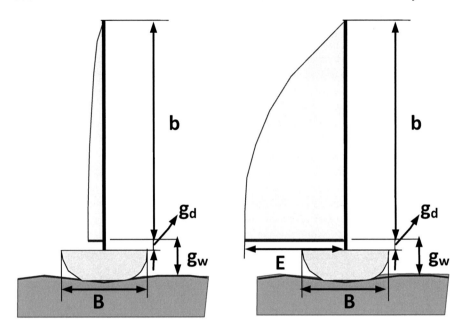

b) small sheeting angle   b) large sheeting angle

Fig. 7.4.42 *Defining gap widths at the foot of a sail*

surface and the distance g_d between the foot of the sail and the deck. It is further clear that when the size of the rig ('b') is much larger than the dimensions of the deck ($b \gg B$), the effective gap width g_{we} will be close to g_w. Similarly we will have $g_{we} \to g_d$ for $b \ll B$. Other parameters involved are the length E of the foot of the sail (Fig. 2.4.1), the sheeting angle δ and the apparent wind angle. The gap width g_{we} will be closer to g_w for larger values of a factor $(E/B)\sin\delta$, where δ is the sheeting angle (Fig. 7.4.43). The reason is that under such conditions a large part of the foot of the sail is directly exposed to the water surface without any shielding through the hull (Fig. 7.4.42b. The apparent wind angle plays a role through the mechanism that it governs the direction in which the trailing vortices from the sail are carried away from the hull (Fig. 7.4.43). For apparent wind angles around 90° the trailing vortices are carried away rapidly from the vicinity of the hull, in particular for large sheeting angles. In such conditions the effect of the hull on the effective gap width will be small. For small apparent wind angles, which usually imply small sheeting angles, the trailing vortices travel along the length of the hull and the effect of the hull will be larger. For apparent wind angles around 180°, with sheeting angles close to 90°, the sail will experience drag only and no lift. As a consequence the notion of effective span is then no longer of significance.

Figure 7.4.44 gives an impression of the trend of the effect of the presence of the hull on the effective span for small apparent wind angles and small sheeting angles.

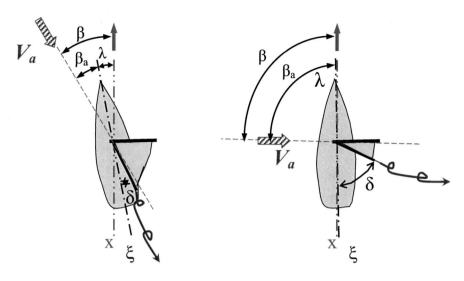

a)   small apparent wind angle   a)   apparent wind angle β ≅ 90°

Fig. 7.4.43 *Trajectory of trailing vortices for different apparent wind angles (qualitatively)*

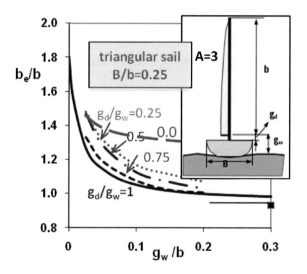

Fig. 7.4.44 *Trend of the effect of the presence of the hull on the effective span of a sail for small apparent wind angles*

Shown is the effective span ratio b_e/b as a function of the distance g_w between the foot of the sail and the water surface for several values of the ratio g_d/g_w of the distance g_d between the foot of the sail and the deck and the distance g_w between the foot of the sail and the water surface. Other, fixed conditions are a sail of triangular planform with a geometrical aspect ratio A=3 and a ratio B/b=0.25 between the width of the deck and the span of the sail. The figure has been constructed by the author on the basis of a simple, approximate, theoretical model for computing the upwash caused by the interaction between the trailing vortices and the deck plus the water surface. The flow about the hull is assumed to be attached in this model.

The figure illustrates that the effect of the presence of the deck on the effective span is significant, that is of the order of 5% for 'normal' values of g_d/g_w (~0.5) and g_w/b (~0.15). This in the sense that the smaller the distance between the deck and the foot of the sail, the better the effective span. With the foot of the sail on the deck ($g_d/g_w = 0$) the effective span settles at about 1.32 times the geometric span for normal values of g_w/b. (Note that the curve for $g_d/g_w = 1$ is the same as that in Fig. 7.4.39 for TR=0).

Using a similar theoretical model, the author has found that for a sail of rectangular planform the effect is about half of that for a triangular sail. For an apparent wind angle of 90° and a large sheeting angle the increase of the effective span due to the presence of the hull was found to be less than a percent. The effect of increasing the beam/span ratio B/b from 0.25 to 0.30 was also found to be modest; the resulting increase of the effective span is of the order of 1%.

In practice, the situation is a little more complex than as described above. The reason is that sailing yachts do, in general, not have a flat deck over the full length of the hull. The roof of the cabin and the spray hood, if present, are usually situated well above the deck and the floor of the cockpit is well below the deck level. This means that the aerodynamically effective value of the gap g_d between the foot of the sail and the deck may be a little smaller or a little larger, depending on the precise configuration of the yacht.

Another complicating factor is that the flow about the hull is, in general, not attached. As already indicated in Sect. 5.16 the flow separates at all sharp edges such as the junction of the deck with the top sides of the hull. This also has its effects on the flow about the sails. We will return to this in Sect. 7.6.

Effect of the Wind Gradient

We have seen already, in Sect. 7.2, that the velocity of the true wind varies with height above the water surface and that this causes the apparent wind velocity to vary also with height, both in magnitude as well as in direction. We have also seen in Sect. 7.3 that the 'twist' of the apparent wind profile can (almost) be neutralized by proper trimming of the geometrical twist of the sail (Fig. 7.3.2).

Neutralizing the effects of the vertical variation of the magnitude of the apparent wind speed is another matter. As already discussed in Sect. 7.2 and shown in Fig. 7.2.4, the difference between the apparent wind speed at the top of the mast and that at the foot of the sail can be as much as 20–40%, depending on the conditions of sailing. The consequences of this are twofold: there is a kinematic and a dynamic part.

The kinematic part is that a sail or foil section that is exposed to a reduced velocity of the free stream generates less circulation. The reason is that less downwash is required to satisfy the Kutta condition of smooth flow detachment at the trailing edge (Sect. 5.14). We recall from Sect. 5.14 that the Kutta condition requires that the component $U_*\alpha$ (α in radians) of the free stream velocity in the direction normal to the sail/foil slope at the trailing edge is neutralized by the local downwash generated by the foil. For high aspect ratios most of this downwash is generated by the bound vortex that represents the circulation about the section. This means that the circulation Γ is approximately proportional to $U_*\alpha$, where U is the (local) wind velocity at the position of the sail section and α the slope of the foil section at the trailing edge.

The second, dynamic part is a consequence of the law of Kutta-Youkowsky (5.47), which states that the local lift per unit span $L(z)$ in a section is proportional to the product of the velocity far upstream and the circulation. In the current notation this can be written as

$$L(z) = \rho U(z)\Gamma(z) \qquad (7.4.16)$$

This means that the local lift in a section is approximately proportional to U^2. This, of course, holds also for the total lift of a foil in uniform flow.

Figure 7.4.45 gives an example of the effect of the wind gradient on the spanwise distributions of circulation and lift for a sail with an aspect ratio of 3 in upwind conditions of sailing with zero aerodynamic twist (See also Fig. 7.2.4). Note that the sail foil section has 12% camber and is attached to a mast with a diameter/sail-chord ratio of 0.08. The width of the gap between the foot of the sail and the water surface is equal to 20% of the span (luff length) of the sail. The latter was set at 10 m. The effective gap width was set at $g_{we}/b = 0.1$. The latter is to be understood as the gap width that, without hull/deck, gives the same effective span as the configuration with hull/deck (see Fig. 7.4.46).

The load distributions in the figure have been determined by means of a modified, so-called 'lifting-line method (Abbott and Von Doenhoff 1949) of the type described in Speer (2010). The (approximate) effect of the wind gradient has been modeled in a way similar to that as proposed by Lissaman (1978).

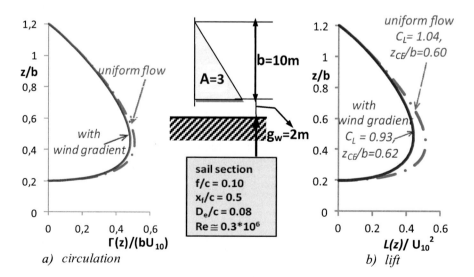

Fig. 7.4.45 Example of the effect of wind gradient on the spanwise distribution of circulation and lift in upwind conditions (calculated, $\gamma = 45°$, $V_b/V_t = 0.6$, $\alpha = 15°$, zero aerodynamic twist)

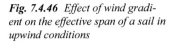

Fig. 7.4.46 Effect of wind gradient on the effective span of a sail in upwind conditions

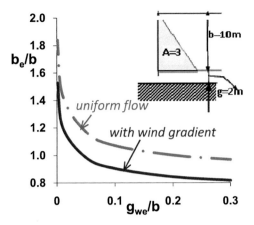

It can be noticed in Fig. 7.4.45 that the effect of the wind gradient on lift is significant, in particular in the lower half of the sail. The loss of the total lift is about 12% compared with the same conditions in uniform flow. There is also a slight change in the shape of the distributions. This has implications for the position of the centre of effort (z_{CE}), the induced drag and the effective span. The latter was found to be about 16% smaller in the case with wind gradient. For a condition of broad reaching ($\gamma = 135°$) the lift was found to be reduced by 16% and the effective span by 20%. It is to be expected that these numbers will be somewhat bigger for shorter rigs and somewhat smaller for taller rigs.

Figure 7.4.46 summarizes the effect of the wind gradient on the effective span of a triangular sail for upwind conditions of sailing. Note that the foot of the sail has been assumed to be at 2 m above the water surface and that the mast length is 10 m. The effective span is given as a function of the effective gap width g_{we}.

An appropriate question in this context is what the influence is of the wind gradient on the minimum induced resistance. According to a theorem of Munk (1923 in Chap. 5) the minimum induced downwash should be constant along the span when there is no constraint and should vary linearly when there is a constraint on bending/heeling moment. This means that in case of a wind gradient the planform should be modified such that the circulation is the same as in uniform flow. It can be shown (Speer 2010) that this requires that the local chord length corresponding with the optimum in uniform flow should be multiplied with a factor U_{10}/U, where, as before, U is the local apparent wind speed and U_{10} the reference apparent wind speed at a height of 10 m. In terms of Fig. 7.4.41 this implies that the optimum planform should have a somewhat wider chord near the foot of the sail than those depicted in the '(b)' figure.

It is further useful to note that the change of the shape of the load distribution (Fig. 7.4.45) also causes an upward shift of the centre of effort or centre of pressure (CE or CP, see Sect. 5.14). For upwind conditions this is from $z_{CE}/b = 0.60$ to 0.62. As we will see later (Sect. 7.8) this implies a corresponding increase of the heeling moment for the same lift force.

Effect of Twist

In the preceding discussions we have assumed the sail to have zero aerodynamic twist. This means that the sail has been assumed to have a geometrical twist that neutralizes the twist of the apparent wind profile (See Fig. 7.3.2). With a larger amount of geometrical twist (increased $\delta_h - \delta_0 \equiv \Delta\delta$ in Fig. 7.3.2) the local angle of attack of the sail decreases towards the tip (head). The opposite is the case when the geometrical twist is decreased. In both cases there is an effect on the spanwise distribution of circulation and, hence, on the effective span.

Figure 7.4.47 illustrates the effect of twist on the spanwise load distribution for the same sail configuration and the same upwind conditions of sailing as in Fig. 7.4.45. The distribution of the geometrical twist has been taken as in Fig. 7.3.1. Three values, $0°$, $3°$ and $6°$, have been chosen for the total geometrical twist angle $\Delta\delta$. Because the aerodynamic twist $\Delta\alpha$ is about zero for $\Delta\delta = 3°$ (see Fig. 7.3.2a this implies aerodynamic twist angles $\Delta\alpha \equiv \alpha_h - \alpha_{f\alpha}$ of $+3°$, $0°$ and $-3°$ respectively. As shown in Fig. 7.4.47, increasing the twist for the same sheeting angle at the foot of the sail causes a reduction of lift and of the effective span. The centre of pressure or centre of effort (z_{CE}) goes a little downward when the twist is increased.

It appears that for every $3°$ additional twist, the effective span decreases by about 9%, that is, the induced drag increases by the same amount (for the same lift). At the same time, the centre of effort (or centre of pressure) goes down by only about 1%. This suggests that increasing the twist of a sail is not a very efficient way to reduce heel. Reducing lift without reducing the luff length is more efficient because it does not affect the effective span. This can, in principle, be done in three different ways. The first is to decrease the angle of attack by increasing the sheeting angle (without increasing twist). Another option is to reduce the amount of camber of the sail. The third possibility is to reduce sail area. For the same luff length this implies a change of sail to one with a shorter chord at the foot. Reducing sail area by reefing at the foot of the sail is less efficient because this also leads to a reduced effective span.

Fig. 7.4.47 *Effect of twist on the load distribution of a sail in upwind conditions of sailing*

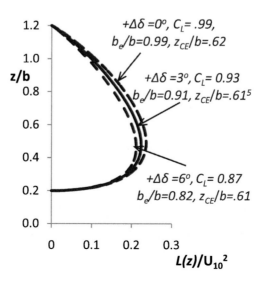

It is further useful to realize that the spanwise load distribution of a sail and, hence, the effective span, can also be changed by varying the camber of the sail sections in the vertical direction. Increasing the amount of camber towards the head of a triangular sail will make the span loading fuller near the top with a corresponding increase of the effective span. This, of course, at the cost of a higher position of the centre of effort and increased heel. For low wind speeds and/or stiff yachts, the increase of heel will, however, be small. Increasing the amount of camber towards the head of the sail may then provide some advantage.

7.5   Jib/Genoa Plus Mainsail

One of the most fervently disputed phenomena in the technology of sailing is the aerodynamic interaction between the jib or genoa of a sailing rig and its mainsail. Many theories have been put forward to explain the perceived superiority of a rig with closely coupled fore- and mainsails. However, it wasn't until the paper (Gentry 1971) published by Gentry[11] in 1971 that the mechanisms involved were fully understood.

In the following sub-sections we will, first, consider the interaction mechanisms for the simplified case of two-dimensional flow. This will be followed by a discussion on three-dimensional effects.

7.5.1   Interaction Mechanisms at Small Apparent Wind Angles

As argued by Gentry the interaction between a mainsail and a foresail is governed by two (or three) different mechanisms. The first (two) are the up/downwash that are generated by each of the sails. The second (or third) is an increased velocity of the flow near the trailing edge or leech of the foresail. Figure 7.5.1 illustrates what is meant.

Shown, schematically, is the flow about a lifting sail section (solid black line) at a small angle of attack. As discussed in Sect. 5.14 the presence of the lifting sail causes upwash upstream and downwash downstream of the section. In addition a region with locally high flow velocity is formed on and above the suction side of the section, just aft of the leading edge.

When another sail section (foresail, dashed black line) is placed in front of the first section, it experiences the upwash induced by the first (mainsail) section as a virtual increase of the angle of attack. The variation of the induced upwash in the

---

[11] Arvel Gentry, aerodynamicist and yachtsman, was, at the time, working as a research scientist at the former Douglas Aircraft Company in Long Beach, California. His boss, A.M.O. Smith, a well-known fluid dynamicist, was to publish, in 1975, a hall-mark paper (Smith 1975) on the aerodynamics of aircraft wings with high lift devices at the leading and trailing edge. The Smith paper describes the mechanisms involved, which are very similar to the mechanisms involved with sails, in substantial detail.

Fig. 7.5.1 *Mainsail/foresail interaction mechanisms*

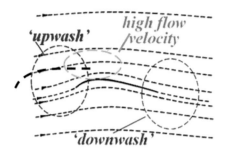

streamwise direction is experienced by the foresail as a virtual increase of camber, in particular towards its trailing edge.

When the trailing edge of the foresail is positioned in the high flow velocity area at a small distance aft of and above the leading edge of the mainsail, which is the case when the sails are sheeted for 'upwind' conditions, even more circulation (resulting in more lift) is required to satisfy the Kutta condition of smooth and tangent flow detachment at the trailing edge (Sect. 5.14). In addition, the boundary layer on the foresail experiences a higher trailing edge 'dumping velocity'. That is, it detaches from the trailing edge of the foresail with a higher velocity than the freestream velocity (see also Sect. 5.13). As shown in Fig. 7.5.2 this implies (Bernoulli's Law) that the pressure at the trailing edge of the foresail is lower (level 'B') than without the presence of the mainsail (level 'A'). As a, positive, consequence separation of the boundary layer on the upper surface of the foresail is postponed to higher levels of lift. In other words, the foresail can carry more lift when it is in the proximity of the mainsail. In summary, one can say that a foresail in the presence of a mainsail

Fig. 7.5.2 *Effect of the presence of a mainsail on the flow about a foresail*

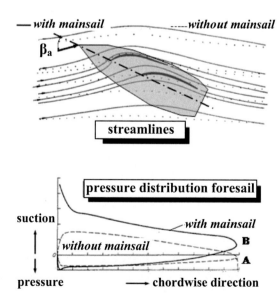

Fig. 7.5.3 *Effect of the presence of a foresail on the flow about a mainsail*

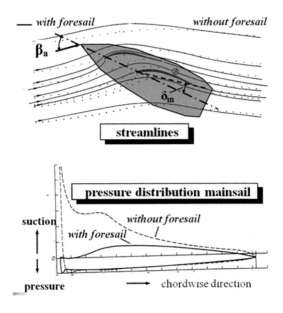

is forced to carry more lift through the upwash induced by the mainsail and that its boundary layer can sustain this higher level of lift by the higher 'dumping velocity'/ lower pressure at the trailing edge.

The other side of the picture is that the mainsail operates in the field of downwash induced by the foresail. This is experienced by the mainsail as a smaller angle of attack but also as some increase of camber, in particular near the leading edge. Figure 7.5.3 shows the effect on the pressure of the mainsail. Because of the reduction of the perceived angle of attack and the increased camber near the leading edge the mainsail carries less lift in the presence of the foresail, in particular near the leading edge.

An additional aspect is that the foresail acts as a sort of 'flow directing device' for the mainsail. Because the flow always leaves the leech of the foresail in the direction of the section slope at the trailing edge, the effective angle of attack of the mainsail hardly varies when, for fixed sheeting angles, the geometrical angle of attack of the combination (apparent wind angle) is varied. As a consequence the lift of the mainsail hardly varies when the apparent wind angle is increased or decreased. This means that most of the variation with angle of attack (apparent wind angle) of the lift and drag of the sail combination takes place on the foresail.

It should be remarked here, that the sail section configurations studied by Gentry (1971) and as shown in Figs. 7.5.2 and 7.5.3, do not contain a mast. This means that the flow near the leading edge of the mainsail will be somewhat different from that shown in the figures. From the discussion in Sect. 7.4 it is clear that, with a mast present, the lift on the mainsail will be smaller anyway (see Fig. 7.4.19) and this will be the case both with and without foresail. Because the effect of the presence of a mast on drag is smaller for low levels of lift (see Fig. 7.4.20), a mast is probably a little less harmful for a mainsail when it operates with a foresail in front. Another aspect is that the mast will create some additional supervelocity due to volume displacement in its vicinity and this may have some benefit for the foresail if its trailing edge is not too far away.

In the absence of any experimental evidence on the effect of the presence of a mast on the interaction between a foresail and a mainsail we may assume that the interaction is not modified in a qualitative sense. However, we should, because of the loss of lift of the mainsail caused by the mast, expect that the mainsail must be set at a higher angle of incidence (smaller sheeting angle) in order to induce the same favourable conditions for the foresail.

7.5.2   Lift and Drag

We have seen above that the aerodynamic interaction between closely coupled main- and foresails causes an increase of lift on the foresail and a reduction of lift on the mainsail. The key question is, of course: what is the net effect in terms of lift and drag? While the precise answer to this question depends on the precise configuration of the sails, about which later, the general answer is that the most important effect is an increase of the total lift of the sail combination. This is illustrated by Fig. 7.5.4. The figure compares the lift and drag characteristics of a mainsail and a foresail section in two-dimensional flow, in isolation and in combination, as measured in a windtunnel (Goovaerts 2000). The '(a)' figure depicts the sail configuration. The mainsail section has 11 % camber with the maximum camber at 30 % of

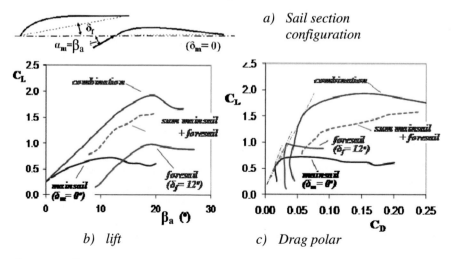

Fig. 7.5.4 *Effect on lift and drag of the interaction between a mainsail and a foresail in two-dimensional flow; coefficients based on the sum of the chord lengths. (From wind tunnel tests* [7.10], *(Goovaerts 2000), Re ≅ 0.2*10[6])*

the chord. The genoa type foresail section has 14.7% camber with its maximum at 40% of the chord. The latter is 3% longer than the chord of the mainsail section.

The genoa section has an overlap with the mainsail section of about 11% of its chord. The sheeting angle δ_m of the mainsail is 0° and that of the genoa (δ_f) is 12°.

The '(b)' figure presents the lift as a function of the angle of attack α_m of the mainsail. Because the sheeting angle of the mainsail section is zero this means that α_m is equal to the apparent heading β_a. Shown is the lift coefficient of the individual sail sections in isolation, the lift coefficient of the combination and the sum of the lift coefficients of the individual sections without aerodynamic interaction. Note that the aerodynamic coefficients are based on the sum of the chord lengths of the individual sections. This means that they can be considered as a measure of the absolute forces. Note also that due to the sheeting angle of the genoa its lift curve is shifted about 12° to the right as compared to the mainsail.

The figure illustrates clearly that the lift of the combination is significantly larger (by about 25%) than the sum of the lift of the individual sails. Although the lift on the individual sections has not been measured in the combination mode, it will be clear from the preceding discussion that the genoa will carry more than this 25% additional lift and the mainsail will have less lift than in isolation.

The figure also shows that the increase of lift with angle of attack of the combination is almost the same as that of the foresail only. This reflects the fact, mentioned in the preceding sub-section, that, at least for small sheeting angles, the lift of the mainsail hardly varies with angle of attack and that the increase of lift with angle of attack manifests itself mainly on the foresail.

The '(c)' figure compares lift-drag polars. Here we can see that the mainsail section by itself has the highest maximum lift/drag ratio (or the smallest drag angle), and that the genoa section has a slightly lower value of $(L/D)_{max}$. The maximum L/D of the combination is also slightly lower, but substantially better than for the summed forces of the individual sections. The main reason for the absence of improvement in terms of L/D is that the (profile) drag of a section increases quadratically, or faster, with lift (Sect. 5.14 and Appendix H). As a consequence there is always a point where an increase of lift causes too much of a drag penalty to improve the lift/drag ratio.

A complementary picture is obtained when the aerodynamic characteristics of the sections are compared in terms of coefficients that have been made dimensionless on the basis of the chord lengths of the individual sections. This is done in Fig. 7.5.5 for the lift-drag polars, where the chord length of the combination has been taken as the distance between the leading edge of the genoa section and the trailing edge of the main for $\delta_m = 0°$ (see Fig. 7.5.4a). It can be argued, that from a yacht/rig design point of view, this is a fairer comparison than that of Fig. 7.5.4c). This because the sections are basically compared for the same chord length. The latter is equivalent with the same rated sail area for real sails. What we can see in Fig. 7.5.5 is that the maximum lift of the combination is still the highest but that the drag is higher than that of the genoa in a small range of lift (around $C_L = 1.7$) only. Recalling, from Chap. 4, that high lift and a high lift drag ratio (or small drag angle) are the most important aerodynamic characteristics of a sail, Fig. 7.5.5 sug-

Fig. 7.5.5 *Comparison of drag polars for individual and interacting sail sections, coefficients based on equal chord lengths (see also Fig. 7.5.4)*

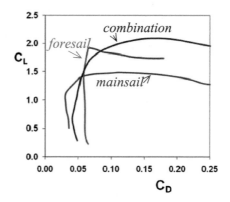

gests that the combination is better than a single genoa of the same area (except for a fairly narrow range of lift around $C_L = 1.7$) and better than a single mainsail for all $C_L > 1.4$.

It is useful to recall at this point, that in two-dimensional flow, any shape, single or multiple foils, has zero total drag in inviscid flow, that is when the Reynolds number is infinitely large (see Sects. 5.11 and 5.14). This implies that under such conditions the sum of the pressure drag of the section segments is always zero; the drag induced on one segment is always fully compensated by thrust induced on the other and there is no way around this. It means that while the pressure drag of a foresail in the presence of a mainsail is negative (implying thrust), due to the perceived up-wash, the pressure drag of the mainsail must necessarily be positive.

We have seen also in Chap. 5 that when the Reynolds number is finite, the 'viscous' or profile drag depends on the pressure distribution. For very thin sections, such as sails, the viscous drag depends primarily on lift and camber. With most of the lift carried by the foresail this means that the foresail also carries most of the viscous drag, (at least when the mainsail is not attached to a mast!). It is to be expected, however, that, for the same lift, the viscous drag of the foresail in the presence of the mainsail will be less than that of the foresail in isolation; the reason being that, for the same lift, the higher 'dumping velocity' at the trailing edge reduces the adverse pressure gradients that the boundary layers have to cope with. In the absence of such favourable conditions for the mainsail the viscous drag of the latter, will, for the same lift coefficient, hardly be influenced by the presence of the foresail. Recalling also, from the sub-section on *'upwash and downwash'*, that most of the variation of lift and drag with angle of attack of the combination is caused by the foresail, this means that the total drag of the combination will, in general, be smaller than that of an isolated foresail of the same size and at the same lift. The only conditions in which, for a given lift, the total drag of the combination may be larger than that of an isolated foresail of the same size is when, in the combination, the lift of the mainsail is relatively low and its drag relatively high. The latter will be the case when the main operates at a level of lift that is below the lift for its maximum lift/drag ratio.

What Figs. 7.5.4 and 7.5.5 suggest, in summary, is that the main advantage of favourably interacting fore- and mainsails is an increase of the level of lift. This at the

cost of some reduction of the maximum lift/drag ratio, i.e, a (small) increase of the
minimum drag angle. From the mechanics of sailing (Chap. 4) we know that a high
level of lift is the prime factor for the driving force when sailing upwind at low wind
speeds. In addition, for small apparent wind angles, a small aerodynamic drag angle
becomes increasingly important. At higher wind speeds it becomes increasingly
important to limit the heeling force. This means (see Fig. 4.3.5) that the importance
of a high level of lift becomes less important and that the importance of a small
aerodynamic drag angle increases progressively with wind speed, in particular for
tender yachts with shallow-draught keels. Hence, properly interacting foresails and
mainsails are more important for low wind speeds than for high wind speeds.

7.5.3   Effects of Gap, Overlap and Sheeting Angles

In the absence of systematic test data in the literature it is not clear to what extent
the results depicted in Figs. 7.5.4 and 7.5.5 depend on the precise configuration of
the sails. However, we know, from the aerodynamics of high lift devices of aircraft
wings (Smith 1975) that the maximum attainable lift increases when the number of
foil segments is increased, but that the maximum lift/drag ratio decreases significantly with increasing number of segments. It is also known from such sources that
the chord lengths of the foil segments should be of the same order of magnitude if
they are to induce the most favourable interaction effects.

For sails this means that the optimum number of segments will not be very large
and that the chord lengths of the leading and trailing segments should be about the
same. It can be argued, since the increase of maximum lift manifests itself on the
foresail and the flow about the latter is not hampered by the presence of a mast, that
the optimal value of the ratio c_f/c_m of the chord lengths of foresail and mainsail is
probably a little larger than 1. However, the difference from 1 is likely to be small, because the improvement of the aerodynamics of the foresail is induced by the mainsail.

For a given number of segments and ratio of chord lengths, the most important
parameters are the gap or slot width, the overlap and the difference of incidence between the segments. We know from systematic studies of the aerodynamics of high-
lift devices of aircraft wings (Smith 1975) that the gap should be as small as possible without causing a direct 'touch' between the boundary layers of the segments.
We also know, as already mentioned above, that, for a given difference of incidence
between the segments, the trailing edge of the leading segment should be positioned
in an area near the trailing segment where the upwash and flow velocity are high.
This is usually not very far from the leading edge of the trailing segment, which
means that the overlap should not be very large. It is further known (Smith 1975)
that with increasing difference of incidence between the segments the positive interaction effect on lift increases but that the maximum lift/drag ratio decreases.

For sails, additional parameters are the amount and the chordwise position of maximum camber. It is also clear that gap, overlap and segment incidence angles (sheeting angles) are not fully independent, as illustrated by Fig. 7.5.6. For a fixed inci-
dence/sheeting angle δ_m of the mainsail, the width 'g' of the gap wil increase or

Fig. 7.5.6 *Defining gap and overlap of sail section segments*

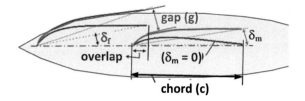

decrease when the sheeting angle δ_f of the foresail is increased or decreased. Similarly, for a fixed sheeting angle of the foresail, the gap width will decrease when the sheeting angle of the mainsail is increased. However, in the case of the latter, the dependence is weaker because the leech (trailing edge) of the foresail is closer to the hinge point of the mainsail section.

For most yachts, there is a minimum for the gap width that is determined by the lateral position of the genoa sheeting rail on the deck and/or the width of the shrouds of the rig. The latter two are, to a large extent, determined by the maximum beam of the yacht. It is further clear that the minimum sheeting angle of the genoa is a function of the beam/length ratio of the hull.

For large sheeting angles, such as utilized at high apparent wind angles, the notions gap and overlap lose their significance. The interaction between a mainsail and a foresail adopts a different nature when the apparent wind angle becomes of the order of 90° or beyond, as we will see shortly.

Effects of Sheeting Angles

What, then, may we expect in terms of changes in lift and drag from changing the sheeting angles of the foresail and mainsail sections?

From the preceding paragraphs it follows that increasing the sheeting angle of the foresail with that of the mainsail fixed, causes a larger gap and a larger difference between the incidence angles (sheeting angles) of the foresail and mainsail section segments. The larger difference in sheeting angles will lead to a larger increase of maximum lift but a decrease of the maximum lift/drag ratio. These effects will, however, be tempered somewhat by the increase of the gap width. Decreasing the sheeting angle of the mainsail with the foresail fixed will, qualitatively, have a similar effect. However it will, quantitatively, be a little more pronounced, because the increase of the gap width is smaller than in the case of a larger sheeting angle of the foresail.

Figure 7.5.7 confirms the expected effect on lift of an increase of the difference $\Delta\delta = \delta_f - \delta_m$ between the sheeting angles. Shown, for the same 2D configuration (Goovaerts 2000) as in Fig. 7.5.4, is the lift coefficient C_L as a function of the angle of attack ($\alpha = \beta_a - \delta_m$) of the mainsail section for two values, 12° and 18°, of the sheeting angle of the foresail section. It appears that the maximum lift is some 10% higher for the larger value of $\delta_f - \delta_m$.

Drag data for $\delta_f = 18°$ are, unfortunately, not available. However, the decrease of the maximum lift/drag ratio (or increase of the minimum drag angle) with increasing $\Delta\delta = \delta_f - \delta_m$ is confirmed by measurements for three-dimensional rigs (Fossati et al. 2008; Marchai and Tanner 1963). An example in the form of (partial) lift-drag polars, based on wind tunnel data from Ref. Fossati et al. (2008) is given

Fig. 7.5.7 *Effect on lift of sheeting angle difference (see also Fig. 7.5.4)*

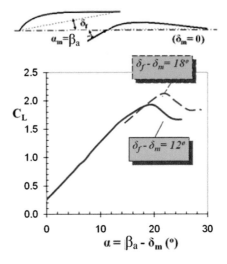

Fig. 7.5.8 *Lift-drag polars for different sheeting angles for a 3-D rig (Fossati et al. 2008) in close-hauled conditions*

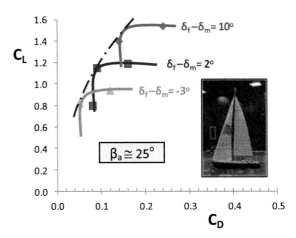

in Fig. 7.5.8. The rig concerned is an IMS-class rig with a non-overlapping jib and an aspect ratio of about 4. The average maximum camber of the jib is about 12%, positioned at about 40% of the chord. The corresponding figures for the mainsail are 10 and 32%, respectively. The average mast-diameter/main-chord ratio D/c is about 0.05. The minimum drag angle is seen to increase with increasing value of $\Delta\delta = \delta_f - \delta_m$ and so does the level of lift.

It is useful to note at this point that the effect of increasing the difference in sheeting angle between the foresail and the mainsail is very similar to that of increasing camber. In both cases it leads to an increase of the maximum lift and an increase of the minimum drag. There is also a difference, however. While camber increases the lift for a given angle of attack, an increase of the difference in sheeting angle causes a reduction of lift at the same apparent heading.

For the yachtsman the implication of the discussion just given is that the difference between the sheeting angles of the mainsail and the foresail should be kept small for small apparent wind angles, in particular at high wind speeds. For larger apparent wind angles the sheeting angle difference should be increased, in particular at low wind speeds when heel does not play a significant role.

Effects of Overlap
Information on the effect of overlap on lift and drag is not abundant in the literature. As already discussed above it can be expected that the amount of overlap should be fairly small to exploit the benefits of favourable aerodynamic interaction. Marchai (2000) reports an 'optimum' overlap of 65 % at the foot of the genoa of a 90 % fractional rig for close hauled conditions ($\beta_a = 30°$). With the local overlap diminishing towards the top of the sail this comes down to an average sectional overlap of about 30 %. The 'optimum' was defined as the overlap giving the best value of V_b/V_t. Assuming that this was determined for low to moderate wind speeds (although Ref. Marchai (2000) does not mention this explicitly) this 'optimum' would represent the maximum driving force condition.

More recent wind tunnel measurements by Fossati et al. (2006), with simulated atmospheric boundary layer, lead to a similar picture. Shown in Fig. 7.5.9 are the maximum driving force coefficient C_{Tmax} and the maximum propulsive efficiency $(C_T/C_H)_{max}$ as a function of the genoa overlap for a 92 % fractional rig in an up-wind ($\beta=27°$) configuration. The rig is similar to that of Fig. 7.5.8, except for the fractionality. It should be noted that the force coefficients in Fig. 7.5.9 have been corrected (Fossati et al. 2006) for the drag of the bare hull and the bare rig and are all based on the same reference sail area (that of the rig without overlap). The figure shows that the maximum driving force is obtained for an overlap of about 45 %, which comes down to an average sectional overlap of about 20 %. The figure also shows that overlap does not pay at all when the maximum propulsive efficiency (C_T/C_H) is the objective. This means that there is no point in sailing with a foresail with overlap in high wind speeds and at very small apparent wind angles. However,

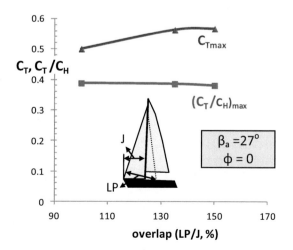

Fig. 7.5.9 *Effect of genoa overlap on maximum driving force (C_{Tmax}) and maximum propulsive efficiency $(C_T/C_H)_{max}$*

overlap does pay for low to moderate wind speeds, provided that the overlap does not exceed about 45 %.

7.5.4 'Optimal' Camber

An interesting question is what the 'optimum' amount of camber is for a foresail in the presence of a mainsail and vice versa. The question has been addressed by Chapin et al (2006, 2008) through computational fluid dynamics for two-dimensional flow and numerical optimization techniques.

The answer depends, of course, on what sort of 'optimum' is required. Figure 7.5.10 shows section shapes and streamlines when the maximum lift/drag ratio or maximum propulsive efficiency is sought for. The optimization was performed (Chapin et al. 2006) for fixed chord lengths ($c_f/c_m = 1.04$), fixed positions of the leading edges, fixed positions of the location (x_f/c) of maximum camber (40 % for the genoa and 30 % for the mainsail), no mast and an apparent heading β_a of 30°. For these conditions[12] the genoa acquired a camber/chord ratio of 19 % and the mainsail of only 4 %. The sheeting angles found were $\delta_f = 31°$ and $\delta_m = 10°$, respectively.

Much larger values of 'optimum' camber ($f_f/c = 30\%$ and $f_m/c = 27\%$, respectively) are found when the objective is chosen to be the maximum driving force (see Fig. 7.5.11). The sheeting angles are, however, of comparable magnitude ($\delta_f = 32°$ and $\delta_m = 3°$). The maximum driving force condition is, as we have seen in Chap. 4, much closer to the condition of maximum lift than the condition for maximum L/D (or minimum drag angle).

It is useful to realize at this point that in real, three-dimensional conditions, for a rig of finite aspect ratio, the apparent heading β_a should be significantly larger than 30° if the same level of lift as in two-dimensional flow is to be realized. Assuming the lift coefficient C_L to be about 1.7 in the case of Fig 7.5.10 (maximum L/D) and about 2.5 in the case of Fig. 7.5.11 (maximum driving force), it follows from lifting surface theory (Chap. 5) that, for a rig of aspect ratio 3.5, the induced angle of attack is about 9° for the lift level of Fig. 7.5.10 and about 13° for the lift level of Fig. 7.5.11. This means that the apparent headings would be about 40 and 45°, respectively, in equivalent 3D conditions.

Fig. 7.5.10 *Example of the amount of camber and sheeting angles required for maximum propulsive efficiency (maximum T/H, or maximum L/D) in upwind conditions. (Two-dimensional flow, from Ref. Chapin et al. (2006))*

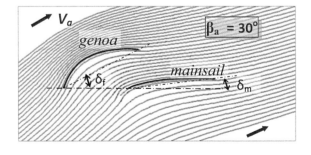

---

[12] Unfortunately, Chapin et al. (2006, 2008) do not mention the Reynolds number for which these studies were performed. An 'educated guess' is that it is of the order of $1.4*10^6$.

Fig. 7.5.11 *Example of the amount of camber and incidence required for maximum driving force in upwind conditions. (Two-dimensional flow, from Ref. Chapin et al. (2008))*

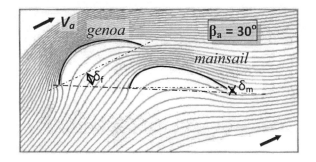

What Figs. 7.5.10 and 7.5.11 suggest, in summary, is that a large amount of camber for both the foresail and the mainsail and a rather large difference between the sheeting angles are required to maximize the driving force in upwind conditions. A much smaller amount of camber, in particular for the mainsail, and a smaller difference between the sheeting angles is required to maximize the propulsive efficiency (T/H).

Information on the optimum position of maximum camber for interacting fore- and mainsails is, unfortunately, not available. We have seen in Sect. 7.4, that, for single sails, a position at 30–40% of the chord gives the best lift/drag ratio (Fig. 7.4.12) and, hence, the best propulsive efficiency. A more rearward position leads, however, to a higher maximum lift (and higher maximum driving force). It can further be argued, that because of the higher trailing edge 'dumping' velocity of the foresail (induced by the mainsail) and the resulting reduced sensitivity to boundary layer separation (Sect. Interaction Mechanisms at Small Apparent Wind Angles), the foresail can cope with a more rearward position of maximum camber than the mainsail. The apparent preference of sail makers to position maximum camber at about 40% of the chord for jibs/genoas and at about 30% for mainsails seems to support this.

7.5.5   Large Apparent Wind Angles

For apparent wind angles approaching 90° and beyond, the interaction mechanisms gradually change. Consider, for example, the case of a sail configuration at an apparent wind angle of about 90° and large sheeting angles (Fig. 7.5.12).

What the figure tries to illustrate is that the bound (or 'lifting') vortex Γ_m of the mainsail induces hardly any upwash or downwash on the foresail, the induced velocity vectors v_{im} being almost parallel to the foresail section. The latter implies a slightly higher lift for the foresail because of an effective increase of the apparent wind speed. The bound vortex Γ_f (lift) of the foresail appears to induce velocity vectors v_{if} that cause some downwash at the trailing edge of the main in addition to a reduction of the effective apparent wind speed. This implies a lower lift of the mainsail. It can be concluded that the intensity of the interaction between a foresail and a mainsail at large apparent wind angles (and corresponding sheeting angles) is much smaller than for small apparent wind angles and small sheeting angles. The main reason is that for large apparent wind angles and large sheeting angles the

Fig. 7.5.12 Illustrating interaction mechanism at large apparent wind angles (foresail to leeward)

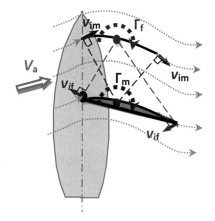

Fig. 7.5.13 Illustrating aerodynamic interaction mechanisms at large apparent wind angles (foresail to weather)

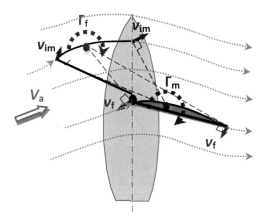

mainsail does not induce much upwash on the foresail. Nevertheless, the foresail will still carry all little more and the mainsail a little less lift than in isolation due to the mutual influence on the effective apparent wind speed.

An alternative for the sail set of Fig. 7.5.12 is to set the foresail to the weather side on the spinnaker pole (Fig. 7.5.13). In this configuration the foresail will receive some more upwash from the mainsail because it is positioned more upstream of the main. For the same reason the mainsail will be subject to some more down-wash from the foresail. This means that a sail configuration like that depicted in Fig. 7.5.13 will behave a little more like upwind sail configurations as shown in Figs. 7.5.2 and 7.5.3 than a configuration like that of Fig. 7.5.12. However, the 'gap' between the foresail and the mainsail in the 'broad reaching' configuration of Fig. 7.5.13 is, obviously, much larger than in upwind conditions (Figs. 7.5.2 and 7.5.3). As a consequence the benefit, in terms of additional maximum lift, will be smaller.

It is worth mentioning at this point that more favourable interaction conditions are obtained if the foresail section is moved aft, thereby reducing the width of the gap between the mainsail and the foresail, as in Fig. 7.5.14. This, in a way, is what we

Fig. 7.5.14 *Illustrating aerodynamic interaction mechanisms at large apparent wind angles (foresail to weather: aerorig)*

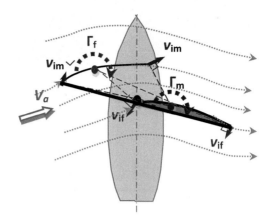

see happening if we look upwards towards the top of the foresail: the gap width becomes zero at the point where the forestay is attached to the mast.

Reducing the gap width, in particular for reaching conditions and as far down to the deck as possible, is a characteristic of the so-called *'aerorig'* concept. The aerorig (to be described in some more detail in Sect. 7.13) is characterized by a large rotating boom that supports the foresail as well as the mainsail. With the luff attached to the front end of the boom rather than the bow, the foresail is in a much better position for beneficial mutual interaction with the mainsail, in particular for reaching conditions.[13]

While the aerorig has also some advantage on off-wind courses, at least for conventional jib/genoa plus mainsail configurations, downwind sailing is really the domain of spinnaker type head sails. We will discuss the latter in some detail in Sect. 7.7.

7.5.6   Additional Factors for Three-Dimensional Rigs: Roach and Fractionality

In the preceding (sub-)sections we have considered the interaction between fore- and mainsails merely from a two-dimensional point of view. That is, we ignored the fact that the vertical dimensions of any practical sail are finite.

For single sails the aspect ratio $A = b^2/S$, where b is the span or distance between head and foot and S is the area of the sail plan without overlap, was seen to be the most important parameter in three dimensions (Sect. 7.4). Multiple sails (main+jib/genoa) can be considered in two different ways:

- As a single sail with the planform of the jib/genoa plus mainsail and the section characteristics of the interacting jib/genoa plus mast/mainsail configuration

---

[13] It can be argued that even better conditions for beneficial mutual interaction would be created if the forward and aft segments of the boom could be rotated independently.

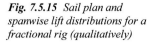

Fig. 7.5.15 *Sail plan and*
spanwise lift distributions for a
fractional rig (qualitatively)

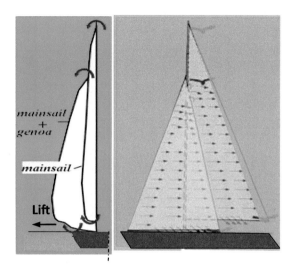

- As two separate, but interacting, three-dimensional sails, each with its own 3-D characteristics but with the interaction modeled through, for instance, the bi-plane theory of Munk (1923)

The first option is a feasible and suitable one for upwind conditions and the associated sail configurations. This means that one can use the descriptions given in Sect. 7.4 with the appropriate jib + mast/mainsail section characteristics to obtain an estimate of the aerodynamic characteristics of a rig in three dimensions. A complicating factor is that the sail planform is rather irregular, in particular for fractional rigs (see Fig. 7.5.15). As a consequence there are discontinuities in the variation of the (multiple) section shape, local chord length and the sectional aerodynamic characteristics. This causes irregularities in the spanwise distribution of the total lift and circulation (mainsail + genoa, as shown in the left side of the figure) with associated concentrations of trailing vorticity. The penalty of this is a reduction of the effective span and, hence, a higher induced drag.

It is also obvious that the benefits in terms of improvement of maximum lift of the mutual interaction between foresail and mainsail disappear when the local chord length of the foresail and/or the mainsail goes to zero, that is near the top of the sail(s).

Figure 7.5.17 gives an impression of the magnitude of the induced drag penalty in terms of a quantity h_e, called the effective rig height. The effective rig height h_e is determined from wind tunnel tests (Claughton et al. 1999, 2008) by fitting a straight line given by

$$C_D = C_{D0} + \frac{C_L^2}{\pi h_e^2 / S} \qquad (7.5.1)$$

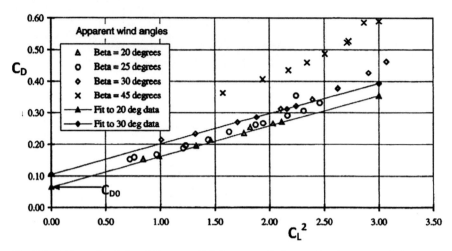

Fig. 7.5.16 *Example of plot of C_D versus C_L^2, for mainsail plus masthead genoa, including windage. (From Ref. Claughton et al. (1999))*

through a plot of measured values, for a given apparent wind angle, of the drag coefficient C_D versus C_L^2 (for an example see Fig. 7.5.16). The quantity C_{D0} is called the parasite drag or zero lift drag and S, as usual, is the sail area. All drag values, including the zero lift drag C_{D0}, incorporate, in principle, the drag of the hull, rig and all other sources of 'windage' (to be addressed in the next section). When measurements without sails are available this 'windage' can be subtracted from the measured drag values to obtain an approximation of the lift and drag of the sails proper. Obviously, all the lift dependent components of drag have been collected in the second term of Eq. (7.5.1). This means that it contains the contributions of the profile drag, the induced drag and, in principle, also the drag due to side edge separation (see Sect. 7.4, Eq. (7.4.12)). For aspect ratios >2 the latter is negligible, however. We have also seen in Sect. 7.4 that, for moderate aspect ratios ($A \cong 3$), the induced drag is much larger, by a factor of the order of 4, or more (see Fig. 7.4.28), than the profile (or viscous) drag due to lift. This means that the induced drag is almost equal to, but smaller than the drag-due-to-lift term in Eq. (7.5.1). Hence we can write, see also Eq. (7.4.12)

$$C_L^2/(\pi A e) = \frac{C_L^2}{\pi(b^2/S)\dfrac{b_e}{b}} \le \frac{C_L^2}{\pi h_e^2/S} \tag{7.5.2}$$

From this it follows that

$$b b_e \ge h_e^2, \tag{7.5.3}$$

or, dividing by $b^2 = h_{mast}^2$, where b is the geometrical span or mast length h_{mast},

Fig. 7.5.17 Effect of rig fractionality on effective rig height (effective span), from windtunnel tests. (Fossati et al. 2006)

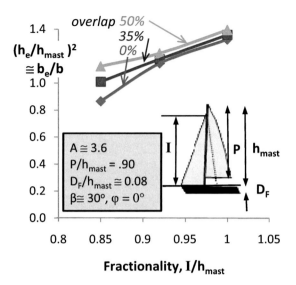

$$b_e/b \geq (h_e/h_{mast})^2 \qquad (7.5.4)$$

In other words the effective rig height is about equal to, but smaller than, the square root of the effective span.

Figure 7.5.17 gives the effective rig height (squared) as a function of the fractionality (ratio I/h_{mast}) as determined from wind tunnel tests (Fossati et al. 2006) on IMS-class rig models similar to that shown in Fig. 7.5.8. The aspect ratio of the rigs is about 3.6. Results are shown for three values of genoa overlap. The figure shows clearly that, in terms of effective span, a masthead rig is superior to a fractional rig.[14]

The figure shows also that the effective span benefits somewhat from genoa overlap. This is probably due to the fact that, with a large overlap, and the genoa down to the deck, a larger portion of the sail combination benefits from a small gap between the foot of the sail and the deck/water surface. The latter, as we have seen in Sub-Sect. Factors Determining the Effective Aspect Ratio (Fig. 7.4.41), is favourable for the effective span.

Roach is another factor with influence on the effective span. As described in Sect. 2.4 the roach of the mainsail is a measure of the deviation of the planform shape of the main from the triangular form. As indicated in Fig. 2.4.1 the amount of roach is usually expressed as the additional area between the actual leech of the mainsail and the basic triangular form as a percentage of the area of the basic tri-

---

[14] There may, however, be other reasons to prefer a fractional rig over a mast head rig. One often heard argument is that fractional rigs offer better possibilities for trimming the sails (adjusting camber). Whether this is sufficient to overcome the reduced effective span of the fractional rig is questionable. The author is not aware of any rational, quantitative data on this matter.

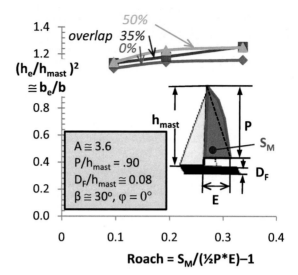

Fig. 7.5.18 *Effect of mainsail roach on effective rig height (effective span), from windtunnel tests. (Fossati et al. 2006; Claughton et al. 2008)*

angular form. Increasing the amount of roach for a fixed area implies that the sail acquires a shorter foot 'E' but longer chord lengths near the head.

The effect of mainsail roach on the effective span is illustrated by Fig. 7.5.18. The figure is based on data from the same set of wind tunnel tests (Fossati et al. 2006; Claughton et al. 2008) as Figs. 7.5.17, 7.5.8 and 7.5.9. The effective span is seen to increase slightly with increasing roach. This is in agreement with Sect. 7.4 in which we saw that for low wind speeds, when the heeling moment does not matter, the best planform shape is one that begins to resemble an elliptical shape (See Figs. 7.4.36, 7.4.37 and 7.4.38).

7.6 Aerodynamics of Other Yacht Components

7.6.1 'Windage' and Parasite Drag

Sails and mast are not the only components of a sailing yacht that are exposed to the wind and that generate aerodynamic forces. The shrouds of the rig, spreaders, boom(s), stays, sheets, halyards, railing, deck equipment, exposed crew members and, of course, the hull top sides and superstructure, are also sources of aerodynamic force, drag in particular. The contributions of all these sources are usually collected in a single term that is called '*windage*'. Windage comprises not only the drag ('*parasite drag*') but also the lift produced by the hull and other air-exposed components of a yacht.

The parasite drag is far from negligible. Its contribution to the total aerodynamic drag can be as much as 25 % in close-hauled conditions. The contribution of the hull in particular is quite important (65 %, or more, of the total parasite drag). The precise magnitude of the parasite drag depends, of course, on the configuration of the yacht and the apparent wind angle.

The lift part of the 'windage' of a yacht is much smaller than the drag, being of the order of 3% or less of the lift produced by the sails in upwind conditions of sailing. Most if not all of this 'windage' lift is due to the hull.

Determination of the contribution of windage to the total aerodynamic forces is not a simple matter. It is, of course, possible to measure the lift and drag force on (a model of) the bare hull with the bare rig (and other sources of windage) without the sails, for example through a wind tunnel test. Figure 7.6.1 presents results of such measurements (Claughton et al. 1999; Hansen et al. 2002, 2006; Fossati 2009; Richards et al. 2006). Shown are the coefficients of drag and lift due to windage as a function of the apparent heading angle β_a for several cruiser-racer/racing yacht configurations. In each case, the lift and drag coefficients are based on the (reference) sail area. That is, C_L and C_D were determined through the following (usual) expressions:

$$C_L = L/(\tfrac{1}{2}\rho V_a^2 S) \qquad (7.6.1)$$

$$C_D = D/(\tfrac{1}{2}\rho V_a^2 S), \qquad (7.6.2)$$

where L and D are the parasitic lift and drag, respectively, V_a is the apparent wind speed and S is the (reference) sail area.

It can be noticed that the drag curves exhibit similar shapes but that the differences in level are appreciable. This holds also for the lift coefficients. The reasons are probably differences between the configurations of the yachts, the ratios between hull dimensions and sail area in particular, and, possibly, differences in the definitions of the reference sail area and the Reynolds number.

One conclusion that can be drawn from Fig. 7.6.1 is that lift and drag of a hull + bare rig without sails behaves, roughly, like

$$C_L \sim c_{wL1} \sin\beta \cos(\beta + c_\beta) \qquad (7.6.3)$$

Fig. 7.6.1 Examples of lift and drag due to windage as measured in wind tunnel tests on models of racing yachts and cruiser-racers

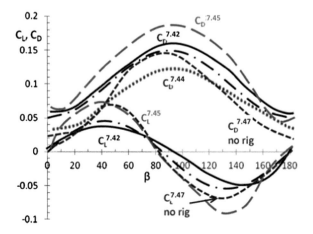

Fig. 7.6.2 Example of the variation with apparent heading of the vertical aerodynamic force coefficient of a sailing yacht hull (10 m IMS cruiser-racer (Hansen et al. 2006))

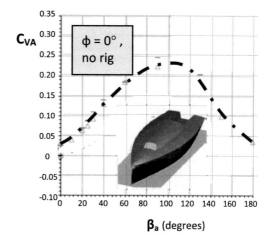

$$C_D \sim c_{wD0} + c_{wD1} \sin^2\beta \qquad (7.6.4)$$

Here, the constants c_{wL1} (~0.1), c_{wD0} (~0.05) and c_{wD1} (~0.11) depend mainly on the ratio between the projected area of the hull and the sail area. The off-set angle c_β (~5°) represents the effect of the fore/aft asymmetry of the hull. For a hull without rig the values of the constants are, roughly, $c_{wL1} \cong 0.075$, c_{wD0} (~0.025) and c_{wD1} (~0.09) More specific formulae can be found in Ref. Richards et al. (2006).

As already mentioned above, the hull and the rig are not the only sources of windage. The contribution of railings, deck equipment and exposed crew members to the parasite drag can be as much as the minimum drag of the hull without the rig.

Windage also causes a vertical aerodynamic force component on the hull. Figure 7.6.2 gives an example from wind tunnel tests (Hansen et al. 2006) on a model of the hull of a 10 m IMS-class cruiser-racer. Shown is the vertical aerodynamic force coefficient C_{VA}, defined as

$$C_{VA} = V_A/(\tfrac{1}{2}\rho V_a^2 S) \qquad (7.6.5)$$

as a function of the apparent heading. It can be noted that the variation with β_a is similar to that of the aerodynamic drag in Fig. 7.6.1 and that the magnitude is even larger than the aerodynamic drag. The greater part of this vertical aerodynamic force is believed to be generated by a 'deck-edge' vortex of the kind already mentioned in Chap. 5.16 (see Fig. 5.16.3).

7.6.2 Interaction Mechanisms

From the point of view of yacht performance, the ultimate interest is, of course, in the aerodynamic forces on the combination of hull plus rig plus sails. Unfortunately the latter are not simply equal to the sum of aerodynamic forces on the sails plus the windage. The reason is that there is aerodynamic interaction between the various sources of windage and the sails as well as between the various sources of windage themselves. One (special) example of this, discussed already in Sect. Effects of a Mast or Head-Foil on Sail Section Characteristics, is the aerodynamic interaction between a sail and the mast to which it is attached: As indicated in Fig. 7.4.19 the lift and drag of a sail plus mast configuration are not equal to the sum of the values of the individual components in isolation. Another example is when an object is situated in the wake of an object that is positioned more upstream. In such a case the object experiences less windage due to the lower effective wind speed in the wake.

Figure 7.6.3 illustrates a further basic mechanism of aerodynamic interaction between objects. For convenience the figure addresses the simplified case of two-dimensional flow in a horizontal plane. On the left side we have a cylinder (for example a rod or wire) subject to an airstream with velocity V. Because the cylinder produces drag only and no lift, the resultant force vector F has the same direction as the free stream velocity vector V and is equal to the drag D.

On the right (b) figure), the cylinder is positioned near a lifting body represented by the curved line with 'lifting vortex'. In the topology considered the lifting body induces a perturbation velocity (for example upwash, red arrow) at the position of the cylinder. Hence, the cylinder experiences an effective onset flow V_e with a direction and magnitude that differs from the undisturbed flow velocity V. As a consequence the resulting force vector F acquires a different direction and a different magnitude also. Decomposition of F into components perpendicular to and parallel to the free stream velocity vector V then gives lift L and drag D.

The example just given illustrates that a body which produces drag only in isolation produces lift and drag when positioned near another body. The drag in the presence of the other body may be smaller or larger than the drag in isolation, depending on the direction of the perturbation velocity relative to the free stream velocity. Whether the other body is lifting or non-lifting is not essential. In both cases it will generate a perturbation of the flow at the position of the first body.

Aerodynamic interaction between two objects is, in principle, always a two-way street: the interaction is mutual. If, however, one object is much smaller than the other and not immediately adjacent, the interaction becomes a one-way street: the effect of the proximity of the big object on the flow about the small object is large but there is hardly any effect of the presence of the small object on the big object.

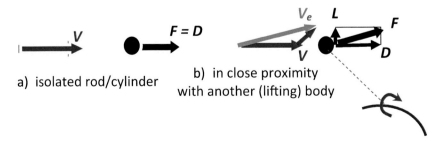

Fig. 7.6.3 *Illustrating aerodynamic interaction in relation to windage*

This is caused by the fact that the radius of aerodynamic influence of a body is proportional to its dimensions.[15] When two objects are immediately adjacent there can be a direct and possibly strong interaction through the boundary layers, for example in the sense that the smaller body triggers boundary layer transition or separation on the bigger body.

A schematic of the main aerodynamic interactions that can be distinguished between the various components of a sailing yacht is given by Fig. 7.6.4. Examples of mutual interaction are those between the hull and the sails and between the mast and the (main)sail. The latter example is special in the sense that mast and sail are immediately adjacent to each other. Most if not all other interactions can be considered to be of the one-way type.

Because of the aerodynamic interaction mechanisms described above, the modeling of the 'windage' of a sailing yacht is not an easy matter. It is, of course, possible to estimate the drag (and lift, if applicable) of single, individual, windage producing components of a yacht. This is usually done by expressing the drag D_n of a component as

$$D_n = \tfrac{1}{2}\rho V_a^{\,2}\, S_n C_{Dn}\,, \tag{7.6.6}$$

where S_n is the cross-sectional area of the component normal to the flow direction and C_{Dn} the related drag coefficient. The latter can, for example, be based on the drag data base contained by Ref. Hoerner (1965). The total parasite drag D_{par} due to windage can then be estimated by summing the distributions of the components:

$$D_{par} = \textstyle\sum_n \tfrac{1}{2}\rho V_a^{\,2} S_n C_{Dn} \tag{7.6.7}$$

Fig. 7.6.4 *Schematic of mutual and one-way aerodynamic interference between the different components of a sailing yacht*

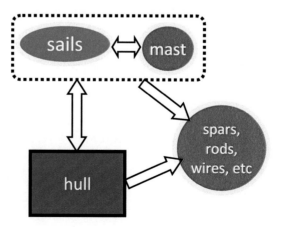

---

[15] There is one exception to this 'rule': when the flow about the large body is unstable, a (very) small disturbance can have a big influence

We know now that, because of the aerodynamic interactions between components, this is not a good procedure. Unfortunately, there is hardly an alternative. The reason being that we do not know the direction and magnitude of the perturbation velocities of the various components as a function of their relative positions and the apparent wind angle. If we would know these it would be possible to take the interactions into account according to the scheme of Fig. 7.6.3.

What is clear from the preceding discussions is that the most important interactions are between the mast and the sails and between the hull and the sails (+ mast). We have seen already in (Sub)-Sect. Effects of a Mast or Head-Foil on Sail Section Characteristics that sails plus mast should preferably be treated like a sail configuration with the sectional characteristics of the sail plus mast combination rather than adding the windage of the mast to the aerodynamic characteristics of the sail. As we saw in Sect. 7.4 this is possibly a feasible procedure, at least for small and moderate apparent wind angles. Determining the effect of the hull on the aerodynamics of the sails (and vice versa) is another matter.

7.6.3   The Flow About the Hull and Its Interaction with the Sails

One effect of the hull is that, by its presence, the effective span of the sails is increased through a reduction of the effective width of the gap between the foot of the sail and the water surface, as already discussed in Sect. 7.4. But there is more to it. The hull also creates its own flow field with its own lift and drag ('windage'), as we saw above, which interacts with the flow about the sails.

As already touched upon in Sect. 5.16, the (air-exposed part of the) hull of a sailing yacht can be considered as some kind of slender body. It can be considered as slender because, for most yacht hulls, the ratio $\sqrt{S_{Xmax}}/L$ is of the order of 0.15, i.e. $\ll 1$. A sailing yacht hull, unlike bodies of revolution, also exhibits several sharp edges, such as at the intersection between the deck and the sides. As already indicated in Sect. 5.16 (see Fig. 5.16.3) such edges trigger the formation of longitudinal vortices due to the rolling-up of the separating shear layer at the edge of the deck. Figure 7.6.5 gives a more realistic illustration of this phenomenon. The figure, taken

Fig. 7.6.5 *Example of the formation of a 'deck-edge vortex'. (From a CFD simulation for a Star class boat (http://www.wb-sails.fi 2008))*

from http://www.wb-sails.fi (2008), shows particle traces (streamlines) from a computational fluid dynamic simulation for a Star class boat.

A consequence of the deck-edge separation is that a sailing yacht hull behaves also somewhat like a (thick) low aspect ratio lifting surface (slender wing) with side edge and/or leading edge separation (Figs. 5.15.12/5.15.15). This is the reason why the lift curve in Fig. 7.6.1 has a slope that is different from zero for $\beta = 0°$. Without side edge separation the slope at $\beta = 0°$ would be, practically, zero (See Sect. 5.16).

As already mentioned, the deck-edge vortex is also believed to be the main cause of the vertical aerodynamic force (Fig. 7.6.2), addressed in a preceding paragraph.

A classical but still very useful experiment on the effect of the presence of the hull on the aerodynamics of the sails is the one by Marchai and Tanner (1963) on a model of a "Dragon" type yacht (Fig. 7.6.6) in the wind tunnel of Southampton University. Measurements were performed for several configurations, including the complete rig as mounted on the hull, the rig alone, without the hull, at the same distance from the (simulated) water surface and the hull alone. The measurements were done for small to moderate apparent wind angles, with zero heel and for fixed sheeting angles of the mainsail ($\delta_m = 5°$) and/or the genoa ($\delta_f \cong 14°$). Sail camber was about 10%. Figure 7.6.7 shows the results of these tests in terms of lift. Shown is the lift of the hull alone (normalized by sail area), the lift of the rig alone and the lift of complete configuration as a function of the apparent heading β_a. Also shown is the sum of the lift of the hull alone plus that of the rig alone.

The important thing to notice is that the lift of the complete configuration (rig with hull) is significantly larger than the lift of the rig alone and the sum of the lift of the rig alone plus the hull alone. This illustrates that there is considerable aerodynamic interaction between the hull and the sails.

When considering the effect of the hull on the sails two or three different mechanisms can be distinguished. One is the increase of the effective span and the effective aspect ratio of the sails due to the fact that the hull reduces the effective width of the gap between the foot of the sail and the water surface (already described in Sect. Factors Determining the Effective Aspect Ratio). This causes an increase of the lift curve slope. The other mechanism is an increase of the angle of attack of the

Fig. 7.6.6 '*Dragon*' *class yacht*

Fig. 7.6.7 Example of the effect of the hull on the aerodynamic lift of a sailing yacht ('Dragon' class), as measured in a wind tunnel. (Marchai and Tanner 1963)

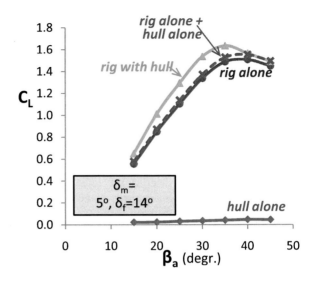

sails plus an increase of the effective wind speed due to additional cross-flow that is generated by the cross-sectional shape of the hull. This additional cross-flow is caused by both the displacement volume of (the air-exposed part of) the hull as well as the deck-edge vortex shown in Fig. 7.6.5. The result is an increase of the lift of the sails for a given apparent wind angle and a shift of the maximum lift of the sails to a smaller apparent wind angle. Because the maximum lift of a lifting surface is almost independent of the effective aspect ratio (Sub-Sect. '3D' Effects: Effective Aspect Ratio the difference between the maximum lift of the complete configuration and the maximum of the sum of the individual components is a measure of the effect of the increase of the effective wind speed. The difference between the apparent wind angles at which maximum lift occurs (about 3°) is a measure of the increase of the effective angle of attack and the lift curve slope. A similar picture for drag is shown in Fig. 7.6.8. The figure shows that for the same apparent heading

Fig. 7.6.8 Example of the effect of the hull on the aerodynamic drag of a sailing yacht ('Dragon' class), as measured in a wind tunnel. (Marchai and Tanner 1963)

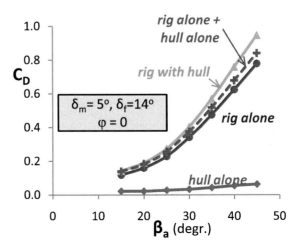

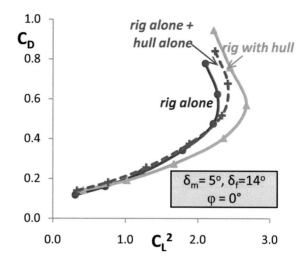

Fig. 7.6.9 *Example of the effect of the hull on the lift-drag characteristics of a sailing yacht ('Dragon' class) in upwind conditions, as measured in a wind tunnel. (Marchai and Tanner 1963)*

the drag of the complete configuration (hull with rig) is larger than the sum of the drag of the hull alone plus that of the rig alone. This is caused by the higher lift of the complete configuration with hull and rig and the associated higher induced drag.

On the other hand it also appears, (Fig. 7.6.9) that the drag for a given lift of the configuration with hull is always smaller than that of the sum of the rig alone plus the hull alone and also smaller than the drag of the rig alone for lift coefficients above ~0.95. The difference is further seen to increase with increasing lift. This behaviour can be ascribed to the increase of the effective span of the rig due to the presence of the hull, as described in Sect. Factors Determining the Effective Aspect Ratio.

When determined from the slope of the curves in Fig. 7.6.9 at low lift coefficients, the effective span of the rig alone is found to be $b_e/b \cong 0.92$. For the rig with hull this is found to be $b_e/b \cong 1.12$, a considerable improvement indeed. Incidentally, this increase, of about 20%, is about equally large as the decrease of the effecting span due to the windgradient (Fig. 7.4.46).

An important consequence of the increase of the effective span is that the minimum drag angle ($\varepsilon_A = \arctan(D/L)$ of the rig with hull is about the same as that of the rig alone, this in spite of the drag increase due to the hull (See Fig. 7.6.10). Moreover, it follows from Fig. 7.6.9, that the drag of the rig with hull is smaller than that of the rig alone for all values of C_L between $\cong 1$ and the maximum lift coefficient of the rig alone. This means a larger driving force for all apparent wind angles <35° (See Fig. 7.6.11). Beyond maximum lift, when, with the sheeting angles fixed, the lift decreases again with increasing apparent wind angle, but the drag increases further, the driving force decreases also.

When considering Figs. 7.6.10 and 7.6.11, it should be emphasized that they represent conditions with the sheeting angles of both mainsail and genoa fixed. Besides, the sheeting angles of the configurations with and without hull are the same. We have seen in Chap. 4 that there is an optimum sheeting angle for every

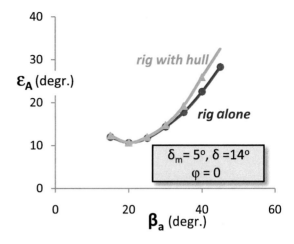

Fig. 7.6.10 *Example of the effect of the hull on the aerodynamic drag angle of a sailing yacht ('Dragon' class), as measured in a wind tunnel. (Marchai and Tanner 1963)*

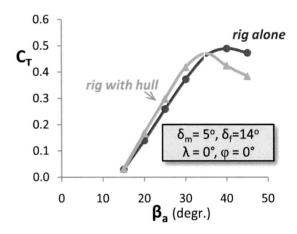

Fig. 7.6.11 *Example of the effect of the hull on the driving force of a sailing yacht ('Dragon' class), as measured in a wind tunnel. (Marchai and Tanner 1963)*

apparent wind angle. What this optimum, and the objective, is, depends on the type of yacht and the conditions of sailing.

When the objective is to minimize the aerodynamic drag angle, or, equivalently, to maximize the propulsive efficiency (applicable to high wind speeds and tender yachts), Fig. 7.6.10 shows that for a sheeting angle of 5° of the mainsail and 14° of the genoa, the drag angle attains its minimum at an apparent heading of about 20° for both configurations. Larger sheeting angles are required if this minimum is to be shifted to higher apparent wind angles.

When the objective is to maximize the driving force it is found that for $\delta_m = 5°$, $\delta_f = 14°$, the maximum driving force is attained at $\beta_a \cong 40°$ for the rig alone and at $\beta_a \cong 35°$ for the configuration with hull (see Fig. 7.6.11).

The other way around is not necessarily the case, as shown by Fig. 7.6.12. The figure gives the driving force coefficient as a function of the sheeting angle of the mainsail for $\beta_a = 35°$. Evidently, the driving force attains its maximum when

Fig. 7.6.12 *Example of the effect of the hull on the driving force of a sailing yacht ('Dragon' class) as measured in a wind tunnel (Marchai and Tanner 1963) as a function of sheeting angle, for a given apparent wind angle*

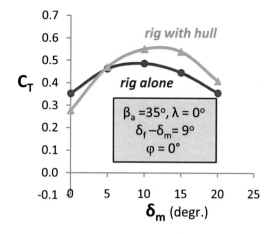

the sheeting angle of the mainsail is about 9° for the rig alone and about 12° for the rig with hull. Note that the maximum driving force is significantly larger for the configuration with hull.

Summarizing the effect of the hull for small to moderate apparent wind angles we can conclude that the hull acts by no means like a drag producing obstruction for the airflow about the sails. On the contrary, its presence causes, in general, an increase of the driving force produced by the sails, provided that the sheeting angles are adjusted to cope with the higher effective angle of attack. Even the minimum drag angle and, hence, the propulsive efficiency (see Sect. 4.3) does not seem to be affected seriously.

A remark to be made with respect to the results discussed above is that there was no simulation of the wind gradient of the atmospheric boundary in the wind tunnel tests. As a consequence the effects of the hull are possibly a little more pronounced than in reality.

For large apparent wind angles very little is known about the effects of the hull on the aerodynamics of the sail. Recalling the three different mechanism that can be distinguished:

1. increased effective span
2. increased effective wind speed
3. increased effective angle of attack,

We have seen already (Sub-Sect. Factors Determining the Effective Aspect Ratio), that the first, i.e., the increase of the effective span of the sails, becomes smaller when the sheeting angle is increased, which is the case for large apparent wind angles. On the other hand, the increase of the effective span will not totally disappear for $\beta=90°$. It can be argued that, for $\beta=90°$, it will still be of the order of B_{max}/L times the value for small sheeting/apparent-wind angles if B_{max} and L are the maximum beam and length of the hull, respectively. This means that the increase of the effective span, which, in a preceding paragraph, was seen to be of the order of 20 % for a 'Dragon' type of yacht at $\beta \cong 30°$, may be expected to be of the order of 3–5 % for large apparent wind angles and the associated large sheeting angles. Figure 7.6.13 gives the effective span of the IMS Class yacht considered earlier in Sect.7.5 for a limited range of small to moderate apparent wind angles, as measured in wind tunnel tests (Fossati et al. 2006). The figure shows that the variation of the effective span with apparent wind angle is quite large. This can be explained partly by the increasing effective gap width between the foots of the sails and the water surface for the associated larger sheeting angles, as mentioned above. Another mechanism involved is probably that the increase of the effective wind speed and the effective angle of attack will be felt mostly by the lower parts of the sails (see the difference in load distribution between uniform flow and with wind gradient in Fig. 7.4.42). This means an increased load in the lower parts of the sails with, possibly, a small increase of the effective span as a consequence. The increase of the effective wind speed will attain its maximum when the hull has maximum exposure to the wind, that is for $\beta = 90°$. The variation with β will be like $\sin\beta$. This means that the curve in Fig. 7.6.13 may be expected to level off for $\beta \to 90°$.

The maximum values of the increase of the effective wind speed and the effective angle of attack are probably a function of the ratio D_F/b between the height D_F of the freeboard of the hull and the span b of the rig and approximately proportional to the ratio D_F/B_{max}. It appears from experimental data (Marchai and Tanner 1963) that the maximum value of the increase of the effective wind speed is of the order of 2.5 %, that is 5 % in terms of dynamic pressure, for $D_F/b \cong .06$ and $D_F/B_{max} \cong .35$.

The increase due to the hull of the effective angle of attack was seen to be of the order of 3° at an apparent heading of about 35–40° for de 'Dragon' Class yacht. This is probably close to the maximum, because the increase due to the hull of the effective angle of attack wil be zero for $\beta_a = 0°$ and $\beta_a = 90°$ and can be expected to attain its maximum around $\beta_a = 45°$. This means that the variation with β will be like $\sin(2\beta)$. For yachts with a higher freeboard the maximum can be expected to be proportionally higher than 3°.

Fig. 7.6.13 *Dependence of effective mast height (effective span) on apparent wind angle, from windtunnel tests. (Fossati et al. 2006)*

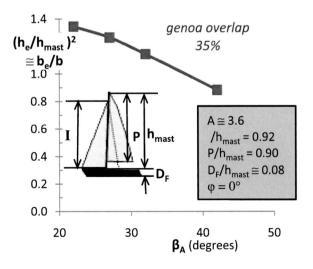

There is no doubt that the aerodynamic interaction between hull and sails is mutual. Unfortunately, there is very little information in the literature about the effect of the sails on the flow about the hull. Hansen et al. (2006) and Richards et al. (2006) form an exception. They give the vertical aerodynamic force with and without rig and sails, as measured in a wind tunnel (with simulation of the atmospheric boundary layer). Figure 7.6.14 presents some data (Hansen et al. 2006). The figure compares the coefficient C_{VA} of the vertical aerodynamic force of the hull without the rig (as in Fig. 7.6.2) with that of the hull with the mast and sails fitted. For $\beta_a > 90°$ the results are given for a mainsail/jib configuration. For $\beta_a > 90°$ the results are for a mainsail/spinnaker configuration.

It is found that the effect of the sails on the hull vertical force is quite large indeed. The vertical force appears to be smaller over the full range of apparent wind angles by an amount that is about equal to half its maximum value. The reason for this is probably that the greater part of the top side of the hull is always exposed to the pressure side (windward side) of the sails. This will cause a higher pressure over most of the deck with a smaller vertical force as a result.

If the mechanism just described is correct, the aerodynamic side force on the hull at small apparent wind angles will be augmented by the presence of the sails. The reason is that the weather side of the hull is also exposed to the pressure side of the sails, implying a higher level of lift as well as drag. The effect is likely to disappear for large sheeting angles and the associated large apparent wind angles. Data from wind tunnel tests (Richards et al. 2006) appear to confirm the occurrence of such phenomena: At small apparent wind angles the lift of a hull has been found to be about twice as large in the presence of sails and the drag about 20 % larger. The effect is also found to disappear for larger apparent wind angles and changes sign for $\beta_a \cong 120°$.

Fig. 7.6.14 *Effect of the sails on the vertical aerodynamic force coefficient of a sailing yacht hull. (10 m IMS cruiser-racer (Hansen et al. 2006))*

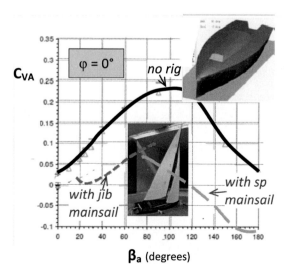

7.7   Centre of Effort, Heeling, Pitching and Yawing Moments

Aerodynamic Heeling and Pitching Moments

The forces acting on a sail, together with the position of the centre of pressure, determine the aerodynamic moments of force of a sailing yacht. In the jargon of sailing the centre of pressure is usually called the Centre of Effort (CE). Its vertical position in particular determines the aerodynamic heeling and pitching moments, as illustrated by Fig. 7.7.1. Together with the hydrodynamic heeling and pitching moments and the hydrostatic righting moments they determine the angles of heel and trim-in-pitch. As already discussed in Chap. 4, heel in particular plays an important role in the performance of a sailing yacht.

It will be clear from Fig. 7.7.1 that the aerodynamic heeling moment M_{Ax} satisfies the formula (see also Sect. 3.4)

$$M_{Ax} = S_A\, z_{CE} - V_A\, y_{CE} \approx H_A\, \zeta_{CE} \approx S_A\, \zeta_{CE}/\cos\varphi \qquad (7.7.1)$$

and that the aerodynamic pitching moment M_{Ay} is given by

$$M_{Ay} = -\left(T\, z_{CE} + V_A\, x_{CE}\right), \quad > 0 \text{ nose up} \qquad (7.7.2)$$

In these expressions S_A, V_A (<0) and H_A are the aerodynamic side force, vertical force and heeling force components, respectively and T is the driving force. If the

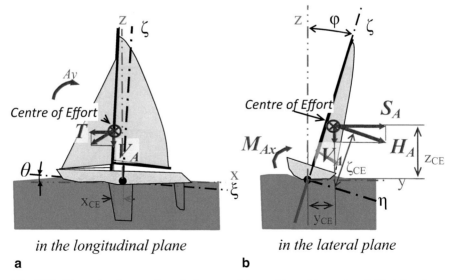

in the longitudinal plane   *in the lateral plane*

a   **b**

Fig. 7.7.1 *Illustrating the aerodynamic forces and the position of the centre of effort that determine the heeling and pitching moments*

origin of the coordinate systems is taken at amidships, in the water surface, the vertical position of the centre of effort is given by z_{CE}, the lateral position by y_{CE} and the longitudinal position by x_{CE} (<0 in Fig. 7.7.1).

The position of the centre of effort can also be expressed in terms of coordinates ξ, η, ζ in the ship coordinate system (Sect. 3.1). It can be shown that for small pitch angles θ and small angles of leeway λ there holds:

$$x_{CE} \approx \xi_{CE} \tag{7.7.3a}$$

$$y_{CE} \approx \eta_{CE} \cos\varphi + \zeta_{CE} \sin\varphi \tag{7.7.3b}$$

$$z_{CE} \approx -\eta_{CE} \sin\varphi + \zeta_{CE} \cos\varphi \tag{7.7.3c}$$

These expressions indicate that there is a strong dependence on the angle of heel φ of the vertical and lateral position of the centre of effort, about which later.

For a given rig and sail configuration the position of the ζ-coordinate ζ_{CE} of the centre of effort in the ship coordinate system is almost independent of the angle of heel φ, but ξ_{CE} and η_{CE} vary with quantities like the sheeting angles and the amount of camber of the sails. Because the vertical aerodynamic force V_A is, in general, relatively small, except for large angles of heel, the vertical position of the centre of effort is the most important geometrical quantity for the heeling and pitching moments. The heeling force and driving force are the most important force components.

Factors Governing the Vertical Position of the Centre of Effort
It will be clear from Fig. 7.7.1 that the most important quantities for the vertical position of the centre of effort are the height or span b of the rig and the height of the foot of the sail above the water surface.

The distribution of lift along the span is another factor of importance. For small to moderate angles of attack (α) the lift distribution depends primarily on the shape of the planform and the twist and camber of the sail. For high angles of attack, when there is an considerable amount of separated flow it depends also on α.

We have seen already in Sect. 5.15 that, for attached flow (small apparent wind angles), the spanwise position of the centre of pressure of half a lifting surface of symmetrical planform with a small taper ratio like a sail, is at around 40% of the (semi-)span (see Fig. 5.15.18). We have also seen in Sect. Factors Determining the Effective Aspect Ratio that for a single sail of triangular planform, of aspect ratio 3 and with the foot of the sail at a distance of 20% of the span above the water surface ($g_w/b = 0.2$), the centre of pressure/effort is positioned at a distance of about 62% of the span above the water surface (Fig. 7.4.42). This corresponds with about 42% of the span above the foot of the sail. Increasing the twist by 6° was seen to lower the centre of effort from 62 to 61% only (Fig. 7.4.47). A substantial amount of twist is, apparently, needed to lower the centre of effort significantly.

In a study of the kind underlying Figs. 7.4.45 and 7.4.47, performed by the author, it was found that the position of the centre of effort at small apparent wind angles, as measured from the foot of a triangular sail (TR=0) with aspect ratio 3, is almost constant (at about 40% of the span of the sail) for a range of practical values of the gap-width/span ratio g_w/b. This means that, as shown in Fig. 7.7.2, the vertical position z_{CE}/b of the centre of effort as measured from the water surface varies almost linearly with g_w/b. It means also that it varies almost linearly with the height $D_F(=g_w-g_d)$ of the freeboard of a yacht (see Fig. 2.3 for the definition of the freeboard height). For a sail with a rectangular planform (TR=1) and A=3 the author has found the centre of effort at about 43% of the span from the foot of the sail. When measured from the water surface it increases also almost linearly with g_w/b.

The effect of camber on the vertical position of the centre of effort was found to be negligible, as long as the distribution of camber along the span was uniform. When the amount of camber and/or the chord-wise position of maximum camber increases towards the head of the sail, the centre of effort moves in the same direction.

In a quantitative sense there is not much known about the effects of large apparent wind angles and the associated larger sheeting angles. As discussed in Sect.7.4 the effective width of the gap between the foot of a sail and the water surface will increase with increasing apparent wind angle and increasing sheeting angles. This means that the lift distribution will change in the sense that the centre of pressure (=centre of effort) will move upwards. On the other hand the increased effective wind speed induced by the cross-flow about the hull at large apparent wind angles will be felt mostly by the lower parts of the sail (as already discussed in Sect. 7.6) and this will tend to lower the centre of effort. Which of the two mechanisms will dominate is hard to say at this stage. We will return to this shortly.

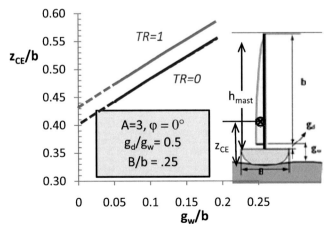

Fig. 7.7.2 *Vertical position of the centre of effort of a single sail as a function of the distance between the foot of the sail and the water surface (small apparent wind angles)*

For Bermuda-type rigs with mainsail and jib/genoa there is a number of additional parameters with effect on the position of the centre of effort. Fractionality, overlap and roach are the most important.

Figure 7.7.3 gives the vertical position of the centre of effort as a function of fractionality for an IMS class rig (the same as in Fig. 7.5.17) for an apparent wind angle of about 30°. The figure is based on results of the same set of wind tunnel tests (Fossati et al. 2006) with simulation of the atmospheric boundary layer. The centre of effort is seen to move upward from about 40% to about 45% of the mast height above the deck when the fractionality is increased from 80 to 100%, not an insignificant amount. This, of course, is not surprising, since the sail area moves also upward with increasing value of the fractionality ratio I/h_{mast}.

Less important is the effect of overlap, which is also shown in the figure. A smaller increase of z_{CE}/h_{mast} from 43 to 45%, is found (Fossati et al. 2006) when the roach is increased from 0.10 to 0.34.

From the wind tunnel tests (Fossati et al. 2006) referred to above, it appears also that there is a significant effect of the apparent wind angle on the vertical position of the centre of effort. This is shown by Fig. 7.7.4, which gives results for the same configuration as in Fig. 7.7.3. The centre of effort is seen to go downward from about $z_{CE}/h_{mast} = 0.45$ to 0.41 when the apparent wind angle is increased from 22 to 45°. This, in all probability, is due to the same, hull-sail, interaction mechanisms that govern the decrease of the effective span shown in Fig. 7.6.13.

In a preceding paragraph it was argued that the effect of the increasing effective width of the gap between the foot of the sails and the water surface tends to move the centre of effort upwards with increasing apparent wind angle but that the increased effective wind speed due to the additional cross-flow induced by the hull tends to lower it. We can now conclude that the latter effect apparently dominates

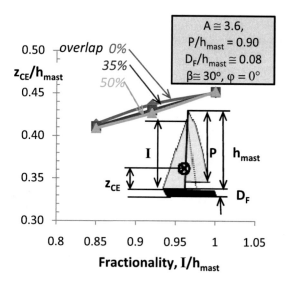

Fig. 7.7.3 *Effect of rig frac-tionality on the vertical position of the centre of effort. (Fossati et al. 2006)*

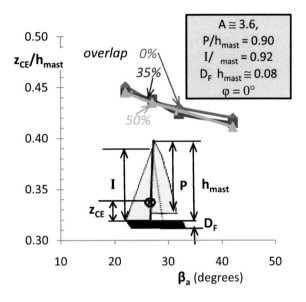

Fig. 7.7.4 *Dependence on apparent wind angle of the vertical position of the centre of effort, from windtunnel tests. (Fossati et al. 2006)*

the former. This means that we may expect the centre of effort to attain its lowest position for $\beta \cong 90°$.

Longitudinal and Lateral Positions of the Centre of Effort

As illustrated by Fig. 7.7.5, the aerodynamic yawing moment M_{Az} of a sailing yacht is determined by the level of the driving force and aerodynamic side force plus the longitudinal and lateral positions of the centre of effort. In the hydrodynamic coordinate system it can be expressed as

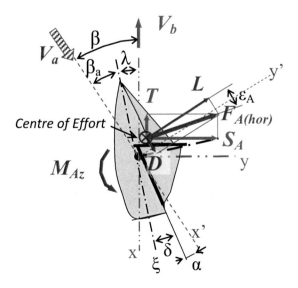

Fig. 7.7.5 *Aerodynamic forces in the horizontal plane and yawing moment*

$$M_{Az} = S_A x_{CE} + T y_{CE}, \tag{7.7.4}$$

where $M_{Az} > 0$ when it tends to turn the bow of the boat to weather. (Note that $x_{CE} < 0$ in Fig. 7.7.5).

We have seen already, that there is a strong dependence on heel of the lateral position of the centre of effort and, through that, of the aerodynamic yawing moment, about which later.

There is also a dependence on sheeting angle. It is relatively easy to see that the centre of effort will move outboard and forward with increasing sheeting angle. For an individual sail the position of the centre of effort varies like

$$\xi_{CE}(\delta) = \xi_{luff} + \{\xi_{CE}(\delta_0) - \xi_{luff}\} \cos\delta/\cos\delta_0 - \eta_{CE}(\delta_0) \tag{7.7.5}$$
$$*(\sin\delta - \sin\delta_0)/\cos\delta_0$$

and

$$\eta_{CE}(\delta) = \eta_{CE}(\delta_0) + \{\xi_{CE}(\delta_0) - \xi_{luff}\}(\sin\delta - \sin\delta_0)/\cos\delta_0 \tag{7.7.6}$$

Here, ξ_{luff} represents the longitudinal position of the luff of the sail and δ_0 is a reference sheeting angle which, conveniently, can be taken as $\delta_0 = 0$.

The quantities $\xi_{CE}(\delta_0)$ and $\eta_{CE}(\delta_0)$ define the position of the centre of effort for the reference sheeting angle. They are functions of the amount of camber and the lift of the sail. Because of the latter, they are also a function of the angle of attack and, hence, the apparent wind angle (recall that $\alpha = \beta_a - \delta$). The reason is that for cambered foil sections, like those of a sail, the position of the centre of pressure is relatively far aft at low and moderate lift but moves forward in the direction of the aerodynamic centre (AC) with increasing angle of attack and lift (see Sect. 5.14). This is illustrated by Fig. 7.7.6, which shows the longitudinal position of the centre of effort as a function of angle of attack for a sail section with 12% camber (data derived from Ref. Milgram (1971)). The centre of effort is seen to move backwards again at angles of attack beyond 15° when, with increasing α, the separation point of the boundary layer on the suction side moves forward, away from the trailing edge. For smaller amounts of camber the centre of effort will, at small to moderate angles of attack, be positioned proportionally closer to the aerodynamic centre. The presence of a mast has a similar effect. As already mentioned in (Sub-)Sect. Effects of a Mast or Head-Foil on Sail Section Characteristics a mast with a diameter/chord ratio(D/c) of 0.10 causes a forward shift of the centre of pressure of about 10% of the chord. For very high angles of attack ($\alpha \rightarrow 90°$), with flow separation from all edges, the centre of effort moves gradually to a position around 50% chord, irrespective of the amount of camber.

The lateral position $\eta_{CE}(\delta_0)$ of the centre of effort of a sail is much less dependent on camber and angle of attack than the longitudinal position. It is readily seen that $\eta_{CE}(\delta_0) = 0$ for a flat sail with sheeting angle zero ($\delta_0 = 0$) and adopts values of the

Fig. 7.7.6 Longitudinal
position of the centre of
effort of a sail section with
12% camber as a function
of angle of attack (2-D)

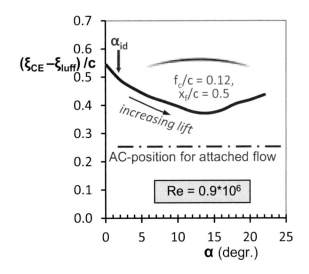

order of the maximum camber (f_c) for sails with camber. Setting $\eta_{CE}(\delta_0) \cong 0.75f_c$ is probably a reasonable approximation.

We have seen in Sect. 5.15 (Fig. 5.15.19) that for a single lifting surface of aspect ratio 3–4 and taper ratio zero, like a sail, the aerodynamic centre is positioned slightly ahead of the 25% chord point at a spanwise position that is slightly outboard of the section at 40% of the span as measured from the root. What this means for the longitudinal position of the centre of effort of a Bermuda type rig with fore- and mainsail is not so easy to say. Not in the least because there is very little experimental or computational data available on this matter.

For a single sail we can obtain an impression of the dependence of the position of the centre of effort on apparent heading and sheeting angle by combining the formulae (7.7.5) and (7.7.6) with the section data given in Fig. 7.7.6. The results of such an exercise are presented in Fig. 7.7.7. Shown are the longitudinal and lateral positions of the centre of effort as a function of sheeting angle at zero heel for a single sail of aspect ratio 3.5 with 12% camber for several apparent headings. The figure indicates that the variation of the longitudinal position with sheeting angle for small apparent headings is highly non-linear, like, but opposite to, that of the basic section with angle of attack (see Fig. 7.7.6).[16] The general trend with increasing apparent heading is that the centre of effort moves forward. For a given side force this implies a smaller moment to weather (or a larger moment to lee). The variation with sheeting angle becomes also less non-linear for higher apparent headings. The general trend for the lateral position is, not surprisingly of course, that the centre of effort moves outboard with increasing sheeting angle. For a given driving force this implies a larger moment to weather (or a smaller moment to lee). Which

---

[16] This is due to the fact that the lift decreases with increasing δ but increases with increasing α ($=\beta_a - \delta$)

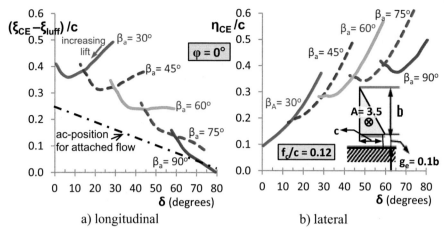

Fig. 7.7.7 *Longitudinal and lateral position of the centre of effort of a single sail as a function of sheeting angle and apparent heading (trend as calculated by means of a theoretical-empirical model)*

of these two opposite trends dominates in terms of aerodynamic yawing moment depends on the variation with angle of attack of lift and drag.

For Bermuda type rigs with mainsail plus foresail the position of the centre of effort of the combination of sails is difficult to determine but will be some weighted sum of the positions of the centre of effort of the individual sails. Factors determining this weighted sum are the relative areas and positions of the fore and main sails, the amounts of camber and the positions of maximum camber. The centre of effort is seen to move a little rearwards when the amount of camber is increased and/or the position of maximum camber is further downstream (Fossati et al. 2008). The position of the combination depends also on the difference between the sheeting angles of the mainsail and the foresail and the twist of the sails. When the sheet of the mainsail is eased, relative to that of the foresail (i.e. decreasing $\delta_f - \delta_m$), the lift of the mainsail will decrease, relative to that of the foresail, and the centre of effort will move forward. The opposite is, of course, the case when $\delta_f - \delta_m$ is increased. This is illustrated by Fig. 7.7.8, which shows the position of the centre of effort, relative to the position of the mast, as a function of $\delta_f - \delta_m$ for the same IMS-type sail configuration as in Fig. 7.5.8 (wind tunnel data from Ref. Fossati et al. (2008)). For this sail configuration the average amount of camber of the mainsail is about 9% and about 11% for the jib. Note that the position of the centre of effort is a little forward of the mast, that the apparent heading is 27° (a close-hauled condition) and that the sheeting angle of the jib is constant at 10°.

It is useful to mention that an engineering way of estimating the position of the centre of effort for small apparent wind angles, that is commonly used in sailing yacht design (Larsson and Eliasson 1996), is the following. First, the geometrical centres of area of the mainsail and foresail are determined separately. The centre of effort of

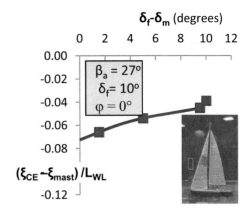

Fig. 7.7.8 Example of the longitudinal position of CE as a function of sheeting angle difference

the combination is then positioned on the line connecting the separate centres while taking into account the difference between the areas of the mainsail and the foresail.

For an individual, triangular sail this positions the centre of effort at 50% of the chord at 1/3rd of the height of the sail. In view of the results shown in Fig. 7.7.7 this seems a little too far to the rear. For a Bermuda type rig the longitudinal position found according to this procedure is usually close to that of the mast. This also seems a little too far backward (see Fig. 7.7.8).

It should further be noted, that when the foresail is sheeted to weather, on the spinnaker pole, as in Fig. 7.5.13, the lateral position of the centre of effort will be close to the centre line of the yacht ($\eta_{CE} \cong 0$). The reason is, of course, that the centres of pressure of the individual sails are then at opposite sides of the centre line. As mentioned earlier, this is the case only for downwind conditions of sailing, which is the domain of spinnaker type foresails.

Aerodynamic Yawing Moment
With respect to the centre of effort the yawing moment of a sail is zero, by definition. With respect to some other point or line the variation of the yawing moment with camber and angle of attack is, of course, coupled to the variation of the position of the centre of effort. This is already implied by Fig. 7.4.7, which shows the moment with respect to the ¼-chord point of the same sail section as that of Fig. 7.7.6 for small to moderate angles of attack. For a three-dimensional sail configuration with fixed sheeting angle(s) this means that the yawing moment with respect to the aerodynamic centre will vary significantly with the apparent heading.

Recalling (see Eq. (7.7.4)), that in the hydrodynamic coordinate system, the yawing moment can be expressed as

$$M_{Az} = S_A x_{CE} + T y_{CE},\qquad(7.7.7)$$

it follows that with Eqs. (7.37) this can also be written as

$$M_{Az} \approx S_A \xi_{CE} + T(\eta_{CE}\cos\varphi + \zeta_{CE}\sin\varphi)\qquad(7.7.8)$$

Dividing by the dynamic pressure $\frac{1}{2}\rho V_a^2$, the sail area S_A and a reference length L_{ref}, Eq. (7.7.8) can also be expressed in dimensionless quantities:

$$C_{MAz}(\varphi) \approx C_{SA}\xi_{CE}/L_{ref} + C_T\{(\eta_{CE}/L_{ref})\cos\varphi + (\zeta_{CE}/L_{ref})\sin\varphi\} \quad (7.7.9)$$

For zero heel ($\varphi = 0$) this reduces to

$$C_{MAz}(0) \approx C_{SA}\xi_{CE}/L_{ref} + C_T\eta_{CE}/L_{ref} \quad (7.7.10)$$

Through substitution of the expressions (7.7.5) and (7.7.6), Eqs. (7.7.9) and (7.7.10) can be used to obtain a qualitative impression of the effect of sheeting angle on the aerodynamic yawing moment. An example, for zero heel, is given in Fig. 7.7.9. The figure has been constructed on the basis of the CE-position for small apparent wind angles of the IMS type yacht (Fossati et al. 2008) of Fig. 7.7.8 in combination with the overall lift and drag data (Marchai and Tanner 1963) of the 'Dragon' type rig of Figs. 7.6.7 and 7.6.8. Hence, it does not represent any particular type of yacht but is merely intended to show the trend of the variation of the aerodynamic yawing moment with sheeting angle for a Bermuda type of rig. Note that the position of the centre of the mast was chosen as the moment reference point and the length of the waterline as the reference length.

The trend reflected by Fig. 7.7.9 is that the aerodynamic yawing moment is <0, i.e. tends to turn the boat to lee, for small apparent wind angles. When the apparent wind angle is increased the yawing moment becomes more positive, that is tends to weather. The trend with sheeting angle for a given apparent wind angle is more complex.

When considering the shape of the curves it is useful to realize that, for a given apparent wind angle, the angle of attack of the sails decreases with increasing sheeting angle. For the rig considered, the maximum lift (C_{Lmax}) is obtained for an angle of attack of about 30°. Since $\delta = \beta - \alpha$, this means a sheeting angle $\delta_m = 0°$ for

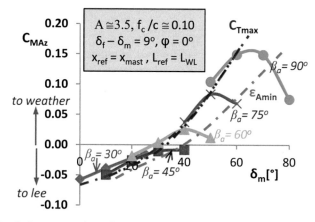

Fig. 7.7.9 *Trend of variation of aerodynamic yawing moment with sheeting angle for a Bermuda-type rig (see Fig. 7.7.8) at zero heel*

$\beta=30°$, $\delta_m=15°$ for $\beta=45°$, etc. For larger apparent wind angles the conditions for the maximum driving force C_{Tmax} are almost equal to the conditions for maximum lift. The conditions corresponding with the minimum drag angle ε_{Amin} (or maximum lift/drag ratio) are met for smaller angles of attack, i.e. larger sheeting angles.

It appears, that for (very) small apparent wind angles the yawing moment increases monotonously when the sheeting angles are increased, but that for larger apparent wind angles the yawing moment decreases again when the sheeting angle is increased beyond the value that corresponds with the angle of attack for maximum lift. The reason for this is that for small apparent wind angles the largest contribution to the yawing moment is due to the side force, which is then dominated by lift (Sect. 4.3). For apparent wind angles around 90°. The driving force has the largest contribution, which is then also dominated by lift. This means that, at zero heel, the yawing moment in upwind conditions is always dominated by lift. The consequence is that the yawing moment will, roughly, decrease in the absolute sense when the lift is decreased. For small apparent wind angles, when the aerodynamic yawing moment is <0, this means a reduction of the (absolute value of the) yawing moment with increasing sheeting angle. When the yawing moment is >0, that is for larger apparent wind angles, a reduction of lift through an increase of sheeting angle implies a less positive yawing moment.

In conditions of broad reaching and running, with the foresail to weather on the spinnaker pole, the aerodynamic yawing moment will, in general, decrease in the absolute sense with increasing apparent wind angle. The reason is that, as mentioned above, the combined centre of effort of the sails comes closer to the centre line of the yacht with increasing sheeting and apparent wind angles. We will come back to this in the next section.

We will see later, in Sect. 7.9, that heel has a large influence on the yawing moment. This is associated with the presence of the term $T\,\zeta_{CE}\sin\varphi$ in Eq. (7.7.8). The sign of this term is such that it implies an additional moment to weather. For large angles of heel its magnitude is such that it overrules all other terms in Eq. (7.7.8).

It is also worth mentioning at this point that in the results of wind tunnel tests the longitudinal position of the centre of effort is usually defined as the ratio M_{Az}/S_A between the yawing moment M_{Az} and the side force S_A. This means that the effect of the driving force T (see Eq. (7.7.8)) is not taken into account. The error involved is small for small sheeting angles and small angles of heel but becomes substantial under heel and for large sheeting angles.

7.8 Foresails for Large Apparent Wind Angles: Spinnakers and Gennakers

Basic Shapes
As already indicated before, sailing at large apparent wind angles is the domain of application of a different class of foresails (or head sails) generally known as spin-

nakers and gennakers. The dimensions of such sails are measured in a way that differs from the way in which jib and genoa type foresails are measured (see Sect. 2.3).

Spinnakers are so-called 'flying' sails, that is, they are 'flown' from (almost) the top of the mast and are not attached to the forestay or the hull but supported by a pole between the tack point and the mast (Fig. 7.8.1, see also Fig. 2.4). Gennakers are not attached to the forestay either but the tack point is attached to the bow or an extendable bowsprit (Fig. 7.8.2). The purpose of spinnakers is to maximize the driving force for conditions of running ($\approx 135° < \beta < \approx 180°$) and (broad) reaching ($\approx 90° < \beta < \approx 135°$). Gennakers, a word constructed by contracting genoa and spinnaker, are meant for close reaching conditions ($\approx 50° < \beta < \approx 100°$.

Since the early twenty-first century still another category of head sails is distinguished: the Code 0 (zero). A Code 0 (see Fig. 7.8.3), is a genoa kind of foresail

Fig. 7.8.1 *Illustrating a spinnaker type of head sail*

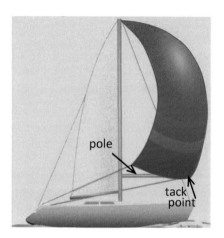

Fig. 7.8.2 *Illustrating a gennaker type of head sail*

Fig. 7.8.3 Illustrating a Code 0 type of head sail

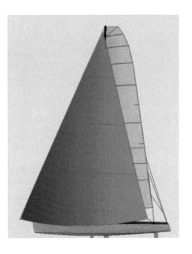

with a large overlap. It is rigged like a genoa (with or without extendable bowsprit) but is measured like a spinnaker. Code 0's are meant for upwind sailing at low wind speeds in a fairly narrow range of apparent wind angles ($\approx 30^\circ < \beta < \approx 50^\circ$).

Spinnakers
Two types of spinnaker are distinguished: symmetrical spinnakers with port-starboard symmetry and asymmetrical ones (see Fig. 7.8.4). The classical form of spinnaker is the symmetrical one. Its voluminous shape, with deep horizontal and vertical profiles, reflects the thought that, when the maximum amount of drag is needed, the shape should resemble that of a hollow semi-circular cylinder or a hollow semi-sphere (Sect. 7.4, Fig. 7.4.32). This would, of course, be perfect for a

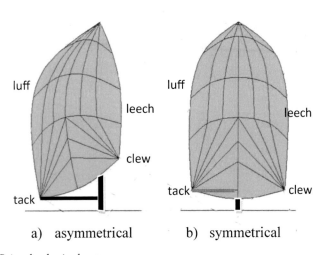

a)   asymmetrical   b)   symmetrical

Fig. 7.8.4 Spinnaker basic shapes

'dead' run (apparent wind angle $\beta = 180°$) without a mainsail. However, we have seen in Sect. 4.6 that a dead run is seldom the fastest course to a waypoint that is positioned exactly downwind. For low to moderate wind speeds, the best downwind VMG (velocity-made-good) was seen to be obtained by gybing downwind at true wind angles γ between 135 and 165°. This corresponds roughly with apparent wind angles β between 90 and 140°. We know also, from Sect. 4.3, Fig. 4.3.1, that, at $\beta = 90°$, all of the lift of a sail is turned into driving force and 0% of the drag (Fig. 4.3.2). At $\beta = 135°$ this is 50% for the lift and 50% for the drag. These figures indicate that lift is more important than drag for $\beta < 135°$. Even at $\beta = 165°$ there is still a significant portion of the lift (25%) that is transformed into driving force. What this means, is that the capability of spinnakers to generate lift is at least as important as that of generating drag. This requires asymmetric sail configurations with asymmetrical camber, at least in principle.

Another reason why symmetrical spinnakers may not embody the optimal downwind shape of a foresail is that spinnakers are usually set in the presence of the mainsail. As a consequence a spinnaker usually operates in an asymmetrical flow field and an asymmetric shape is required to take optimal advantage of the interaction with the mainsail.

While a spinnaker should, in principle, be asymmetric from the point of view of performance, a symmetrical one is easier *to* handle for the crew. The reason is that, because of the port-starboard symmetry, the luff and leech can interchange their roles when gybing. For this reason symmetric spinnakers are still quite popular in the cruising yacht community.

Asymmetric spinnakers are, roughly, distinguished in two categories: 'runners' and 'reachers' (Fig. 7.8.5). A nomenclature that is sometimes used for further classification distinguishes Code 1 to Code 6 asymmetric spinnakers. Runners (Code 2, 4 and 6) are bulbous, with up to 45% camber and are meant for very large apparent

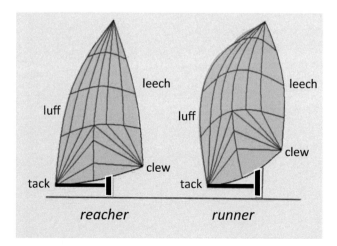

Fig. 7.8.5 Asymmetric spinnakers

wind angles ($\approx 135° < \beta < \approx 180°$). They are shaped so as to combine a large drag with a reasonable amount of lift and reasonable handling qualities. Reachers (Code 1, 3 and 5) are flatter (35 % or less camber) and are meant for reaching conditions ($\approx 90° < \beta < \approx 135°$). They are shaped so as to generate a lot of lift with a moderate amount of drag. Code 1 and 2 sails are meant for light wind, Codes 3 and 4 for moderate winds and Codes 5 and 6 for heavy weather conditions. The reaching kind of asymmetric spinnakers are sometimes also called gennakers. A Code 0 is sometimes considered as the flattest of the reaching asymmetric spinnakers or the flattest of the gennakers.

Spinnaker Sheeting and Basic Trimming Mode
While the sheeting of a gennaker type of head sail is similar to that of a genoa, the sheeting of a spinnaker and the definition of the sheeting angle is distinctly different. As already indicated above and in Sect. 2.4, a spinnaker is set to weather, similar as in Fig. 7.5.14, with the tack point fixed at the end of an approximately horizontal pole that is attached to the mast and with 'sheets' attached to the tack point as well as the clew. While the 'sheet' attached to the clew is still called the sheet, the other line, attached to the end of the spinnaker pole is called the 'guy' or tack line. The sheeting angle δ_p of the spinnaker pole is defined as the angle between the pole and the yacht's longitudinal axis (Fig. 7.8.6). With this convention it is not so easy to define the angle of attack of a spinnaker. For mainsail and jib/genoa the angle of attack α was conveniently defined as the angle between the apparent wind vector and the chord of a sail section. This implies the relation $\alpha = \beta - \delta$. For a spinnaker this is obviously no longer applicable. Besides, the shape of a spinnaker, an asymmetrical one in particular, is such that it is not obvious how to define the chord. A reasonable choice for the foot section would be the line between the tack and clew points. However, the angle of attack for a given orientation of the pole would then still depend strongly on the position of the clew.

Fig. 7.8.6 *Defining spinnaker pole sheeting angle*

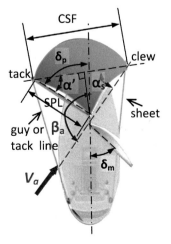

For symmetrical spinnakers it is possible to make a rough estimate of the geo-metrical angle of attack at the foot. Because of the symmetry, the tack point and the clew will have about the same distance to the mast. It is also known among yachts-men, as a classical rule of the thumb, that in the basic trimming mode of a spinnaker, the pole, as well as the boom of the mainsail, makes an angle of about 90° with (i.e. is square to) the apparent wind vector. This implies that

$$\delta_p \approx \delta_m \approx \beta_a - 90 \text{ (degrees)} \tag{7.8.1}$$

Trigonometry then teaches that the angle (α_s) between the foot of the spinnaker and the apparent wind vector is equal to

$$\alpha_s = 90 - \alpha' \text{ (degrees)}, \tag{7.8.2}$$

where

$$\alpha' \approx a\cos(\text{CSF}/2/\text{SPL}) \tag{7.8.3}$$

is the angle between the spinnaker pole and the foot chord. Here, CSF is the chord length of the foot of the spinnaker and SPL the length of the spinnaker pole. For most spinnakers the foot chord length CSF appears to be of the order of 2/3 of the foot length SF with the latter of the order of 1.75 (or less) times the length SPL of the pole (see also Fig. 2.4.2). This gives that α' is of the order of 50°. Hence, α_s is of the order of 40°. This does not mean that α_s is 40° for all apparent wind angles. Because high lift and low drag is required for reaching conditions and high drag but low lift while running, α_s will have to be $<40°$ (and $\alpha'>50°$) for $\beta_a\downarrow90°$ and $\alpha_s>40°$ or $\alpha'<50°$ for $\beta_a\uparrow180°$. The first, $\alpha'>50°$, means that the sheet attached to the clew has to be eased. The second, $\alpha'<50°$, that the sheet has to be pulled. Obviously, $\alpha'\cong0$ for $\beta_a=180°$.

Combining Eqs. (7.8.1) and (7.8.2) gives a relation between α_s and δ_p:

$$\alpha_s = \beta_a - \delta_p - \alpha' \text{ (degrees)} \tag{7.8.4}$$

or

$$\alpha_s \approx \beta_a - \delta_p - 50 \text{ (degrees)} \tag{7.8.5}$$

At this point it is interesting to note that the dimensions of spinnakers are usually such that the aspect ratio of the maximum projection in a vertical plane is of the order of 1.3–2. This would mean, see Figs. 7.4.33 and 7.4.35, that spinnakers attain their maximum lift as well as a local maximum in drag at an angle of attack of about 40°.[17] This is also in reasonable agreement with experimental data (Lasher et al. 2003; Richards and Lasher 2008) for isolated spinnakers and corresponds well with

---

[17] The absolute maximum of the drag is, of course, obtained for $\alpha=90°$.

the geometrical angle of attack implied by the basic trimming mode of a spinnaker described above. It also means that this angle of attack will not be very far from the optimum for an apparent wind angle of 135°. This because the maximum driving force at $\beta = 135°$ is obtained when the sum of lift and drag attains it maximum (see Sect. 4.3).

Applying the same reasoning to the mainsail, with its higher aspect ratio (≈ 4) suggests that the mainsail attains its maximum lift and a local maximum of the drag when the angle of attack is in the order of 30°. Here also, the absolute maximum of the drag is, of course, obtained for $\alpha = 90°$. Note that trimming the mainsail for maximum lift would require a sheeting angle of the mainsail that is some 60° larger than that of the classical 'square' rule of the thumb value given by Eq. (7.8.1). The latter is obviously meant to realize maximum drag. We will come back to this point shortly.

Interaction with the Mainsail

As already discussed in Sect. Factors Determining the Effective Aspect Ratio, there is aerodynamic interaction between a foresail and the mainsail. With the foresail poled-out to weather a lifting main will generate some upwash and some super-velocity at the position of the foresail and a lifting foresail will cause some downwash and some sub-velocity at the position of the main. This applies also to spinnaker type foresails, at least as long as the apparent wind angle is such that both sails produce a significant amount of lift. We have seen above that this should be the case for all, except the very large, apparent wind angles, that is except when β_a approaches 180°.

Because of the upwash due to the mainsail, the foresail/spinnaker will require a smaller geometric angle of attack α_s ($< 40°$) for the same lift when the mainsail carries lift. This means that under reaching conditions the spinnaker pole shall be pulled further to windward than square to the wind ('over-squared' pole, $\delta_p > \beta_A - 90°$) with the clew sheet eased correspondingly. This comes on top of the reduction of α, or the corresponding increase of δ_p, that, as discussed in the preceding paragraph, is required to increase the lift and reduce the drag at smaller apparent wind angles.

At the same time the mainsail will require a larger angle of attack (smaller sheeting angle δ_m) because of the downwash induced by the circulation around the foresail. This compensates, to some extent, the reduction of α and associated increase of δ_m, to a value $> \beta_a - 90°$, that is required to increase the lift and reduce the drag at the smaller apparent wind angles.

When β_a is close to 180° both sails will mainly produce drag, with flow separations from all edges that form a big wake of 'dead air' behind each sail. In such conditions the virtual 'displacement body' formed by the mainsail and its 'dead air' wake may cause some increase of the effective wind speed felt by the foresail, at least as long as the foresail is out of the 'wind shadow' of the mainsail. Because of the latter it is important to keep the foresail/spinnaker out of the wake of the mainsail by setting the spinnaker pole as far out as possible and, if necessary, by easing the sheet. It is, amongst others, for this purpose that asymmetric, runner type of spinnakers have a relatively short leech (Fig. 7.8.5).

The mainsail, at the same time, may experience some reduction of the effective wind speed due to the stagnation of the flow approaching the concave side of the spinnaker.

Aerodynamic Forces and Optimal Sheeting Angles

With the general characteristics and mechanisms described above in mind, it is interesting to look at some measured (wind tunnel) data (Richards and Lasher 2008) of sail configurations with spinnakers. Figure 7.8.8 shows lift and drag for a generic spinnaker configuration (Fig. 7.8.7) for an IACC type of yacht. The spinnaker is relatively flat and symmetric with an aspect ratio of about 1.5. The aspect ratio of the mainsail is about 4.5. The Reynolds number of the tests, based on average sail chord, was about $0.1*10^6$, which is relatively low, compared with full scale conditions.

The figure presents lift and drag coefficients, based on total (mainsail + spinnaker) sail area, as a function of mainsail sheeting angle with the spinnaker pole fixed at $\delta_p = 40°$ for an apparent heading of $\beta_A = 120°$. The latter implies that the spinnaker pole is 'oversquared' by 10°.

Fig. 7.8.7 *Wind tunnel model of a generic sail configuration with symmetric spinnaker for an IACC type yacht. (From Ref. Richards and Lasher (2008))*

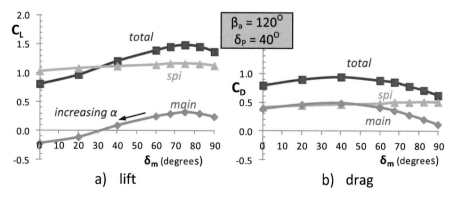

Fig. 7.8.8 *Lift and drag, from wind tunnel tests (Richards and Lasher 2008), for a generic sail configuration with symmetric spinnaker (Fig. 7.8.8), at an apparent heading of 120° as a function of mainsail sheeting angle*

The lift and drag coefficients are given for the individual sails (in each other's presence) as well as the 'total' values for the combination. The mainsail is seen to attain its maximum lift for a sheeting angle δ_m of about 75°. This is about 45° more than the 'square' rule of the thumb value of $120-90=30$° and about 15° less than is obtained when the angle of attack for maximum lift is assumed to be 30° $(120-30=90)$. The mainsail attains its maximum drag for a sheeting angle δ_m of about 35°, or an angle of attack of 85°. This is only some 5° smaller than the 'normal' value of 90°. The sign of these differences is in agreement with the expectations given in the preceding paragraph. It is to be expected that the effect of the spinnaker on the mainsail in terms of downwash is larger when the mainsail is closer to the spinnaker, that is for $\delta_m = 75$°. However, the large difference (15° versus 5°) between the, presumed, spinnaker induced downwash at maximum lift and that at maximum drag is a bit of a surprise.

It can also be noticed that, in spite of the fixed sheeting angle of the spinnaker pole, there is a small, but significant, increase of lift and drag of the spinnaker with increasing lift of the mainsail. This reflects the upwash at the spinnaker induced by the mainsail.

Streamline patterns for some of the flow conditions covered by Fig. 7.8.8 are shown in Fig. 7.8.9. They represent results of Computational Fluid Dynamic simulations from Richards and Lasher (2008). Note that for a sheeting angle δ_m of 0° the mainsail is completely stalled with a wide 'dead air' region on the leeside. For the larger

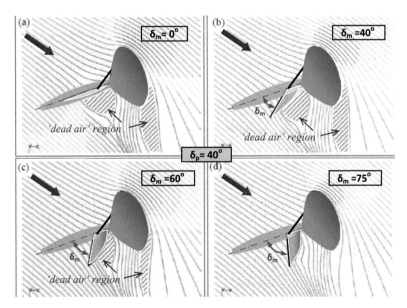

Fig. 7.8.9 *Streamlines from a computational fluid dynamic simulation in a plane about halfway up the mast for the sail configuration of Figs. 7.8.8 and 7.8.9. (From Ref. Richards and Lasher (2008), adapted)*

sheeting angles the 'dead air' region decreases in size and seems to have almost disappeared for $\delta_m = 75°$. The streamline patterns suggest that there are also indications for some separated flow in a, thinner, region just ahead and downstream of the leech of the spinnaker. The size of this also seems to decrease with increasing sheeting angle of the mainsail. It reflects that lift on the mainsail has a positive effect, in the sense of an increase of circulation, on the flow around the spinnaker.

Figure 7.8.10 presents lift and drag as a function of the spinnaker pole sheeting angle δ_p with the mainsail sheeting angle fixed at $\delta_m = 75°$. The spinnaker is seen to attain its maximum lift at a pole sheeting angle of about 42°. According to Eqs. (7.8.4/7.8.5) this means a geometrical angle of attack α_s of about $\beta_a - \delta_p - 50 = 28°$ if the angle α' (see Fig. 7.8.6) is set at 50°. This is significantly smaller than the 40° or so that is suggested by Fig. 7.4.33. It implies that the spinnaker sheet is to be eased so that $\alpha' \cong 38°$ in order to realize maximum lift. Part of this must be due to the upwash induced by the mainsail. Spinnaker maximum drag seems to be attained for $\delta_p < 20°$, which, according to Eq. (7.8.5) implies an angle of attack $> 50°$. The latter is still quite far from the 'normal' value of about 90°. The difference of 40°, or less, may not be completely due to the upwash from to the mainsail. Part of or even the greater part may be caused by the trimming of the clew sheet of the spinnaker. If the clew sheet was pulled when the pole sheeting angle was decreased, the angle α' (see Fig. 7.8.6) will have been $< 50°$, with a corresponding increase of α.

A surprising feature of Fig. 7.8.10 is further that lift (and drag) of the mainsail increase slightly with increasing spinnaker pole sheeting angle. Since the lift of the spinnaker increases when δ_p is increased from 20 to 40°, one would expect the lift of the mainsail to decrease slightly because of the increased downwash from the spinnaker. This, apparently, is overruled by another mechanism that tends to

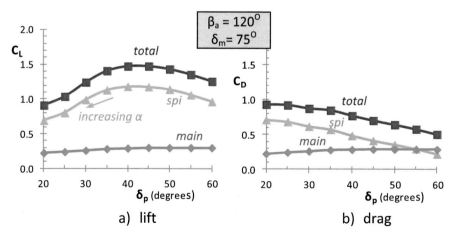

Fig. 7.8.10 *Lift and drag, from wind tunnel tests (Richards and Lasher 2008), for a generic sail configuration with symmetric spinnaker (Fig. 7.8.8), at an apparent heading of 120°, as a function of spinnaker pole sheeting angle*

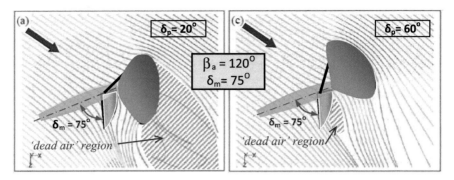

Fig. 7.8.11 *Streamlines from a computational fluid dynamic simulation in a plane about halfway up the mast for the sail configuration of Figs. 7.8.7 and 7.8.10. (From Ref. Richards and Lasher (2008))*

increase the lift of the mainsail with increasing spinnaker pole sheeting angle. The explanation is possibly also in the trimming of the spinnaker clew sheet. If the latter is eased when the pole sheeting angle is increased by pulling the guy, the distance between the spinnaker and the mainsail will increase, with a reduced downwash at the mainsail and more lift as a result.

Figure 7.8.11 shows streamline patterns for different pole sheeting angles with the mainsail sheeting angle fixed at 75° and, again, an apparent heading of 120° (for a pole angle of 40° see Fig. 7.8.9d). The figure illustrates that for $\delta_p = 20°$ there is massive flow separation on the lee side of the spinnaker, while the flow seems to be almost fully attached for $\delta_p = 60°$. With the latter pole angle there is some indication of separated flow on the lee side of the mainsail. This probably because, due to reduced downwash from the spinnaker, the mainsail now operates slightly beyond the angle of attack for maximum lift.

Figure 7.8.12 gives the driving force and side force coefficients for the conditions described above. The driving force is seen to attain its maximum for a spinnaker pole sheeting angle δ_p of about 38° and a mainsail sheeting angle of $\delta_m \cong 68°$. Note that these angles are slightly smaller than the values for which the lift attains its maximum. This is caused by the contribution of the drag, which increases when the angles of attack are increased (sheeting angles decreased) beyond those for maximum lift.

It appears from the (b) and (d) figures that for the apparent heading concerned, i.e. 120°, the condition for the maximum driving force just mentioned coincides with the condition for zero side force. Note that this is in agreement with the findings in Sect. 4.3.

It is also worth noting that, at the optimum condition, about 75 % of the driving force comes from the spinnaker and only 25 % from the mainsail.

Comparing Fig. 7.8.12 with Figs. 7.8.8 and 7.8.10 it is further clear from the shape of the curves that, for $\beta = 120°$, the driving force is dominated by lift and the side force by drag. This in agreement, of course, with the considerations of Chap. 4.

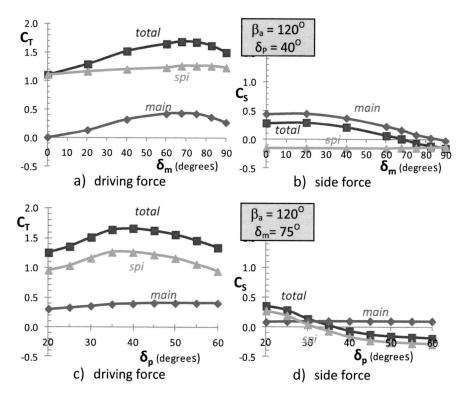

Fig. 7.8.12 *Driving force and side force, from wind tunnel tests (Richards and Lasher 2008), for the sail configuration of Fig. 7.8.7 and the same conditions as in Figs. 7.8.8 and 7.8.10*

Richards and Lasher (2008) also contains data for an apparent heading of 100°. It appears that the maximum driving force for a spinnaker pole angle δ_p of about 20° is then obtained with a mainsail sheeting angle δ_m of 58°. This implies a geometrical angle of attack of the mainsail of about $\alpha_m = 42°$, versus about $120 - 68 = 52°$ for $\beta_a = 120°$ and $\delta_p = 40°$. These numbers indicate a smaller angle of attack and less drag of the mainsail for $\beta_a \downarrow 90°$, which is also in agreement with the general considerations in Chap. 4. For the same reason it is to be expected that the optimum angle of attack of the spinnaker at $\beta_a = 100°$ should also be smaller than at $\beta_a = 120°$. In case of the latter α_s was seen to be about 28° (effectively probably about 40°) for maximum lift and about $(120 - 38 - 50 =)$ 32° for the maximum driving force condition. This would suggest a spinnaker pole angle of about $100 - 30 - 50 = 20°$ at $\beta_a = 100°$ for the maximum driving force.

We have seen already that for $\beta_a = 135°$ the maximum driving force is obtained when the sum of lift and drag attains its maximum. This appears to be the case for $\alpha_s \cong 37°$ or a spinnaker pole angle $\delta_p \cong 135 - 37 - 50 = 48°$ and $\alpha_m \cong 52°$ or a mainsail sheeting angle of $\delta_m \cong 135 - 52 = 83°$.

Figure 7.8.13 gives the sheeting angles for the maximum driving force as a function of the apparent wind angle. The figure illustrates that the 'square' rule of the thumb

Fig. 7.8.13 Measured/estimated sheeting angles for maximum driving force, from wind tunnel tests (Lasher et al. 2003), sail configuration with spinnaker of Fig. 7.8.7

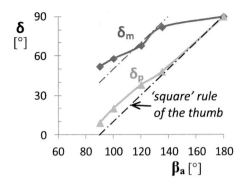

$\delta = \beta_a - 90$ (degrees) is not so bad for the spinnaker, at least for apparent headings $\geq 140°$. For smaller apparent headings the spinnaker pole should be over-squared progressively up to about $+10°$ for $\beta_a = 90°$. For the mainsail sheeting angle the 'square' rule of the thumb appears to be way off, except for apparent wind angles close to $180°$. For $\beta_a \leq 140°$ a much better rule of the thumb for the mainsail seems to be

$$\delta_m \approx \beta_a - 50 \text{ (degrees)}, \tag{7.8.6}$$

with some additional under-trimming (about $10°$) required when the apparent wind angle approaches $90°$.

Figure 7.8.13 also implies that for $\beta_a > 140°$ the boom of the mainsail should be set out as far as possible.

It is emphasized that Fig. 7.8.13 should not be interpreted as a precise recipe for sail trim with spinnaker. The precise numbers will depend on the geometry of the sail configuration. It is to be expected, for example, that for a rig with a lower aspect ratio (<4.5), the sheeting angle of the mainsail at $\beta_a \cong 90°$ will have to be somewhat smaller than suggested by Fig. 7.8.13. This because maximum lift will be obtained at a higher angle of attack (see Fig. 7.4.33).

There is probably also an effect of Reynolds number on the optimum sheeting angle. For a given amount of camber, maximum lift at a higher Reynolds number will be higher and will be attained at a higher angle of attack. This points also towards somewhat smaller sheeting angles for reaching conditions at higher Reynolds numbers (higher wind speeds/bigger yachts). For running conditions, when drag dominates and the flow is fully separated, the effect of Reynolds number will be negligible.

Driving Force and Side Force as a Function of Apparent Wind Angle

As illustrated in the preceding paragraphs it is possible to determine the conditions for the maximum driving force by a proper combination of the sheeting angles of foresail and mainsail. This can be done for every apparent wind angle of interest.

Figure 7.8.14 gives examples of how the maximum driving force and the associated side force of spinnaker-mainsail configurations may vary as a function

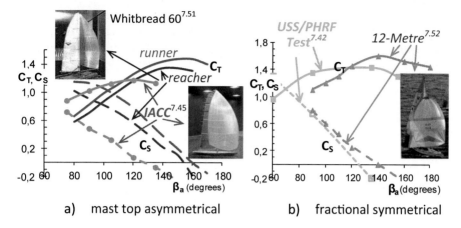

a)   mast top asymmetrical   b)   fractional symmetrical

Fig. 7.8.14 *Examples of the variation with apparent wind angle of the maximum driving force and the associated side force for spinnaker configurations (zero heel)*

of the apparent wind angle. The left, (a), figure gives results of wind tunnel tests (Van Schaftinghen 1999) for a 'runner'-type (Code 2) and a 'reacher'-type (Code 1) asymmetric spinnaker for a Whitbread 60 Class of yacht and an asymmetric mast top spinnaker for an IACC type of yacht (Fossati 2009). The right, (b), figure gives results of wind tunnel tests (Claughton et al. 1999; Flay and Vuletich 1995) for fractional, symmetric spinnaker configurations.

Considering these figures it should first of all be said that, because of different choices for the reference sail area, the absolute values of the force coefficients have little significance. For the Whitbread 60 configurations the reference sail areas are the same. It can be noted that the force level for the 'reacher' is lower than that of the 'runner'. This, by itself, is not surprising, because (Van Schaftinghen 1999), in this case, the actual area of the 'reacher' is smaller than that of the 'runner'.

A closer inspection reveals that the reacher has a better driving-force/side-force ratio. This means that the reacher is more suitable for higher wind speeds.

Note also that there are considerable differences in the values of the apparent wind angle for which the driving force attains its maximum. The range is, roughly, between $\beta_a = 120$ and 160°. This holds also for the apparent wind angles at which the side force goes through zero. More precisely, it can be noticed that, for all configurations, the side force is zero when the driving force attains its maximum. This in perfect agreement with the general discussion on the mechanics of sailing in Chap. 4. There we saw, in Sect. 4.3, that the side force is always zero when the driving force attains its absolute maximum. Part of the differences between the tests results may be due to the fact that some of the wind tunnel tests were done in uniform flow (Flay and Vuletich 1995), while others (Fossati 2009; Flay and Vuletich 1995) were done with simulated wind gradient. The differences in the sail configurations will also have played a role.

At this point it is useful to recall from Sect. 4.3 that the apparent wind angle corresponding with the condition of the absolute maximum driving force is given by

$\beta = 90 + \varepsilon_A$ (degrees). Because the aerodynamic drag angle ε_A is a function of the lift/drag ratio (tan $\varepsilon_A = D/L$), the absolute maximum driving force is attained at a relatively small value of the apparent wind angle for a high lift/drag ratio and at a relatively high value of β when L/D is low. This means also (Sect. 7.4) that a high aspect ratio rig will attain its absolute maximum driving force at a smaller apparent wind angle than a low aspect ratio rig. This is also evident from Fig. 7.4.35.

Vertical Force and Moments

While the vertical force components of 'ordinary' sails is usually not of great interest, at least for zero heel, it is of significant magnitude for spinnaker type foresails. It has been found (Van Schaftinghen 1999) that, under reaching conditions, it is of the order of 40 % of the maximum lift for 'runner' type asymmetric spinnakers and of the order of 20 % for the, flatter, 'reacher' type of spinnaker. A level of about 25 % of the maximum lift is reported (Flay and Vuletich 1995) for symmetrical spinnakers. For running conditions the same percentage seems to apply, but then in terms of maximum drag.

The orientation of this vertical force is upward. This is due to the almost horizontal orientation of the part of the sail near the head (see Fig. 7.8.1). Because of the highly swept edges of the sail near the head there is also edge vortex formation like on a highly swept delta wing (See Fig. 5.15.15). This increases the vertical force. Because the horizontal orientation of the top of the sail is more pronounced for runner type spinnakers than for reachers, the vertical force is larger for the former.

Because the orientation of the vertical force is upward, it tends to lift the hull out of the water which means a reduced displacement and reduced wave-making resistance (Sect. Wave Making Resistance). Because of the forward position of the centre of effort of the spinnaker in downwind conditions it also reduces the (bow down) pitching moment. This is, usually, also beneficial for the wave-making resistance.

It is appropriate to mention at this point that jib/genoa type foresails also acquire significant inclination for large sheeting angles. This is due to the fact that they are attached to the forestay. The latter makes, usually, an angle of 30–40° with the mast. Hence, jib/genoa type foresails also produce a vertical force (even at zero heel) for large sheeting angles, albeit not as large as in the case of spinnakers.

About the aerodynamic moments there is very little information available in the literature. However, it appears that reasonably accurate values of the heeling and pitching moments of a fractional spinnaker (Flay and Vuletich 1995) can be obtained from the side force and the driving force if the centre of effort is assumed at 43 % of the mast height (see Eqs. (7.7.1) and (7.7.2)). For the mast top runner and reacher configurations of Fig. 7.8.14a the centres of effort were found at 62 and 56 % of the mast height, respectively. Because the wind tunnel tests (Van Schaftinghen 1999) were done in uniform flow these values may be a little too small (see Fig. 7.4.42).

For the heeling moment this implies that the variation with apparent wind angle is similar to that of the side force in Fig. 7.8.14, except for a scale factor that is proportional to the height of the centre of effort. The behaviour of the pitching moment is similar to that of the driving force.

With respect to the yawing moment it can be mentioned that the available experimental data (Van Schaftinghen 1999; Flay and Vuletich 1995) indicates that the aerodynamic yawing moment of sail configurations with spinnaker at large apparent wind angles is an order of magnitude smaller than the heeling and pitching moments. This is caused by the fact that, with the spinnaker pole to weather, the lateral position of the centre of effort is close to the centreline of the yacht so that there is only a small contribution of the driving force to the yawing moment (cf. Eq. (7.7.8)). The contribution of the side force is also small, because the latter goes through zero somewhere in the range $120 < \beta < 160$.

Figure 7.8.15 gives an example of the variation with the apparent wind angle of the aerodynamic yawing moment of sail configurations with spinnaker. The conditions are the same (maximum driving force) as those in Fig. 7.8.14a for the Whitbread 60 configuration. Note that the moment coefficient is given with respect to the centre of the mast and that the reference length is the mast height. The yawing moment is seen to act to lee (<0 in the coordinate system of sailing) for apparent wind angles around 90° and to disappear completely ($=0$) for $\beta \cong 160°$.

Optimal Draft or Camber
An interesting question to consider is "what is the optimal draft or camber of a spinnaker?" Obviously the answer depends on the apparent wind angle because high lift is required for reaching and a large drag for running.

We have seen in Sect. 7.4 that the maximum lift of a sail of high ormoderate aspect ratio is about equal to the maximum lift of the (average) section of the sail. The latter was found to attain its largest value for a camber/chord ratio of about 30%. We saw also that for small aspect ratios ($A < \cong 2$) there is some additional lift at high angles of attack due to side edge separation.

It is important to realize that the values given above refer to a situation with a given length of the chord or a given projected area of the sail. The answer to the question posed is different when the size of the sail is measured in terms of the length of the girth, that is the arc-length of a cross-section of the sail. The latter is, in general, the case, as indicated in Sect. 2.4 (see Fig. 2.4.2).

We have seen in Sect. 5.2 that the total aerodynamic forces acting on a lifting body like a sail are proportional to the projected area. What this means is that it is more appropriate to look at the product of maximum lift coefficient and chord

Fig. 7.8.15 Example of the variation with apparent wind angle of the aerodynamic yawing moment of sail configurations with spinnaker. (Van Schaftinghen 1999; see also Fig. 7.8.14)

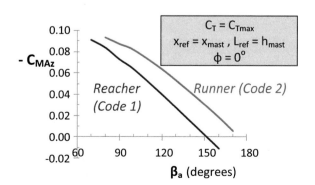

length for a given length of the girth or arc-length of a sail section if the objective is to maximize lift for a given rated sail area or length of the girth. Similarly, the product of maximum drag and chord length is a more appropriate quantity if the maximum drag for a given girth length is the objective.

Figure 7.8.16 shows how these quantities vary as a function of the camber/chord ratio f_c/c of a sail section. Shown are the products of the maximum lift and drag coefficients with the chord/girth length ratio c/lg as a function of the camber/chord ratio. The maximum lift coefficients correspond with the sectional data given in Fig. 7.4.13 and the drag coefficients are those for sails with an aspect ratio A=1 at an angle of attack of 90° in Fig. 7.4.34.

The figure indicates that for a given girth length the lift is maximized for a camber chord ratio of about 18% and the drag for $f_c/c \cong 12\%$. Note that these values are much smaller than the 30 and 50%, respectively, that were found in the commonly considered situation of a fixed length of the chord!

The reason for these large, if not enormous differences is that the variation of the maximum values of the aerodynamic coefficients with camber/chord ratio (drag in particular) is much smaller than the progressive decrease with camber of the chord/girth length ratio.

The message contained by Fig. 7.8.16 seems to be that, for a fixed rated size and area, the optimal amount of camber of spinnakers is much smaller than the values of 25–45% that seem to be commonly adopted (Richards et al. 2006). This is also suggested by spinnaker test results presented in Hansen et al. (2006).

Gennakers and Code 0

Gennakers and Code 0 type foresails are, as already mentioned, meant for (close) reaching conditions and upwind sailing in light air. This is illustrated by Fig. 7.8.17 which compares wind tunnel data (Claughton et al. 2008) for a 150% genoa, a Code 0 and a gennaker type headsail for an IMS class cruiser-racer model, all with the same mainsail. Shown are the driving force T and the side force S_A, scaled by the dynamic pressure q, as a function of the apparent wind angle, at zero heel and assuming zero leeway. It appears that the Code 0 type headsail has the best driving force for apparent wind angles between 20 and 45° and the gennaker for β>45°.

Fig. 7.8.16 *Maximum lift and drag of sail sections for a fixed girth length or arc-length, as a function of camber/chord ratio*

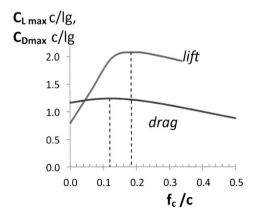

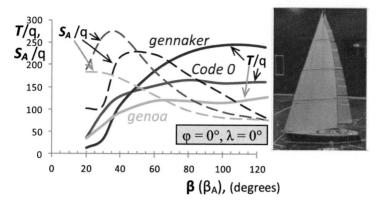

Fig. 7.8.17 *Comparison of driving force and side force for a 150% genoa, Code 0 and gennaker type foresails at zero heel. (From wind tunnel data Claughton et al. 2008)*

Hence, the Code 0 and the gennaker are a good choice for low wind speeds, each in its appropriate range of apparent wind angles. A closer inspection reveals that the genoa has a better lift/drag ratio (or a smaller drag angle). This means that the genoa is more suitable for higher wind speeds.

7.9   Effects of Heel and Trim-In-Pitch

Introduction
Heel has, as we have seen already in Chap. 4, a negative effect on the performance of a sailing yacht. This in particular in upwind conditions, when the heeling force is relatively large.

In Sect. 6.7 we saw in some detail that it reduces the hydrodynamic side force and lift/drag ratio of the underwater body. In this section we will see that heel is equally harmful for the driving force and the aerodynamic drag angle of the sails. The effects of trim-in-pitch, while significant for the hydromechanics of the hull, are usually quite small and negligible for the aerodynamics of the sails.

Prior to going into the details of the mechanisms involved it is useful to recall from Sect. 3.1 that the angle of heel φ of a sailing yacht is defined as the angle between the vertical (z-) axis and the plane of symmetry of the yacht (or ξ-ζ plane. The trim-in-pitch angle θ is defined as the angle between the longitudinal (ξ-) axis of the yacht and the horizontal plane.

Apparent Wind Angle and Angle of Attack
We have seen already in Chap. 5 and Sect. 7.4 that the aerodynamic forces acting on a sail of given dimensions are determined by the apparent wind speed and the angle of attack α of the sail. For zero heel the angle of attack α is the angle between the apparent wind vector V_a and the reference plane of the sail. The latter is defined as the plane through the mast (or luff) and the chord line of the foot of the sail. The

angle of attack α of the sail and the apparent wind angle β were seen (Sect. 3.4) to satisfy the following relation:

$$\alpha = \beta_a - \delta\ (= \beta - \lambda - \delta),\qquad\qquad(7.9.1)$$

where β_a is the apparent heading, δ the sheeting angle (defined in the ξ-η plane of the ship coordinate system, through rotation about the ζ-axis) and λ the angle of leeway (see also Fig. 7.9.1).

It can be shown, making use of the theory of rotation matrices (http://en.wikipedia.org/wiki/Rotation_matrix in Chap. 3), that, under heel and pitch, this relation takes the more complex form

$$\tan(\alpha + \delta) = (\cos\varphi\sin\beta_a - \sin\varphi\ \sin\theta\cos\beta_a)/(\cos\theta\ \cos\beta_a)\qquad(7.9.2)$$

For small pitch angles θ this reduces to

$$\tan(\alpha + \delta) \approx \cos\varphi\ \sin\beta_a/\cos\beta_a = \cos\varphi\ \tan\beta_a\qquad(7.9.3)$$

Note that the expressions (7.9.2/7.9.3) are, in essence, the same as those (6.7.1/2) for the angle of attack of the keel of a yacht under heel and pitch. Equation (7.9.3) implies that, for a given sheeting angle and apparent heading, the angle of attack under heel is a little smaller than at zero heel. This means that the sheeting angle must be decreased in order to realize the same angle of attack. That is, for $\beta_a < 90°$. It turns out that δ must be increased for $\beta_a > 90°$. The difference appears to be about 4° or less for $\varphi = 30°$.

Effects of Heel on the Aerodynamic Force Components of the Sails
The reduction of the angle of attack implies that, for a given apparent heading and a given sheeting angle, the aerodynamic forces will also be smaller under heel, as we will see shortly. There is, however, a more important effect.

Fig. 7.9.1 *Aerodynamic forces and moments in the horizontal plane*

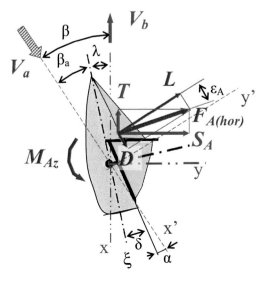

Heel also causes a redirection of the aerodynamic forces in the coordinates of sailing. We have seen in Sect. 5.2 that when the friction forces are small, the total aerodynamic force F_A acting on a thin body, like a sail, is approximately normal to the reference plane of the body. This means that for the total lift L' acting on a sail (in the lateral aerodynamic (y'-z') plane, perpendicular to the apparent wind vector V_a, there holds

$$L' = F_N \cos\alpha - F_T \sin\alpha \approx F_A \cos\alpha, \qquad (7.9.4)$$

where F_N is the normal force, F_T the tangential force and α is the angle of attack.

It also means that the total aerodynamic force F_A makes an angle φ_e with the horizontal plane. Considering the lift force on a sail (Fig. 7.9.2), this implies that the component L' of the total aerodynamic force component in the lateral aerodynamic (y'-z') plane makes an angle (φ_e) with the horizontal plane, so that for the horizontal component L of the lift there holds

$$L = L' \cos\varphi_e \qquad (7.9.5)$$

We will call φ_e the effective angle of heel. Using the calculus of rotation matrices (http://en.wikipedia.org/wiki/Rotation_matrix in Chap. 3), it can be shown that there holds for φ_e

$$\tan\varphi_e = (\cos\delta\sin\varphi\cos\theta - \sin\delta\sin\theta)/ \qquad (7.9.6)$$
$$\{(\sin\delta\cos\theta + \cos\delta\sin\varphi\sin\theta)\sin\beta_a + \cos\delta\cos\varphi\cos\beta_a)\}$$

For small pitch angles $\theta \ll 1$ this reduces to

Fig. 7.9.2 *Aerodynamic forces in the aerodynamic lateral plane (small βa, small δ)*

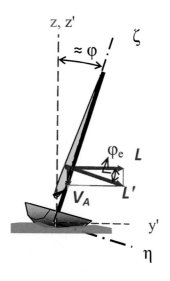

$$\tan(\varphi_e) \approx (\cos\delta\sin\varphi)/\{\sin\delta\sin\beta_a + \cos\delta\cos\varphi\cos\beta_a\} \qquad (7.9.7)$$

For small sheeting angles ($\delta \ll 1$) this reduces further to

$$\tan(\varphi_e) \approx \tan\varphi/\cos\beta_a \qquad (7.9.8)$$

Equations (7.9.7/7.9.8) imply that, for a given geometrical angle of heel φ, the effective angle of heel φ_e increases with increasing apparent heading β_a and decreases with increasing sheeting angle δ (see Fig. 7.9.3). For sheeting angles approaching 90°, ($\sin\delta \approx 1$, $\cos\delta \approx 0$), we get

$$\tan(\varphi_e) \approx \cos\delta \ \sin\varphi/\sin\beta_a \approx 0 \qquad (7.9.9)$$

This is related to the fact that, for $\delta \approx 90°$, the plane of the sail is almost parallel to the vertical, irrespective of the angle of heel and the apparent heading.

Note in Fig. 7.9.3 that the effective angle of heel φ_e for a fixed geometrical heel angle φ depends strongly on the apparent heading for small sheeting angles. The dependence on β_a is much smaller for sheeting angles of 30° and beyond.

Because the drag is, by definition, the component of the total aerodynamic force in the direction of the apparent wind vector, its direction is not effected by heel. Its magnitude is a function of heel because of the dependence of the induced drag on lift.

The vertical aerodynamic force V_A on the sails is, see Fig. 7.9.2, given by

$$V_A = L' \sin\varphi_e \qquad (7.9.10)$$

It is easily verified that for the heeling force H_A, that is the component of F_A normal to the plane of symmetry of the yacht, there holds

$$H_A \approx F_N \cos\delta + F_T \sin\delta, \qquad (7.9.11)$$

where F_N is the normal force and F_T the tangential force.

Fig. 7.9.3 *Effective heel angle as a function of apparent heading, for $\varphi = 30°$ and constant sheeting angle δ*

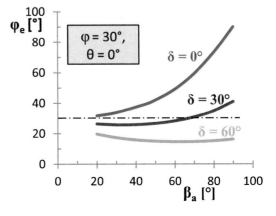

Apart from the direct effects of heel on the direction and magnitude of the aero-dynamic forces there are also secondary effects through changes in the effective geometry of the sail. One manifestation of this is a reduction of the effective span due to the fact that the mirror image of the sail with respect to the water surface, including its tip vortex, is in closer proximity to the real sail. This mechanism was already encountered in Sect. 6.7 (see also Appendix E) when considering the effects of heel on the side force acting on the keel of a sailing yacht (see Fig. E.4). The smaller effective span causes a further reduction of lift for a given angle of attack and an increase of induced drag for a given lift due to increased downwash.

Another geometrical phenomenon is that the planform of a sail acquires, effec-tively, a sweep angle (Λ) through heel, at least for non-zero apparent wind angles.[18] This can be seen readily for the case of $\beta_a \approx 90°$ (see Fig. 7.9.4). Using the calculus of rotation matrices (http://en.wikipedia.org/wiki/Rotation_matrix in Chap. 3), it can be shown, that, for arbitrary values of the apparent heading and the sheeting angle, the effective angle of sweep of a rig is given by

$$\tan \Lambda_e = (\cos\varphi \sin\theta \cos\beta_a \cos\delta + \sin\varphi \cos\theta \sin\beta_a \sin\delta)/(\cos\varphi \cos\theta) \quad (7.9.12)$$

For small pitch angles θ this reduces to

$$\tan \Lambda_e = \sin\beta_a \tan\varphi \sin\delta \quad (7.9.13)$$

Sweep, as we have seen in Sect. 5.15, causes a reduction of lift for a given angle of attack (represented through the edge correction factor 'E'). It also reduces the effective camber of a sail by a factor $\cos\Lambda_e$. However, the consequence for the ef-fective span is not necessarily unfavourable. As shown by Fig. 5.15.11, sweep has,

Fig. 7.9.4 *Illustrating that heel causes sweep for $\beta_a \neq 0$*

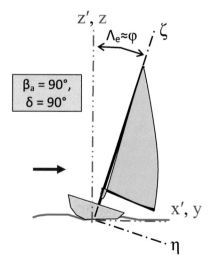

---

[18] Recall from Sect. 5.15 that the angle of sweep Λ of a lifting surface is defined, at zero angle of attack, as the angle between a line perpendicular to the direction of the undisturbed flow and a line connecting points of equal chord percentage. For the latter, the line through the centre of the mast is usually taken in case of a sailing rig type of lifting surface.

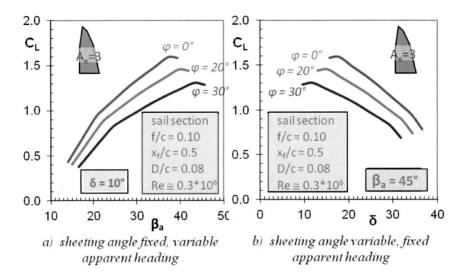

a) sheeting angle fixed, variable apparent heading

b) sheeting angle variable, fixed apparent heading

Fig. 7.9.5 *Example of the effect of heel on the lift of a sail + mast configuration (calculated, zero aerodynamic twist)*

for small taper ratios (TR), a favourable effect on the aerodynamic efficiency factor 'e' and, through that, on the effective span (see Eq. (5.15.21)). Another effect of heel for sheeting angles around 90° is that the foot of the sail comes closer to the water surface, which is favourable for the effective span.

Figure 7.9.5 gives examples of the effect of heel on lift for a single Bermuda type sail (plus mast). The figures are based on calculated data using the formulae given above and the aerodynamic model described in Sub-Sect. '3D' Effects: Effective Aspect Ratio. The figure on the left shows the coefficient C_L of lift in the horizontal plane as a function of the apparent heading β_a for three angles of heel φ and a sheeting angle δ of 10°. It can be noticed that the loss of lift due to heel can, for a given apparent wind angle, be as much as 25% for $\varphi=30°$. A similar loss of lift is found for a fixed apparent heading β_a and variable sheeting angle.

The effect of heel on the drag of a sail is relatively small. The reason is that the 'viscous' or profile drag is a function of, primarily, Reynolds number, camber, mast-diameter/chord ratio, 'normal' lift L' and sail planform (aspect ratio, sweep angle). These quantities are hardly affected by heel and this applies also to the profile drag. The main, but still modest, effect is on the induced drag through the slightly de-creased effective span. There is, however, a substantial effect on the lift-drag polar when plotted against the horizontal component L of the lift (See Fig. 7.9.6). This is caused by the fact that the latter is strongly affected by heel, as already evident from Fig. 7.9.5.

It is further evident from Fig. 7.9.6 that there is a also significant effect of heel on the minimum drag angle and on the (horizontal component of) lift at which the drag angle attains its minimum. It appears that the increase of tyhe minimum drag angle

Fig. 7.9.6 *Effect of heel on the lift-drag polar of a sail + mast (upwind, zero aerodynamic twist)*

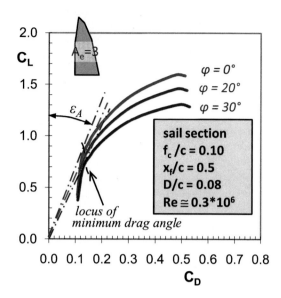

is about 7% for a heel angle of 20° and about 15% for a heel angle of 30°. This, as discussed in Chap. 4, has consequences for the minimum apparent wind angle.

Because, as we have seen in Sect. 4.3, the driving force T is related to the horizontal component L of the lift like

$$T = L\sin\beta - D\cos\beta \qquad (7.9.14a)$$

Equation (7.9.5) implies that heel also reduces the driving force. This is also the case for the aerodynamic side force, which was seen to be given by

$$S_A = L\cos\beta + D\sin\beta \qquad (7.9.14b)$$

The effect on the driving force is illustrated by Fig. 7.9.7. The figure gives the driving force coefficient of the same sail configuration as a function of the sheeting angle and the apparent wind angle (assuming zero leeway) for several angles of heel. The figure shows that at 30° heel the maximum driving force is reduced by some 25%. The (a) figure also illustrates that, under heel, for a fixed sheeting angle and variable apparent wind angle, the maximum driving force is obtained at a larger apparent wind angle. Similarly, the (b) figure shows that the maximum driving force for a given apparent wind angle is obtained for a smaller sheeting angle.

The loss of driving force under heel as shown by Fig. 7.9.7, is accompanied by a reduction of the heeling force. We have seen above (Eq. (7.9.11)), that the latter can be expressed as

$$H_A = F_N \cos\delta + F_T \sin\delta \qquad (7.9.15a)$$

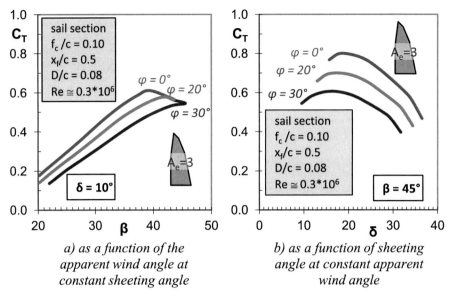

Fig. 7.9.7 *Examples of the effect of heel on the driving force of a sail + mast configuration (calculated, zero aerodynamic twist, zero leeway)*

This can also be written as

$$H_A = (L'\cos\alpha + D'\sin\alpha)\cos\delta + (-L'\sin\alpha + D'\cos\alpha)\sin\delta \quad (7.9.15b)$$

Although the maximum heeling force appears to be almost independent of heel (see Fig. 7.9.8), the heeling force for a given sheeting angle and apparent wind angle at 30° of heel is about 13 % smaller than at zero heel. This implies that, in equilibrium conditions, the underwater body experiences somewhat less resistance due to side force. This compensates, partly, for the loss of driving force. For multiple sails, i.e. a mainsail with jib/genoa, the mechanisms involved and the magnitude of the effects of heel are very similar, as we will see shortly.

Effects on the Flow About the Hull and Its Interaction with the Sails
Heel also modifies the flow about the hull and the interaction with the flow about the sails. Because the lateral projected area of a hull increases with heel, the lift and drag forces acting on the hull will, in general, also be larger under heel.

We have seen in Sect. 7.6 that the so-called deck-edge vortex, (Fig. 7.6.5), is an important feature of the flow about a sailing yacht hull. It is particularly important for the pressure level on the deck and, hence, for the force component in the direction perpendicular to the deck.

It appears (Hansen et al. 2006), that heel enhances the strength of the deck-edge vortex. The reason is that heel increases the incidence of the deck with respect to the incoming, horizontal flow. The stronger deck-edge vortex causes a lower pressure on the deck, and, through that, a larger component $F_{\zeta A}$ of the force acting on the hull

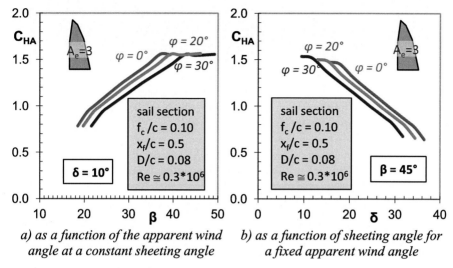

a) as a function of the apparent wind
angle at a constant sheeting angle

b) as a function of sheeting angle for
a fixed apparent wind angle

Fig. 7.9.8 *Examples of the effect of heel on the heeling force of a sail + mast configuration (cal-culated, zero aerodynamic twist, zero leeway)*

that is perpendicular to the plane of the deck (or, roughly, in the ζ-direction 'along the mast'). This is shown in Fig. 7.9.9, which gives the force component $F_{\zeta A}$ (or rather its coefficient $C_{\zeta A}$) as a function of the apparent wind angle for several angles of heel. The figure is based on wind tunnel data (http://www.wb-sails.fi 2008) for the same IMS class cruiser-racer model as in Fig. 7.6.14. Note that, therefore, the curve for $\varphi = 0°$ is the same as that in Fig. 7.6.14 for $\beta < 90°$. Note also that the mea-surements were done with jib and mainsail present.

The main message of Fig. 7.9.9 is that the component of the force on the hull normal to the plane of the deck of a yacht doubles or even triples for heel angles

Fig. 7.9.9 *Example of the effect of heel on the com-ponent of the force on the hull normal to the plane of the deck, in the pres-ence of jib and mainsail (10 m IMS cruiser-racer (Hansen et al. 2006))*

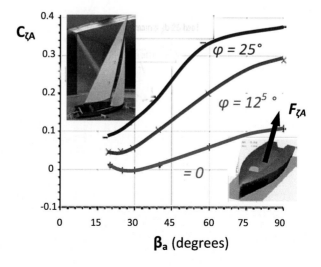

approaching 30°. This, of course, is quite significant, because it means that, at large heel angles, it is of the same order of magnitude as the driving force at small apparent wind angles (see, for example Figs. 7.6.11 and 7.9.7).

Because of the orientation of the deck, the increase of the force component $F_{\zeta A}$ perpendicular to, and upwards from, the deck implies a similar, but perhaps slightly smaller, increase of the vertical force on the hull (see also Fig. 7.6.13). This compensates, partly, the downward component of the force on the sails. It also implies an increase of the aerodynamic side force on the hull.

This author is not aware of any information on the effects of heel on the aerodynamic interaction between the hull and the sails, except for the statement in Ref. Hansen et al. (2006) that "the increase with heel (of the normal force on the hull/deck) with the sails present is even more significant than without sails". As to the effect of the hull on the flow about the sails, described in some detail for zero heel in Sect. 7.6, it can be argued that this will be amplified by heel. The reason is that a hull forms a greater obstruction for the incoming wind when heeled, which means a stronger cross-flow over the topside. This will be the case in particular for hulls with a large beam/draft ratio. The latter will tend to raise their 'weather' parts further out of the water than hulls with a small beam/draft ratio.

Overall Effect of Heel on Aerodynamic Forces

We have seen above that, in general, heel reduces the lift forces on the sails of a yacht but increases the forces due to the hull. Because the lift forces on the sail are, in general, much larger than those due to the hull, heel also reduces the total lift force as well as the driving force acting on a sailing yacht.

Figure 7.9.10 gives an example of the magnitude of this reduction. Shown is the ratio of the maximum driving force coefficient $C_{Tmax}(\varphi)$ at a heel angle φ divided

Fig. 7.9.10 *The variation with heel angle of the maximum driving force in upwind conditions for an IMS class yacht. (From wind tunnel tests (Fossati and Muggiasca 2008))*

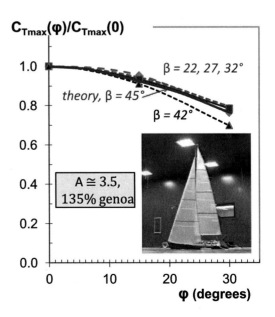

by the same coefficient $C_{Tmax}(0)$ at zero heel, as a function of the angle of heel, as measured in a wind tunnel (Fossati and Muggiasca 2008) for an IMS class cruiser-racer type of yacht. Data are shown for upwind conditions at four different apparent wind angles. Also shown is the same quantity, for $\beta = 45°$, as calculated according to the theoretical model for a single sail as shown in Fig. 7.9.7. The general conclusion to be drawn from Fig. 7.9.10 is that 30° of heel reduces the maximum driving force by some 23 % at small apparent wind angles (but, of course, progressively less at smaller heel angles).

It appears further, that, as a rule of the thumb, the ratio $C_T(\varphi)/C_T(0)$ is approximately equal to the cosine of the effective angle of heel φ_e given by Eqs. (7.9.7/7.9.8), i.e.

$$C_T(\varphi)/C_T(0) \approx \cos\varphi_e \qquad (7.9.16)$$

Bearing in mind that the lift of a sail is also proportional to $\cos\varphi_e$ (see Eq. (7.9.5)), this is not a big surprise, because lift is the main contributor to the driving force at small apparent wind angles.

Because the effective angle of heel increases with increasing apparent heading en decreasing sheeting angle (see Fig. 7.9.3) the loss of driving force due to heel also increases with increasing apparent heading en decreasing sheeting angle. This means, amongst others, that the loss of driving force due to heel at the condition of maximum lift/drag ratio or maximum T/H will be somewhat smaller than at the condition of maximum driving force. The reason being that the condition of maximum T/H requires a smaller angle of attack and, hence, a larger sheeting angle than the condition of maximum driving force. A representative number is that for $\varphi = 30°$ a loss of driving force of 15 % is incurred at the maximum T/H condition for the sail configuration of Fig. 7.9.7, rather than the 23 % that is found for T_{max}.

Effect on Yawing Moment
As already described in Sect. 7.7, the aerodynamic yawing moment M_{Az} of a sailing yacht can be expressed as

$$M_{Az} \approx S_A \xi_{CE} + T(\eta_{CE} \cos\varphi + \zeta_{CE} \sin\varphi), \qquad (7.9.17)$$

or, in coefficient form

$$C_{MAz} \approx C_S \xi_{CE}/L_{ref} + C_T\{(\eta_{CE}/L_{ref})\cos\varphi + (\zeta_{CE}/L_{ref})\sin\varphi\} \qquad (7.9.18)$$

In these expressions S_A is the side force and T is the driving force. ξ_{CE}, η_{CE} and ζ_{CE} are the longitudinal, lateral and 'vertical' positions of the centre of effort, respectively, in the ship's coordinate system. L_{ref} is a suitably chosen reference length, such as the mast height or the length of the waterline. The dependence on heel is evident from the presence of the $\cos\varphi$ and $\sin\varphi$ terms. The sine term in particular indicates that the aerodynamic yawing moment changes immediately when a yacht acquires heel. This is caused by the outboard displacement of the centre of effort when a yacht heels (see, e.g., Fig. 7.7.1b. In addition there is, of course, the dependence of the side force S_A and the driving force T on the angle of heel φ. The position of the centre of effort (ξ_{CE}, η_{CE}, ζ_{CE}) in the ship's coordinate system can be taken as independent of heel for most if not all practical purposes.

Figure 7.9.11 illustrates how strongly the aerodynamic yawing moment can vary with heel. Shown is the aerodynamic yawing moment (coefficient) as a function of the angle of heel for three values of the apparent wind angle (upwind conditions). The solid lines represent the conditions of maximum driving force, the chain lines those of the maximum propulsive efficiency. Note that the difference between the two is small.

The yacht configuration considered is the same as that of Fig. 7.7.9. That is, the position of the centre of effort is the same as that of the IMS type yacht (Fossati et al. 2008) of Fig. 7.7.8 but the lift and drag data (Marchai and Tanner 1963) are those of the 'Dragon' type yacht of Figs. 7.6.7 and 7.6.8. Hence, Fig. 7.9.11 does not represent any particular type of yacht but is merely intended to show the trend of the variation of the aerodynamic yawing moment with heel angle for a Bermuda type of rig. Note that the position of the centre of the mast was chosen as the moment reference point and the length of the waterline as the reference length.

There are several things to be noted in Fig. 7.9.11. Firstly, the level of the yawing moment is more positive (more 'to weather') at larger apparent wind angles. This, of course, in agreement with Fig. 7.7.9. Secondly, the yawing moment is seen to increase with increasing angle of heel. This is due mainly by the more outboard

Fig. 7.9.11 *Trend of variation of aerodynamic yawing moment with heel angle for a Bermuda-type rig*

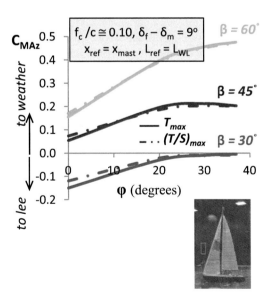

position of the centre of effort when a yacht heels (represented by the $\sin\varphi$ term in Eq. (7.9.17). The slope is a function of the effective angle of heel φ_e that was introduced in the beginning of this section (see, e.g., Eq. (7.9.7)). For small sheeting angles, which are characteristic for upwind sailing, the effective angle of heel was seen to increase with increasing apparent wind angle (Fig. 7.9.3). This is why the curves in Fig. 7.9.11 have a higher slope for larger apparent wind angles.

It is finally noted that the curves in Fig. 7.9.11 level off for heel angles around 25°. This is due to the fact that the lift of the sails, the driving force and the side force are decreasing progressively with increasing angle of heel (Figs 7.9.5 and 7.9.7. This counteracts the effect due to the more outboard position of the centre of effort.

Effect of Heel on Apparent Heading as Measured by a Wind Vane
Heel is also known to introduce an error in the reading of the apparent heading of a wind vane. As discussed in some detail in Gentry (1981a), this is caused by the fact that, under heel, the axis of a wind vane is no longer perpendicular to the apparent wind vector. It can be shown (Fossati and Muggiasca 2008) that the relation between the indicated apparent heading β_a' and the real apparent heading β_a is given by

$$\tan\beta_a' \approx \cos\varphi\sin\beta_a/\cos\beta_a = \cos\varphi\tan\beta_a \qquad (7.9.19)$$

(see also Eq. (7.9.3)).

Figure 7.9.12 shows the relation graphically for different angles of heel φ. The error can be seen to be as much as 5° for an angle of heel φ of 30°.

Fig. 7.9.12 *Effect of heel on apparent heading as indicated by a wind vane*

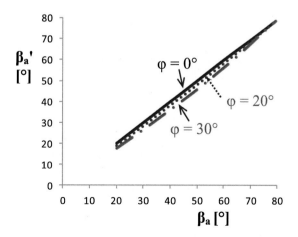

7.10   Unsteady Aerodynamic Effects

As already indicated in Sects. 3.1 and 5.11 the (turbulent) flows about the sails of a sailing yacht are unsteady, that is time dependent, by nature. We have also seen that for many practical purposes, such as performance analysis, the flow can be treated as if it were (quasi-)steady through the introduction of a time-averaging process (Sect. 3.1). There are, however, (see also Sect. 5.18) a number of aerodynamic phenomena with unsteady characteristics that can not be neglected. One is the effect on the aerodynamic forces acting on the sails of a yacht in periodic motion in waves. Another phenomenon, in which time-varying effects in the aerodynamic forces acting on the sails are essential, is rolling in downwind conditions. In the following we will consider both in some further detail.

Effects of Motion in Waves on the Aerodynamic Forces
When a yacht is sailing in waves it is subject to periodic, wave induced pitching, rolling and yawing ((Sub)Sects. Added Resistance in Waves and 6.8). For the sails this means that they are exposed to periodic variations in angle of attack and apparent wind speed. In principle there will also be periodic variations in the effective sweep of the rig, but the effect of this on the aerodynamic forces is, in general negligible compared to the effects of the variations in angle of attack and apparent wind speed.

Because a yacht rolls and pitches usually around the centre of gravity, or thereabout, the periodic variations in angle of attack and apparent wind speed are felt mostly near the top of the sail. It is also clear that the frequency of encounter with the waves and the wave direction angle play an important role, because these determine the kind of motion (pitching and rolling) and their frequency.

An estimate of the frequency and magnitude of the periodic variations in angle of attack and apparent wind speed due to the pitching oscillation of a yacht sailing upwind in a seaway can be obtained as follows. Assuming a coastal/estuary type of environment, a typical (but relatively short) wave length is about 15 m (Sect. 5.19). The corresponding frequency of encounter with the waves in upwind conditions ($\gamma_w \cong 45°$) is then $f_e \cong 0.45$ s^{-1} for a boat speed of about 6 kts (see Fig. 6.5.40). A typical value for the maximum wave slope θ_w is $\theta_w \cong 0.27$ radians (or about 15°, see Fig. 5.19.7). Because of the true wind angle of 45° the effective maximum wave slope and maximum pitch angle θ_{max} experienced by the yacht will be about $15 * \cos\gamma_w \cong 11°$. Hence, the pitch angle of the yacht will, in regular waves, vary in time like

$$\theta(t) = \theta_w \cos\gamma_w \sin(\omega_e t), \qquad (7.10.1)$$

where $\omega_e = 2\pi f_e$ is the circular frequency of encounter with the waves.

As already mentioned in Sect. 5.18, the so-called reduced frequency k_f, defined as $k_f = \omega_e c/(2U)$, where c is the chord length and U the (average) flow speed, is an important measure for the unsteady aerodynamic effects. Based on the numbers given above, an assumed chord length of 2 m, and taking for U the apparent wind speed V_a at 25 kts, it is found that $k_f \cong 0.2$. This means that unsteady effects (lift deficiency and phase lag) can be expected to be significant.

Because of this pitching motion a sail section at a height 'ζ_{CER}' above the water surface will oscillate backward and forward with the same frequency. The local oscillatory speed V_b' of the sail section is then

$$V_b' = -\dot{\theta}(t)\zeta_{CER} = -\zeta_{CER}\,\omega_e\theta_w\,\cos\gamma_w\,\cos(\omega_e t) \qquad (7.10.2)$$

Assuming an average boat speed of 30% of the true wind speed and a true wind angle of 40°, we can determine the corresponding variation in time of the apparent wind speed V_a and apparent wind angle β_{CER} of the sail section according to the formulae given in Sect. 3.3. The result of such a calculation, assuming zero heel and zero leeway, for a section at a height $\zeta_{CER} = 8$ m above the water surface is given in the top graph of Fig. 7.10.1. Shown is the pitch angle θ and the local apparent wind angle β_{CER}. The local apparent wind speed V_a of the section, divided by its value $(V_a)_{stat}$ in steady flow, as a function of time over one period of the pitching motion is shown in the bottom figure. The figures show that the amplitude of the variation of the apparent wind angle is about the same as that of the pitch angle and that the amplitude of the variation of the apparent wind speed squared is about 60% of the average apparent wind speed.

Fig. 7.10.1 *Estimated trend of the variation, with time, of the aerodynamic characteristics of a sail of aspect ratio 4, pitching in waves (coastal/estuary environment), at a constant true wind angle and zero heel*

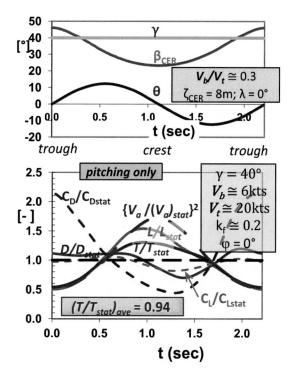

It should be realized that, because of Eq. (7.10.1), the oscillatory parts of the apparent wind speed and apparent wind angle increase in, roughly, a linear way with the height above the water surface. Because of the quadratic dependence on the apparent wind speed of the aerodynamic forces, a height of 8 m is probably a reasonably representative value for estimating the unsteady effects on the sails due to pitching of a yacht with its mast top at about 17.5 m above the water surface.

Based on the unsteady conditions assumed and described above, it is possible to make an estimate of the resulting unsteady aerodynamic forces acting on a sail. This can be done, for example, on the basis of measured characteristics for steady flow and unsteady characteristics as calculated according to the models described in Sect. 5.18. The bottom graph of Fig. 7.10.1 presents results of such an estimate for a rig with aspect ratio 4. The estimate is based on the steady flow data of Fig. 7.6.7. This implies that the sails will experience some mild dynamic stall phenomena (Sect. 5.18). Shown is the variation, with time, over one period, of various aerodynamic quantities, each divided by its value in steady (stationary) flow. The most noticeable features are the following:

- The variation of the drag coefficient is much larger than that of the lift coefficient. This is due to the quadratic, or even stronger, dependence of the drag on lift (through aspect ratio).
- The variation of the lift and drag coefficients is overruled by the variation of the apparent wind speed. The latter is half a period out of phase with the lift and drag coefficients (recall that $L \div C_L V_a^2$ and $D \div C_D V_a^2$)
- The effect on the time-averaged driving force is significant and negative; the time-averaged driving force is about 5% less than the driving force in steady flow.

This, however, is only a fraction of the complete story. The real life situation is much more complex. A yacht in a seaway does not oscillate in pitch only but also in roll. The latter is also influenced by the time-varying heeling force. This is also the case for the angle of yaw. In addition there is the effect of the orbital motion of the water particles in the waves on the motion of the yacht, and, last but not least, the effects of inertia. The latter will tend to cause a time lag between the variation of the wave slope and the motion of the yacht. In the example given above this time lag was assumed to be zero.

It is interesting to consider the effects of the wave-induced rolling motion. We can make a crude estimate of the latter by assuming an average heel angle of, say, 20° and a rolling motion that conforms instantaneously to the wave shape and the wave direction. This means that the effects of the varying heeling force and the inertial time lag is not taken into account. The heel/roll angle then varies in time like

$$\varphi(t) = \theta_w \sin \gamma_w \sin(\omega_e t) \qquad (7.10.3)$$

The oscillatory rolling motion has an additional kinematic effect on the effective apparent wind angle, the angle of attack and the effective apparent wind speed experienced by the sail. It can be calculated in a similar way as described above for the zero heel/roll case.

Fig. 7.10.2 Estimated trend
of the variation, with time,
of the aerodynamic characteristics of a sail of aspect
ratio 4, pitching and rolling
in waves (coastal/estuary
environment), at a constant
true wind angle

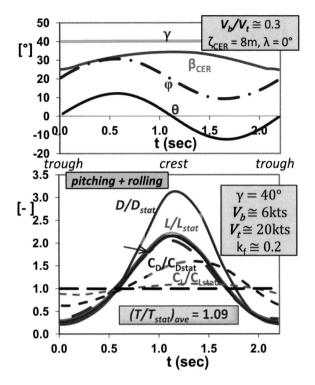

Figure 7.10.2 gives the results of such calculations. The main difference with Fig. 7.10.1 is in the variation of the (effective) apparent wind angle of the sail, which implies a similar difference in angle of attack. While in the case of pitching only, at zero heel, the effective apparent wind angle attains its minimum at the crest of a wave it attains its maximum at the crest when pitching and rolling under heel. As a consequence the variation with time of the aerodynamic forces is also distinctly different. The net effect is that the time averaged driving force is now some 9 % larger than in steady flow at zero heel.[19] The explanation of this phenomenon is that, under heel and rolling, the sail develops additional, dynamic driving force through the same heaving type of mechanism that is used by birds to generate a propulsive force by flapping its wings (Sect. 5.18).

Inclusion of the effect of the orbital motion of the water in the waves is relatively simple. Figure 7.10.3 summarizes the results. The main difference with Fig. 7.10.2 is a reduction of the aerodynamic forces when the yacht is near the crest of a wave. This is caused by the reduction of the apparent wind speed when the yacht is pushed downwind by the orbital motion near the crest. The opposite occurs near a trough. Another aspect is that the time history of the pitching and rolling motion is less sinusoidal. This is caused by the fact that due to the orbital motion the yacht spends more time near the crests, where the VMG over the bottom is decreased, than near

---

[19] When compared with steady flow at the average heel angle of 20° this is even 20 %!

Fig. 7.10.3 Estimated trend of the variation, with time, of the aerodynamic characteristics of a sail of aspect ratio 4, pitching and rolling in waves (coastal/estuary environment), at a constant true wind angle, including the effect of the orbital motion of water

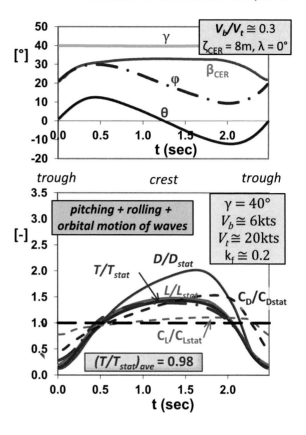

the troughs, where the VMG over the bottom is increased. The instantaneous pitch and roll rates, that is the slope of the φ- and θ-curves, near the wave crests are smaller, while they are higher near the troughs. This causes a further reduction of the unsteady part of the aerodynamic forces near the wave crests but an increase of the unsteady part of the aerodynamic forces near the troughs. The net effect is a reduction of the time-averaged driving force; the ratio $(T/T_{stat})_{ave}$ reduces from 1.09 to 0.98 when the effects of the orbital motion of the water particles in the waves in taken into account. When the apparent heading β_{CER} rather than the true wind angle γ is kept constant the ratio $(T/T_{stat})_{ave}$ is found to be 0.99.

The values just given suggest that there is hardly a loss in time-averaged driving force in waves when a yacht is pitching as well as rolling under heel (recall that with pitching only, at zero heel, the ratio $(T/T_{stat})_{ave}$ was 0.95). The author is not aware of any information in the available literature (Fossati and Muggiasca 2011; Gerhardt 2010) on the subject that is in conflict with this (preliminary) finding.

The orbital motion of the water particles in waves has also another, more direct effect on the velocity-made-good to windward of a yacht. Because the yacht spends relatively more time near the wave crests, where the VMG is reduced by the orbital motion and relatively less time near the troughs, where the VMG is increased, there is a loss in time-averaged VMG. We will come back to this aspect in Chap. 8.

It is emphasized that the examples given above embody only a crude, qualitative indication of the effects of waves on the aerodynamic forces in upwind conditions. This in particular because the effects of the time-varying forces on the motion and leeway of the yacht were not taken into account. Recall also that the estimates were made for yachts with a length of about 10 m, sailing upwind in relatively short and steep waves that are representative for a coastal/estuary environment. The effects will be proportionally smaller in the open sea with longer waves and smaller wave slopes and lower frequencies of encounter. However, larger yachts with higher boat speeds might experience a performance loss of comparable magnitude in longer waves. This because resonance phenomena in pitching motion, which lead to higher amplitudes, depend on wave-length/boat-length ratio (Sect. 6.5). Another factor is that longer and higher waves imply higher orbital velocities (Sect. 5.19). The effect of the latter is, however, less pronounced for higher boat speeds.

It is, further, good to realize that loss of time-average driving force, if any, is not primarily caused by the reduced frequency dependent, unsteady aerodynamic phenomena. The important mechanism, that is the quadratic (or even stronger) dependence of drag on the (varying) lift, is equally present in quasi-steady conditions, that is for (very) low frequencies. The amplitudes of the motion induced variations of the angle of attack and the apparent wind speed seem to be the most important parameters. As indicated above, these are functions of the wave slope as well as the frequency of encounter and the height of the rig.

Self-Excited Rolling
A phenomenon well known to most yachtsman is that a yacht sailing (almost) dead downwind, with the sheets (almost) fully eased, can suddenly be subject to wild oscillations in roll without any obvious cause. The phenomenon was studied in detail and explained for the first time through the classical experiments of Marchai (2000).

Figure 7.10.4 illustrates the main mechanism involved. Let us assume that a yacht sailing dead downwind with a sheeting angle of about 90° starts to roll, for whatever reason, and swings to port, as in the figure on the left. The sail will then experience a cross-wind component v_r equal but opposite to the speed of the swing motion. The same, but opposite, happens during a swing to starboard.

The roll-induced cross-wind velocity depends, of course, on the height above the water surface. The magnitude will increase linearly with the distance from the centre of rotation of the rolling motion. The latter will, in general, be close to the centre of gravity.

Before the swing, the angle of attack α of the sail will be about 90° and the sail will produce drag only (which is then equal to the driving force). However, during the swing, the swing-induced velocity component v_r causes a change of the angle of attack. The resultant apparent wind velocity V_R will cause the sail to experience an angle of attack varying between $90 + \delta\alpha_r$ and $90 - \delta\alpha_r$, (or an apparent wind angle between $180 + \delta\alpha_r$ and $180 - \delta\alpha_r$) where $\delta\alpha_r$ is the (maximum) roll-induced angle of attack:

$$\delta\alpha_r = -\arctan(v_r/V_a), \qquad (7.10.4)$$

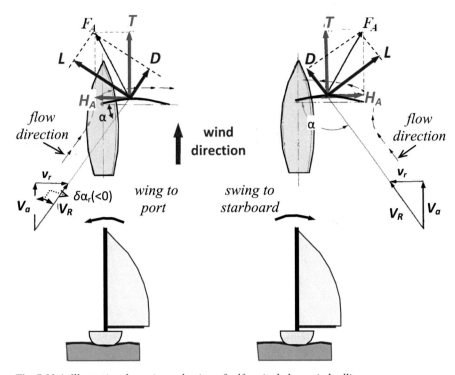

Fig. 7.10.4 *Illustrating the main mechanism of self-excited, downwind rolling*

V_a being the apparent wind speed. At $\alpha = 90°$ the sail will, in general, not produce any lift. It is the change $\delta\alpha_r$ of the angle of attack due to rolling that causes the sail to generate lift, while the drag hardly changes, at least for small values of $\delta\alpha_r$ (see, e.g. Figs. 7.4.35 and 7.4.36). As indicated in Fig. 7.10.4 a consequence is that the sail now also produces a heeling force component H_A. The orientation of this heeling force is such that it enhances the swinging motion. This is the case irrespective of the direction of the swinging motion. Potentially, this is an unstable situation.

In the situation just described there is a strong coupling between the aerodynamics of the sails and the hydro-mechanic forces on the underwater body. We know from Sect. 6.2 that the hull of a yacht in rolling motion begins to develop a hydrostatic restoring force as soon as it develops heel, because of its lateral hydrostatic stability. This causes the yacht to swing back to the other side under the influence of inertia and the production of the aerodynamic heeling force. As a consequence the yacht will begin to oscillate in roll, in a preferred, natural frequency. As indicated in Sect. 6.8 this frequency is of the order of 0.25 cycles per second.

The behaviour just described is very much like that of a 'dynamic' (mass-spring-damper) system of the type described in Sect. 5.18. It will be described in some more detail in Chap. 8. At this point it suffices to note that the role of the spring is played by the hydrostatic restoring forces while the damping function is embodied mainly by the hydrodynamic forces on the appendages of the hull, the keel in particular. The sails can also be considered to generate damping, but, in this case, of the wrong (negative) sign. The 'net' damping (sum of the positive hydrodynamic and negative aerodynamic damping) determines whether the 'system' as a whole is dynamically stable or unstable. In the latter case the roll amplitude will increase to, potentially, catastrophic values.

Considering the aerodynamics in some further detail we note first that the velocity v_r induced by the rolling motion is approximately given by

$$v_r = -\dot{\varphi}(t)\zeta_{CER} = -\zeta_{CER}\,\omega_r\varphi_{max}\,\cos(\omega_r t), \qquad (7.10.5)$$

where $\dot{\varphi}$ is the angular velocity, ω_r the circular frequency of the rolling motion and φ_{max} the amplitude of the rolling motion. 'ζ_{CER}' is the distance between the sail section considered and the centre of rotation. The circular frequency ω_r corresponding with the 'real' frequency of 0.25 cycles per second is $\omega_r \cong 2\pi*0.25 \cong 1.5$ radians/second. For an assumed height ζ_{CER} of the centre of effort of the sails in the rolling motion of 10 m and a maximum roll angle of 30° this gives $v_{rmax} \cong 7.5$ m/s. With an assumed (real) apparent wind speed of 12 kts $\cong 6$ m/s, that is a true wind speed of the order of 20 kts, this gives a maximum roll-induced change $\delta\alpha_r$ in angle of attack of about 50°. This is a large amount indeed, which can easily lead to a situation in which the lift is temporarily larger than the drag, as in Fig. 7.10.4 (see also Figs. 7.4.35 and 7.4.36).

It will be clear that during the swing there is also an increase of the resulting effective apparent wind speed. For the situation assumed above, with a steady V_a of 12 kts and $v_{rmax} \cong 7.5$ m/s $\cong 15$ kts we have a maximum apparent wind speed $(V_a)_{max}$ of $\sqrt{(12*12 + 15*15)} \cong 19$ kts. Because aerodynamic forces are proportional with the square of the apparent wind speed this means an increase of lift and drag of up to 150 % as compared to the steady situation. This is a very large amount that leads to a substantially larger time-averaged driving force as compared to the steady situation.

The maximum roll-induced angle of attack $\delta\alpha_r$ and apparent wind speed are, of course, smaller for smaller values of the maximum roll angle and/or lower roll frequencies and/or shorter masts. The figures assumed above are meant to give an indication of how large the effects can be.

It should also be mentioned that the assumed frequency of the rolling motion of about 0.25 cycles per second implies that the reduced frequency of the unsteady flow experienced by the sails is about $k_f = 2\pi f_r c/(2V_a) \cong 0.17$ (assuming a chord length 'c' of 3 m). Because the sails, in downwind running conditions, are operating in their deep stall regime, this means that unsteady, dynamic effects, such as time lag and 'overshoot' phenomena (see Sect. 5.18) can be significant. During a cycle, vortices will be shed periodically from the leading and trailing edges of the sail, as indicated qualitatively in Fig. 7.10.5 and described also in Ref. Bethwaite (2010, pp. 323–324). The (red) vortices will be shed, roughly, when the swing speed attains its maximum, with the trailing edge producing the strongest vortex. The latter

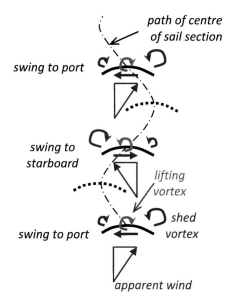

Fig. 7.10.5 *Formation of alternating, shed vortices at sail edges during downwind rolling oscillations*

is a consequence of the vortex theorems of Helmholtz (Prandtl and Tietjens 1934b in Chap. 1). These require that the total circulation around the shed vortices plus the lifting vortex (blue) or circulation around the sail section is zero. When the sail section swings to port, the orientation of the lifting vortex (circulation) is clockwise, because the lift is also directed to port. This means that the anti-clockwise vortex shed from the trailing edge must be stronger that the clockwise vortex shed by the leading edge.

The resulting pattern is somewhat similar to that of the Von Kármán vortex trail described in Sect. 5.18. The latter was seen to form a stable pattern when the Strouhal number fL/V is about 0.25, where f is the frequency, L the characteristic dimension of the object and V the wind speed. For the rolling sail situation with $f \cong 0.25$ this means $V/L \cong 1$ [s^{-1}]. Taking for L the effective chord length of the sail, say 3 m, gives an apparent wind speed of about 3 m/s or 6 kts (or about 11 kts true wind). This is much lower than the 12 kts apparent wind (or 20 kts true wind) of the example described above. It means that the unsteady vortex wake will, in this case, not be structured as in the case of the Von Kármán vortex trail but will be more 'fuzzy' (blown apart) due to the relatively high wind speed. For low wind speeds and/or large yachts (big 'L') the wake is probably more structured. The latter may imply a stronger periodicity of the aerodynamic forces.

It is important to realize that the phenomenon sketched above can happen in every imaginable sea state. Once triggered into rolling, through whatever incident, the self-exciting mechanism sets off, even in completely flat water, if the hydrodynamic damping from the keel is not sufficient to neutralize the negative aerodynamic damping from the sails. Waves can, of course, enhance the phenomenon, in particular when they meet the yacht with a frequency close to the natural frequency of the rolling motion.

The phenomenon is not limited to ordinary, single sails. It can also happen with spinnaker type sails (Marchai 2000, pp. 658–675).

An important question is if, and how, downwind rolling can be avoided. We have seen above that the basic mechanism is the orientation of the heeling force with respect to the direction of motion. It is easily verified that there is negative aerodynamic damping when the rate of change $\dfrac{dH_A}{d\alpha}$ of the aerodynamic heeling force with angle of attack is positive, that is when

$$\frac{dH_A}{d\alpha} > 0 \tag{7.10.6}$$

Because, see Sect. 4.3,

$$H_A \approx L\cos\beta + D\sin\beta \tag{7.10.7}$$

and

$$\beta = \alpha + \delta \approx 90 + \delta\alpha_r + \delta \tag{7.10.8}$$

Equation (7.10.6) can be rewritten as

$$\frac{dL}{d\alpha}\cos\beta - L\sin\beta + \frac{dD}{d\alpha}\sin\beta + D\cos\beta > 0 \tag{7.10.9}$$

For $\beta \cong 180°$ and $\alpha \cong 90°$, i.e. small $\delta\alpha_r$ and $\delta \cong 90°$, we have $\cos\beta < 0$ and $dD/d\alpha \approx 0$ (see Fig. 7.4.36). Equation (7.10.9) then reduces to

$$(\frac{dL}{d\alpha} + D)\cos\beta - L\sin\beta > 0 \tag{7.10.10}$$

or

$$\frac{dL}{d\alpha} < L\tan\beta - D \tag{7.10.11}$$

or, in coefficient form,

$$\frac{dC_L}{d\alpha} < C_L\tan\beta - C_D \tag{7.10.12}$$

We know that D is always >0 and that $L>0$ for $\alpha < \cong 90°$ and $L<0$ for $\alpha > \cong 90°$. For $\delta \cong 90°$ this means that $L<0$ for $\beta > 180°$ and $L>0$ for $\beta < 180°$. We also know that $\tan\beta < 0$ for $90° < \beta < 180°$ and $\tan\beta > 0$ for $180° < \beta < 270°$. Hence, for the term

$L \tan\beta$ there holds $L \tan\beta \leq 0$, irrespective of the value of β. Equations (7.10.11/7.10.12) therefore implies that negative aerodynamic damping can occur when the lift curve slope $\dfrac{dC_L}{d\alpha}$ is sufficiently negative $(< -C_D)$.[20] Because $C_D \cong 1.25$ for $\alpha \cong 90°$ (see Fig. 7.4.36) this means that

$$\frac{dC_L}{d\alpha} < \cong -1.25 \text{ per radian} \cong -.02 \text{ per degree} \qquad (7.10.13)$$

and that there can be no self-excited rolling when

$$\frac{dC_L}{d\alpha} > \cong -1.25 \text{ per radian} \cong -0.02 \text{ per degree} \qquad (7.10.14)$$

It appears from Fig. 7.4.36 that this requires roughly that the angle of attack α should be smaller than that for maximum lift (α_{CLmax}). As shown in Fig. 7.4.35 this depends on the aspect ratio of the sail. For small aspect ratios A, say $A \cong 1$, we have $\alpha_{CLmax} \cong 45°$, decreasing to $\alpha_{CLmax} \cong 25°$ for $A \cong 4$.

It is useful to recall at this point that, under normal conditions of sailing, $\alpha < \alpha_{CLmax}$ implies apparent wind angles $\beta < \cong 90°$.

Reducing the angle of attack at a fixed apparent wind angle means, for a mainsail, increasing the sheeting angle, since $\alpha = \beta - \delta$. For a spinnaker type head sail this means releasing the sheet and/or over-squaring the pole (Fig. 7.8.6). For a conventional, Bermuda rig mainsail this is not possible for $\delta \cong 90°$, due to the presence of the shrouds. The way out is to make use of the phenomenon that the slope of the lift curve reverses again for angles of attack beyond, say, about 150°, as illustrated by Fig. 7.10.6 for an aspect ratio $A = 4$ rig. For an apparent wind angle of 180° this means a sheeting angle of only about 20°, depending on aspect ratio. In practice the allowable sheeting angles will be larger due to the positive hydrodynamic damping of the rolling motion contributed by the underwater appendages of a yacht.

Fig. 7.10.6 *Lift as a function of angle of attack upto $\alpha = 180°$. (From wind tunnel tests (Marchai 2000) on an $A = 4$ Bermuda rig)*

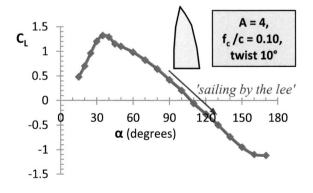

Although such 'sailing by the lee' ($\alpha > 90°$) reduces the risk of self-excited rolling, it increases, of course, the risk of an unintentional gybe.

It is also clear from Eqs. (7.10.11/7.10.12) that any deviation of the time averaged apparent wind angle from the $\beta \cong 180°$ condition has some influence on the conditions for (negative) aerodynamic damping. Equation (7.10.12) implies that the more negative the product $C_L \tan\beta$ the smaller the risk of self-excited rolling. This means that if $\tan\beta < 0$, that is $\beta < 180°$, C_L should be > 0 and as large as possible. The latter implies $\alpha < 90°$. It follows from Eq. (7.10.7) that this is always the case for a fixed sheeting angle $\delta \geq 90°$. It is similarly found that if $\tan\beta > 0$, that is $\beta > 180°$ (or $< -180°$), C_L should be as much negative as possible. This means $\alpha > 90°$, which, for $\beta > 180°$, is always the case for $\delta \leq 90°$. It follows that, in general, the further away the apparent wind angle is from $180°$ the smaller the risk of self-excited rolling. However, it also follows, because $\delta \geq 90°$ is not feasible, at least for conventional rigs, that 'sailing by the lee' is the only simple option to reduce the risk of self-excited rolling.

It is useful to note, at this point, that the lift of a sail is usually not zero at precisely $90°$ angle of attack (see Fig. 7.10.6). The reason is asymmetry due to the presence of the mast, position of maximum camber twist and planform shape. In the case of the sail of Fig. 7.10.6 the lift is zero for $\alpha \cong 105°$. This means that, for $\beta < 180°$, α should be $< 105°$ or $\delta \geq 105°$ (not feasible) to reduce the risk of self-excited rolling. Similarly α should be $> 105°$ or $\delta \leq 105°$ for $\beta > 180°$. This means, again, that 'sailing by the lee' is a feasible option to reduce the risk of self-excited rolling.

A further point to be mentioned with respect to downwind rolling is the following. We have seen above that downwind rolling is accompanied by a substantial increase of the time-averaged driving force. Rolling causes also additional hydrodynamic resistance of the underwater body. However, the latter seems to be quite moderate (Marchaj 2000, pp. 658–675). This, presumably, is due to the relatively small distance between the centre of lateral resistance (CLR) of the underwater body and the rolling axis and the resulting relatively small values of the roll-induced cross-flow velocity. The substantial increase of the time-averaged driving force plus the moderate increase of the time-averaged hydrodynamic resistance means that there is a potential for a performance gain, provided that the crew is prepared to accept the discomfort associated with rolling.

7.11   Indicators: 'tell tales'

Perhaps the most frustrating (or interesting?) aspect of sailing that it is, in general, not possible to judge in a direct way whether the sails are operating like they should for given conditions of sailing. The reason is, of course, that there are no instruments available that give an indication of, for example, the (maximum) driving force or the (maximum) lift/drag ratio of the sails. The information we do have available in general, in the form of boat speed or velocity-made-good (through a log and/or GPS), is, although indispensable, of an indirect nature as far as the aerody-

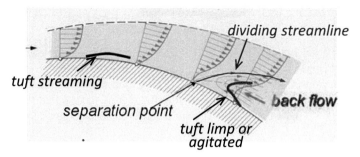

Fig. 7.11.1 *Illustrating how tufts (tell tales) work*

namics of the sails in concerned. There is, however, still another way to obtain some information about certain properties of the flow about the sails: 'tell tales'.

Tell tales are small woollen tufts or small ribbons (made, for example, of thin spinnaker tissue) that are attached to a sail in a number of strategic positions. Their intention is to indicate the local direction of the flow and to provide information on the condition of the boundary layer. As illustrated by Fig. 7.11.1 tufts are 'streaming' in the direction of the local flow when the boundary layer at their position is attached. They are either 'limp' or agitated when they are situated in an area of separated flow. The 'limp' condition is indicative of a thick 'dead water' area. In a thin area of separated flow with large-scale turbulent fluctuations, a tuft can move around (flutter) in an 'agitated' way.

By strategic positioning tell tales can give useful information about the following aerodynamic properties:

- The 'ideal' angle of attack of a jib or genoa
- A crude indication about the angle of attack for maximum lift
- The proper amount of camber of a sail
- The proper twist of the sail

Information about the 'ideal' angle of attack of a sail section can be obtained by so-called 'Gentry' tales.[21] As illustrated by Fig. 7.11.2, Gentry tales are a set of three or four tufts positioned in tandem aft of the leading edge in the first 15–20% of the chord. At the so-called 'ideal' angle of attack α_{id} when the flow enters the leading edge of a sail section smoothly, see Sect. Flow and Force Characteristics of Sail Sections, the boundary layers on both sides of the sail near the leading edge are fully attached and all tufts are streaming. At smaller angles of attack, $\alpha < \alpha_{id}$, a laminar separation bubble develops on the pressure (windward) side of the sail. Depending on the length of the separation bubble, the first two or three tufts on this side are now limp or agitated. The fourth tuft in the figure is streaming, indicating that the free shear layer has reattached between the third and fourth tuft.

---

[21] Named after the aerodynamicist (Douglas and Boeing) and yachtsman Arvel E. Gentry, who was the first to propose them (Gentry 1981b).

Fig. 7.11.2 *Illustrating 'Gentry tales'*

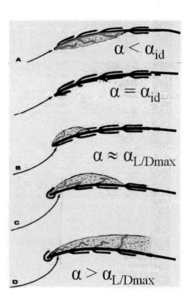

For $\alpha > \alpha_{id}$ a similar picture develops on the leeward (suction) side of the sail. When the laminar separation bubble is small, that is when the last tuft, or the last two tufts, are streaming, the angle of attack is close to that for the maximum lift/ profile-drag ratio of the section ($\alpha \cong \alpha_{L/Dmax}$). For still larger angles of attack ($\alpha > \alpha_{L/Dmax}$) the separation bubble extends beyond the last Gentry tuft or the boundary layer does not reattach at all.

It is important to realize that the condition $\alpha_{L/Dmax}$ refers, as mentioned, to the maximum lift/profile-drag ratio of the sail section. For a sail of finite aspect ratio this is not necessarily the same as the condition for the maximum ratio between the lift and the total drag, as discussed in Sect. 7.4. Because of the phenomenon of in-duced drag the maximum lift/drag ratio of a sail with a finite aspect ratio is obtained at a lower lift than in two-dimensional flow. For a sail with an aspect ratio $A \cong 3$ this implies (see Figs. 7.4.28 and 7.4.29) that the angle of attack for the maximum L/D is about 3–4° larger than the angle of attack at which the section ('2D') attains its maximum profile L/D. This is about 2° smaller in terms of effective section angle of attack ('2D'). It means, see also Fig. 7.4.5, that the tuft picture for the maximum lift/ drag ratio of a sail with finite aspect ratio should, in terms of Fig. 7.11.2, be closer to that for $\alpha \cong \alpha_{id}$ rather than that for $\alpha \cong \alpha_{(L/D)max}$.

Tell tales can also provide information on boundary layer separation at other locations, on foresails as well as mainsails. This can be used, for example, to give a crude indication of the condition of maximum lift of a sail. Figure 7.11.3 gives an example. We have seen in Sect. 5.14 that the maximum lift of a foil is usually attained when the boundary layer on the suction surface separates at about 50 % of the chord. With tufts positioned around 50 % and at 90 % of the chord we have a crude indication for maximum lift when the tuft at 90 % is limp or agitated and the one near 50 % is still streaming. This is the preferred picture when trimming for the maximum driving force at apparent wind angles approaching 90°.

Fig. 7.11.3 *Tell tales indicating separated flow near the trailing edge (leech) of a sail*

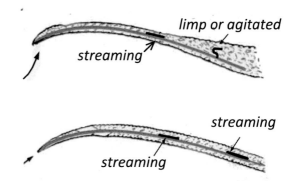

With both tufts streaming we have fully attached flow over the rear part of the section. In this case the profile drag is still low. This is the preferred situation when trimming for maximum L/D. When this cannot be realized for any sheeting angle it is necessary to reduce the amount of camber of the sail (see also Sect. 7.4). This situation can occur at (very) low Reynolds number that is (very) low wind speed.

In three-dimensional, separated flows there can be a substantial amount of cross-flow in the separation zone. This is the case in particular near a swept leading or trailing edge. The mechanism is illustrated in Fig. 7.11.4. Shown, schematically, is the leading edge area of a swept lifting surface or foil with a separation bubble. When the aspect ratio of the foil is sufficiently large, only the component V_n of the undisturbed flow V normal (perpendicular) to the leading edge will be subject to perturbation by the foil. Within the separation bubble the normal flow is almost stagnant ($V_n \approx 0$). Hence, only the tangential component V_s is retained within the bubble. A tuft positioned in the bubble will therefore tend to adopt the direction of V_s.

For a foresail this means that, as shown in Fig. 7.11.5, there remains a flow component towards the head of the sail inside a leeside leading edge separation bubble (if there is one). Tufts in this area will tend to point upwards towards the head of the sail when the cross-flow is sufficiently strong to overcome the gravitational pull.

For a mainsail with separated flow near the trailing edge it means that there remains a flow component towards the foot in the separated flow region.

Fig. 7.11.4 *Illustrating the cross-flow mechanism in a separated flow*

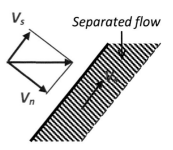

Fig. 7.11.5 Cross-flow in areas with separated flow on sails

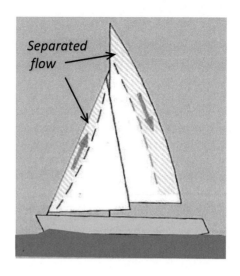

Figure 7.11.6 gives an example of strategically positioned tufts on a complete set of sails. On the jib/genoa Gentry tales are positioned at about 1/3 of the height (span) above the deck. Other sets of tufts, at 10, 50 and 90% of the local chord are positioned near the foot, at about half of the span and at about 80% of the span above the deck.

On the mainsail we have tufts at 50 and 90% of the local chord at 25, 50 and 75% of the span above the boom.

A tuft system like this allows to monitor flow separation over almost the entire sail area. It also provides information for the setting of the twist of the sails. The twist should be (near-)optimal when the tuft picture is the same in all sections of the sail.

Fig. 7.11.6 Example of strategic tuft positioning on a set of sails

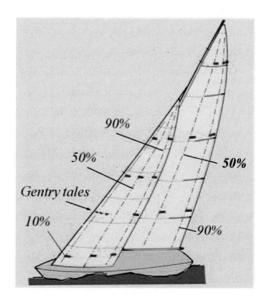

7.12   Sail Trim: Adjusting Angle of Attack, Twist and Camber

In this section we will indicate briefly how the angle of attack, twist and camber of a sail, of a Bermuda type of rig, can be adjusted or 'trimmed'. It is not about boat handling. That is, it is not about the most efficient way(s) to handle and trim the sails when beating upwind, reaching, running, tacking and gybing. There are several books (Dedekam 2006; Gladstone 2003), by experienced yachtsmen, that deal extensively with boat handling and sail handling in particular. The purpose of this section is merely to indicate, for those with no practical experience of sailing, the mechanical means that are available for adjusting the sails.

Mainsail
The angle of attack α of the mainsail, or any sail for that matter, is controlled by two quantities, the apparent heading β_a and the sheeting angle δ. (Recall that $\alpha = \beta_a - \delta$). The apparent heading is controlled by pointing the yacht through the rudder. The sheeting angle is controlled by means of the main sheet (see Fig. 7.12.1). The main sheet is a multiple line (rope) attached between blocks on the boom and the main sheet traveller. The traveller is a small car on a lateral rail spanning the cockpit. The sheeting angle can be increased by easing the sheet and/or moving the traveller to lee.

The sheeting angle is decreased by tightening the sheet and/or moving the traveller to the windward side of the boat.

The main sheet and the traveller are also instrumental in controlling the twist of the sail. An additional element for controlling the twist, in particular for large sheeting angles is the (boom) vang or downhaul. The vang (see Figs. 7.12.2 and 7.12.3) is a multiple line (or hydraulic cylinder on bigger yachts) between blocks on the boom and at the foot of the mast that can be adjusted to bring and keep the boom down. When the sheeting angle is small it is usually possible to control the amount

Fig. 7.12.1 *Means to control the sheeting angle of a mainsail*

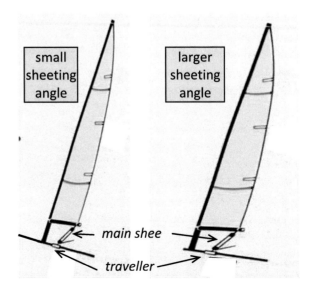

small sheeting angle

larger sheeting angle

← main shee

traveller

Fig. 7.12.2 *Controlling the twist of a mainsail*

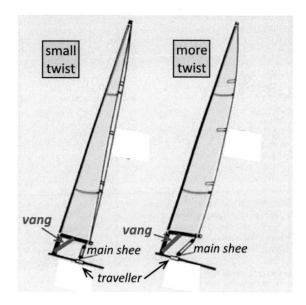

of twist through the sheet and traveller only. The twist is decreased by tightening the sheet while moving the traveller to lee. An increase of twist requires easing of the sheet while the traveller is moved to windward.

For large sheeting angles it is no longer possible to control the twist by the main sheet and traveller. Because of the limited range of the traveller the vang has now to be tightened to keep the boom down and reduce the twist (Fig. 7.12.3).

The camber of the mainsail is controlled by a number of other provisions. The amount of camber is controlled primarily by the outhaul and mast bending (Fig. 7.12.4). The outhaul is a line attached to the clew of the mainsail that stretches the foot of the sail when tightened. It controls the amount of camber in the lower parts of the sail in particular. Mast bending, through tightening of the backstay(s), flattens the sail in the upper reaches.

The chordwise position of maximum camber (or draft) can be varied through the tension of the luff of the sail. The latter can be controlled in two different ways: through the tension of the halyard and/or, if available, through the so-called

Fig. 7.12.3 *Use of the vang to keep the boom down and control twist at large sheeting angles*

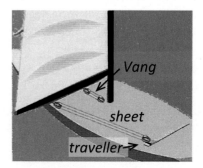

Fig. 7.12.4 *Camber control through outhaul and mast bending*

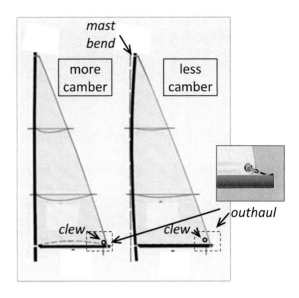

'cunningham'. The cunningham, (Fig. 7.12.5), is a line through an eye near the tack point of the sail that stretches the luff when tightened. Increasing luff tension causes the position of maxium camber (draft) to move forward. Additional aft camber can be induced by stretching the leech line. The latter is a thin line in the leech of the sail that can be tightened through a clamping cleat near the foot.

Genoa

The main means of control of a genoa are the sheet tension and the position of the lead, plus the tensions of the forestay and the luff of the sail. The genoa lead is a block, guiding the sheet, on a slide on a longitudinal rail (See Fig. 7.12.6). Some, bigger yachts have double, parallel rails. Genoa control is complex in the sense that variation of one parameter causes changes in several quantities. For example, easing the sheet for a fixed position of the lead block gives a larger sheeting angle but increases camber, in particular in the lower parts of the sail, and twist at the same time.

Increasing the tension of the sheet causes a smaller sheeting angle, less camber and less twist.

The longitudinal position of the lead, in combination with the proper amount of tension of the sheet, can be used to control the amount of camber, in particular in the lower parts of the sail. The amount of camber can be increased by moving the lead forward and decreased by positioning it more rearward (Fig. 7.12.7).

The longitudinal position of the lead is also the main parameter for controlling the twist of the genoa (Fig. 7.12.8). Positioning the lead to the rear provides opportunity for more twist. Moving it forward reduces the twist.

In principle, the sheeting angle can also be controlled through variation of the lateral position of the lead (Fig. 7.12.9). This offers better possibilities for controlling sheeting angle and camber independently. However, the possibilities are usually very limited. Multiple, parallel rails are an option on some yachts. Another possibility is

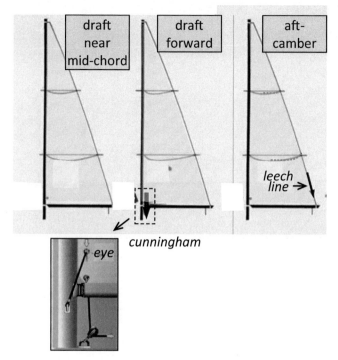

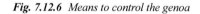

Fig. 7.12.5 *Control of chordwise position of maximum camber through luff and leech tension*

Fig. 7.12.6 *Means to control the genoa*

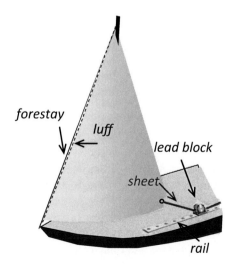

Fig. 7.12.7 *Control of camber in the lower parts of the sail through longitudinal positioning of the lead*

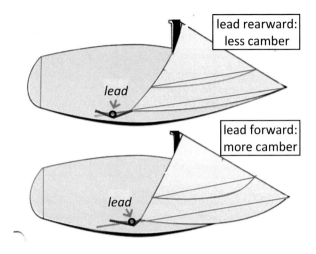

Fig. 7.12.8 *Control of genoa twist through longitudinal positioning of the lead*

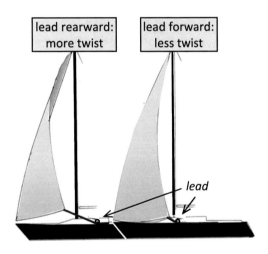

to install an additional block and tackle that pulls the clew of the genoa sideways. This is known as a 'barberhauler'. If a larger sheeting angle is required the block can be attached to the foot rail at the lee side of the boat. If a smaller sheeting angle is the objective, the block should be attached at some point to windward of the clew (the foot rail at the windward side is usually an option).

A barberhauler can be effective for apparent wind angles up to, say, 40°. For larger apparent wind angles and the associated larger sheeting angles it becomes increasingly difficult to avoid that the genoa adopts excessive camber. Setting the spinnaker pole to lee with the genoa clew attached to its far end can, in the author's experience, be a useful option in such conditions.

While, as described above, the amount of camber is controlled primarily by the tension of the sheet and the longitudinal position of the lead, the chordwise position of

Fig. 7.12.9 *Control of genoa sheeting angle through lateral positioning of the lead or a barberhauler*

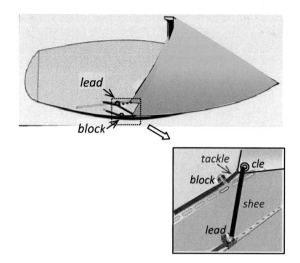

the maximum draft is governed by the tension of the forestay and the luff tension. The latter is controlled through the halyard. The tension of the forestay is controlled indirectly through the tension of the backstay(s). Increasing the tension of the backstay increases the tension of the forestay also. When luff and forestay are tight, the position of maximum camber (draft) is relatively forward, at least, when the sail is properly cut. With a slack luff and forestay the position of maximum draft is further to the rear (Fig. 7.12.10). A sagging forestay also increases the profile depth (amount of camber), in particular in the upper and middle reaches of the sail.

Spinnaker

The sheeting system of a spinnaker and its basic trimming mode (sheeting angle and horizontal pole angle) have already been described in Sect. 7.8 and do not need to be repeated here. Another factor of importance is the vertical position of the pole. This

Fig. 7.12.10 *Control of position of genoa maximum camber through luff/forestay tension*

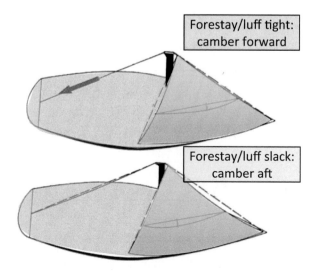

is the main quantity for controlling the camber of a spinnaker. A high position of the
pole causes more vertical camber but less horizontal camber in the upper part of the
spinnaker (Fig. 7.12.11). The reason for the latter is that luff and leech of the sail
come closer when stressed, which increases the draft. The other way around is that
a low position of the pole leads to less vertical camber and more horizontal camber.

The horizontal camber in the lower part of the sail can be controlled by the position of the lead of the spinnaker sheet. (Fig. 7.12.12). When the lead is aft, the lower

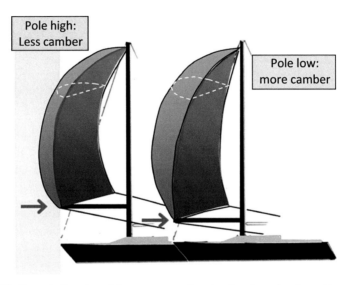

Fig. 7.12.11 *Control of camber of the upper part of a spinnaker through pole position*

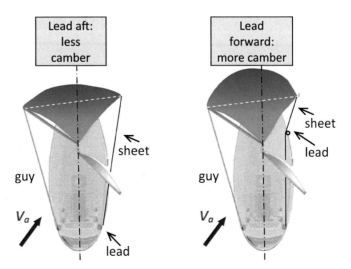

Fig. 7.12.12 *Controlling camber of the lower part of a spinnaker through positioning of the lead
of the sheet*

part of the spinnaker will be relatively flat, that is it will have a relatively small amount of camber. With the lead in a forward position the sail will acquire more horizontal camber.

7.13   Examples of Other Types of Rig

Although the vast majority of keel yachts is equipped with a Bermuda (sloop) type of rig there are some alternatives that are worth mentioning.

Gaff Rigs
In its original form (seventeenth century), the Bermuda rig had a gaffed mainsail, as in Fig. 7.13.1. In terms of aerodynamics the gaffed mainsail may have a slight advantage over a triangular sail in light weather in the sense of a little less induced drag for the same rig height. This, because the wider chords near the top of the sail will produce a span loading that is a little closer to the elliptical one. The gaff will, however, disturb the flow about the top of the mainsail in a similar way as the mast does for the larger part of the mainsail.

A variant is the gaff-topsail rig shown in Fig. 7.13.2.

Another, modern variant is formed by the square-topped mainsails sported, amongst others, by the IACC and VOR 70 class racing yachts (See Fig. 7.13.3 for a schematic picture).

AeroRig®
A further modern variant of the Bermuda rig is the so-called AeroRig (McDonald and Roberts 1996), reportedly first proposed by the British yacht designer Ian Howlett in the 1990s. An aerorig is characterised by a large, integral boom, attached to and extending for and aft of a, usually rotating, free standing mast (Fig. 7.13.4). The rig is fitted with a normal, reefable mainsail and a non-overlapping jib. The jib is self-tacking through a short, lateral sheeting rail mounted on the boom.

In terms of aerodynamics there are two main differences with a conventional rig. The first is that the coupled sheeting angles of fore- and mainsail are controlled by only one parameter, that is by rotating the complete boom. In addition there are limited possibilities, usually not instantaneous, to fix the sheeting angle of the self-tacking jib relative to the boom.

The second important difference is that the gap (or slot) between the leech of the jib and the luff of the mainsail is independent of the sheeting angle. In this way the gap/slot can be kept small to promote optimal aerodynamic interaction between the jib and the mainsail at all sheeting angles and apparent wind angles (as already discussed in Sect. 7.5, see Fig. 7.5.14).

The sheeting arrangement described above does not have much advantage for close hauled conditions but is superior for larger apparent wind angles, say $>40°$ (Fig. 7.13.5). For downwind courses (Fig. 7.13.6) the aerorig has the advantage that the foresail is not, or at least less, blanketed by the mainsail. This, however, is

Fig. 7.13.1 *Gaffed Bermuda rig*

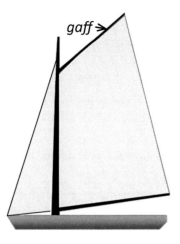

Fig. 7.13.2 *Gaff-topsail rig*

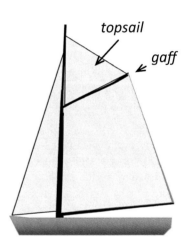

Fig. 7.13.3 *Modern sloop rig with square-topped mainsail (schematic)*

Fig. 7.13.4 *Large yacht with AeroRig.*
(McDonald and Roberts 1996)

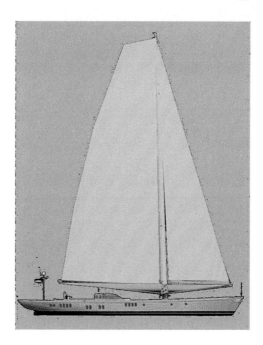

also the case, at least partly, when the foresail is set to weather on a spinnaker pole
(Fig. 7.5.14).

A disadvantage of the aerorig, at least in close-hauled conditions, is that it is not
possible to close the gap between the foot of the sail and the deck. For a given size
of the sail this means a reduction of the effective span and effective aspect ratio.

Fig. 7.13.5 *Aerorig, close reaching*

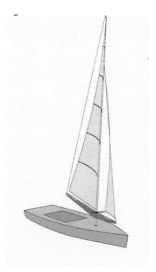

Fig. 7.13.6 *Aerorig, running*

Further disadvantages are the limited possibilities for setting the proper twist of the sails and for varying the difference between the sheeting angle of the mainsail and that of the jib. We have seen in Sect. 7.5 that it is desirable to have a small sheeting angle difference in close-hauled conditions when it is important to minimize the aerodynamic drag angle. For reaching conditions, when it is important to maximize the lift of the sail, it is desirable to be able to apply a large difference between the sheeting angle of the mainsail and that of the jib.

Another disadvantage is that the rig is relatively heavy in order to realize sufficient stiffness and strength.

An interesting variant of the aerorig could be one in which the front and rear parts of the boom are hinged separately around the mast. It would then be possible to vary the sheeting angle difference while maintaining the proper (small) gap/slot width between the leech of the jib and the luff of the mainsail (see Fig. 7.13.7). A disadvantage of such a segmented boom would be, of course, that strength and stiffness requirements would lead to more weight.

Fig. 7.13.7 *Aerorig with segmented boom: better options for varying the sheeting angles of jib and mainsail (conceptual)*

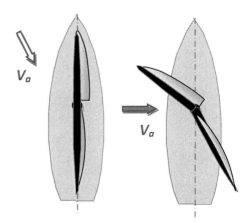

Fig. 7.13.8 *'Balanced Rig™'. (http://www.balancedrig.com/)*

The author has stimulated application of the concept in model sailing boats, with success, but is not aware of any full-scale application.

Balanced Rig™
The so-called Balanced Rig (http://www.balancedrig.com/) can be considered as another variant of the aerorig. As such, it shares most of the advantages and drawbacks of the aerorig. The balanced rig is characterized by a boom at the foot as well as the top of the sails (Fig. 7.13.8). These are shaped so as to act as 'end-plates'. The 'end-plates' are claimed to reduce the induced drag. This may be the case to some limited extent (at least when the plates do not have holes for weight reduction as in the figure). A more significant reduction of the induced drag can be expected when the 'end-plates' are shaped so as to act like winglets (see Sect. 5.15).

Of more importance for the performance of the rig is probably that the taper ration of the sails is about 0.6. This is probably not far from the optimum for light weather conditions (Sect. 7.4). Since both the foresail as well as the main sail have a finite, non-zero chord length at the top, an additional advantage is that the, profitable, foresail-mainsail aerodynamic interaction (see Sect. 7.5) extends over the full length of the sail. The latter is, of course not the case for rigs with a triangular sail planform, fractional rigs in particular.

DynaRig™
Another option, for (very) large yachts, is the so-called 'DynaRig', the concept of which was first proposed in the 1960s by the German engineer Wilhelm Prölls. At first sight it would seem (see Fig. 7.13.9) that the dynarig can be considered as a modern version of the classical square rig of the famous tea-clippers of the nineteenth century. There are, however, significant differences, not only in hardware but also in terms of aerodynamics. The most important differences in terms of aerodynamics are the horizontally curved yards and the absence of slots between the sails and the yards. The curved yards allow the sails to have camber in the proper

Fig. 7.13.9 *'Super' yacht with DynaRig: Maltese Falcon. (http://www.symaltesefalcon.com/)*

direction. This, together with the possibility to vary the sheeting angle by rotating the mast(s) and the rigidly attached yards, creates a rig with a high aspect ratio that is efficient at most points of sail, from close-hauled to broad reaching.

For dynarigs with multiple masts, the dead run is probably the only condition where the rig scores poorly, because of the blanketing of the foresail(s) by the rear sail. Another disadvantage, for any number of masts, is that, due to the fixed yards, the twist of the sails cannot be varied.

Reefing is possible by furling the (top)sails into the mast.

A final remark to be made for multiple mast configurations is that the downwash generated by the leading sail(s) and the upwash generated by the following sail(s) requires different sheeting angles, decreasing from front to rear, for the individual rigs.

Wing Sail

Although hard wing sails have been used before, it would seem, since the 2010 edition of the America's Cup, that the ultimate racing rig is one with a hinged and articulated wing sail as sported by the Oracle multi-hull (http://en.wikipedia.org/wiki/USA_17_(yacht)). For ocean-going and cruising yachts this is not an option because of the absence of a possibility for reefing. The success of the AC-type wing sail rig is due to its extremely high aerodynamic efficiency (L/D). For very fast yachts, such as cat/trimarans (or ice yachts) this can be utilized at almost all true wind angles because very fast yachts sail almost always at small apparent wind angles (Sect. 3.3). The high L/D is realized by a very high aspect ratio (about 7), together with an efficient, two-element, single-slotted high-lift airfoil system. (Fig. 7.13.10). The airfoil system comprises symmetrical front and rear sections.

Fig. 7.13.10 *AC45 Class wing sail. (http://en.wikipedia.org/wiki/USA_17_(yacht))*

The rigid front section, with integral mast, can be rotated as a whole. The rear or flap section can be deflected relative to the front section, to port as well as to starboard, up to something like 35°. The slot between the sections is narrow, as between an aircraft wing and its flap when the latter is deflected.

The front section is relatively thick, with a large leading edge radius for high maximum lift. The rear section is thinner and with a smaller leading edge radius (Fig. 7.13.11). When deflected, it rotates, by means of external brackets, about a hinge point ahead of the trailing edge of the front section. This in such a way that the trailing edge of the front section is always close to the position of maximum suction near the leading edge of the flap. In this way it is possible to realize almost optimal aerodynamic interaction, as in aircraft high-lift systems (Smith 1975), at all points of sailing.

When a high lift/drag ratio is required, the flap angle is set at a relatively small value (10–15°). When maximum lift is the objective, the flap angle is set close to its maximum of about 30°.

Fig. 7.13.11 *Wing sail airfoil section with flap*

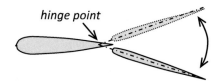

hinge point

Estimating the maximum lift coefficient of the segmented section at about 2.5,[22] suggests that this is about as large as that of a single, highly cambered, thin sail section without mast (or with a thin wing mast), see Fig. 7.4.12. The major advantage is, however, in the reduction of drag and, hence, a much better maximum lift/drag ratio, through the absence of flow separation and the high aspect ratio.

A disadvantage of hard wing sail rigs, apart from the price and the reefing/stowing problem, is that the possibilities for adapting the twist are modest. With the front part of the wing sail rigid (although rotatable) the only possibility for introducing some form of geometrical twist is by variation, through segmentation, of the flap deflection in the spanwise direction.

Although wing sails have much to offer for very fast yachts like catamarans and trimarans, it is questionable whether this holds also for, slower, keel yachts. The reason is that the latter will have to sail with (much) larger apparent wind angles for a considerable part of the time. In such conditions it is maximum lift (and drag) rather than the lift/drag ratio (L/D) that governs boat speed.

References

Abbott IH, Von Doenhoff AE (1949) Theory of wing sections. McGraw Hill, New York

Bethwaite F (2010) High performance sailing, 2nd edn. Adlard Coles Nautical, London, p 194, pp 209–215, 323–324

Chapin VG, Jamme S, Neyhousser R (2004) Aérodynamique grand voile—mât', 2nd workshop Ecole Navale

Chapin VG et al (2005) Sailing yacht rig improvements through viscous computational fluid dynamics. The 17th Chesapeake sailing yacht symposium, Anapolis

Chapin VG, Neyhousser R, Dulliand G, Chassaing P (2006) Analysis, design and optimization of Navier-Stokes flows around interacting sails. MDY06 international symposium on yacht design and production, Madrid, 30–31 March 2006

Chapin VG et al (2008) Design optimization of interacting sails through viscous CFD. International conference on innovation in high performance sailing yachts, Lorient, RINA, May 2008

Claughton AR, Shenoi RE, Wellicom JF (eds) (1999) Sailing yacht design: theory (Chapter 7). Prentice Hall, New York. ISBN 0582368561

Claughton A, Fossati F, Battistin D, Muggiasca S (2008) Changes and development to sail aerodynamics in the ORC international handicap rule. 20th international HISWA symposium on yacht design and yacht construction

Collie SJ, Jackson PS, Gerritsen M, Fallow JB (2004) Two-dimensional CFD-based parametric analysis of downwind sail designs. RINA, London

Dedekam I (2006) Illustrated sail & rig tuning. Dedekam Design, Oslo. ISBN 978 18 9866 067 5

Eiffel G (1910) La resistance de l'air: examen des formules et des experiences. H. Dunod et E. Pinat, Paris

Flay RGJ, Vuletich IJ (1995) Development of a wind tunnel test facility for yacht aerodynamic studies. J Wind Eng Aerodyn 58:231–258

Fossati F (2009) Aero-hydrodynamics and the performance of sailing yachts. International Marine/ McGraw-Hill, New York, p 251. ISBN 978-0-07-162910-2

Fossati F, Muggiasca S (2008) Influence of heel on yacht sailplan performance. 6th international conference on high-performance marine vehicles, HIPER 08, Naples, Sept 2008

---

[22] Based on data from Abbott and Von Doenhoff (1949).

Fossati F, Muggiasca S (2011) Experimental investigation of sail aerodynamic behavior in dynamic conditions. J Sailboat Technol, Article 2011-02. ISSN 1548-6559

Fossati F, Muggiasca S, Viola IM (2006) An investigation of aerodynamic force modelling for IMS rule using wind tunnel techniques, HISWA

Fossati F, Muggiasca S, Martina F (2008) Experimental database of sails performance and flying shapes in upwind conditions. International conference on innovation in high performance sailing yachts, Lorient, RINA

Gentry A (1971) The aerodynamics of sail interaction. Paper presented at the 'Ancient Interface III', 3rd AIAA symposium on sailing, Redondo Beach, Nov 1971

Gentry AE (1976) Studies of mast section aerodynamics. Presented at The Ancient Interface VII, 7th AIAA symposium on sailing, Long Beach, Jan 1976

Gentry AE (1981a) Sailboat performance testing techniques. The Ancient Interface XI, 11th AIAA symposium on sailing, 12 Sept 1981, Seattle

Gentry AE (1981b) A review of modern sail theory. The Ancient Interface, 11th AIAA symposium of sailing, 12 Sept 1981, Seattle

Gerhardt FC (2010) Unsteady aerodynamics of upwind sailing and tacking. Thesis, University of Auckland

Gladstone B (2003) North U. Trim. North U, Madison. ISBN 0-9724361-1-1

Goovaerts FEG (2000) Development of a high power sail with the aid of gurney flaps. Master's thesis Delft University of Technology, Laboratory of Aero- and Hydrodynamics, Sub-faculty of Applied Physics, Rept. No. MEAH-199

Haddad B, Lepine B (2003) Etude expérimentale de l'interaction mât-voile. Rapport de stage de fin d'étude, Ecole Navale

Hansen H, Jackson P, Hochkirch P (2002) Comparison of wind tunnel and full-scale aerodynamic sail force measurements. High performance yacht design conference Auckland, 4–6 Dec 2002

Hansen H, Richards P, Jackson P (2006) An investigation of aerodynamic force modeling for yacht sails using wind tunnel techniques. 2nd high performance yacht design conference Auckland

Hoerner SF (1965) Fluid dynamic drag. Hoerner Fluid Dynamics, Bakersfield

http://www.wb-sails.fi. Accessed 2008

http://en.wikipedia.org/wiki/Calculus_of_variations. Accessed 2012

http://www.balancedrig.com/. Accessed 2014

http://www.symaltesefalcon.com/. Accessed 2013

http://en.wikipedia.org/wiki/USA_17_(yacht). Accessed 2012

Jones RT (1950) The spanwise distribution of lift for minimum induced drag of wings having a given lift and a given bending moment, NACA TN 2249

Larsson L, Eliasson RE (1996) Principles of yacht design. Adlard Coles Nautical, London, pp 141–142. ISBN 0-7136-3855-9

Lasher WC et al (2003) Experimental force coefficients for a parametric series of spinnakers. The 16th Chesapeake sailing yacht symposium, Annapolis, March 2003

Liebeck RH (1973) A class of airfoils designed for high lift in incompressible flow. J Aircr 10(10) :610–617

Lissaman PBS (1978) Lift in a sheared flow. Proceedings of 16th AIAA/SNAME symposium on the aero and hydrodynamics of sailing

Marchai CA (1964) Sailing theory and practice. Adlard Coles Ltd, Southampton

Marchai CA (2000) Aero-hydrodynamics of sailing. Adlard Coles Nautical, London, p 240, 444,553, 587, 652, pp 302–325, 328–338, 510–516, 604–606, 658–675

Marchai CA, Tanner T (1963) Wind tunnel tests on a ¼ scale dragon rig. Southampton University Yacht Research, Rep. 14

McDonald N, Roberts D (1996) The AeroRig®—rig of the future. Paper presented at HISWA symposium, Amsterdam

Milgram JH (1971) Section data for thin, highly cambered airfoils in incompressible flow, NASA CR-1767

Milgram JH (1978) Effects of masts on the aerodynamics of sail sections. Mar Technol 15(1): 35–42

Munk MM (1923) General bi-plane theory, NACA Report 151

Olejniczak J, Lyrintzis AS (1994) Design of optimized airfoils in subcritical flow. J Aircr 31(3):680–687

Paton JS, Morvan HP (2007) The effect of mast rotation and shape on the performance of sails. Int J Marit Eng 149

Richards P, Lasher W (2008) Wind tunnel and CFD modelling of pressures of downwind sails. BBAA VI international colloquium on bluff bodies aerodynamics & applications, Milano, Italy

Richards P et al (2006) The use of independent supports and semi-rigid sails in wind tunnel studies. 2nd high performance yacht design conference, Auckland (N-Zl)

Simiu E, Scanlan RH (1986) Wind effects on structures, 2nd edn. Wiley, New York

Smith AMO (1975) High-lift aerodynamics, 37th Wright brothers lecture. J Aircr 12(6):501–530

Sparenberg JA, Wiersma KA (1976) On the maximum thrust of sails by sailing close to wind. J Ship Res 20(2):98–106

Speer T (2010) Minimum induced drag of sail rigs and hydrofoils. http://www.tspeer.com. Accessed 2011

Sugimoto T (1992) A first course in optimum design of yacht sails. 11th Australasian fluid mechanics conference, University of Tasmania, Hobart, Australia, 14–18 Dec 1992

Van Schaftinghen N (1999) Experimental investigation and optimisation of asymmetric spinnakers. Thesis, Von Karman Institute for Fluid Dynamics

Wiersma AK (1977) On the maximum thrust of a yacht by sailing close to wind. J Eng Math 11(2):145–160

Wilkinson S (1984) Partially separated flow around 2D masts and sails. PhD Thesis, University of Southampton

Chapter 8
Encore: Sailing; Aerodynamics Plus Hydromechanics

8.1 The Problem of Optimizing Boat Performance

Introduction

We have seen already in Chap. 4, that the determination and realization of optimal boat performance (expressed in terms of so-called dependent variables like boat speed, VMG, etc.) is quite a complex affair. The essence of the problem is that boat performance for a given yacht depends on a large number (about 25) of independent variables. Realization of the combination of adjustments/values of these 25 variables that maximizes boat performance, is far from an easy matter. About six of the independent variables are external variables, determined by wind and waves, and cannot be influenced by the crew. The other 20 or so are variables that can be manipulated by the crew. They are, sometimes, called decision variables. In principle, these decision variables are independent of each other. Determining what values they should have, for given values of the six external variables, is one part of the problem. Realizing these values through proper trimming of the boat is the other part.

Table 8.1, presented in the next two pages, provides an overview of the variables. It is useful to note that rudder deflection is not considered to be an independent variable. The equilibrium of yawing moments determines the amount of helm that is needed to balance a yacht in yaw.

Establishing what the optimal values of the independent, adjustable variables are for a given set of external variables, can, in principle, be done in two different ways. Measurements of boat performance for different trim settings through sailing trials is one way. The other way is through the application of a so-called Velocity Prediction Program (VPP), about which later.

© Springer International Publishing Switzerland 2015
J. W. Slooff, *The Aero- and Hydromechanics of Keel Yachts,*
DOI 10.1007/978-3-319-13275-4_8

Table 8.1 *Summary of the main variables of sailing*

External independent variables

true wind	speed	
	direction	
	profile/gradient	
waves (spectrum)	direction	
	length	
	height	
total number	6	

Dependent variables

Dependent variables	determined by	dependent of	**Independent variables** can be influenced through	# of independent variables
above the water surface	aerodynamic forces and moments	external variables (wind)	–	–
	equilibrium of forces and moments	heel/pitch/yaw		
	apparent wind speed/angle	true wind speed/angle, boat speed	steering	1
	sail geometry and trim			
		size/type foresail	choice (by crew)	1
		size/type mainsail	choice/reefing (by crew)	1
		genoa sheeting angle	position of lead	1
			sheet tension	1
		mainsail sheeting angle	(barberhauler)	(2)
			position traveller	1
			sheet tension	1
		genoa twist	position lead	1
		mainsail twist	sheet tension	
			vang	1
		genoa camber	sheet tension	1
			leech line	1
			forestay tension	1
			halyard tension	1
		mainsail camber	outhaul	1
			leech line	1
			halyard tension/cunningham	1
			backstay(s) tension	1
		parasatic drag	position/exposure crew	1
			Σ	16

Table 8.1 (continued)

below the water surface	hydromechanic forces and moments	equilibrium of forces and moments	heel/pitch/yaw		
	frictional resistance	equilibrium of forces	surface roughness	sanding/polishing	1
	induced resistance			(trimtab deflection)	(1)
	wave-making resistance				
	added resistance in waves	external variables / sea state	mass moments of inertia	longitudinal position of movable mass	1
			relative wave direction	steering	
	resistance due to side force		rudder deflection	sail trim	
					Σ 2
common					
	heel angle	equilibrium of forces		lateral position of movable mass	1
	pitch angle	equilibrium of forces		longitudinal position of movable mass	
	leeway/yaw angle	equilibrium of forces			
					Σ 1
					ΣΣ 19

Measuring Boat Performance Through Sailing Trials

Measuring boat performance through sailing trials ('building experience') is, of course, the classical and ultimate way to achieve a useful picture of the performance characteristics of a sailing yacht. Unfortunately this is also quite a complex and time consuming affair.

One part of the problem is that there is no control over the external variables. Because of the continuously varying conditions of wind and water and the related response to this of the helmsman, it is impossible to perform measurements of even only two different settings of a single adjustable variable under identical conditions of wind and waves. This means that the results of the measurements cannot be compared properly, at least not directly. Comparing measured data is useless, unless the data have been corrected in some way for the differences in the conditions of wind and waves. In general, such corrections require some form of interpolation between data obtained at a sufficiently large number of different external conditions.

The other, and probably the greater part of the problem, is that the number of independent variables to be considered is of the order of 20. Assuming three values per variable gives 3^{20} combinations, requiring an impossibly large number of measurements. Even when it is realized that not all arbitrary combinations of values of these variables make sense, or are feasible, the number is still impossibly large. In practice, the number of variables is therefore limited to a small number (5–10) of the most important ones, with 'default' values, based on experience, of the other variables.

The large number of variables also requires extensive instrumentation and registration of data. Furthermore, statistical and curve-fitting techniques have to be applied to the data base in order to extract useful information.

Needless to say that determining the optimal performance of a sailing yacht through sailing trials is very time consuming. Sailing trials with two identical yachts in close proximity can help to reduce the total amount of time needed. Not in the least because the external conditions of the boats can be assumed to be identical when they are sailing in close proximity.

Velocity Prediction Programs (VPP)

As already mentioned, another way of determining (optimal) boat performance is through the application of a so-called Velocity Prediction Program (VPP). A Velocity Prediction Program is a computer program based on a mathematical model of the mechanical properties of a yacht. A 'VPP' usually consists of several modules (See Fig. 8.1.1).

The core of the model is formed by the equations describing the equilibrium of forces and moments as discussed in Sect. 4.1. For convenience we recall these as follows:

- Equilibrium of forces in the direction of motion

$$T = R \tag{8.1.1},$$

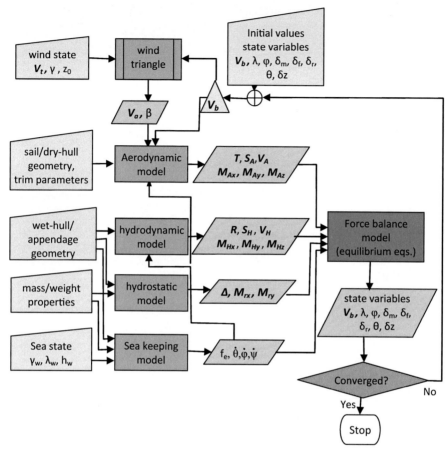

Fig. 8.1.1 *Flow diagram of a velocity prediction program (generic)*

where T is the (aerodynamic) driving force and R is the hydrodynamic resistance

- Equilibrium of forces in the lateral direction (side force)

$$S_A = S_H \qquad (8.1.2),$$

where S_A and S_H are the aerodynamic and hydrodynamic side force, respectively.

As an alternative to Eq. (8.1.2) one can use Eq. (4.2.4), expressing the directional balance of forces (Lanchester's Law, Sect. 4.2).

- Equilibrium of forces in the vertical direction

$$V_H + \Delta = V_A + \nabla \qquad (8.1.3),$$

where V_A and V_H are the aerodynamic and hydrodynamic vertical force, respectively, $V_H + \Delta = V_A + V$ the weight of the yacht and Δ the hydrostatic buoyancy force.

- Equilibrium of moments around the longitudinal (x-) axis

$$M_{Hx} + M_{Ax} = M_{rx} \qquad (8.1.4),$$

where M_{Ax} and M_{Hx} are the aerodynamic and hydrodynamic rolling moments, respectively, and M_{rx} is the (hydrostatic) righting moment.

- Equilibrium of moments around the lateral (y-) axis

$$M_{Ay} + M_{Hy} = M_{ry} \qquad (8.1.5),$$

where M_{Ay} and M_{Hy} are the aerodynamic and hydrodynamic pitching moments, respectively, and M_{ry} is the longitudinal (hydrostatic) righting moment.

- Equilibrium of moments around the vertical (z-) axis

$$M_{Az} = M_{Hz} \qquad (8.1.6),$$

where M_{Az} and M_{Hz} are the aerodynamic and hydrodynamic yawing moments, respectively.

The equilibrium equations given above are supplemented with the relations between true wind speed, boat speed, apparent wind speed, apparent wind angle and true wind angle that can be derived from the wind triangle (Sect. 3.3).

We have seen in Chap. 7 that the aerodynamic forces and moments are a function of the geometry of the sail(s), the apparent wind speed, the apparent heading and the various trimming parameters like sheeting angles, etc. This means that, for a given type of yacht, the aerodynamic forces can be expressed as

$$T = \tfrac{1}{2}\rho_A V_a^2 S_s C_T = \tfrac{1}{2}\rho_A V_a^2 S_s (C_L \sin\beta - C_D \cos\beta) \qquad (8.1.7a)$$

$$S_A = \tfrac{1}{2}\rho_A V_a^2 S_s C_{SA} = \tfrac{1}{2}\rho_A V_a^2 S_s (C_L \cos\beta + C_D \sin\beta) \qquad (8.1.7b)$$

$$V_A = \tfrac{1}{2}\rho_A V_a^2 S_s C_{VA} \qquad (8.1.7c),$$

where S_s is the sail area and the apparent wind speed

$$V_a = V_a(V_t, V_b, \beta)$$

is a function of the true wind speed V_t, boat speed V_b and the apparent wind angle β or the true wind angle γ (see Eqs. (3.3.3) and (3.3.4)).

For a given configuration of sails and hull, the lift and drag coefficients C_L and C_D and the vertical force coefficient C_{VA} must be known as a function of the apparent heading β_a $(=\beta-\lambda)$, the angle of attack α $(=\beta_a-\delta)$ and the angles of heel and pitch φ and θ, respectively.[1] In principle lift and drag also depend, though weakly, on Reynolds number and, hence, on the apparent wind speed V_a. i.e.:

$$C_L = C_L(\beta_a, \; \delta, \; \varphi, \; \theta, \; (V_a)) \qquad (8.1.8a)$$

and

$$C_D = C_D(\beta_a, \; \delta, \; \varphi, \; \theta, \; (V_a)) \qquad (8.1.8b)$$

A similar kind of relation is also required for the vertical force aerodynamic coefficient C_{VA}. Because of the relatively strong dependence of the vertical aerodynamic force on the flow about the hull (see Sect. 7.6) the dependence of

$$C_{VA} = C_{VA}(\beta_a, \; \delta, \; \varphi, \; \theta, \; (V_a)) \qquad (8.1.8c)$$

will, in general, be of a somewhat different form.

Relationships of the kind as indicated above are also required for the aerodynamic moments and their dimensionless coefficients. For small pitch angles and small angles of leeway they are of the form (see Sect. 7.7):

$$M_{Ax} = S_A \zeta_{CE} / \cos \varphi \qquad (8.1.9a)$$

$$M_{Ay} = T(\eta_{CE} \sin \varphi - \zeta_{CE} \cos \varphi) - V_A \xi_{CE} \qquad (8.1.9b)$$

$$M_{Az} = -S_A \xi_{CE} - T(\eta_{CE} \cos \varphi + \zeta_{CE} \sin \varphi) \qquad (8.1.9c),$$

with T, S_A and V_A given by Eqs. (8.1.7) and (8.1.8). The position $(\xi_{CE}, \eta_{CE}, \zeta_{CE})$ of the centre of effort in the ship coordinate system for a given sail configuration is a function of the sheeting angles δ and the apparent heading β_a, as also indicated in Sect. 7.7.

The relationships given above form what is called the aerodynamic model of the yacht.

Similar relationships are also required for the hydrodynamic forces and moments. For the hydrodynamic forces this can be summarized as follows (see Chap. 6):

$$R = \tfrac{1}{2} \rho_H V_b^2 S_H C_R \qquad (8.1.10a)$$

---

[1] For multiple sails there will, in general, be a sheeting angle δ and angle of attack α for each individual sail.

$$S_H = \tfrac{1}{2}\rho_A V_a{}^2 S_H C_{SH} \tag{8.1.10b}$$

$$V_H = \tfrac{1}{2}\rho_A V_a{}^2 S_H C_{VH} \tag{8.1.10c},$$

For a given yacht, the dimensionless coefficients C_R, C_{SH}, and C_{VH} are functions of the angle of leeway λ, the heel and pitch angles φ and θ and the deflection angle δ_r of the rudder.[2] To a lesser extent they are also a function of boat speed V_b, through Reynolds number. Symbolically this can be expressed as

$$C_R = C_R(\lambda,\ \delta_r,\ \varphi,\ \theta, V_b) \tag{8.1.11a}$$

$$C_{SH} = C_{SH}(\lambda,\ \delta_r,\ \varphi,\ \theta, V_b) \tag{8.1.11b}$$

$$C_{VH} = C_{VH}(\lambda,\ \delta_r,\ \varphi,\ \theta, V_b) \tag{8.1.11c}$$

The hydrodynamic moments can be expressed as

$$M_{Hx} = S_H \zeta_{CLR}/\cos\varphi \tag{8.1.12a}$$

$$M_{Hy} = R(\eta_{CLR}\sin\varphi - \zeta_{CLR}\cos\varphi) - V_A \xi_{CLR} \tag{8.1.12b}$$

$$M_{Hz} = -S_H \xi_{CLR} - R(\eta_{CLR}\cos\varphi + \zeta_{CLR}\sin\varphi) \tag{8.1.12c},$$

The hydrostatic forces and moments are those described in Sect. 6.2.

Most VPP's seem to be inspired by the requirement for a 'fair' rating system, such as the IMS and ORC rating systems (International Measurement System 2011; ORC Rating Systems 2011 in Chap. 2), for races between yachts of different types (ORC-VPP Documentation 2012). For this purpose they solve only a sub-set of the equilibrium Eqs. (8.1.1) to (8.1.6). The minimum requirement seems to be equilibrium of the forces in the longitudinal ('thrust and drag') and lateral (heeling force) directions and equilibrium of moments around the longitudinal axis (Claughton et al. 1999). In this case, the equilibrium equations are solved for a reduced set of state variables consisting of boat speed V_b and the heel angle φ. These are determined for given hydrostatic, hydrodynamic and aerodynamic models of the yacht, a given true wind speed V_t and a given true wind angle γ.

The aerodynamic models (Kerwin 1978; Hazen 1980) are usually simplified in the sense that, apart from the sail area, the span of the sails and the vertical position of the centre of effort, they contain the envelopes of the maximum lift and 'viscous' or parasitic drag coefficients of the individual sails as a function of the apparent

---

[2] Additional variables are involved for yachts with a trim tab, water ballast or a canting keel.

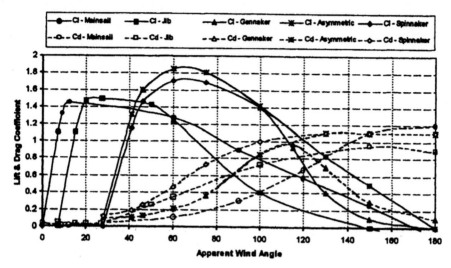

Fig. 8.1.2 *Example of sail force coefficient envelopes for several types of individual sails. (From IMS VPP, Ref. Claughton et al. (1999))*

wind angle. Figure 8.1.2 gives an example for different types of individual sails. In general, this is supplemented with a model for the windage of the hull topsides, rigging, etc.

The drag due to lift is modeled through the introduction of an effective span or effective aspect ratio (see Sects. 7.4 and 7.5). The effect of heel is modeled by relating the envelope curves for lift and drag to the angle between the apparent wind vector and the heeled plane through the mast rather than the apparent wind angle in the horizontal plane (see Sect. 7.9). In case of multiple sails the interaction in terms of lift is modeled through 'blanketing' factors that weigh the contributions of the individual sails. The induced drag (drag-due-to-lift) of multiple sails is calculated according to bi-plane theory (Munk 1923). Sail trim modeling options are usually limited to factors ('reef' and 'flat') (Hazen 1980) that adapt sail area and the amount of camber.

The hydrodynamic force model usually contains viscous and 'residuary' resistance plus 'resistance due to heel' and 'resistance due to heeling force' (see Sect. 6.5).

In terms of Eqs. (8.1.8)/(8.1.11) this means that the drag coefficient C_D of the sails is expressed as a function of the lift coefficient C_L rather than as in Eq. (8.1.8b) and the hydrodynamic resistance coefficient C_R as a function of the heeling force coefficient C_H or the side force coefficient C_S rather than as Eq. (8.1.11a).

Note that the simplified models just mentioned do not explicitly contain dependence of forces on angle of attack, i.e., explicit dependence on apparent heading, sheeting angle and angle of leeway is not modeled. For this reason the equations are solved with the additional condition that boat speed is maximized for the given true wind angle. It is this additional condition that determines the level of lift of the sails (within the given envelope of lift as a function of the apparent wind angle) and the corresponding side force on hull and appendages. With this procedure, explicit knowledge of angles of attack or sheeting angles is not required.

A weak point of the procedure is that the effect of leeway on the aerodynamics of the sails and the hull is not, or not fully, accounted for. This applies in particular to the direct and indirect aerodynamic effects of the hull, which are a function of the apparent heading rather than the apparent wind angle. The difference will, however, only be significant when the angle of leeway is large, that is in very close-hauled conditions.

The philosophy behind the class of VPPs that follow the procedure sketched above, is that they should provide a fair comparison of what is achievable, in terms of performance, for different types of yachts. This determines the rating of a yacht. It is, however, up to the crew to approach the performance potential as closely as possible.

There are other examples of VPP mentioned in the literature (Van Oossanen 1993) that do contain explicit modeling of the dependence of the aero- and hydrodynamic forces on the angle of leeway. VPPs of this category are, at least in principle, better suited for simulation of sailing performance. The most recent developments are aiming at time-accurate simulation (Proceedings 2012).

A VPP developed by the author, to be applied later in this chapter, handles 4°–6° of freedom, that is

* Directional equilibrium of forces in the horizontal plane (Lanchester's Law)
* Equilibrium of aero- and hydrodynamic forces in the direction of sailing
* Equilibrium of heeling moments
* Directional balance of aero- and hydrodynamic yawing moments,

plus, optionally, equilibrium in pitch and vertical forces.

The aerodynamic and hydrodynamic models are based on the descriptions given in Chaps. 6 and 7, with distinction between induced drag and viscous drag due to lift.

The program solves for the maximum boat speed for a given apparent wind angle, true wind speed and sea state (average wave period and significant wave height). The dependent variables are boat speed V_b, leeway angle λ, heel angle φ, true wind angle γ, sail sheeting angles (δ_m, δ_f) and rudder deflection δ_r.

8.2 Upwind Performance, Dependence on Shapes of Hull and Appendages

Introduction
Although a VPP type of program is required for a full evaluation of the aero-hydrodynamic interaction for a given type of sailing yacht, some general trends can

be identified more clearly through simpler models. In this section we will do so for several of the aspects of the mechanics of sailing discussed in Chap. 4.

As we have seen in Sect. 4.5, determination of the maximum velocity-made-good to windward:

$$V_{mgw} = V_b \cos \gamma \qquad (8.2.1)$$

is a rather complex affair. For a given true wind speed, it requires, see Fig. 4.5.2, a small true wind angle γ and a high boat speed V_b. As shown by Fig. 4.5.3 this implies also a small apparent wind angle β.

   The requirements of a high boat speed and a small apparent wind angle form the simple part of the problem. The complexity comes in through the (non-linear) dependence of V_b on γ or β (V_b decreases rapidly when the apparent wind angle β becomes smaller than about 30°), plus the fact that there is a minimum value attached to the apparent wind angle β. The latter was seen to be a consequence of Lanchester's Law (Sect. 4.2) which states that the apparent wind angle is equal to the sum of the aerodynamic drag angle ε_A of the sails (plus hull top sides, etc.) and the hydrodynamic drag angle ε_H of the under-water part of the hull plus appendages like keel and rudder. We have also seen, that, as a direct consequence of Lanchester's Law, the directional equilibrium or balance of aerodynamic and hydrodynamic forces can, for zero heel, be written as

$$\varepsilon_A + \varepsilon_H = \beta = \beta_a + \lambda = \alpha + \delta + \lambda \qquad (8.2.2)$$

Here, as before, β_a is the apparent heading, λ the angle of leeway, α the angle of attack of the sail(s) and δ the sheeting angle. For non-zero angles of heel the geometrical relation $\beta_a = \alpha + \delta$ is to be replaced by

$$\beta_a \approx \arctan(\tan(\alpha + \delta) / \cos \varphi) \qquad (8.2.3),$$

see Sect. 7.9.

Minimizing the Apparent Wind Angle
Realizing a low value of the apparent wind angle in equilibrium conditions means that ε_A and ε_H must be small at the same time. In the normal practice of sailing ε_A can, for a given type of rig, be minimized through proper setting of the angle of attack of the sails, that is proper setting of the apparent heading and the sheeting angle (recall that, at zero heel, $\alpha = \beta_a - \delta$). However there is no direct control over ε_H. The hydrodynamic drag angle settles at a value that results from the equilibrium of aero- and hydrodynamic forces.

   We have seen in Sect. 4.4 that this process can be illustrated through a β-λ diagram, as illustrated, qualitatively, by Figs. 4.4.1-3. There we saw already that higher wind speeds require larger sheeting angles when the objective is to minimize the apparent wind angle β. The reason was seen to be the increase of ε_H with boat speed V_b due to wave-making resistance and the increase of both ε_A and ε_H with heel (recall that both boat speed and heel increase with wind speed).

We are now (after consulting Chaps. 6 and 7) in a position to consider the mechanisms involved with minimizing the apparent wind angle in some further detail. First of all we note that any increase of either ε_A and/or ε_H due to an increase of aerodynamic drag or hydrodynamic resistance requires larger sheeting angles when sailing to windward. This is, for example, also the case when there is a lot of added resistance due to waves in a seaway. The mechanism is, of course, the same as that illustrated by Fig. 4.4.3.

It is also useful to consider the role of the hydrodynamic characteristics of the underwater body on the outcome of the equilibrium process. For this purpose we consider β-λ diagrams of hypothetical yachts with identical rigs but different hull-appendage configurations. The rig has been assumed to have a minimum drag angle of 12.5°. This is a representative value for a yacht with a rig of moderate aspect ratio (about 3.5) at a heel angle of 20° (see Sects. 7.5, 7.6 and 7.9). The centre of effort has been assumed at 36% of the length of the waterline. The hull-appendage configurations are the same as those in Fig. 6.7.11 for the medium depth hull. Figure 8.2.1 illustrates the effect of keel configuration on the directional equilibrium of horizontal forces, including balance in yaw. The keels have all the same area, aspect ratios A_k of 0.5, 1 and 2, respectively, and are fitted on the same hull. The curves labeled $\varepsilon_A + \varepsilon_H$ correspond with the assumed aerodynamic (minimum) drag angle ε_A of 12.5° plus the side force versus resistance polars shown in Fig. 6.7.19. These polars

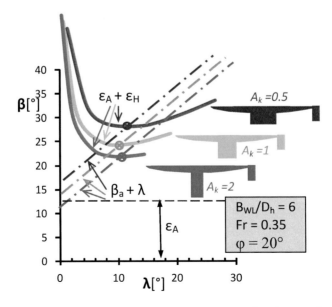

Fig. 8.2.1 *Effect of keel configuration on minimum apparent wind angle in conditions of directional equilibrium*

include the effect of yaw balance through rudder deflection. The curves reflect the earlier finding that deep draft, high aspect ratio keels imply a smaller minimum apparent wind angle.

The lines labeled $\beta_a + \lambda$ represent the variation of $\beta_a + \lambda$, as a function of λ, that is required to intersect the $\varepsilon_A + \varepsilon_H$ curves at their minimum value. These intersection points represent equilibrium conditions. The intersections with the vertical axis indicate the corresponding values of the apparent heading β_a. Figure 8.2.1 shows that yachts with a deep, high aspect ratio keel require a smaller value of the apparent heading β_A to attain their minimum apparent wind angle β than yachts with a shallow draft, low aspect ratio keel. Recalling Eqs. (8.2.2) and (8.2.3) and that the same aerodynamic drag angle ε_A means the same angle of attack α, a smaller value of β_a implies a smaller sheeting angle δ. Hence, yachts with a deep, high aspect ratio fin keel require a smaller apparent heading and smaller sheeting angles to attain their minimum apparent wind angle than yachts with a shallow draft, low aspect ratio keel.

Figure 8.2.2 gives a similar picture for different hull shapes. The curves shown represent the three different hull shapes of Fig. 6.7.11, all fitted with similar appendages (the medium aspect ratio ($A_k \approx 1$) keel of Fig. 8.2.1) but with the span adapted to give the same total draft. The beam/draft ratios B_{WL}/D_h are 3, 6 and 9, respectively. The corresponding hydrodynamic resistance polars, which include the effect of yaw balance through rudder deflection, are also shown in Fig. 6.7.19. It follows from Fig. 8.2.2 that 'beamy', shallow draft hulls require a slightly larger ap-

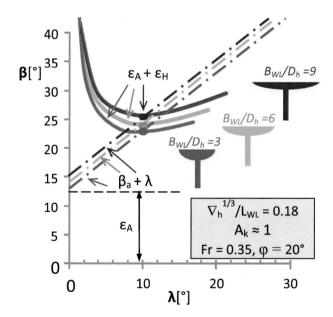

Fig. 8.2.2 *Effect of hull configuration on minimum apparent wind angle in directional equilibrium conditions, at 20° of heel*

parent heading and associated larger sheeting angles, than narrow, deep draft hulls in order to realize the minimum apparent wind angle.

A remark to be made here is that, in practice, the difference will be smaller. The main reason is that 'beamy' hulls will, in general, adopt less heel and narrow hulls more heel than the 'average' value of 20° assumed for the figure. We will come back to this point shortly.

When considering Figs. 8.2.1 and 8.2.2 it is also clear (again) that a smaller aerodynamic drag angle leads to a smaller minimum apparent wind angle with, correspondingly, smaller apparent heading and smaller sheeting angles. As we have seen in Chap. 7, rigs with a high (effective) aspect ratio, small cross-sectional dimensions of the mast, sails with a modest amount of camber (in particular for the mainsail), a small amount of aerodynamic twist and a small difference between the sheeting angles of the foresail and the mainsail are favourable in this respect. Other favourable factors are a small parasite drag and, of course, little heel.

Maximizing Boat Speed for a Given Apparent Wind Angle
As already discussed in Chap. 4, boat speed, for a given true wind angle or apparent wind angle, is governed mainly by the equilibrium of aerodynamic and hydrodynamic forces in the direction of sailing. It follows from Eqs. (8.1.1), (8.1.7a) and (8.1.10a) that this equilibrium can be expressed as

$$\tfrac{1}{2}\rho_A V_a^2 S_s C_T = \tfrac{1}{2}\rho_H V_b^2 S_H C_R \qquad (8.2.4)$$

From this it follows that

$$\left(\frac{V_b}{V_t}\right)^2 = \frac{\rho_A S_s C_T}{\rho_H S_k C_R}\left(\frac{V_a}{V_t}\right)^2 \qquad (8.2.5),$$

where the keel area S_k has been taken as the hydrodynamic reference area S_H. Equation (8.2.5) expresses that, for a given apparent wind speed V_a, boat speed is proportional to the square root of the ratio C_T/C_R between the driving force coefficient C_T of the sails and the hydrodynamic resistance coefficient C_R of the underwater body. We will call this quantity the boat speed factor (BSF $= \sqrt{(C_T/C_R)}$).

From the directional equilibrium of forces (8.14 and 8.15)

$$\varepsilon_A + \varepsilon_H = \beta = \beta_a + \lambda = \ \arctan\{\tan(\alpha + \delta)/\cos\varphi\} + \lambda \qquad (8.2.6),$$

we can, for a given rig and a given angle of heel, determine the aerodynamic drag angle ε_A for a chosen value of the apparent heading β_a and the sheeting angle δ. For given hydrodynamic characteristics of the hull and appendages we can then determine the values of λ and ε_H for which the directional equilibrium condition (8.2.6) is satisfied. The process is basically the same as that illustrated by Figs. 8.2.1/2, except for the fact that the lines representing $\beta_a + \lambda$ should now be considered for

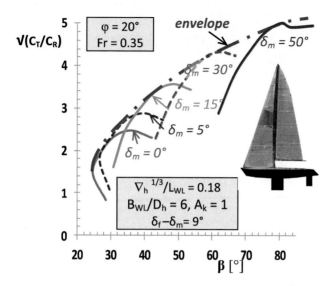

Fig. 8.2.3 *Trend of the effect of sheeting angle on boat speed factor (BSF = √(C$_T$/C$_R$)) in (approximately) equilibrium conditions of sailing*

other values of β_a. From the intersection with the $\varepsilon_A + \varepsilon_H (= \beta)$ curves and the corresponding values of ε_A and ε_H we can determine the equilibrium values of C_T and C_R and the ratio C_T/C_R for every chosen value of β_a.

Figure 8.2.3 gives the results of such a process for several values of the sheeting angle δ_m of the mainsail of a hypothetical yacht configuration. The under-water body is the same as the medium hull ($B_{WL}/D_H = 6$) plus medium keel ($A_k = 1$) configuration of Fig. 8.2.1 (see also Figs. 6.7.19 and 6.7.11). This has been combined with the aerodynamic characteristics (Fossati et al. 2008) of the IMS rig considered in Sects. 7.5, 7.6 and 7.7 and shown in Fig. 7.7.8. The camber of the sails is 12% for the jib and 10% for the mainsail. Note that the difference between the fore- and mainsail sheeting angles is fixed at 9°, that the heel angle is 20°and that the Froude number has been assumed constant at Fr=0.35.

Figures 8.2.3 gives the boat speed factor $\sqrt{(C_T/C_R)}$ as a function of the apparent wind angle β for constant values of the sheeting angle δ_m of the mainsail. Also shown is the envelope of these curves, representing the maximum value of $\sqrt{(C_T/C_R)}$ as a function of β for the conditions as indicated. Inspecting the underlying data it is found that the angle of attack corresponding with $(C_T/C_R)_{max}$ is close to, but a fraction smaller than the angle of attack for the maximum driving force. It is, therefore, not surprising that the figure reflects the earlier finding (Sect. 4.4) that the sheeting angle required for the maximum driving force increases with increasing apparent wind angle.

In considering Fig. 8.2.3 it should be further be realized that the minimum apparent wind angle would be smaller for a smaller difference between the sheeting angles of the foresail and the mainsail. The driving force and boat speed factor would be

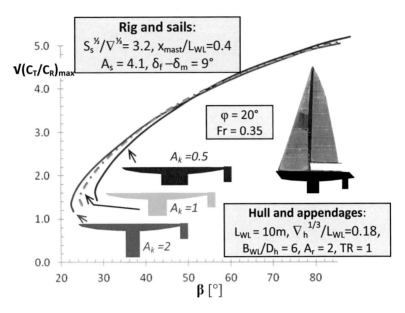

Fig. 8.2.4 Boat speed factor envelopes for hypothetical yacht configurations with different keels and identical hulls, all fitted with the same rig, at 20° of heel

larger for the range of apparent wind angles approaching 90° if the sheeting angle difference $\delta_f - \delta_m$ and the camber of the sails would be increased (see Sect. 7.5). In real sailing conditions the driving force and boat speed factor would, of course, also benefit from the reduced heel at the larger apparent wind angles.

Figure 8.2.4 compares the boat speed factor envelopes for the hypothetical yacht configurations also considered in Fig. 8.2.1, all fitted with the same rig (also shown in Fig. 8.2.3). The figure illustrates that yachts with a deep, high aspect ratio keel have a significant advantage in close-hauled conditions, both in terms of the minimum attainable apparent wind angle as well as in terms of boat speed for a given (attainable) apparent wind angle. However, this advantage disappears for apparent wind angles above, say, about 50°. In fact, the low aspect ratio keel appears to have a (very) small advantage for the higher apparent wind angles. This is due to the fact that the low aspect ratio keel has a little less frictional resistance due to its higher Reynolds number.

Boat speed factor envelopes of the hypothetical yacht configurations with different hulls, already considered in Fig. 8.2.2 are compared in Fig. 8.2.5. As already mentioned the hulls have the same displacement/length ratio but different beam/draft ratios (B_{WL}/D_h = 3, 6 and 9), and are fitted with similar appendages ($A_k \approx 1$) giving the same total draft and have the same rig, see also Fig. 6.7.11. The first thing to note is that, as already shown in Fig. 8.2.2, the configuration with the deep, narrow hull has the smallest minimum apparent wind angle. As already mentioned, this is

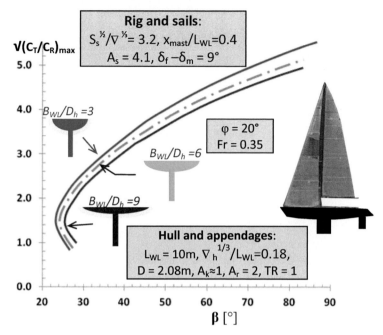

Fig. 8.2.5 *Boat speed factor envelopes for hypothetical yacht configurations with different hulls, fitted with similar appendages with the same total draft and the same rig, at 20° of heel*

due to the less favourable, keel-hull hydrodynamic interaction for the wider, more shallow hulls under heel (Sect. 6.7). It should, however, also be mentioned, as before, that the picture is flattered for the deep, narrow hull. This because the more 'beamy' hulls will, in general, adopt less heel. We will come back to this point in a following paragraph.

The other point to note is that the deep, narrow, $B_{WL}/D_h = 3$ (blue) hull has also a higher boat speed factor $\sqrt{(C_T/C_R)}$, over the full range of apparent wind angles, than the 'beamy', more shallow hulls. Analysis of the underlying data teaches that, for the fixed Froude number of 0.35, this is due mainly to the smaller wetted area and the associated smaller viscous resistance of the deep, narrow hull. A second factor is the smaller wave-making resistance of the $B_{WL}/D_h = 3$ hull.

Leeway

It is instructive to consider the variation of the angle of leeway at the $(C_T/C_R)_{max}$ condition as a function of the apparent wind angle. This is done in Fig. 8.2.6 for the medium hull with the three different keels and the same rig as in Fig. 8.2.4. The figure shows that the angle of leeway increases rapidly when the apparent wind angle is reduced below, say, 45°. This can, of course, also be observed in Fig. 8.2.1 by considering horizontal cuts for different levels of the apparent wind angle β. The angle of leeway can be seen to adopt a considerable value, of the order of 10°, when the apparent wind angle attains it minimum and increases further beyond this point. Sailing under such conditions is known as 'pinching', as already noted in Chap. 4.

Fig. 8.2.6 *Angle of leeway as a function of apparent wind angle at the $(C_T/C_R)_{max}$ condition of hypothetical yacht configurations with identical hulls, fitted with different keels but the same rig. (See also Fig. 8.2.4)*

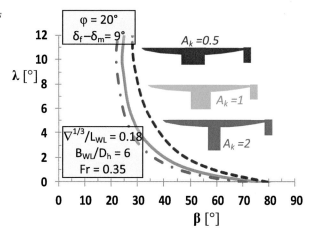

Pinching is met first by the configuration with the shallow draft keel, at an apparent wind angle of about 28° with an angle of leeway of about 12°. The corresponding apparent heading is about $28-12=16°$ The corresponding figures for the deep draft keel are $\beta_{min} \cong 22°$ with $\lambda \cong 11°$ and $\beta_a \cong 11°$, respectively. For the medium $(A_k=1)$ keel the numbers are somewhere in between.

Figure 8.2.6 also teaches that keels are hardly active for apparent wind angles beyond, say, 60°. For such conditions the angle of leeway appears to be smaller than 1°, with, of course, a correspondingly small side force.

It will be clear, by comparing Figs. 8.2.1 and 8.2.2, that a similar behaviour is found for the leeway characteristics of the configurations with the different hulls.

A frequently discussed subject in the world of yachting is the importance of leeway for boat importance. While almost every yachtsman will rapidly agree that it is a non-issue for large apparent wind angles (as illustrated above), there is often divergence of opinion on the importance of leeway in upwind, close-hauled conditions.

At first sight it would seem, from comparing Figs. 8.2.6 and 8.2.4, that, indeed, more leeway implies poorer performance. Obviously, the configuration with shallow-draft keel exhibits more leeway and a smaller boat speed factor $\sqrt{(C_T/C_R)}$ than the configuration with the deep keel. However, there is more to it. The configuration with the shallow-draft, low aspect ratio keel has also more induced resistance and, hence, a larger hydrodynamic drag angle than the configuration with the deep-draft, high aspect ratio keel. Lanchester's Law (Sect. 4.2), which states that the apparent wind angle is equal to the sum of the aerodynamic and hydrodynamic drag angles $(\beta = \varepsilon_A + \varepsilon_H)$, teaches that this holds also for the minimum apparent wind angle. In other words, it is the larger minimum hydrodynamic drag angle that causes the poorer close-hauled performance of the configuration with the shallow-draft keel and not the larger amount of leeway.

For a given apparent wind angle there is, of course, also the direct effect of the higher induced resistance on the boat speed factor $\sqrt{(C_T/C_R)}$.

An explicit way of investigating the importance of leeway is by comparing boat speed factor envelopes for, for example, the medium hull, medium keel configuration described above, but with different settings of the keel incidence with respect to the centre plane of the hull. Although such an 'all-flying' keel is not a very practical contraption from the point of view of vulnerability of the construction, it can suitably serve the purpose of investigating the importance of leeway explicitly. This, because different incidence settings imply almost equally large differences in the angle of leeway while scarcely affecting the resistance for a given side force.[3] When the boat speed factor envelopes are determined for different, positive and negative, incidence settings of the keel it is found (but not shown here) that, as expected, they are almost identical, in spite of large differences in the angle of leeway.

Boat Speed and Heel
In the preceding paragraphs we have seen the effect of keel and hull shapes on the Boat Speed Factor ($\sqrt{(C_T / C_R)}$) envelopes for a constant angle of heel ($\varphi = 20°$). In order to determine the effects on the actual boat speed, the dependence of aero- and hydrodynamics on heel and the effect of boat speed on the apparent wind speed have to be taken into account. A Velocity Prediction (type of) Program is needed for this purpose. In this section we will do this by applying a VPP to the yacht configurations considered in the preceding section.

The VPP used is the one mentioned in Sect. 8.1, developed by the author. In the version utilized here it handles four degrees of freedom, that is

- Directional equilibrium of forces in the horizontal plane (Lanchester's Law)
- Equilibrium of aero- and hydrodynamic forces in the direction of sailing
- Equilibrium in heel
- Balance of aero- and hydrodynamic yawing moments

The aerodynamic model has been tuned to match the wind tunnel data (Fossati et al. 2008) for the IMS rig (upwind configuration) shown in Fig. 7.7.8. The camber of the foresail is 12 % and that of the mainsail 10 %. The aerodynamic twist was assumed to be zero. The model does not, at this stage, explicitly model the effect of Reynolds number on the aerodynamic coefficients.

It follows from Ref. Fossati et al. (2008) that the wind tunnel data were obtained at Reynolds numbers that are about 1/3 of those in reality, at full scale (about $0.4 * 10^6$ versus about $1.1 * 10^6$). This means that, for attached flow conditions, the aerodynamic forces are probably a little pessimistic for the higher apparent wind speed conditions.

The VPP program solves for the maximum boat speed for a given apparent wind angle, true wind speed and sea state (average wave period and significant wave height). The dependent variables are boat speed V_b, leeway angle λ, true wind angle γ, main sail sheeting angle δ_m, foresail sheeting angle δ_f and rudder deflection δ_r. Note

---

[3] If the incidence of the keel would be set such so as to reduce the angle of leeway of the centre plane of the hull, the resistance of the hull would be marginally smaller. There will also be an effect on the yawing ("Munk") moment of the hull and an associated change in the induced resistance of the rudder due to the different balance in yaw.

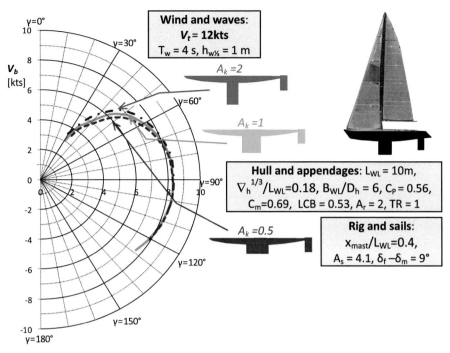

Fig. 8.2.7 *Comparison of upwind boat speed polars for hypothetical yacht configurations with identical hulls, fitted with different keels but with the same rig ($S_s^{1/2} / \nabla^{1/3} = 3.2$). (See also Fig. 8.2.4)*

that in several of the applications to be discussed hereafter the sheeting angle of the foresail was kept at $\delta_f - \delta_m = 9°$ relative to that of the mainsail.

Figure 8.2.7 compares the results of simulations for the three configurations with different keels but the same hull already shown in Fig. 8.2.4. The comparison is presented in the form of speed polar diagrams for a true wind speed of 12 kts and a sea-state corresponding with North Sea conditions (see Sect. 5.19).

Because of the identical hulls, the lateral stability characteristics of the configurations are not very different (see the table below).

B_{WL}/D_h	$\nabla_{TOT}(m^3)$	A_k	gm/L_{WL}	$gb(30)/L_{WL}$	ζ_{CG}/L_{WL}
6	6.3	0.5	0.119	0.040	0.027
6	6.3	1	0.133	0.047	0.013
6	6.3	2	0.152	0.056	-0.006

The slightly larger values of the metacentric height, **gm** at zero degrees heel and the righting arm **gb(30)** at 30 degrees of heel, of the deeper keels are due to differences in the vertical position (ζ_{CG}) of the centre of gravity. The latter has been estimated on the basis of information contained by Larsson and Eliasson (1996) and Gerritsma et al. (1977). The hydrostatic characteristics have been determined according to the estimation 'rules' of Sect. 6.2 and Appendix D. The values given in the table are believed to be representative for modern, light to medium displacement yachts with a ballast percentage of 40%.

As a consequence of the comparable stability characteristics the differences in heel (not shown here) are quite small and boat speed performance with balance in heel in Fig. 8.2.7 is qualitatively similar to the Boat Speed Factor envelopes for 20 degrees of heel shown in Fig. 8.2.4. The configuration with deep keel is seen to have an advantage in close-hauled conditions (6% improvement of VMG as compared to the medium keel configuration) but loses its advantage for larger apparent and true wind angles. The shallow draft keel configuration loses a similar amount of VMG.

Not shown in the figure is that a winged keel with the same draft as the shallow keel and a span/draft ratio of about .85 was found to have an improvement in VMG of 4% when compared with the shallow draft keel without winglets.

In contrast with the picture for different keels on the same hull, the effect of balance in heel is quite large for the configurations with a similar keel but different hulls shown in Fig. 8.2.5. Figure 8.2.8 compares the results of VPP simulations for the three configurations with different hulls, all fitted with the same rig and sails as shown in the Fig. 8.2.7.

The hydrostatic characteristics for these configurations, also determined according to the estimation 'rules' of Sect. 6.2 and Appendix D, are summarized in the table below:

B_{WL}/D_h	$\nabla_h^{1/3}/L_{WL}$	$\nabla_{tot}(m^3)$	gm/L_{WL}	$gb(30)/L_{WL}$	ζ_{CG}/L_{WL}
3	0.18	6.3	0.050	0.043	-0.024
6	0.18	6.3	0.133	0.093	+0.013
9	0.18	6.3	0.256	0.176	+0.039

Note that the righting arm at 30 degrees of heel $gb(30)$, of the narrow hull $(B_{WL}/D_h = 3)$, is less than half of that of the medium beam $(B_{WL}/D_h = 6)$ hull and that $gb(30)$ of the wide $(B_{WL}/D_h = 9)$ hull is almost twice that of the medium hull.

Figure 8.2.8 compares the boat speed polar diagrams of the three configurations for the same conditions as in Fig. 8.2.7.

It appears that, in contrast with the Boat Speed Factor envelopes for 20 degrees of heel (Fig. 8.2.5), the tender, small beam/draft ratio configuration now has the poorest upwind performance of all and the stiff, high beam/draft ratio configuration the best upwind performance. Furthermore, the medium and high beam/draft ratio configurations attain about the same maximum boat speed at a true wind angle γ of about 90° (which corresponds with an apparent wind angle of about 50°, see Fig. 3.3.3). The yacht with the small beam/draft ratio has a small advantage for true wind angles above 110° only.

The reason for the poor performance of the $B_{WL}/D_h = 3$ yacht is apparent from Fig. 8.2.9: the boat is too tender. It adopts excessive heel for true wind angles below

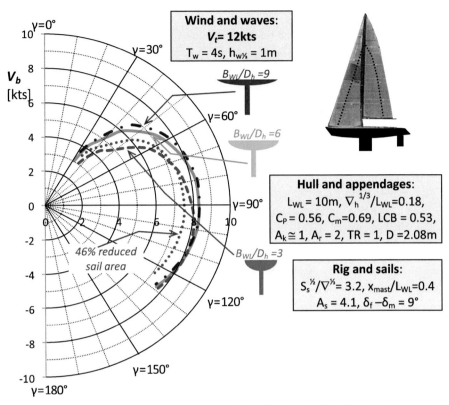

Fig. 8.2.8 *Comparison of upwind boat speed polars for hypothetical yacht configurations with different hulls, all fitted with similar keels and the same rig. (See also Fig. 8.2.5)*

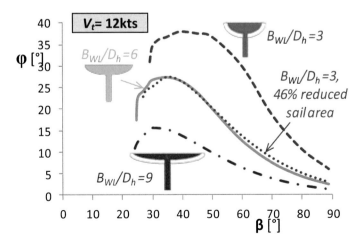

Fig. 8.2.9 *Heel angle as a function of apparent wind angle for the configurations of Fig. 8.19*

about 100° (which corresponds with apparent wind angles below about 65°) with severely degraded aero- and hydrodynamic characteristics as a consequence.

Also shown in the figures is the amount of heel and boat speed of the narrow $(B_{WL}/D_h = 3)$ hull with a reduced sail area that gives about the same angles of heel as the medium beam/draft ratio configuration (dotted line). This is seen to lead to better performance in close-hauled conditions but also to a big speed loss for larger apparent wind angles. Note that the sail area was reduced by as much as 46% through a 27% reduction of the vertical and longitudinal dimensions of the sails.

Figure 8.2.8 does not mean to say that yachts with a small beam/draft ratio are always inferior to more 'beamy' yachts. For example, they will heel much less at lower wind speeds with a smaller loss of boat speed. They will also profit more from the reduced wetted area in low wind speed conditions. Both points are illus-trated by Fig. 8.2.10, which compares boat speed polars for a true wind speed of only 6 kts. The small beam/draft ratio yacht is now seen to exhibit the best perfor-

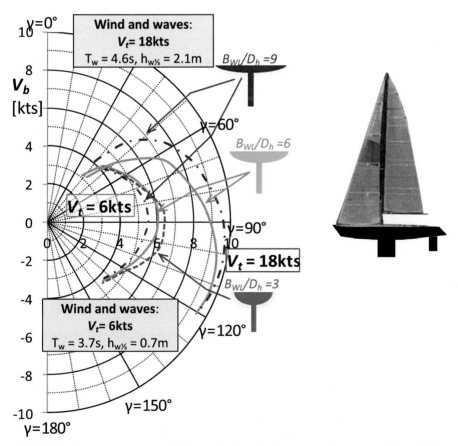

Fig. 8.2.10 *Comparison of upwind boat speed polars for the yacht configurations of Fig. 8.2.8 at low wind speed and high wind speed*

mance in reaching conditions. Close-hauled the differences between the three hulls are small, although the heel angle for the narrow configuration (not shown here), is found to be still quite appreciable (up to 20°). It should also be mentioned that narrow yachts will, in general, be designed to carry more low positioned ballast with more displacement in order to reduce heel.

Figure 8.2.10 also gives boat speed polars for a true wind speed of 18 kts. The high beam/draft ratio yacht now has an even bigger advantage than at 12 kts true wind. The medium beam/draft ratio yacht is now seen to lose a lot of boat speed in close-hauled conditions. Results for the small beam/draft ratio hull are not shown because the VPP did not produce feasible results for this configuration at this wind speed.

The big speed loss of the medium beam/draft ratio yacht is due to excessive heel. The angles of heel at 18 kts true wind (not shown here) are up to 28° for the medium beam/draft ratio hull and up to 23° for $B_{WL}/D_h=3$, with larger sheeting angles in both cases. The corresponding figures for the low wind speed ($V_t = 6$kts) are 8° and 4°, respectively.

The important point to note here is, of course, that heel can be the decisive factor for boat speed.

Is There an 'optimum' Angle of Heel?
The discussions given above raise the question whether there is something like an 'optimum' angle of heel. The answer is 'yes' but it is not a simple 20- or 30- something degrees. From a pure aerodynamics point of view the answer is simple: 0°. However, there is (a lot) more to it, as the reader may have guessed already from the preceding paragraph.

We have seen earlier that the aerodynamic forces and moments acting on a sailing yacht are proportional to the sail area and the square of the apparent wind speed. This is also the case for the heeling force. As a consequence, the heel angle of a yacht at a given apparent wind angle increases with wind speed and sail area. We have seen also, that the rate of increase of heel with wind speed and sail area is determined by the static lateral stability characteristics (righting moment) of the yacht as well as by the ('vertical') distance between the centre of effort of the sails and the centre of lateral resistance of the underwater body. Heel, as we have seen earlier, reduces the driving force for a given sail area and wind speed. Hence, there is a certain, 'critical', angle of heel (φ^*) beyond which any increase of driving force due to an increase of wind speed or sail area is overruled by the negative effect on the driving force of the increased heel. The other way around is also true. Beyond φ^* any decrease of driving force due to a decrease of wind speed or sail area is compensated by the positive effect on the driving force of the decreased heel. Figure 8.2.11 illustrates the point. Shown is the rate of change $\dfrac{d\mathrm{T}_{max}}{d(q\mathrm{S}_s)}$ of the maximum driving force T_{max} per unit change of the so-called dynamic pressure force $q\mathrm{S}_s$ on the sails as a function of the angle of heel. The critical angle of heel φ^* is attained when $\dfrac{d\mathrm{T}_{max}}{d(q\mathrm{S}_s)} = 0$. The curves shown are representative for, respectively, an average

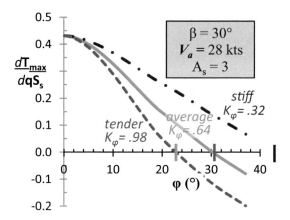

Fig. 8.2.11 *Rate of change of the maximum driving force per unit dynamic pressure force on the sails as a function of the angle of heel*

cruising yacht, a stiff yacht and a tender yacht that heels easily. The rig considered is the same, of aspect ratio 3, for which the effects of heel on driving and heeling force are shown in Figs. 7.9.7 and 7.9.8. Note that the apparent wind angle is 30° and that the apparent wind speed is 28 kts. The latter corresponds, roughly, with a true wind speed of 22 kts.

The line for the tender yacht can be seen to intersect the horizontal axis at about φ $(=\varphi^*)=23°$. For the average cruising yacht and the stiff yacht the values of the critical angle of heel φ^* are about 31° and 43°, respectively.

As discussed in further detail in Appendix N, the level and shape of the curves in Fig. 8.2.11 depends mainly on a 'heel forcing factor' K_φ, values of which are indicated in the figure. This heel forcing factor is given by

$$K_\varphi = qS_s \frac{(\zeta_{CE} - \zeta_{CLR})}{\nabla \, gm} \tag{8.2.7}$$

Here, $q(=\frac{1}{2}\rho V_a^2)$ is the dynamic pressure, S_s is the sail area, ζ_{CLR} and ζ_{CE} are the values of the vertical position of the centre of lateral resistance of the keel and the centre of effort of the sails, respectively, ∇ the displacement weight and **gm** the metacentric height. Stiff yachts have a small value of K_φ/q, tender yachts a large value.[4]

Figure 8.2.12 shows the critical angle of heel φ^* as a function of the heel forcing factor. The lower curve is for the maximum driving force *(T_{max})* condition, the upper one for the condition of maximum propulsive efficiency (or *$(T/H)_{max}$* or *$(L/D)_{max}$*). The figure indicates that for tender yachts and high wind speeds it does not make sense to sail at heel angles above, say, 20° when the objective is to maximize the driving force. When the objective is to maximize the propulsive efficiency the corresponding figure is about 30°. For stiffer yachts the critical angle of heel

---

[4] For q=279, the expression for the heel forcing factor K_φ becomes equal to the so-called Dellenbaugh angle, a rough measure of the stiffness of a yacht (Larsson and Eliasson 1996).

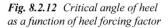

Fig. 8.2.12 *Critical angle of heel as a function of heel forcing factor*

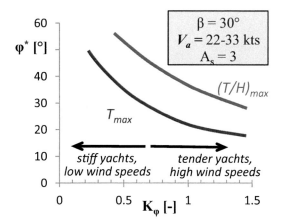

increases progressively with decreasing value of the heel forcing factor K_φ. This means that, from a performance point of view, stiff yachts should, somewhat counter-intuitively, and if wind speeds allow, be sailed at considerably larger angles of heel than tender yachts. Reefing for stiff yachts can be justified only for reasons of crew comfort and/or loads on the rigging.

We have seen in Chap. 6 that there is still another factor involved in the performance mechanisms associated with heel. Heel was seen to increase the hydrodynamic resistance of the hull, keel and rudder. This comes on top of the aerodynamic limitations on the angle of heel described above. It means that when boat speed, rather than driving force, is taken as the criterion for the critical angle of heel φ^*, the values of the latter will be somewhat smaller than as indicated by Fig. 8.2.12. A VPP type of program is required to determine φ^* more precisely for any specific yacht and specific sailing conditions.

It is of some interest to take notice of the value of the heel forcing factor K_φ for the yacht configurations and conditions of sailing considered in Fig. 8.2.8. For the medium and high beam/draft ratio configurations the values are about $K_\varphi = 0.24$ and 0.12, respectively. This is way below a level that corresponds with a critical angle of heel of practical interest. However, the value of K_φ for the narrow, tender hull is about 0.63. This, according to Fig. 8.2.12, gives a critical heel angle of about 30° for maximum driving force conditions. Figure 8.2.9 shows that for the tender yacht this is attained at an apparent wind angle of about 60°. This, in all probability, explains the rapid loss of boat speed of the narrow, tender hull configuration for apparent wind angles below 60° (or true wind angles of about 110°, see Fig. 8.2.8) when the angle of heel acquires values significantly beyond 30° (Fig. 8.2.9).

In this context it is also useful to recall that yachts with a large beam/draft ratio of the hull and shallow draft keel and rudder were seen to pick up more additional hydrodynamic resistance under heel than 'narrow' yachts with deep draft keel and rudder. It means that the reduction of φ^* through additional hydrodynamic resistance is larger for 'beamy', shallow draft yachts then for 'narrow', deep draft vessels.

Sheeting Angles

From the point of view of the trimming of the sails it is interesting to consider the sheeting angles required to realize the maximum boat speed conditions given in Figs. 8.2.7 and 8.2.8. Figure 8.2.13 does so for the mainsail sheeting angle δ_m for the configurations with the different hulls for a true wind speed of 12 kts. Note that δ_m is given as a function of the apparent heading β_a. The figure shows that, in close-hauled conditions, the tender, narrow yacht has to be sailed with larger sheeting angles than the medium beam/draft ratio $(B_{WL}/D_h = 6)$ and the stiff, wide yacht $(B_{WL}/D_h = 9)$ with smaller sheeting angles. A closer inspection reveals that the larger sheeting angles are required when the angle of heel has increased to about 25° and beyond. Figure 8.2.9 teaches that this is the case for $\beta \leq 65°$ for the tender, narrow boat and $\beta \leq 45°$ for the medium beam yacht. For the configurations with the medium beam/draft ratio hull but different keels (Fig. 8.2.7) it is found, but not shown here, that the required sheeting angles are almost the same, with the deep keel requiring just a few degrees less and the shallow keel just a few degrees more sheeting angle than the medium keel configuration.

Figure 8.2.14 shows the sheeting angles for 6 and 18 kts true wind speed. For $V_t=6$ kts the required sheeting angles are practically independent of hull shape and vary almost linearly with the apparent wind angle. Larger sheeting angles are required for the higher wind speed in order to control heel, in particular for the medium beam/draft ratio hull.

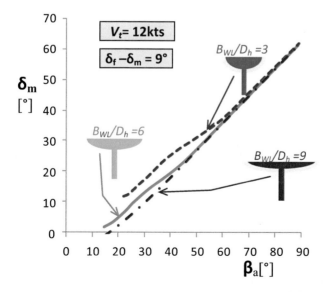

Fig. 8.2.13 *Comparison of sheeting angles for maximum boat speed conditions for the yacht configurations of Fig. 8.2.8*

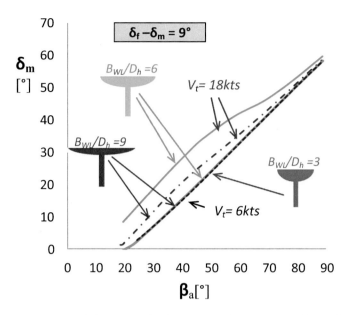

Fig. 8.2.14 *Comparison of sheeting angles for maximum boat speed conditions for the yacht configurations of Fig. 8.2.8 at high and low wind speeds*

Rudder Deflection

Figure 8.2.15 gives an impression of the amount of helm that is required for balance in yaw in upwind conditions of the hull and keel configurations considered above. Shown is the load (lift coefficient C_{Lr}') on the rudder as a function of the apparent heading β_a for a true wind speed of 12 kts.

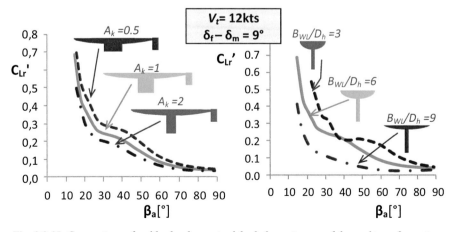

Fig. 8.2.15 *Comparison of rudder loads required for balance in yaw of the yacht configurations of Figs. 8.2.7 and 8.2.8*

The configuration with low aspect ratio, shallow draft keel is seen to require a higher load on the rudder, that is more helm, than the high(er) aspect ratio keels. Analysis of the underlying data teaches that this is mainly due to the driving-force—resistance $(T\text{-}R)$ couple (see Fig. 4.1.1), the higher induced resistance of the shallow draft keel in particular. More weather helm is required to balance this couple.

For the configurations with different hulls (figure on the right) the narrow, tender hull ($B_{WL}/D_h = 3$) is seen to require the highest load on the rudder for balance in yaw. This is not surprising, because the driving force-resistance $(T\text{-}R)$ couple is dominated by heel. What is, perhaps, surprising at first sight is the dip in the curve for $B_{WL}/D_h = 3$ around $\beta_a \cong 40°$. A closer inspection of the data reveals that this is caused mainly by the reduced wave-making resistance when the boat speed drops to below 8 kts. For $\beta_a < 35°$ the load on the rudder rises again, but this is caused by the increasing induced resistance and increasing "Munk" moment of the hull due to increased leeway when the apparent wind angle is decreased further.

Because balance in yaw is dominated by heel (and sheeting angles), smaller rudder loads are required at lower wind speeds and higher rudder loads (more weather helm) at higher wind speeds.

8.3 Upwind Performance, Dependence on rig and Sails

We have seen in Chap. 7 that sail area, the planform of the sails (aspect ratio, fractionality, overlap and roach), sheeting angles and the amount of camber of the sails are important parameters for the aerodynamics of the rig and sails of a sailing yacht. In this section we will try to get an impression of the magnitude of the effect on boat speed of variation of some of these quantities for a Bermuda rig of the type considered in the preceding section. We will do so for the medium beam/draft ratio hull with medium keel also considered in the preceding section (See Figs. 8.2.7 and 8.2.8).

Sheeting Angles
In the examples on the effects of keel shape and hull shape, discussed in the preceding section, the sheeting angle δ_f of the jib (foresail) was coupled to that (δ_m) of the mainsail in the sense that the difference $\delta_f - \delta_m$ was kept fixed at 9°. This, of course, is not fully realistic. We have seen in Chap. 7 that a small difference is required to minimize the attainable apparent wind angle and a relatively large value to maximize the driving force in reaching conditions.

Figure 8.3.1 compares maximum boat speed polars for conditions with the sheeting angle difference $\delta_f - \delta_m$ kept fixed at 9° with those obtained with δ_f and δ_m varied independently. The difference is found to be small but significant, if not substantial, in particular at low wind speeds. For the true wind speed of 12 kts, the increase of boat speed is found to be less than 2%. For the lower true wind speed

(6 kts) the maximum increase is about 5 %. For 18 kts true wind the increase in boat speed is not much different from that at 12 kts.

A closer inspection of the detailed VPP data (not given here) teaches that there are two factors at play. The first is that for all wind speeds the increase of the aerodynamic drag of the sails with increasing sheeting angle difference limits the increase of the effectiveness of propulsion by the increased lift. The second factor is that the higher boat speeds and the associated dominance of the wave-making resistance limit the increase of boat speed at the higher true wind speeds. A general conclusion that can be drawn from Fig. 8.3.1 is that optimal sheeting is more important for low wind speeds than for high wind speeds.

A further point to note is that the best VMG to windward at 18 kts true wind is considerably smaller than that at $V_t = 12$ kts. One reason, that applies to all configurations, is the increased added resistances in waves at the high true wind speeds. Another reason, already mentioned above, is that, at 18 kts true wind, the medium hull configuration is overcanvassed, causing too much heel for apparent wind angles below about 70° (or $\gamma < 100°$). For a true wind speed of 12 kts this was seen to be the case for $\beta \leq 45°$ (or $\gamma < 75°$).

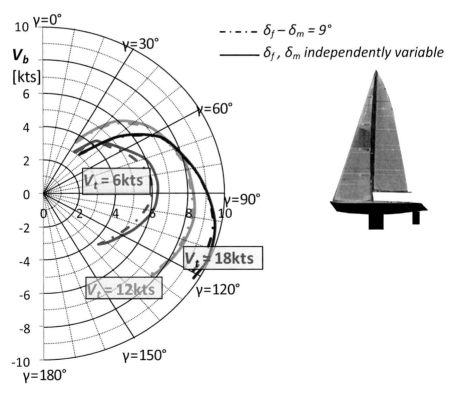

Fig. 8.3.1 *Comparison of upwind boat speed polars for the medium hull/medium keel configuration of Fig. 8.2.8, with, respectively, fixed and optimum sheeting angle difference between the jib and the mainsail. (For sea states see also Fig. 8.2.10)*

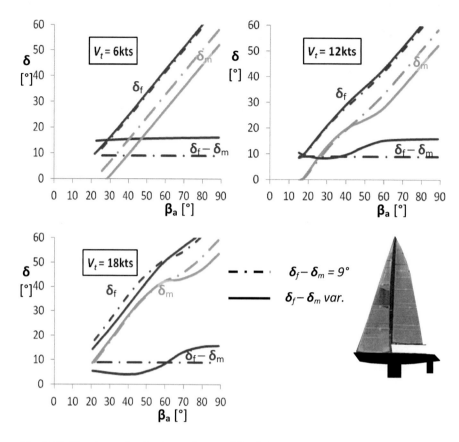

Fig. 8.3.2 *Sheeting angles for the conditions of Fig. 8.3.1*

Figure 8.3.2 gives the sheeting angles corresponding with the conditions of Fig. 8.3.1. It can be noticed that at a true wind speed of 6 kts the sheeting angle difference $\delta_f - \delta_m$ should be somewhat larger than 9° for all apparent headings (β_a). At 12 kts true wind this is the case above $\beta_a \cong 40°$, while $\delta_f - \delta_m$ should be a about equal to 9° for $\beta_a < 40°$. At the same time, the sheeting angle of the mainsail should be a similar amount smaller for $\beta_a > 40°$. This implies, remarkably, that the optimum value of the sheeting angle δ_f of the foresail is almost independent of the sheeting angle of the mainsail. It seems to confirm the rule of the thumb, used by some yachtsmen, that good trimming practice requires that the foresail is trimmed first, before the mainsail.

The picture for the lower true wind speed of 6 kts is similar to, but more regular then, that for $V_t = 12$ kts. This is due mainly to the reduced importance of heel at the lower wind speed. The main difference is a larger value of the sheeting angle difference $\delta_f - \delta_m$ and a smaller value of the mainsail sheeting angle δ_m for all apparent headings.

For a true wind speed of 18 kts the variation of the sheeting angles is even more non-linear than at 12 kts. The reason is that, as already mentioned above, the yacht configuration is overcanvassed at $V_t = 18$ kts and tends to adopt excessive heel for apparent wind angles below 70°. This is compensated by a depowering of the sails through increase of the sheeting angles.

Dependence of Boat Speed on Sail Camber

Figure 8.3.3 shows the effect on boat speed of different amounts of camber of the sails for the yacht configuration shown in Fig. 8.3.1 with medium hull and medium keel. The figure compares speed polar diagrams for three levels of camber and true wind speeds of 6, 12 and 18 kts. The camber levels for the foresail and mainsail are 8 and 6%, 12 and 10%, 16 and 14%, respectively. For the jib, the chordwise position of maximum camber is at 40%. For the mainsail this is at 32%.

The first thing to note is that, in a broad sense, the picture is similar to Fig. 8.3.1. The reason is, as mentioned earlier, that the (hypothetical) yacht is overcanvassed at the higher wind speeds for small apparent wind angles. In combination with the

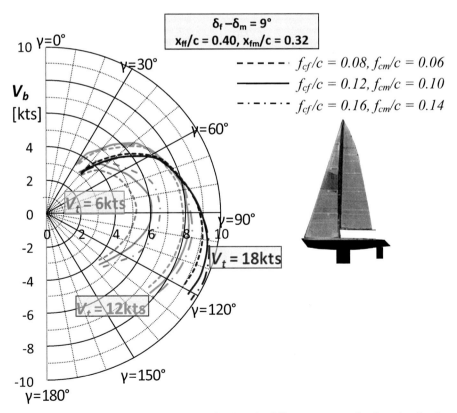

Fig. 8.3.3 *Comparison of upwind boat speed polars for different amounts of sail camber for the medium hull/medium keel configuration of Fig. 8.2.8. (For sea states see also Fig. 8.2.10)*

much larger added resistance in waves at $V_t = 18$ kts this causes a rather dramatic loss of VMG at small wind angles.

The important point to note is that it obviously pays to employ a large amount of camber at low wind speeds. At $V_t = 6$ kts the difference in boat speed is as large as 10–20%. Moreover, even for small true (and apparent) wind angles more camber means more boat speed. The latter is no longer the case at the higher wind speeds. The advantage of high camber disappears in close hauled conditions at 12 and 18 kts true wind. A closer inspection of the data teaches that the medium camber sails have the best VMG at $V_t = 12$ kts and the small camber sails the best VMG at $V_t = 18$ kts. The differences in VMG are of the order of 3%. For large(r) apparent wind angles camber pays at all wind speeds.[5]

A remark to be made in this context is that the profit of increasing camber as shown above is probably a little smaller in reality. The reason is that the chord lengths of the sail were supposed to be constant while they will be shorter with increased camber for a fixed girth length (see also Sect. 7.8, Fig. 7.8.16). This means that the effective sail area will be smaller when the amount of camber is increased (about 4.5% for the 18/16% camber case) and a little larger when camber is decreased (about 2% for the 8/6% camber case). In terms of boat speed this means that, roughly, the increase of boat speed due to increased camber is about 2% less and the decrease of boat speed due to reduced camber about 1% less.

As discussed in Sect. 7.4 there is also a dependence of lift and drag on the chordwise position of maximum camber. This in the sense that the maximum lift increases but the maximum lift/drag ratio decreases when the position of maximum camber is more rearward. As explained earlier, this is advantageous for reaching conditions and low wind speeds. On the basis of Figs. 7.4.10 and 7.4.12 we may expect the effects on boat speed to be similar, but a little smaller than in Fig. 8.3.3 when the position of maximum camber is varied between 30 and 50% of the chord.

Sail Twist

The boat speed data presented so far have all been determined on the assumption that the sails had zero aerodynamic twist. This means, see Sect. 7.4, that the geometrical twist of the sails was assumed to be such that it compensates for the twist in the apparent wind profile to which the sails are exposed (see also Sect. 7.2). As shown in Fig. 7.4.47 an increase of the twist, that is an increase of the difference between the sheeting angles of the foot and of the head of the sail, causes a reduction of lift, a reduction of the effective span and a lower position of the centre of effort (CE).

When this is accounted for in the aerodynamic model in the VPP it is found (but not shown here) that the effect on boat speed of increasing the geometrical twist of the sails within reasonable bounds ($< \cong 10°$) is not very large. For the higher apparent wind angles it is found that an increase of the twist of the sails always leads to a loss of boat speed (of the order of 2%).

---

[5] Note that this behaviour was already anticipated in Sect. 7.4.

For close-hauled conditions there is also a loss of VMG at low and moderate wind speeds. For the higher true wind speed of 18 kts the VMG in close-hauled conditions is found to be almost independent of the amount of twist. This seems to be due to the fact that the lowering of the centre of effort (less heel) due to an increase of twist is counteracted by a reduction of the effective span (more induced drag and a correspondingly larger drag angle).

The message of this seems to be that increasing the twist of the sails is not a very efficient way to control heel.

Sail Area (Reefing)

Reducing sail area offers a more efficient way to control heel in close-hauled conditions and its negative effects on boat speed, at least in principle. This was already illustrated briefly in Sect. 8.2 (See Fig. 8.2.8). For large apparent wind angles, when heel is not important, a reduction of sail area always means a loss of boat speed.

From an aerodynamic viewpoint sail area can, hypothetically, be reduced in several ways. One is to reduce the vertical dimensions of the rig. This results in a smaller aspect ratio. Reducing only the longitudinal dimensions, which gives a larger aspect ratio, is another.

In the hypothetical case shown in Fig. 8.2.8 sail area was reduced by multiplying both the vertical and longitudinal dimensions of the rig and sails by a factor 0.7, while keeping the mast in the same position. This implies that the aspect ratio of the sails was kept constant. In a practical situation this is not always possible.

For a given rig, sail area can be reduced by setting a smaller jib/genoa, or furling-in the genoa, and by setting a smaller mainsail or reefing/furling the mainsail. For cruising yachts with a limited wardrobe of sails the furling/reefing options are often the only possibilities.

In the case of reefing or furling the vertical and longitudinal dimensions of the sails are reduced by factors that depend on the precise planform shape of the sail. For a triangular mainsail that is reefed at the boom the aspect ratio remains unchanged. For a mainsail with roach it is mainly the vertical dimension of the sail that is reduced. This means a smaller aspect ratio. When furled into the mast the aspect ratio also remains unchanged for a triangular sail. However, the aspect ratio now increases when the mainsail has roach.

Replacing the foresail by a smaller genoa or jib usually also implies a smaller overlap and, possibly, a change of fractionality (smaller value of the ratio $I/(P + BAD)$, see Fig. 2.3.1). A smaller overlap and change of fractionality result also when the area of the foresail is reduced by furling. An additional factor in this case is the effect of the furling roll on the aerodynamic characteristics of the foresail. As argued in Sect. Effects of a Mast or Head-Foil on Sail Section Characteristics, such effects are similar to those of a mast in front of a mainsail. Other negative factors in the case of furling-in the jib are a larger gap between the foot of the jib and the deck and a higher position of the centre of effort.

Figure 8.3.4 shows the effect of reefing on boat speed for the medium hull/medium keel configuration considered before, for true wind speeds of 18 and 12 kts. The

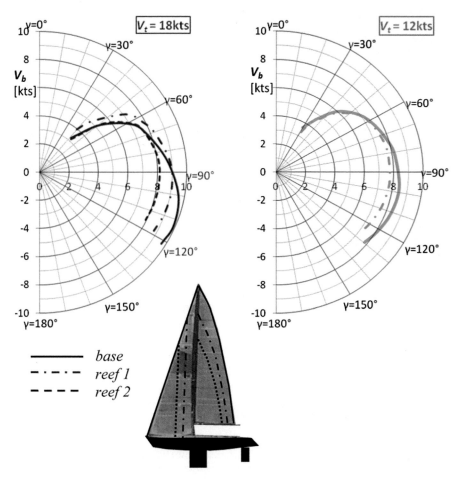

Fig. 8.3.4 *Effect of reefing on upwind boat speed for the medium hull/medium keel configuration of Fig. 8.2.8. (For sea states see also Fig. 8.2.10)*

base rig and sails are again the same as before, with 12% camber of the foresail and 10% of the mainsail. The reefing option considered is one with a smaller jib/genoa of the same aspect ratio and the main reefed at the boom. This means that the overlap of the foresail is reduced (to negative values) for the smaller jibs. The fractionality of the rig was kept unchanged. The aspect ratio of the complete rig is hardly changed. The changes in overlap, effective span and the additional drag due to the bare, exposed parts of the mast have been taken into account in the aerodynamic model.

Two levels of reefing are considered, one with the dimensions of the sail reduced to a factor of about 0.8 (reef 1) and one with a reduction factor of about 0.6 (reef 2). The figure shows that, at 18 kts true wind, reefing (reef 1) pays for true wind angles below 90°, that is for apparent wind angles below about 60°. With reef 2 the yacht configuration is obviously undercanvassed at 18 kts true wind. Higher wind speeds are required to obtain a profit in case of reef 2.

The figure on the right in Fig. 8.3.4 shows that at 12 kts true wind reef 1 does not yet offer any advantage.

Aspect Ratio

The aspect ratio of rig and sails is an important parameter for the aerodynamic efficiency, as already discussed in some detail in Sect. 7.4. In this paragraph we will try to get an impression of the sensitivity of boat speed for variation of aspect ratio while keeping the sail area constant. We will do so for the same hull and appendage (medium hull, medium keel) configuration considered above.

For this purpose VPP computations were done for rig and sail configurations with different mast heights. At the same time the foots of the sails were varied to maintain the same sail area. With the forestay attached to the bow this means that the position of the mast is more rearward when the mast is shorter and more forward when the mast is taller (see Fig. 8.3.5). In order to maintain the same directional balance properties, the position of the keel was moved forward or rearward by a similar amount. In yacht design jargon (Larsson and Eliasson 1996) it is said that the 'lead', that is the difference in the longitudinal position ξ_{CE} of the centre of effort of the sails and the centre of lateral resistance ξ_{CLR} of the keel plus hull is, approximately, kept constant.

When plotted in a speed polar diagram it is found (but not shown here) that there is surprisingly little difference in boat speed (less than 5 %) between the different configurations, in particular at the higher apparent wind angles. The reason is that a change of aspect ratio implies counteracting effects on boat speed. The main factors at play are the induced drag of the sails, and the angle of heel. A higher aspect ratio implies less aerodynamic drag (positive effect on boat speed) but increased heel (negative effect on boat speed). The latter is due to the higher position of the centre of effort of the sails. A lower aspect ratio means more drag but less heel. Factors of secondary importance are the drag due to the mast (see Sect. 7 and a (small) change in the vertical position of the centre of gravity. Both have a small positive effect on boat speed when the aspect ratio is reduced.

Which of these factors dominates depends primarily on wind speed and the lateral stability ('stiffness') of the yacht. For stiff yachts at low wind speeds the induced drag will be dominant. For tender yachts at high wind speed it is more important to control heel.

Figure 8.3.5 also shows the best VMG to windward as a function of the aspect ratio (A_s) for a true wind speed of 12 kts as determined through the VPP. It appears that for the yacht configuration and conditions considered a moderate aspect ratio of about 3.3 gives the best upwind performance. There is no doubt that the optimum aspect ratio will be higher for lower wind speeds and/or for a 'stiffer' yacht. For a more tender yacht and/or higher wind speeds the optimum aspect ratio will be smaller.

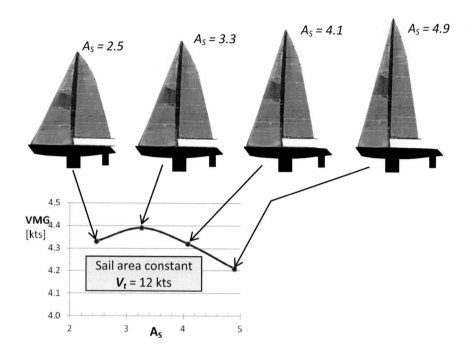

Fig. 8.3.5 *Effect of aspect ratio of the sails on the best maximum VMG to windward, at 12 kts true wind, for the medium hull/medium keel configuration of Fig. 8.2.8*

8.4   Directional Stability

We have seen in Sect. 6.4 that many if not most sailing yacht configurations are directionally unstable in the hydrodynamic sense. That is, without sails, they do not return to their original heading after a directional disturbance without rudder action.

Fortunately the situation is different under sail. Sailing yachts are considered to be directionally stable under sail if the combined aerodynamic plus hydrodynamic yawing moments tend to turn the yacht to weather when there is a sudden increase of the apparent wind angle β (or to lee when there is a sudden decrease of β). When this is the case, the steering of the yacht will be less tiring and will require less attention from the helmsman. It may even steer itself with the helm lashed.

Upwind
It seems that most sailing yachts are, to varying extents, directionally stable, at least in close-hauled conditions. Figure 8.4.1 summarizes the forces at play (see also Chap. 4). The yawing moment is governed by the driving-force—resistance (*T-R*) and side-force (S_A-S_H) couples plus the aerodynamic and hydrodynamic 'Munk' moments of the hull. The lever 'arms' d_y and d_x, of the couples can be written as

$$d_x = x_{CLR} - x_{CE} \tag{8.4.1}$$

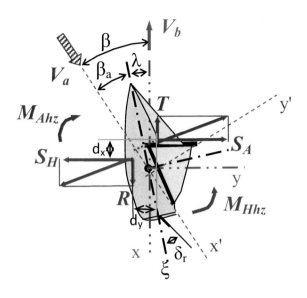

Fig. 8.4.1 *Forces and moments in the horizontal plane*

and

$$d_y = (\zeta_{CE} - \zeta_{CLR}) \sin\varphi \qquad (8.4.2)$$

(see Sect. 7.7). In equilibrium conditions the total yawing moment M_z is zero. This can be written as:

$$M_z = Td_y - S_A d_x - M_{Ahz} + M_{Hhz} = 0 \qquad (8.4.3),$$

where we have defined M_z to be >0 when it acts to weather. The calculus of differentiation teaches that a small change dM_z due to a sudden change in apparent wind angle can be expressed as

$$dM_z = d(Td_y) - d(S_A d_x) + dM_{Hhz} - dM_{Ahz} \qquad (8.4.4)$$

This can be expanded to read

$$dM_z = d_y dT + Td d_y - d_x dS_A - S_A dd_x + dM_{Hhz} - dM_{Ahz} \qquad (8.4.5)$$

Letcher (1965), in a classical, analytical study, has shown, qualitatively, that the *T-R* driving force-resistance couple (the term $d(Td_y)$ in Eq. (8.4.4) is the most important heading-restoring term. The important destabilizing term is $-d(S_A d_x)$. In close-hauled conditions the *T-R* couple develops an additional contribution dT to weather to the yawing moment because the driving force increases through the increase of the apparent wind angle (see Fig. 8.4.1). The restoring *T-R* couple increases also when the lever arm d_y of the moment is increased ($dd_y > 0$). It follows from the expression (8.4.2) that this is the case when there is an increase of heel due to a larger heeling/side force. However, the latter is the case only for apparent wind angles below, say, 35° (see Fig. 8.4.1). This means that the term $+ Tdd_y$ is destabilizing for $\beta > \approx 35°$, $(dd_y < 0)$.

The contribution of the term $-d_x dS_A$ of the side force couple is less important in close-hauled conditions. This because the side force attains its maximum usually at an apparent wind angle around 35°, i.e. $dS_A \approx 0$ for $\beta \approx 35°$. The term $-S_A dd_x$ is destabilizing because the moment arm d_x increases through the forward displacement of the centre of effort of the sails with increasing angle of attack (see Fig. 7.7.6), at least as long as the sail operates at an angle of attack below that for which the lift attains its maximum. This, as indicated earlier, is, in general the case for $\beta < \approx 90°$.

It is further useful to note that the hydrodynamic 'Munk moment' term dM_{Hhz} is stabilizing as long as the angle of leeway and the angle of heel are increasing as a result of a change of the apparent wind angle. The opposite is the case for the aerodynamic 'Munk moment' term dM_{Ahz}. As discussed in Sect. 7.6 the aerodynamic 'Munk moment' varies roughly like $\div \sin 2\beta$ in the aerodynamic coordinate system. This implies that $dM_{Ahz} \div \cos 2\beta$. Hence, $-dM_{Ahz}$ is <0 for $\beta < \approx 45°$ and $\beta > \approx 135°$, and >0 for $\approx 45° < \beta < \approx 135°$. The magnitude is, however, probably not very large.

It is possible to obtain a quantitative assessment of the (static) directional stability of a sailing yacht by running the velocity prediction program described in the preceding sections in a none-equilibrium mode. Starting out from an equilibrium condition it is possible to determine the response of a yacht to a small change in apparent wind angle while the sheeting angles and the rudder angle are kept fixed. The resulting non-zero yawing moment is then a measure of the static directional stability.

The directional stability appears to depend quite strongly on the apparent wind angle β. This is shown by Fig. 8.4.2 which presents the rate of change $\dfrac{dC_{Mz}}{d\beta}$ of the total (aerodynamic plus hydrodynamic) yawing moment coefficient[6] per degree change of the apparent wind angle β, as a function of β for the medium-hull/medium-keel configuration considered earlier (depicted, for example, in Figs. 8.27/8). It is clear that $\dfrac{dC_{Mz}}{d\beta} > 0$, implying directional stability, for close-hauled conditions. It appears also that the directional stability decreases rapidly with increasing apparent wind angle β for $\beta > \cong 30°$. For $\beta > \approx 45°$ the static directional stability is, roughly, neutral ($\dfrac{dC_{Mz}}{d\beta} \approx 0$). The main reason for this behaviour is in the term Tdd_y, which, as mentioned above, has a destabilizing contribution ($dd_y < 0$) for $\beta > \cong 35°$.

---

[6] Made dimensionless by hydrodynamic quatities (made dimensionless by hydrodynamic quatities (ρH, V_b, S_k, L_{WL})).

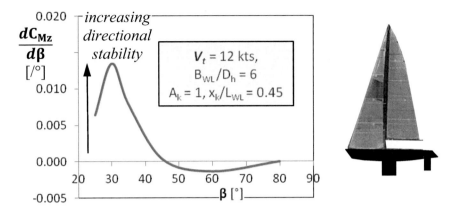

Fig. 8.4.2 *Dependence of the directional stability quantity* $\dfrac{dC_{Mz}}{d\beta}$ *on the apparent wind angle*
for the medium-hull/medium-keel configuration with medium (12%/10%) sail camber in upwind
conditions

The reduced directional stability for $\beta > \approx 45°$ manifests itself in, for instance, the fact that for $\beta > \approx 45°$ an autohelm will have to work harder to keep the yacht on course than in close-hauled conditions, a phenomenon that many yachtsmen may recognize.

Figure 8.4.3 illustrates the dependence of the directional stability on the longitudinal position x_k of the keel[7] for the medium beam/medium keel/medium rig configuration. Shown is the rate of change of the directional stability quantity $\dfrac{dC_{Mz}}{d\beta}$ for close-hauled conditions ($\beta = 30°$), as a function of x_k/L_{WL}. It is clear that $\dfrac{dC_{Mz}}{d\beta} > 0$, implying directional stability, for the range of x_k/L_{WL} considered. It also appears that a more rearward position of the keel, i.e. more 'lead', has a small positive effect on the directional stability. Because more 'lead' implies less weather helm this is in agreement with the qualitative findings of Ref. Letcher (1965).

Figure 8.4.4 gives an impression of the effect of keel aspect ratio on the directional stability in close-hauled conditions for the medium hull configuration. The figure shows that the directional stability quantity $\dfrac{dC_{Mz}}{d\beta}$ decreases fairly rapidly with increasing aspect ratio of the keel. The main reason for this is not immediately obvious but seems to be due to a smaller positive contribution of the term Td_y due to a reduced sensitivity for the angle of heel for the apparent wind angle (reduced rate of change $\dfrac{d\varphi}{d\beta}$) for configurations with the higher aspect ratio keels.

---

[7] Recall at this point that a larger value of x_k implies an increased 'lead' (distance) between the centre of effort (CE) of the sails and the centre of lateral resistance (CLR) of the underwater body.

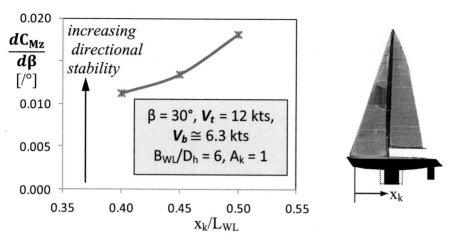

Fig. 8.4.3 *Effect of longitudinal position of the keel on the directional stability quantity* $\dfrac{dC_{Mz}}{d\beta}$ *for the medium hull/medium keel configuration of Figs. 8.2.7 and 8.2.8 in close-hauled conditions*

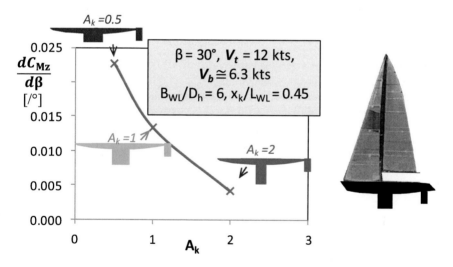

Fig. 8.4.4 *Effect of keel aspect ratio on the directional stability quantity* $\dfrac{dC_{Mz}}{d\beta}$ *for the medium hull configurations of Fig. 8.2.7 in close-hauled conditions*

The dependence on beam/draft ratio of the hull is shown in Fig. 8.4.5.[8] The directional stability is seen to decrease with increasing beam/draft ratio of the hull. This is also in agreement with the qualitative finding of Letcher (1965) that, close-hauled, tender yachts are directionally more stable than stiff yachts. The reason is that tender yachts experience a larger increase of heel for a given increase of

---

[8] Note that the apparent wind angle β is 35° rather than the 30° in the preceding figures.

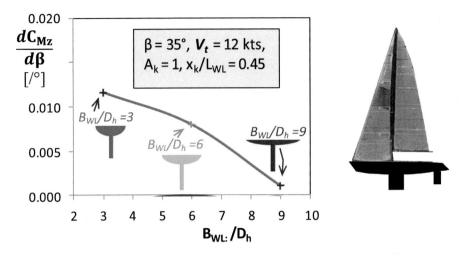

Fig. 8.4.5 *Effect of beam/draft ratio of the hull on directional stability quantity* $\dfrac{dC_{Mz}}{d\beta}$ *for the medium hull configurations of Fig. 8.2.8 in close-hauled conditions*

side force causing a larger lever arm for the restoring **T-R** couple. It also explains (Letcher 1965) why catamarans are directionally unstable.

The directional stability of a sailing yacht depends, of course, also on the characteristics of rig and sails. Yachts with a tall, high aspect ratio rig usually have a good directional stability in close-hauled conditions because they tend to be more tender than yachts with a short, low aspect ratio rig.

There is also a small effect of the camber of the sails. Figure 8.4.6 shows the directional stability quantity $\dfrac{dC_{Mz}}{d\beta}$ as a function of the amount of camber of the

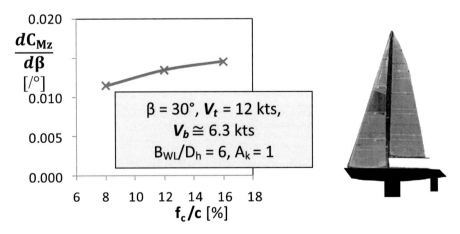

Fig. 8.4.6 *Effect of sail camber on the directional stability quantity* $\dfrac{dC_{Mz}}{d\beta}$ *for the medium-hull/ medium-keel configurations of Fig. 8.2.7 in close-hauled conditions*

foresail for the medium-hull/medium-keel configuration. In all cases the camber of the mainsail is 2% less than that of the foresail.

The directional stability in the close-hauled condition is seen to improve slightly with increasing amount of camber of the sails. The reason is that more camber implies a larger side force and more heel which results in a larger restoring moment through the term $d_y dT$ in Eq. (8.4.5).

Downwind

With a spinnaker type head sail and apparent wind angles beyond 90° the static directional stability of a sailing yacht is a different matter. There are several reasons for this.

With the foresail set to weather and without significant heel (small side force), the lever arm term d_y is practically zero ($d_y \approx 0$). Because the driving force T attains its maximum for $\beta \approx 140°$ (see, e.g., Fig. 7.8.14), its variation dT and, consequently, the term $d_y dT$ in Eq. (8.4.5) are ≈ 0 for $\beta \approx 140°$. Because the side force S_A is also ≈ 0 for $\beta \approx 140°$, this applies also to the term $-S_A dd_x$.

For $\approx 90° < \beta < \approx 140°$, dT will be > 0 and so is the lever arm d_y because the boat will still heel towards lee due to the (non-zero) side force. This means that the term $d_y dT$ is stabilizing for $\beta < \approx 140°$. With $dT < 0$ and $d_y \approx 0$ the term $d_y dT$ is probably negligible for $\beta > \approx 140°$.

Because the centre of effort of a sail that operates beyond maximum lift travels rearward with increasing angle of attack, the term dd_x is < 0 for $\beta > 90°$. Hence, the term $-S_A dd_x$ is > 0 (stabilizing) for $\approx 90° < \beta < \approx 140°$ but destabilizing for $\beta > \approx 140°$ when $S_A < 0$.

Of the remaining terms the term Tdd_y will be destabilizing, because $dd_y < 0$. The latter is due to the fact that, beyond $\beta \approx 35°$, the side force S_A decreases monotonously with increasing apparent wind angle (see Fig. 7.8.14). With $dS_A < 0$, the term $-d_x dS_A$ will be > 0, i.e. stabilizing, for all $\beta > 35°$.

The hydrodynamic 'Munk' moment term dM_{Hhz} in Eq. (8.4.5) requires some further attention. Since it depends on leeway and the angle of heel (see Sect. 6.7), and, hence, on side force, this term can be expanded to read

$$dM_{Hhz} = \frac{\partial M_{Hhz}}{\partial \varphi} d\varphi + \frac{\partial M_{Hhz}}{\partial \lambda} d\lambda \qquad (8.4.6)$$

Because $d\varphi < 0$, at least for $\beta > \approx 35°$, and $\dfrac{\partial M_{Hhz}}{\partial \varphi} > 0$ (Sect. 6.7) the first term is destabilizing. Since $d\lambda$ is also < 0 because of the decreasing side force and $\dfrac{\partial M_{Hhz}}{\partial \lambda}$ is also > 0, the second term is also destabilizing.

The effect of the force couples and moments as described above on the change dM_z of the total yawing moment due to an instantaneous increase of the apparent wind angle is summarized in the Table 8.4.1.

We have seen already that sailing yachts are usually directionally stable in close-hauled conditions and have approximately neutral directional stability in close-

Table 8.4.1 *Summary of contributions to the change in yawing moment due to a (sudden) increase dβ(>0) of the apparent wind angle*

β-range	$Td\,d_y$	$d_y\,dT$	$-S_A\,d\,d_x$	$-d_x\,dS_A$	$-dM_{Ahz}$	dM_{Hhz}	dM_z
β < ≈35°	> 0	> 0	< 0	< 0	< 0	> 0	> 0
≈35° < β < ≈90°	< 0	> 0	< 0	> 0	> 0	< 0	≈ 0
≈90° < β < ≈140°	< 0	> 0	> 0	> 0	> 0	< 0	>0/ ≈0
≈140° < β < ≈180°	< 0	≈ 0	< 0	> 0	< 0	< 0	< 0

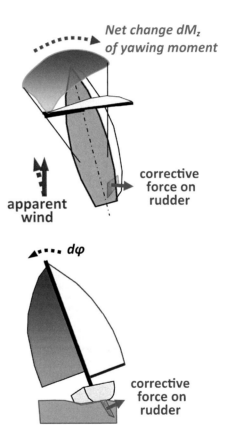

Fig. 8.4.7 *Directional instability and corrective rudder force in downwind conditions with a sudden increase of the apparent wind angle (risk of unintentional gybe)*

Net change dM_z of yawing moment

corrective force on rudder

apparent wind

$dφ$

corrective force on rudder

reaching (≈45°<β<≈90°). The latter seems to be the case also in broad reaching (≈90°<β<≈140°). Most sailing yachts are probably directionally unstable when the apparent wind is in the stern quarter (≈140°<β<≈180°).

As mentioned earlier rudder action is needed to restore the course of a yacht that is directionally unstable. This would not be too bad as long as it is not too tiring for the helmsman. Unfortunately there is more to it. As illustrated by Fig. 8.4.7 the

Fig. 8.4.8 *Directional instability and corrective rudder force in downwind conditions with a sudden decrease of the apparent wind angle (risk of broaching)*

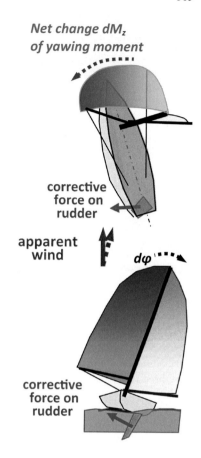

corrective force on the rudder required to restore the net change dM_z of the yawing moment due to an increase of the apparent wind angle tends to turn the yacht to lee. The corrective force on the rudder then also amplifies the change $d\varphi$ of the angle of heel, towards the weather side. This, in turn, increases the yawing moment to lee, etc. We obviously have a risky situation in which the yacht can become uncontrollable with diverging course and heel followed by an unintentional gybe.

A similar but opposite situation arises when there is a sudden decrease of the apparent wind angle (Fig. 8.4.8). The change of the yawing moment and its components is then opposite to that summarized in Table 8.4.1 and tends to decrease the apparent wind angle further. A corrective rudder action now increases the angle of heel towards the lee side, which in this case causes further luffing. Again there is a risky situation in which the yacht can become uncontrollable with course and heel diverging. In this case followed by broaching.

Whether a broaching or gybing calamity will actually happen or not depends on the configuration of the yacht and the conditions of sailing in more detail. Considering Table 8.4.1 it is clear that the risk will be greater for high wind speeds (large *T*, large

Table 8.4.2 *Summary of contributions to the change in yawing moment due to a gust (sudden increase of the wind speed)*

β-range	$Td d_y$	$d_y dT$	$-S_A d d_x$	$-d_x dS_A$	$-dM_{Ahz}$	dM_{Hhz}	dM_z
$\beta < \approx 140°$	> 0	> 0	≈ 0	< 0	< 0	> 0	> 0
$\approx 140° < \beta < \approx 180°$	< 0	≈ 0	≈ 0	> 0	< 0	< 0	< 0

S_A and relatively large dM_{Ahz}) and tender yachts (large $d\varphi$ and, hence, large, negative dd_y and large, negative dM_{Hhz}).

An important aspect is further, that, as discussed in Sect. 5.18, static stability is a prerequisite for stability in dynamic (unsteady) conditions. We will come back to this in the next section.

Gusts
It is also useful to see what happens when a sailing yacht encounters a gust, that is a sudden increase of wind speed. It will be clear that an increase in wind speed will immediately cause an increase dT of the driving force if T itself is > 0. This is also the case for dS_A when S_A is > 0. For $S_A < 0$, that is for $\approx 140° < \beta < \approx 180°$, dS_A will be < 0. $dS_A > 0$ implies $d\varphi$ (and dd_y) > 0 and $d\lambda > 0$. Because the angle of attack of the sails does not change, the position of the centre of effort of the sails does not change either. Hence, $dd_x = 0$. Table 8.4.2 summarizes what this means, qualitatively, for the static directional stability.

A quantitative indication for upwind conditions, obtained by running the VPP in a non-equilibrium mode, is given by Fig. 8.4.9 for the medium hull/medium keel configuration. Shown is the change dC_{Mz} of the yawing moment coefficient, for a 1 kts increase of the true wind speed, as a function of the apparent wind angle. The first

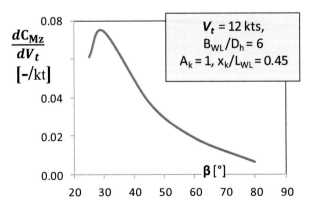

Fig. 8.4.9 *Variation of the directional stability quantity $\dfrac{dC_{Mz}}{dV_t}$ with the apparent wind angle for the medium-hull/medium-keel configuration with medium (12%/10%) sail camber in upwind conditions*

thing that strikes is that the range of apparent wind angles in which a yacht yaws to weather due to a gust is much larger than when subject to a change of wind direction (compare with Fig. 8.4.2). This suggests that, in terms of directional stability, a sailing yacht in upwind conditions is less sensitive for gusts than for wind shifts. This in particular also in reaching conditions. Nevertheless, Table 8.4.2 indicates that the directional stability in a gust is probably also a problem when running downwind ($\beta > \approx 140°$).

8.5 Dynamic Interactions

Dynamic Directional Stability
As already discussed in some detail in Sect. 6.8, the dynamic directional stability of a yacht is already a fairly complex affair without sails, in particular in waves. Apart from the static stability requirement, the dynamic stability was seen to depend on the characteristics of the coupling between the yawing and swaying motions. A ship or yacht was seen to behave like a 'dynamic mass-spring-damper system' with periodic, wave-induced external forces. Moreover, we saw that, in following waves, the wave-induced yawing moment can be so large, that it overrides any directional-stability-in-calm-water property that a ship or yacht may have.

With sails it is even more complicated. Due to the aerodynamic forces on the sails there is now also a strong coupling between the yawing and rolling motions (as already indicated in the preceding section).

A yacht under sail can still be considered to behave like a 'dynamic mass-spring-damper system' (Fossati 2009). For example, the equilibrium of moments around the yawing axis in flat water can be expressed as (see also Sect. 6.8, Eq. (6.8.5))

$$(I_{zz} + I'_{zz})\ddot{\psi} + b_\psi \dot{\psi} + k_\psi \psi = 0 \tag{8.5.1}$$

Here, the first ($\ddot{\psi}$-) term represents the moment due to inertia, the second ($\dot{\psi}$-) term the fluid dynamic damping and the third (ψ-) term the restoring moment. The 'spring stiffness' coefficient k_ψ is, in the coordinate system of sailing, given by (Sect. 6.8)

$$k_\psi = -\frac{dM_z}{d\psi} \tag{8.5.2}$$

This can also be written as

$$k_\psi = \frac{dM_z}{d\beta} \tag{8.5.3}$$

The quantity M_z is now the total, hydrodynamic plus aerodynamic, yawing moment. As mentioned before it is a prerequisite for dynamic stability that the stiffness coefficient $k_\psi > 0$. As shown in the preceding section this is the case only for close-hauled conditions.

The damping coefficient b_ψ represents the hydrodynamic damping only when a yacht moves without sails (Sect. 6.8). With sails it also contains the contribution of the sails. It can then be written as

$$b_\psi = \frac{dM_z}{d\dot\psi}$$
(8.5.4)

$\dfrac{dM_z}{d\dot\psi}$ is the rate of change of the total yawing moment M_z with the rate of yaw $\dot\psi$ (the angular speed of rotation about the z-axis).

For a sailing yacht the two important contributions to the hydrodynamic damping in yaw are those from the hull and the rudder. Both imply positive damping. That is they oppose the direction of rotation (Sect. 6.8). The effect of (constant) heel is positive on the contribution of the hull but negative on the contribution by the rudder. The effect of the keel is, in general, small, because its location is close to the centre of rotation.

The aerodynamic contributions can be distinguished into a contribution of the hull and one of the sails. The aerodynamic contribution of the hull will also be counteracting the rotation and act like positive damping. Heel will have a negative effect.

Little is known about the contribution of the sails. We have seen in Sect. 5.18 that a lifting surface in pitching motion (or a sail in yawing motion) can develop negative damping, that is enhance the pitching motion, when the axis of rotation is located between 50 and 75 % of the chord. This is already the case in attached flow, i.e. for small to moderate angles of attack. It suggests that a sail in yawing motion at small angles of attack (i.e. in close-hauled conditions) could also generate negative damping. What this means for the total damping depends on the relative magnitudes of the hydrodynamic and aerodynamic contributions. It is the author's impression that the total damping of the yawing motion of a sailing yacht in close-hauled conditions is usually positive.

Even less is known about the behaviour at large angles of attack, with fully separated flow. However, it can be argued that in this case (representative for downwind conditions) the damping will be positive because it is caused by drag rather than by lift forces. Unfortunately this does not help for the total directional stability because the restoring moment was already found to be negative (Sect. 8.4). The latter implies negative directional stability, irrespective of any damping.

A complicating factor is further that the yawing motion of a sailing yacht is strongly coupled with the rolling motion, as we have seen already in the preceding section. This is the case in particular in downwind conditions. It is therefore useful to revisit the phenomenon of downwind rolling, the aerodynamics of which were already discussed in Sect. 7.10.

Dynamics of Rolling Motion

As discussed in Sect. 7.10 the aerodynamic forces on the sails of a yacht in rolling motion can enhance the rolling motion in downwind conditions with large sheeting angles. This as a result of the fact that the aerodynamic forces are in phase with the rolling motion. The phenomenon was seen to lead to a rolling oscillation with a frequency about equal to the natural rolling frequency (Sect. 6.8) of the yacht.

It will be clear that the eventual resulting motion will not only be determined by the aerodynamics and the hydrostatics of the yacht but that, as discussed in Sect. 6.8 the hydrodynamic characteristics of the hull and appendages play an important role. This can be illustrated by considering the equivalent dynamic 'spring-mass-damper' system (Sect. 5.18) for the rolling motion.

The equilibrium of moments around the rolling (x-) axis in flat water can be written as (see Sect. 6.8)

$$(I_{xx} + I'_{xx})\ddot{\varphi} + b_\varphi \dot{\varphi} + k_\varphi \varphi = 0 \tag{8.5.5}$$

As before, φ, $\dot{\varphi}$ and $\ddot{\varphi}$ are the angle, angular velocity and angular acceleration, respectively, of the rolling motion. I_{xx} and I'_{xx} are the mass moments of inertia. The 'torsional spring constant' k_φ is mainly determined by the (restoring) lateral hydrostatic stability characteristics of the yacht. The damping coefficient b_φ has hydrodynamic and as well as aerodynamic contributions. One can write

$$b_\varphi = b_{\varphi H} + b_{\varphi A} \tag{8.5.6}$$

A stable rolling motion requires that $b_\varphi > 0$.

The hydrodynamic damping in roll $b_{H\varphi}$ is given by

$$b_{\varphi H} = \frac{dM_{Hx}}{d\dot{\varphi}} \tag{8.5.7}$$

It has already been discussed in Sect. 6.8. There we saw that the hydrodynamic damping is, in general, positive (except, perhaps, for combinations of (very) low boat speeds and high roll rates). Most of the hydrodynamic damping was seen to be provided by the keel.

The aerodynamic damping can be expressed as

$$b_{\varphi A} = \frac{dM_{Ax}}{d\dot{\varphi}} \tag{8.5.8}$$

The rolling moment due to the sails can be approximated as

$$M_{Ax} \approx H_A \zeta_{CER} \tag{8.5.9}$$

In Eq. (8.5.9) H_A is the heeling force and ζ_{CER} the vertical position of the centre of effort of the sails in rolling motion. It is noted here that the centre of effort due to rolling is situated higher than that in steady sailing conditions because of the higher wind velocities due to rolling that are experienced by the upper parts of the sail. Half-way up the mast, or slightly higher, is probably a reasonable estimate for a conventional Bermuda-type rig.

Equation (8.5.8) can also be written as

$$M_{Ax} \approx H_A\zeta_{CER} = (S_A/\cos\varphi)\zeta_{CER} \tag{8.5.10}$$

Assuming ζ_{CER} to be constant, the damping coefficient (8.5.8) can, with (8.5.10), be approximated as

$$b_{\varphi A} = \frac{dM_{Ax}}{d\dot\varphi} \approx \left\{\frac{dS_A}{d\dot\varphi}/\cos\varphi\right\}\zeta_{CER} \tag{8.5.11}$$

Setting $S_A \equiv \frac{1}{2}\rho_A V_a^2 S_s C_{SA}$, this can, utilizing the calculus of differentiation, also be written as

$$b_{\varphi A} \approx \frac{1}{2}\rho_A V_a^2 S_s\left\{\frac{dC_{SA}}{d\dot\varphi}/\cos\varphi\right\}\zeta_{CER} \tag{8.5.12}$$

The derivative in Eq. (8.5.12) can be written as

$$\frac{dC_{SA}}{d\dot\varphi} = \frac{dC_{SA}}{d\alpha}\frac{d\alpha}{d\dot\varphi} \approx \frac{dC_{SA}}{d\alpha}\frac{\zeta_{CER}}{V_\alpha} \tag{8.5.12}$$

Since $\beta = \alpha - \delta$, we can, with fixed sheeting angle δ, replace α by β in Eq. (8.5.13). Hence, Eq. (8.5.13) can also be written as

$$\frac{dC_{SA}}{d\dot\varphi} \approx \frac{dC_{SA}}{d\beta}\frac{\zeta_{CER}}{V_\alpha} \tag{8.5.14}$$

This means that the aerodynamic damping is >0 as long as $\dfrac{dC_{SA}}{d\beta} > 0$. This is certainly the case for $\beta < \cong 45°$, as we have seen in Chaps. 4 and 7. For larger apparent wind angles the aerodynamic roll damping was seen to become negative for apparent wind angles somewhere beyond $90°$ (Sect. 7.10).

It is interesting to consider the total roll damping of a yacht as a function of the apparent wind angle. Figure 8.5.1 serves this purpose. The figure shows the total damping coefficient b_φ (for small roll amplitudes) as a function of the apparent wind angle β for normal conditions of sailing (normal sheeting angles). The configuration considered is the 'average' medium-hull/medium-keel cruising yacht considered

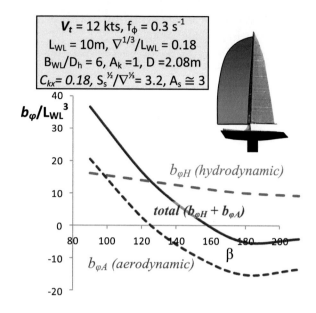

Fig. 8.5.1 *Trend of the variation of roll damping with apparent wind angle for an 'average' cruising yacht configuration with spinnaker type head sail (calculated)*

earlier, now with a spinnaker type head sail. The rolling frequency $f_\varphi = 0.3$ s^{-1}, which is about equal to the natural rolling frequency.

Shown in the figure are the damping coefficients b_φ, normalized by the third power of the length L_{WL} of the waterline, as a function of the apparent wind angle. The curves are based on calculations of the hydrodynamic and aerodynamic forces in rotary motion about the x-axis. For this purpose the modified, 'lifting line' method already mentioned in Sect. 6.8 has been used. The method is semi-empirical in the sense that it uses measured aero/hydrodynamic characteristics of the sail profiles and keel section in steady flow as well as sheeting angles as described in Sect. 7.8. It also contains a correction factor for the sectional lift curve slope to account for low aspect ratio effects. Unsteady flow effects have been taken into account through the theoretical effect of the reduced frequency on the section characteristics (see Sect. 5.18).

The figure shows that the hydrodynamic damping is positive over the range of β considered but that the aerodynamic damping is negative for $\beta > \cong 125°$. As a result the total damping is negative for $\beta > \cong 160°$. The results given are for a true wind speed of 12 kts. Calculations for 18 kts true wind (not shown here) give a similar picture, with increased absolute values of the separate hydrodynamic and aerodynamic damping coefficients but little change in the total damping.

It is emphasized that because of the approximate character of the calculations Fig. 8.5.1 should be interpreted as an indication of the trend for small roll amplitudes rather than an accurate reflection of the roll damping characteristics as a whole.[9] This not in the least because, as already described in Sect. 7.10, a yacht is

---

[9] Partly also because the effect of the rudder on the hydrodynamic damping was not taken into account. It is estimated that, with rudder, the hydrodynamic damping would increase by $\cong 25\%$.

likely to attain large amplitudes of roll with associated non-linear behaviour under
conditions with negative damping.

Roll-Yaw Coupling

Another aspect that has not been taken into account in the calculation model underlying Fig. 8.5.1, is that the hull develops a hydrodynamic yawing moment to starboard
when the boat heels to port and vice versa (Sect. Effects of Heel on Hydrodynamic
Yawing Moment, Static Directional Stability and Yaw Balance). As a consequence
a yacht tends to yaw to starboard when it heels to port and the other way around
(Fig. 8.5.2). This, probably, with some time lag, due to inertia and with some corrective
rudder action by the helmsman (see also Sect. 8.4). This rotation in yaw causes an
increase of the angle of attack of the sail when it heels to port and a decrease when
it heels to starboard. Hence, it softens the aerodynamic effects of rolling during the
part of the cycle that the angle of heel increases (in the absolute sense) but augments
the effects of rolling when the sail swings back.

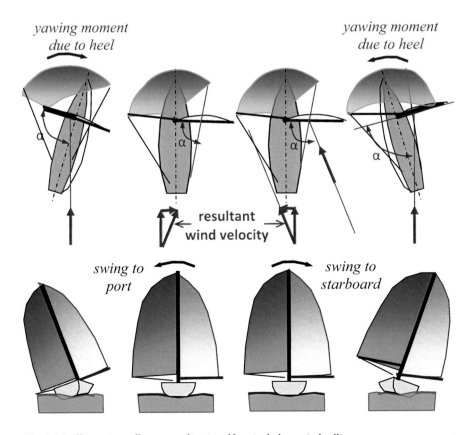

Fig. 8.5.2 *Illustrating roll-yaw coupling in self-excited, downwind rolling*

As we have seen above and before, heel causes yaw. We have seen in Sect. 8.4 that the other way around is also true: yaw causes heel. All together, it will be clear that there is a strong, non-linear coupling between the rolling and yawing motions of a sailing yacht, in particular on downwind courses.

8.6 Sailing in Waves

Pitching, Rolling and Yawing
We have seen in Sect. 6.8 that the motion of a yacht in waves depends on several factors. Without sailes, the most important factors, apart from the wave direction and the associated excitation factor, the wave slope and the mass characteristics of the yacht, were seen to be the (hydrostatic) restoring force and the (hydrodynamic) damping.

Under sail, the aerodynamic damping of the sails is another factor of importance, as already indicated in the preceding section. This is the case for the rolling motion in particular. For the pitching motion the aerodynamic damping of the sails is much smaller than the hydrodynamic damping due to the hull, at least in upwind conditions. The reason is that the aerodynamic pitch damping is approximately proportional to the rate of change with time of the driving force. The latter is probably relative small (10% or less of the average driving force), as we have seen in Sect. 7.10.

For the rolling motion in waves we can obtain an impression of the amplitude of the oscillatory motion by application of the mass-spring-damper model described in Sects. 5.18, 6.8 and 8.5. Some results are shown in Figs. 8.6.1 and 8.6.2.

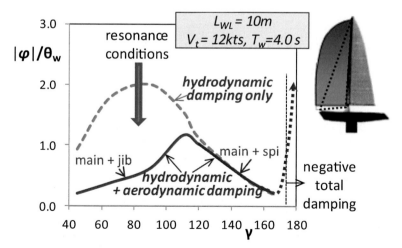

Fig. 8.6.1 *Amplitude of the rolling motion of an 'average' cruising yacht, as a function of true wind angle and North-Sea-like wave conditions (qualitatively, normalized by wave slope)*

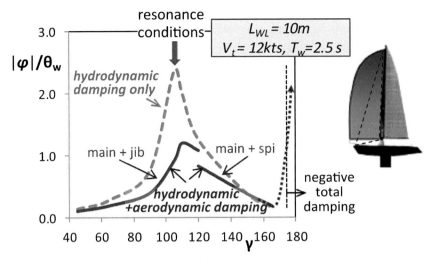

Fig. 8.6.2 *Amplitude of the rolling motion of an 'average' cruising yacht, as a function of true wind angle in estuary-like wave conditions (qualitatively, normalized by wave slope)*

Figure 8.6.1 gives the roll amplitude, normalized by the wave slope, as a function of the true wind angle for the 'average' cruising yacht considered before, (see, for example, Fig. 8.2.7 and Sect. 6.8), in time-averaged equilibrium conditions of sailing at 12 kts true wind in North-Sea-like wave conditions (see Sect. 5.19).

The curves shown give the response in rolling of the yacht with hydrodynamic damping only and with hydrodynamic plus aerodynamic damping. Note that resonance conditions (frequency of wave encounter equal to the natural rolling frequency) occurs at a true wind angle γ of about 85°. Note also that the total damping becomes < 0, implying a dynamic instability (amplitude becomes unbounded), for $\gamma > 175°$.

Another point to note is the large difference between the curves for true wind angles below, say, $\gamma \cong 110°$. This reflects the fact that the aerodynamic damping becomes (much) larger than the hydrodynamic damping for $\gamma < 110°$.

A similar picture for a coastal, estuary-like wave environment (Sect. 5.19) is shown in Fig. 8.6.2. Note that the resonance conditions have shifted to $\gamma \cong 110°$ as a consequence of the shorter wave length.

Comparing Figs. 8.6.1 and 8.6.2 it appears further that, in both cases, the maximum response in roll occurs at $\gamma \cong 115°$. This is probably due to the rapid change of the aerodynamic damping with varying true wind angle γ for $\gamma \cong 115°$.

As already discussed in the preceding sections there is little information available about the dynamics of the yawing motion of a yacht under sail. There is, as discussed in the preceding sections, reason to believe that, in general, a sailing yacht is dynamically stable in yaw in close-hauled conditions but unstable for apparent wind angles beyond, say, $\beta \cong 140°$, and possibly already for, $\beta > \cong 45°$. Besides, it has been found, as already discussed in Sect. 6.8, that the magnitude of the destabilizing, wave-induced yawing moment in the downwind, stern-on-crest/bow-in-trough situation is, in general, so large, that it overrides any directional-stability-in-calm-water property that a ship or yacht may have. The conclusion in Sect. 6.8 that a large and effective rudder and heavy rudder action is required to control a yacht in such situations is, therefore, probably equally valid under sail.

'Snaking' Through Waves

As described in Sect. 7.10 (see Fig. 7.10.3), the combined effects of pitching, rolling and the orbital motion of the water particles in waves, seem to cause hardly a difference in the time-average driving force as compared with flat water conditions.

However, the orbital motion of the water particles in waves has also another, more direct effect on the velocity-made-good to windward of a yacht. Because the yacht spends relatively more time near the wave crests, where the VMG is reduced by the orbital motion and relatively less time near the troughs, where the VMG is increased, there is a loss in time-averaged VMG. Figure 8.6.3 shows, qualitatively, the corresponding track of the yacht over the bottom when the true wind angle is kept constant.

As discussed in some more detail in Bethwaite (2010), the loss in time-average VMG can be reduced significantly (5–30 %) by 'snaking' over the waves in a way so that less time is spent near the crests and more near the troughs. This can be realized by luffing near the crests and bearing away near the troughs, as indicated schematically in Fig. 8.6.4.

Fig. 8.6.3 *Bottom track of a yacht sailing upwind in waves at constant true wind angle γ (qualitatively)*

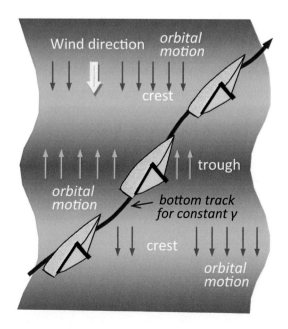

Fig. 8.6.4 *'Snaking' upwind: bottom track of a yacht sailing upwind in waves with the true wind angle varying between 35° (crests) and 55° (troughs)*

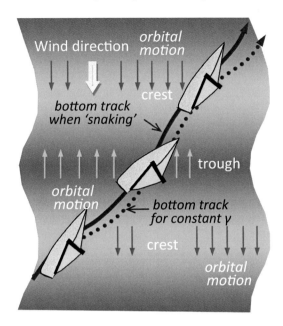

A similar, but opposite technique can, in principle, be used when sailing downwind in waves. Obviously one should, in this case, try to increase the time spent near the crests and reduce the time spent near the troughs. This, however, is relatively complex because of the small difference, possibly, between boat speed and the propagation speed of the waves. In addition there is the risk, or, perhaps, rather the (positive) potential, of surfing (see Sect. 6.5).

Frank Bethwaites' book (2010) contains a lot more information and tips about sailing in waves and many other aspects of "High Performance Sailing". For the racing and dedicated cruising yachtsmen it is highly recommendable if not indispensible.

References

Bethwaite F (2010) High performance sailing, 2nd edn. Adlard Coles Nautical, London, pp 311–314

Claughton AR, Shenoi RE, Wellicom JF (eds) (1999) Sailing yacht design: theory, Chap 7. Prentice Hall, Upper Saddle River. ISBN 0582368561

Fossati F (2009) Aero-hydrodynamics and the performance of sailing yachts. International Marine, New York, p 179. ISBN 978-0-07-162910-2

Fossati S, Muggiasca S, Martina F (2008) Experimental database of sail performance and flying shapes in upwind conditions. International conference on innovation in high performance sailing yachts, Lorient, 29–30 May

Gerritsma J, Moeyes G, Onnink R (1977) Test results of a systematic yacht hull series. 5th HISWA symposium yacht architecture

Hazen GS (1980) A model of sail aerodynamics for diverse rig types. New England yacht symposium, New London

Kerwin JE (1978) A velocity prediction program for ocean racing yachts (Rept 78-11). MIT, Cambridge, July

Larsson L, Eliasson RE (1996) Principles of yacht design. Adlard Coles Nautical, London. ISBN 0-7136-3855-9

Letcher JS Jr (1965) Balance of helm and static directional stability of yachts sailing close-hauled. JRAeS 69(652):241–248

Munk MM (1923) General bi-plane theory. NACA Report 151

ORC-VPP Documentation (2012) Off-shore racing congress

Proceedings 22nd HISWA symposium on yacht design and yacht construction, Amsterdam, Nov. 2012

Van Oossanen P (1993) Predicting the speed of sailing yachts. SNAME Trans 101:337–397

Retrospection

'Mine is a long, wet tail' the author said, in a variation on a theme from Alice in Wonderland. *'Long I can see',* the printer said, counting the pages. *'But why wet? The ink is dry!"*

When I started out on this project I didn't have the slightest idea that it would take me some ten years to finish the job. What started out as an effort to create a full text for a 'Power Point' presentation on the aero-and hydrodynamics of sailing, evolved gradually into a book with a much wider scope and depth. (Too) many questions requiring answers popped-up in the author's mind during the process of writing. The answers to some if not most of these questions required a substantial amount of research and often triggered new questions.

The author feels that this process of questions and answers has increased his knowledge on the subject substantially. Whether it has also improved the accessibility of the book is another matter. It is, of course, up to the reader to decide if it was worth the effort. The author hopes that the reader may benefit from the answers to at least some of the questions.

March 2014   Joop Slooff

© Springer International Publishing Switzerland 2015
J. W. Slooff, *The Aero- and Hydromechanics of Keel Yachts,*
DOI 10.1007/978-3-319-13275-4

Appendices

Appendix A

Some Basic Mathematical Notions

Trigonometric Functions

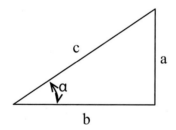

The sine (or sinus), *sin* α, of an angle α is defined as (see figure)

$$\sin \alpha = a/c$$

The cosine (or cosinus), *cos* α, of an angle α is defined as

$$\cos \alpha = b/c$$

The tangens, *tan* α, of an angle α is defined as

$$\tan \alpha = a/b$$

The cotangens, *cotan* α, of an angle α is defined as

$$\text{cotan } \alpha = b/a,$$

© Springer International Publishing Switzerland 2015
J. W. Slooff, *The Aero- and Hydromechanics of Keel Yachts*,
DOI 10.1007/978-3-319-13275-4

where a and b are perpendicular and a is the leg of a rectangular triangle opposing the angle α.

It follows that

$$\sin \alpha / \cos \alpha = \tan \alpha$$

and

$$\cos \alpha / \sin \alpha = \cotan \alpha$$

Inverse Trigonometric Functions

The arcsine (or arcsinus), *arcsin*p, of a quantity p is defined as the angle, the sine of which is equal to p. Hence (see figure)

$$\arcsin (b/a) = \alpha$$

The arccosine (or arccosinus), *arccos*q, of a quantity q is defined as the angle, the cosine of which is equal to q. Hence

$$\arccos (b/c) = \alpha$$

The arctangens, *arctan*r, (also denoted as *atan*r) of a quantity 'r' is defined as the angle, the tangens of which is equal to r. Hence

$$\arctan (a/b) = \alpha$$

The arccotangens, *arccotan*(1/r), of a quantity 1/r is defined as the angle, the cotangens of which is equal to 1/r. Hence

$$\arccotan (b/a) = \alpha$$

Units of Angle

Angles are usually measured in *degrees* [°], with 360° circumscribed by a full circle.

A non-dimensional unit of angle is the *radian* [rad], with 2π radians circumscribed by a full circle, where π = 3.141…. (the ratio between the circumference and the diameter of a circle).

It follows that

$$1 \text{ rad} \equiv 360/(2\pi) \approx 57.3\ldots \text{degrees}$$

Logarithms

The value 'y' of the common logarithm *logx* or $^{10}logx$ of a number 'x' is defined as

$$10^y = x,$$

where 10 is the base number of the common logarithm

The value 'y' of the natural logarithm *lnx* of a number 'x' is defined as

$$e^y = x,$$

where e=2.178... is the base number of the natural logarithm as it appears in the calculus of differentiation.

Rate of Change, Derivatives and Integrals

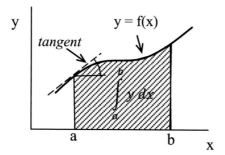

When a quantity 'y' is a function f(x) of another quantity x, the *derivative* or *rate of change* of 'y(x)' with respect to x is usually denoted as $\frac{dy}{dx}$ or y'(x). The value of $\frac{dy}{dx}$ at the point y(x=a) is equal to the slope (in radians) of the tangent to the curve representing y(x) at y(x=a), see figure.

When 'y' is a function of more than one, e.g. two quantities 'p' and 'q', the rate of change of y(p, q) with respect to p for constant q or *partial derivative* with respect to p is denoted as $\frac{\partial y}{\partial p}$ and the rate of change or *partial derivative* with respect to q as $\frac{\partial y}{\partial q}$.

In case 'p' and 'q' are linked in the sense that the change *dp* of 'p' and the change *dq* of 'q' are coupled, the *total derivative* $\frac{dy}{dp}$ of 'y' with respect to 'p' is given by

$$\frac{dy}{dp} = \frac{\partial y}{\partial p} + \frac{\partial y}{\partial q}\frac{dq}{dp}$$

Similarly, the *total derivative* $\dfrac{dy}{dq}$ of 'y' with respect to 'q' is given by

$$\frac{dy}{dq} = \frac{\partial y}{\partial q} + \frac{\partial y}{\partial p}\frac{dp}{dq}$$

The area between the curve representing $y = f(x)$ and the abscissa (x-axis) over the interval $a \le x \le b$ is usually denoted as $\int_a^b y(x)dx$. This is known as the *integral* of the function $y = f(x)$ over the interval $a \le x \le b$.

Appendix B

Formulae for the Induced Angle of Attack of a Lifting Surface

Induced Angle of Attack, Effective Span
As discussed in Sect. 5.15, the induced angle of attack of a lifting surface, due to the trailing vorticity ('tip vortices') can be expressed as

$$\alpha_i = C_L/(\pi Ae) \quad \text{(radians)} \tag{B.1}$$

It has been found by the author, using information from wing theory (Abbott et al. 1949; Küchemann 1978), that a suitable, empirical approximation for the efficiency factor e ($=$ ratio b_e/b of the effective span to the geometric span) of a lifting surface is given by

$$e = 1 - f_1(A) - f_2(A)(TR - TR_{opt})^2/\{1 + 2(TR - TR_{opt})^2\} \tag{B.2}$$

Here, TR_{opt} is the 'optimum' taper ratio for minimum induced drag as given in Fig. 5.15.9. TR_{opt} is suitably approximated by

$$TR_{opt} = 0.45\,(1 - \sin\Lambda)^2/\cos^2\Lambda \tag{B.3}$$

$f_1(A)$ and $f_2(A)$ are empirical factors that depend on aspect ratio and taper ratio:

$$f_1(A) = (0.011\,A)^2 \tag{B.4}$$

$$f_2(A) = f_0(TR)f_2(A,0) + f_1(TR)f_2(A,1) \tag{B.5}$$

with

$$f_2(A,0) = -0.00774\,A^2 + 0.185\,A \tag{B.6}$$

$$f_2(A,1) = -0.00316\,A^2 + 0.0118\,A \qquad (B.7)$$

and

$$f_0(TR) = 1 - TR / (a + TR) \qquad (B.8)$$

$$f_1(TR) = TR^2 / \{(TR^2 + b)f_2(A,1)\} \qquad (B.9)$$

where

$$a = 0.0105A \qquad (B.10)$$

$$b = 1/(f_2(A,1)) - a / \{1 + a\,f_2(A,0)\} - 1 \qquad (B.11)$$

References

Abbott IH, Von Doenhoff AE (1949) Theory of wing sections. McGraw Hill, New York
Küchemann D (1978) The aerodynamic design of aircraft. Pergamon Press, Oxford

Appendix C

Fluid Dynamic Forces on Lifting Surface Configurations

C.1 Lift and Induced Drag of Wing-body Configurations in Steady Flow

There are many 'classical' theoretical/empirical models for the estimation of lift and drag on wing-body configurations (Schlichting and Truckenbrodt 1979). It seems, however, that few, if any, of these are based on the physics based concept (Slooff 1984, 1985) of lift carry-over introduced in Sect. 5.17. A simple model based on this concept will therefore be worked out in this appendix.

Starting out from Kutta-Youkovski's Eq. (5.14.1) for the relation between the lift and the circulation of a wing section, the lift of a symmetrical wing-body configuration can be written as

$$z = -D/2$$
$$L = 2\int \rho V \Gamma(z)\,dz + \rho V \Gamma_R D \qquad (C.1)$$
$$z = -(b/2 + D/2)$$

In Eq. (C.1), b is the geometrical span of the exposed wing, D is the body diameter and $\Gamma(z)$ represents the span-wise distribution of circulation. Γ_R is the circulation at the root of the wing or fin.

Obviously, the integral in (C.1) represents the lift on the exposed part of the wing or fin. This means that Eq. (C.1) can also be written as

$$L = \rho V \Gamma_{avw} b + \rho V \Gamma_R D = L_w \{1 + (\Gamma_R / \Gamma_{avw}) D/b\} = \rho V \Gamma_{av} (b + D) \quad \text{(C.2)}$$

Here, Γ_{avw} is the average circulation on the wing or fin and Γ_{av} is the average circulation on the wing plus body. Dividing by $\frac{1}{2} \rho V^2 S$, where S is the area of the (two) wing (halves), gives

$$C_L = C_{Lw} \{1 + (\Gamma_R / \Gamma_{avw}) D/b\} \quad \text{(C.3)}$$

As described in Sect. 5.15 the lift acting on a wing or fin without a body can be expressed as

$$C_{Lw} = 2\pi E(\alpha - \alpha_i) \quad \text{(C.4)},$$

where α_i is the induced angle of attack. The latter can be written as

$$\alpha_i = 2L/(\rho\pi V^2 b b_e) = 2(\Gamma_{avw}/V)/(\pi b_e) = C_{Lw}/(\pi Ae) \quad \text{(C.5)},$$

When attached to a body, the downwash is determined by the circulation and the effective span of the wing plus body. This means, that (C.5) should be replaced by

$$\begin{aligned}
\alpha_i &= 2(\Gamma_{av}/V)/\{\pi(b_e + D)\} \\
&= 2(\Gamma_{avw}/V)[\{1 + (\Gamma_R/\Gamma_{avw}) D/b\}/(1 + D/b)]/\{\pi(b_e + D)\} \quad \text{(C.6)} \\
&= C_{Lw}[\{1 + (\Gamma_R/\Gamma_{avw}) D/b\}/(1 + D/b)]/\{\pi A(e + D/b)\}
\end{aligned}$$

Substitution of (C.4) and (C.6) into (C.3) gives

$$\begin{aligned}
C_L &= 2\pi E(\alpha - \alpha_0)\{1 + (\Gamma_R/\Gamma_{avw}) D/b\} \\
&/[1 + 2\pi E\{1 + (\Gamma_R/\Gamma_{avw}) D/b\}/(1 + D/b)/\{\pi A (e + D/b_e)\}]
\end{aligned} \quad \text{(C.7)}$$

The quantity $b_e + D$ in Eq. (C.6) can be associated with the notion 'effective span of a wing-body configuration' (b'_e) in the sense that

$$b'_e = b_e + D \quad \text{(C.8)},$$

where b_e is the effective span of the exposed wing.

As already indicated above, Γ_R / Γ_{avw} represents the ratio between the circulation at the root (wing-body intersection) and the average circulation on the exposed wing or fin. It can be shown (Katz and Plotkin 1991) that for an elliptic distribution of circulation this ratio is equal to $4/\pi$. It has been verified by the author, using the data contained by (DeYoung and Harper 1948), that for non-elliptical distributions of circulation the ratio Γ_R / Γ_{avw} can be approximated empirically by

$$\Gamma_R / \Gamma_{avw} = (4/\pi) F_R \quad \text{(C.9)},$$

with F_R given by

$$F_R = c\left[1-(A/3)f(TR)/\cos^{3/2}\Lambda/\left\{1+(A/3)f(TR)/\cos^{3/2}\Lambda\right\}\right] \qquad (C.10)$$

In Eq. (C.10)

$$c = 1+0.633(-0.072\sin^2\Lambda+0.015\sin\Lambda+0.05)A/(1+0.1A^2) \qquad (C.11)$$

and

$$\begin{aligned}
f(TR) = {} & (-0.00501\,\Lambda^2 +0.0123\,\Lambda -0.0126)\,TR^2 \\
& +(-0.0376\,\Lambda^2 +0.0386\,\Lambda+0.155)\,TR \qquad (C.12) \\
& + (-0.00514\,\Lambda^2 +0.0657\,\Lambda -0.066)
\end{aligned}$$

with Λ given in radians.

It is further noted that in the derivation given above it has, silently, been assumed that the spanwise distribution of circulation and, hence, the efficiency factor e is not influenced by the presence of the body. This, however, is highly unlikely.

Unfortunately there is, to the author's knowledge, no simple theoretical or empirical model for the effects of the body on the efficiency factor of a wing. For this reason the efficiency factor of the wing plus body is taken as that of the exposed wing alone.

The model described above is valid only for small to moderate values of the ratio of body diameter to wing span. This can be seen in the fact that Eqs. (C.6), (C.7) do not reflect that for very large ratios of body diameter to wing span, the side of the body begins to act as a plane of symmetry for the flow about the wing or fin (as in Fig. 5.17.1). That is, for very large ratios D/b of body diameter to wing span, the effective span of a wing-body should tend to that of the exposed wing. This can be modeled by replacing the term D/b in the denominator $(e+D/b)$ of Eq. (C.6) by a term D_e/b given by

$$D_e/b = (D/b)/\{1+(0.1D/b)^2\} \qquad (C.13),$$

with a similar replacement in Eq. (C.7).

As a further refinement, intended to take into account that the body also induces upwash at the position of the wing (Schlichting and Truckenbrodt 1979), one can apply a multiplication factor F_α to the expression (C.7) for the lift curve slope. It has been found by the author that the upwash effect is reasonably modeled by choosing

$$F_\alpha = 1+0.07(D/B)(D/b)[\ln\{1+0.07(D/b)\}-\ln\{0.07(D/b)\}] \qquad (C.14),$$

where ln is the natural logarithmic function.

Other aspects not covered are the effects of non-circular cross-section and finite body length. The effect of a non-circular cross-section can, tentatively, be modeled by replacing the body diameter or width D in Eq. (C.13) by 2R, where R is the radius of curvature of the body at the position of the wing (see also Fig. 5.17.5). R can be approximated by

$$R = (B^2/4 + D^2/4)/D \qquad (C.15)$$

The effect of the length (L) of the body (relative to the span of the wing) seems to be small, except, possibly, for relatively large values of D/b in combination with small values of the ratio L/b. In such a case there can be a significant contraction of the trailing vortex system due to the flow field about the body (Greeley and Cross-Whiter 1989). Although this does not have an effect on the active span b' (= b + D), it causes a reduction of the effective span b_e and the efficiency factor e.

It is emphasized, that the model described above is necessarily crude. The most important shortcoming is probably that it does not model the effect of the body on the span-wise distribution of circulation and the effective span or efficiency factor of the wing. It is believed, nevertheless, that it is sufficiently representative to model the main trends in the fluid dynamics of wing-body configurations.

Figure C.1 gives a comparison for straight tapered wings between results of the present analytical model with those of the 'classical' empirical models of Vladea (1934) and Flax (1973) as presented in Oossanen (1981). The comparison is given in terms of the ratio between the lift of the isolated exposed wing and the lift of the complete wing-body configuration. It is to be noted that, unlike the present model, the models of Vladea and Flax do not distinguish between wings of different

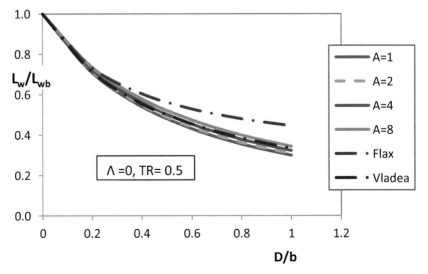

Fig. C.1 *Ratio of lift of exposed wing alone to lift of wing-body as a function of body diameter to exposed wing span ratio*

planforms in terms of interference; the body diameter to wing span ratio is the only parameter determining wing-body interference effects.

It appears that, for the configuration considered ($\Lambda=0$, TR$=0.5$) the results of the present method agree reasonably well with those of Vladea's method. The weak dependence on aspect ratio that can be noticed in Fig. C.1 reflects small changes in the spanwise lift distribution due to the dependence of the 'root factor' F_R (see Eqs. (C.9)–(C.12)) on variation of planform shape.

C.2 Additional Lift and Drag of Low Aspect Ratio Wings at High Angle of Attack

As described in Sect. 5.15 lifting surfaces of small aspect ratio develop additional lift as well as additional drag at high angle of attack due to side edge and/or leading separation. The additional lift δC_L is approximately proportional to the square of the angle of attack and depends on the aspect ratio and the angle of sweep of the leading edge. Using available experimental data, the author has found that for low to moderate angles of attack the additional lift can be approximated by the following empirical relation:

$$\delta C_L \cong [1.8\{A^4/(1+A^5)\}\tan\Lambda_0 + 6.5\{A/(1+A^4)\}\,TR/(1+\tan\Lambda_0)]\alpha^2 \quad (C.16),$$

where the angle of attack α is measured in radians.

The additional drag can be approximated by (Katz and Plotkin 1991)

$$\delta C_D = \delta C_L \tan\alpha \tag{C.17}$$

The additional maximum lift can be estimated by an expression similar to (C.16):

$$\delta C_{Lmax} \cong 0.1\,[1.8\{A^4/(1+A^5)\}\tan\Lambda_0 + 6.5\{A/(1+A^4)\}\,TR/(1+\tan\Lambda_0)] \quad (C.18)$$

These expressions reflect that the additional lift is negligible for lifting surfaces that combine a small taper ratio (TR≈ 0) with small leading edge sweep ($\Lambda\approx 0$).

C.3 Fluid Dynamic Moments in Unsteady Flow

For the moment M with respect to the axis of rotation it has been found (Katz and Plotkin 1991) that this can be expressed like Eq. (5.18.15):

$$M = M_0(\alpha_e) + M_1(\dot\alpha_e) + M_2(\ddot\alpha_e) \tag{C.3.1},$$

with

$$M_0(\alpha_e) = \tfrac{1}{2}\rho V^2 S\,c\,\{C_M(\alpha_{ave}) + \frac{dC_M}{d\alpha}C'(k_f)|\alpha|\sin\omega t\} \tag{C.3.2a}$$

$$M_1(\dot\alpha_e) = \tfrac{1}{2}\rho V^2 S\,c(f_{1Mh} + f_{1M\alpha})k_f\cos\omega t \tag{C.3.2b}$$

and

$$M_2(\ddot{\alpha}_e) = \tfrac{1}{2}\rho V^2 S \; cf_{2M} k_f^2 \sin\omega t \qquad\qquad (C.3.2c)$$

Here, $C_M(\alpha_{ave})$ is the moment coefficient at the mean angle of attack in steady flow and $\dfrac{dC_M}{d\alpha}$ is the rate of change of the moment coefficient C_M in steady flow with the angle of attack. Note that this depends on the position of the reference point (axis of rotation). $\dfrac{dC_M}{d\alpha} > 0$ when the axis of rotation is aft of the aerodynamic centre ($x_a/c > \frac{1}{4}$, see Fig. 5.18.7 and Sect. 5.15), $\dfrac{dC_M}{d\alpha} = 0$ when $x_a/c = \frac{1}{4}$ and $\dfrac{dC_M}{d\alpha} < 0$ when $x_a/c < \frac{1}{4}$.

The factors f_{1M} and f_{2M} depend on the amplitudes of the surging, heaving and pitching motion, the position of the axis of rotation of the pitching motion and the lift deficiency factor $C'(k_f)$. The other quantities are the same as before (Eqs. (5.18.16–5.18.19)).

In a pure heaving motion $f_{1M\alpha}=0$ and the factor f_{1Mh} is <0 for $x_a/c > \frac{1}{4}$, $f_{1Mh}=0$ for $x_a/c = \frac{1}{4}$ and $f_{1Mh}>0$ for $x_a/c < \frac{1}{4}$. The factor f_{2M} is found to be >0 for $x_a/c > \frac{1}{2}$ in heaving mode, 0 for $x_a/c = \frac{1}{2}$ and <0 for $x_a/c < \frac{1}{2}$.

In a pure pitching motion we have $f_{1Mh}=0$. The factor $f_{1M\alpha}=0$ for $x_a/c = \frac{3}{4}$ and $x_a/c \approx \frac{1}{2}, >0$ for $\frac{1}{2}\leq x_a/z < \frac{3}{4}$ and <0 for $x_a/c \leq \frac{1}{2}$, $x_a/c > \frac{3}{4}$. The latter means that the M_1 part of the unsteady moment opposes the pitching motion for $x_a/c \leq \frac{1}{2}$, $x_a/c > \frac{3}{4}$. The factor f_{2M} is, in the pitching mode, also found to be 0 for $x_a/c = \frac{3}{4}$ and $x_a/c \approx \frac{1}{2}, <0$ for $\frac{1}{2}\leq x_a/c < \frac{3}{4}$ and >0 for $x_a/c \leq \frac{1}{2}$, $x_a/c > \frac{3}{4}$.

From the preceding paragraphs it follows that, in terms of a 'dynamic system', the quasi-steady moment $M_0(\alpha_e)$ is restoring when the axis of rotation is forward of the $\frac{1}{4}$-chord point and that the rotation is damped through the $M_1(\dot{\alpha}_e)$ moment when the axis of rotation is forward of the $\frac{1}{2}$-chord point or aft of the $\frac{3}{4}$-chord point. It also follows that a dynamic instability in pitching motion occurs when the axis of rotation is located between the $\frac{1}{2}$-chord point and the $\frac{3}{4}$-chord point. This is when the contribution of the $M_1(\dot{\alpha}_e)$ moment to the damping is negative.

References

Flax AH (1973) Simplification of the wing-Body Problem. J Aircraft 10(10, Oct. 1973):640

Katz J, Plotkin A (1991) Low-speed aerodynamics. McGraw-Hill, New York. ISBN 0-07-05040446-6

Oossanen Peter van (1981) Method for the calculation of the resistance and side force of sailing yachts, paper presented at Conference on 'calculator and computer aided design for small craft—the way ahead', R.I.N.A., 1981

Slooff JW (1984) On wings and keels. Int Shipbuild Prog 31(356):94–104

Slooff JW (1985) On wings and keels (II), AIAA 12th Annual Symposium on Sailing ('The Ancient Interface'), Seattle, Wa, 21–22 Sept 1985

Schlichting H, Truckenbrodt E (1979) Aerodynamics of the airplane. McGraw-Hill, New York

Vladen J (1934) Fuselage and engine nacelle effects on airplane wings, NACA TM No. 736

Appendix D

Approximate Formulae for the Position of the Transverse Metacentre of the Hull of a Sailing Yacht

As mentioned in Sect. 6.2, the distance **bm** between the centre of buoyancy and the metacentre is given by

$$bm = \frac{I_T}{\nabla} \tag{D.1a}$$

or

$$bm/L_{WL} = \frac{I_T}{\nabla L_{WL}} \tag{D.1b}$$

In this expression ∇ is the displacement volume. The quantity I_T is given by an integral along the length of the waterline (Lewis 1988):

$$I_T = \int_0^L y^3 \, dx \tag{D.2},$$

which can also be written as

$$I_T = B_{WL}{}^3 \, L_{WL} \int_0^1 \left(\frac{y}{B_{WL}}\right)^3 d\left(\frac{x}{L_{WL}}\right) \tag{D.3}$$

With (D.3), Eq. (D.1) can be written as

$$bm/L_{WL} = \frac{(B_{WL}/L_{WL})^3}{\nabla/(L_{WL})^3} \int_0^1 \left(\frac{y}{B_{WL}}\right)^3 d\left(\frac{x}{L_{WL}}\right) \tag{D.4}$$

The author has found that for many, if not most, sailing yachts the value of the integral in (D.4) can be approximated by

$$\int_0^1 \left(\frac{y}{B_{WL}}\right)^3 d\left(\frac{x}{L_{WL}}\right) \approx .125(C_{WP})^3 \tag{D.5},$$

Where C_{WP} is the waterplane area coefficient (see Sect. 2.2).

A formula for the approximation of the distance $b0_h$ between the centre of buoyancy CB of the hull and the water surface at zero heel, also used by the author, reads

$$b0_h/D_h = 1.12 \, C_P \left(-5.2C_m{}^2 + 8.1C_m - 2.4\right)\left(\frac{C_m}{.766}\right)^{1.3} /(2.36 \, C_P - 0.34) \tag{D.6},$$

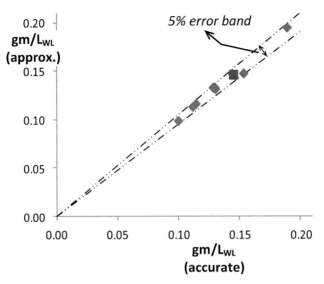

Fig. D.1 *Correlation for metacentric height*

where C_p is the prismatic coefficient and C_m the mid-ship section coefficient. The effect of the keel on the position of the CB can be taken into account through the formula

$$b0/D_h = \{(b0_h/D_h)\nabla_h + (\nabla - \nabla_h)(1 + 0.33\, b_k/D_h)\}/\nabla \qquad (D.7)$$

The metacentric height at zero heel **gm** can then be expressed as

$$gm = bm - b0 + g0 \qquad (D.8)$$

Figure D.1 gives an impression of the correlation between the gm_0 values determined according to the approximation formula and more accurate values obtained through numerical integration for ten different yacht configurations (Larsson and Eliasson 1996; Gerritsma et al. 1977). It appears that the error is, in general, less than 4%.

A reasonably accurate approximation for the variation with heel $\delta gm(\varphi)$ of the metacentric height is, in the author's experience, given by

$$\delta gm(\varphi) = bm\, k(\varphi) \qquad (D.9),$$

where

$$k(\varphi) = \left\{-1.39\left(C_m - C_m^*\right) - 0.096\right\}(1 - \cos\varphi)/(1 - \cos(30^\circ)) \qquad (D.10)$$

Here, C_m^* is the section coefficient of a circle segment section with the same value of B_{WL}/D_h as the actual mid-ship section of the yacht. The difference is a rough measure of the effect of flare or tumblehome. $\left(C_m - C_m^*\right)$ will be >0 in case of tumblehome and <0 in case of flare.

In terms of the description in Sect. 6.2 the righting moment can be written as

$$M_{rx} \approx \Delta\{gm\ \sin\varphi + bm\ k(\varphi)\} \tag{D.11}$$

References

Gerritsma J, Moeyes G, Onnink R (1977) Test results of a systematic yacht hull series. 5th HISWA symposium yacht architecture

Lewis EV (ed) (1988) Principles of naval architecture, vol II. SNAME, Jersey City. ISBN 0-9397703-01-5

Larsson L, Eliasson RE (1996) Principles of yacht design. Adlard Coles Nautical, London. ISBN 0-7136-3855-9

Appendix E

Heuristic Model of the Effects of a Free Surface on the Side Force and Induced Resistance of a Surface Piercing Body

Surface Piercing Fin

Consider the situation of a surface piercing fin with constant chord of length c and zero sweep as in Fig. 6.3.1.

The hydrodynamic side force on the fin can be expressed as

$$S = \int_{z=-b}^{Z_{h-}(x,0)} \int_{x=0}^{c} \Delta p\ dx\ dz + \int_{z=Z_{h-}(x,0)}^{Z_{h+}(x,0)} \int_{x=0}^{c} \Delta p\ dx\ dz \tag{E.1},$$

where Δp is the pressure difference between the pressure $(+)$ and the suction $(-)$ side of the fin. $Z_{h-}(x,0)$, <0, is the surface elevation at the suction side and $Z_{h+}(x,0)$, >0, the elevation at the pressure side.

It follows from Bernoulli's equation (Eq. (5.19.5)), that, for $z < Z_{h-}(x,0)$, Δp can be expressed as

$$\Delta p = \Delta p(x,z) = p_+(x,z) - p_-(x,z)$$
$$= \tfrac{1}{2}\rho\ \{V_-^2(x,z_0) - V_+^2(x,z_0)\} + \rho\ g\{Z_{h-}(x,z_0) - Z_{h+}(x,z_0)\} \tag{E.2}$$

Here, V_- and V_+ represent, respectively, the local surface velocity at the suction and pressure sides of the fin. $Z_{h-}(x,z_0)$ and $Z_{h+}(x,z_0)$ represent the local elevation of the streamlines below the water surface where z_0 is the vertical distance above the water surface of the streamline at a large distance ahead of the fin. Obviously

$$z_0 = z - Z_h(x, z_0) \tag{E.3a}$$

and

$$dz_0 = dz - (\partial Z_h / \partial z_0)\, dz_0 \tag{E.3b}$$

or

$$dz = (1 - \partial Z_h / \partial z_0)\, dz_0 \tag{E.3c}$$

For $Z_{h-}(x, 0) < z < Z_{h+}(x, 0)$ we have

$$p_-(x, z) = p_a \tag{E.4}$$

and

$$p_+(x, z) = p_a + \tfrac{1}{2}\,\rho\{V_b^2 - V_+^2(x, z_0)\} - \rho g\, Z_{h+}(x, z_0) \tag{E.5}$$

so that

$$\Delta p(x, z) = \tfrac{1}{2}\,\rho\{V_b^2 - V_+^2(x, z_0)\} - \rho g\, Z_{h+}(x, z_0) \tag{E.6}$$

It follows from Sect. 5.19 (Eq. (5.19.9)) that, at the air-water interface ($z_0 = 0$), the surface (= streamline) elevation can be expressed as

$$\rho g\, Z_{h+/-}(x, 0) = \tfrac{1}{2}\,\rho\,\{V_b^2 - V_{+/-}^2(x, 0)\} \tag{E.7}$$

A key question is how the streamline elevation Z_h varies with the distance from the air-water interface. It is suggested that an answer to this question can be obtained by considering the deformation of the free surface at the waterline as a traveling surface wave. We have seen in Sect. 5.19 that the amplitude of the orbital motion of surface waves decreases exponentially with depth (see Eq. (5.19.20)). This suggests that the variation of the streamline elevation of a surface piercing fin also varies exponentially with depth:

$$Z_h(x, z_0) = Z_h(x, 0)\, e^{2\pi z_0 / \lambda_w}, \quad z_0 \le Z_h(x, 0) \tag{E.8},$$

Adopting Eq. (E.8) as the basic expression for the variation of the streamline elevation with depth, the next question is what value should be taken for the wavelength λ_w.

According to the theory of simple gravity waves, the wavelength λ_w and the propagation speed c_w of a traveling free surface wave are coupled according to (see Sect. 5.19, Eq. (5.19.21)):

$$\lambda_w = 2\pi c_w^2 / g \tag{E.9},$$

With the fin and the associated disturbance traveling at boat speed V_b we can replace c_w by V_b so that Eq. (E.8) can also be written as

$$Z_h(x, z_0) = Z_h(x, 0) e^{z_0 g / V_b^2} \tag{E.10}$$

or

$$Z_h(x, z_0) = Z_h(x, 0)\, e^{Fr_c^{-2} z_0 / c} \tag{E.11}$$

where Fr_c is the Froude number based on the chord length c of the fin:

$$Fr_c^2 = V_b^2 / c\, g \tag{E.12}$$

It can also be argued that the characteristic length of the surface disturbance caused by the fin is the length c of the chord, irrespective of boat speed or Froude number, and that, therefore, one should use Eq. (E.8) with $\lambda_w = c$ rather than (E.10) or (E.11). This gives

$$Z_h(x, z_0) = Z_h(x, 0)\, e^{2\pi\, z_0 / c},\ z_0 \le 0 \tag{E.13a}$$

Because a surface piercing body does not generate infinitely long waves normal to the direction of motion, there is, further, reason to believe that the streamline elevation will decay somewhat faster with depth than in the case of free running ocean waves. This can be taken into account by multiplying the power of the exponential function in (E.13a) with a factor κ (> 1):

$$Z_h(x, z_0) = Z_h(x, 0)\, e^{2\pi\, \kappa z_0 / c},\ z_0 \le 0 \tag{E.13b}$$

It then follows that the gradient of the variation of the streamline elevation with depth is given by

$$\partial Z_h / \partial z_0 = 2(\pi \kappa / c) Z_h(x, 0)\, e^{2\pi\, \kappa z_0 / c},\ z_0 \le 0 \tag{E.14}$$

It is the author's impression that an appropriate value for κ is $\kappa \approx 1.5$.

For small absolute values of z_0 Eq. (E.13b) can, using the series expansion

$$e^{2\pi\, \kappa z_0 / c} = 1 + 2\pi \kappa z_0 / c + \ldots \tag{E.15}$$

be approximated by

$$Z_h(x, z_0) = Z_h(x, 0)(1 + 2\pi \kappa z_0 / c + \ldots) \tag{E.16}$$

and Eq. (E.14) by

$$\partial Z_h / \partial z_0 = 2(\pi \kappa / c)\, Z_h(x,0)\,(1 + 2\pi \kappa z_0 / c + \ldots), \quad z_0 \leq 0 \tag{E.17}$$

It is further noted that for $Fr^2 = 1/2\pi$, that is $Fr \approx 0.4$, the alternative expressions (E.11) and (E.13a) become identical.

The pressure difference Δp on the fin Eq. (E.2) can now, for $z \leq Z_{h-}(x,0)$, or $z_0 < 0$, be written as

$$\Delta p(x,z) = \tfrac{1}{2}\rho \left\{ V_-^2(x,z_0) - V_+^2(x,z_0) \right\} + \rho g \left\{ Z_{h-}(x,0) - Z_{h+}(x,0) \right\} e^{2\pi \kappa z_0 / c} \tag{E.18}$$

or, utilizing Eq. (E.7), as

$$\Delta p(x,z) = \tfrac{1}{2}\rho \left\{ V_-^2(x,z_0) - V_+^2(x,z_0) \right\} - \tfrac{1}{2}\rho \left\{ V_-^2(x,0) - V_+^2(x,0) \right\} e^{2\pi \kappa z_0 / c} \tag{E.19}$$

For small absolute values of z_0 this can be approximated by

$$\Delta p(x,z) = \tfrac{1}{2}\rho \left\{ V_-^2(x,0) - V_+^2(x,0) \right\} (-2\pi \,\kappa z_0 / c + \ldots) \tag{E.20}$$

For $Z_{h-}(x,0) < z < Z_{h+}(x,0)$ the expression for Δp takes the form

$$\Delta p(x,z) = \tfrac{1}{2}\rho \left\{ V_b^2 - V_+^2(x,z_0) \right\} - \tfrac{1}{2}\rho \left\{ V_b^2 - V_+^2(x,0) \right\} e^{2\pi \kappa z_0 / c} \tag{E.21}$$

For small values of z_0/c this can be approximated by

$$\Delta p(x,z) = \tfrac{1}{2}\rho \left\{ V_b^2 - V_+^2(x,0) \right\} (-2\pi \kappa z_0 / c + \ldots) \tag{E.22}$$

Considering first the integrals $\int \Delta p(x,z)\,dx$ in Eq. (E.1) we note that the local surface velocities V_+ and V_- are related to the local circulation $\Gamma(z)$ around the fin. It can be shown (Prandtl and Tietjens 1957) that, at least for small disturbances, there holds

$$\int_{x=0}^{c} \left\{ V_-^2(x,z) - V_+^2(x,z) \right\} dx = 2 V_b \Gamma(z) \tag{E.23}$$

and, similarly,

$$\int_{x=0}^{c} \left\{ V_b^2(x,z) - V_+^2(x,z) \right\} dx = V_b \Gamma(z) \tag{E.24}$$

Equations (E.23), (E.24) are obtained by expressing $V_-(x,z)$ and $V_+(x,z)$ as

$$V_-(x,z) = V_b + v_-(x,z) \tag{E.25a}$$

and

$$V_+(x,z) = V_b + v_+(x,z) \tag{E.25b},$$

where v_- and v_+ are the local perturbation velocities on the suction and pressure sides of the fin, respectively.

The integrant in (E.23) can, neglecting squares of the perturbation velocity, be approximated by

$$V_-^2(x,z) - V_+^2(x,z) = 2V_b\{v_-(x,z) - v_+(x,z)\} \tag{E.26}$$

which gives

$$\int\limits_{x=0}^{c} \{V_-^2(x,z) - V_+^2(x,z)\}dx = \int\limits_{x=0}^{c} 2V_b\{v_-(x,z) - v_+(x,z)\}dx = 2V_b\Gamma(z) \tag{E.27}$$

Substituting (E.19) with (E.27) into the first integral (I_1) of the expression (E.1) for the side force, then gives, assuming $|Z_h(x,0)| \ll c$, and using the identity $2\pi\kappa z_0/c = 2\pi\kappa A_k z_0/b$ with $A_k = b/c$

$$I_1 = \rho V_b \int\limits_{z_0=-b}^{0} \left\{\Gamma(z_0) - \Gamma(0)\ e^{2\pi\kappa\ Akz_0/b}\right\}dz_0 \tag{E.28}$$

Obviously, the first term of the integrant corresponds with classical wing theory. The second term represents the pressure relief effect of the free surface.

The second integral

$$I_2 = \int\limits_{z=Z_{h-}(x,0)}^{Z_{h+}(x,0)} \int\limits_{x=0}^{c} \Delta p\ dx\ dz \tag{E.29a},$$

can be worked out to give

$$I_2 = -\tfrac{1}{2}\rho\pi \int\limits_{x=0}^{1} \left\{V_b^2 - V_+^2(x,0)\right\}\left[\left\{Z_{h-}^2(x,0) - Z_{h+}^2(x,0)\right\}\right]d(x/c) \tag{E.29b}$$

It is easily verified that this is proportional to Fr^4 (see Eq. (5.19.14) in Sect. 5.19) and, hence, for low Froude numbers can be safely neglected for most practical purposes.

Returning to the first integral it is noted that the integration cannot be performed without knowledge of the spanwise distribution of circulation $\Gamma(z_0)$. Assuming a (semi-)elliptic distribution, that is

$$\Gamma(z_0) = \Gamma(0)\sqrt{\left\{1 - (z_0/b)^2\right\}} \tag{E.30},$$

the integration can be performed to give

$$S = \rho\ V_b\Gamma(0)\ b\left[\pi/4 - \{1/(2\pi\kappa A_k)\}\ (1 - e^{-2\pi\kappa A_k})\right] \tag{E.31}$$

It can also be shown (Gerritsma and Keuning 1984) that for an elliptical distribution of circulation

$$\Gamma(0) = 4\, bV_b(\lambda - \lambda_0)/(1 + A_k) \tag{E.32a}$$

where λ is the angle of attack or angle of leeway in radians and b and A_k are the geometric (not the effective!) span and aspect ratio, respectively, of the submerged part of the fin.

In the derivation (Gerritsma and Keuning 1984) leading to Eq. (E.32a), the 'edge factor' E, introduced in Sect. 5.15 to improve the accuracy of lift prediction for small aspect ratios, has not been taken into account. When 'E' is taken into account it gives

$$\Gamma(0) = 4EbV_b(\lambda - \lambda_0)/(E + A_k) \tag{E.32b}$$

Substituting (E.32b) into (E.31) gives

$$S = 4E\rho V_2^b b^2 \left[\pi/4 - \{1/(2\pi\kappa A_k)\}\left(1 - e^{-2\pi\kappa A_k}\right) \right] (\lambda - \lambda_0)/(E + A_k) \tag{E.33}$$

Dividing by $\tfrac{1}{2}\rho V_b^2\, S$ gives

$$C_S = 2\pi E \left[1 - \{2/(\pi^2\kappa A_k)\}\left(1 - e^{-2\pi\kappa A_k}\right) \right](\lambda - \lambda_0)/(1 + E/A_k) \tag{E.34}$$

This can also be written as

$$C_S = C_{S_0} \left[1 - \{2/(\pi^2\kappa A_k)\}\left(1 - e^{-2\pi\kappa A_k}\right) \right] \tag{E.35}$$

Here,

$$C_{S_0} = 2\,\pi E(\lambda - \lambda_0)/(1 + E/A_k) \tag{E.36}$$

is the side force coefficient without free surface effect for an elliptic distribution of circulation and zero angle of sweep.

At this point it must be mentioned that the assumptions of a constant chord fin and an elliptic distribution of circulation are not (fully) compatible, except, to some extent, for (effective) aspect ratios smaller than about two. As indicated in Chap. 5, lifting surfaces of small aspect ratio always have an almost elliptic distribution of circulation, irrespective of the taper ratio.

Effects of Sweep Angle and Taper

It follows from Sect. 5.15 and 6.3 and Appendices B and C that the effects of sweep and taper on the side force C_{S_0} of a fin-keel without free surface effects can be taken into account by extending the expression (E.36) to

$$C_{S_0} = 2\pi E(\lambda - \lambda_0)/\{1 + E/(A_k e)\} \tag{E.37}$$

Here it is to be noted that the edge factor E implied by Fig. 5.15.10 and the efficiency factor e (or effective span ratio b_e/b, see Appendix B, Eq. (B.1) and Fig. 5.15.11) should be taken for an (effective) aspect ratio that is twice the value of the geometrical aspect ratio A_k of the submerged part of the fin. This means, for example, that the aspect ratio A in Fig. 5.15.10 should be taken as twice the geometric aspect ratio of the fin-keel and that $A = 2A_k$ should be substituted for A in all equations leading to the expression for $e = b_e/b$ in Appendix B.

The effect of taper on the free surface effect is another matter. It can be argued that for a tapered fin the root chord c_R is the governing quantity in the exponential function that models the free surface effect. This means that the exponent $2\pi\kappa z_0/c$ in (E.19) should be taken as $2\pi\kappa z_0/c_R$. For a straight-tapered fin this can, introducing the taper ratio $TR = c_T/c_R$, be written as $\pi\kappa A_k(1+TR)z_0/b$, so that a more general version of (E.28) is

$$I_1 = \rho V_b b \int_{z_0/b=-1}^{0} \left\{ \Gamma(z_0) - \Gamma(0) \, e^{\pi\kappa \, A_k(1+TR)\, z_0/b} \right\} d(z_0/b) \tag{E.38}$$

It can further be argued that the loss of side force due the free surface effect at a given depth z_0 must be proportional to the local chord length $c(z_0)$ of the fin. This suggests that Eq. (E.38) should be extended to read

$$I_1 = \rho V_b b \int_{z_0/b=-1}^{0} \left[\Gamma(z_0) - \Gamma(0) \{ c(z_0)/c(0) \} \, e^{\pi\kappa \, A_k(1+TR)\, z_0/b} \right] d(z_0/b) \tag{E.39}$$

where $c(0) = c_R$ is the chord length at the waterline. For a straight-tapered fin this can be written as

$$I_1 = \rho V_b b \int_{z_0/b=-1}^{0} \left[\Gamma(z_0) - \Gamma(0) \{ 1 - (z_0/b)(TR-1) \} \, e^{\pi\kappa \, Ak(1+TR)z_0/b} \right] d(z_0/b) \tag{E.40}$$

It can also be argued that the effect of sweep on the free surface effect can be approximated by replacing the factor z_0/b in the exponential function by $z_0/(b\cos\Lambda)$. In this way the increase due to sweep of the absolute distance between a point on the fin and the intersection with the water surface is, roughly, taken into account. Introducing this in (E.40) gives

$$I_1 = \rho V_b b \int_{z_0/b=-1}^{0} \left\{ \Gamma(z_0) - \Gamma(0) \left[1 - (z_0/b)(TR-1) \right] e^{\pi\kappa \, A_k(1+TR)\, z_0/(b\cos\Lambda)} \right\} d(z_0/b) \tag{E.41}$$

This can be integrated to give

$$S = \rho V_b \Gamma(0) \, b[F_0 - F_1 - F_2] \tag{E.42}$$

where, for an elliptic distribution of circulation, $\Gamma(0)$ is given by (E.32) and the basic term F_0 takes the value

$$F_0 = \pi/4 \tag{E.43}$$

It follows from Appendix B that Eq. (E.42) can also be written as

$$S = \rho\, V_b \Gamma_{av} b \left[1 - \{\Gamma(0)/\Gamma_{av}\}(F_1 + F_2)\right] \tag{E.44},$$

where $\Gamma(0) = \Gamma_R$ is the circulation at the root and Γ_{av} the average circulation. As described in Appendix C, the ratio $\Gamma(0)/\Gamma_{av}$ or Γ_R/Γ_{av} can be written as

$$\Gamma_R/\Gamma_{av} = (4/\pi)F_R \tag{E.45}$$

with the factor F_R given by Eq. (C.10)). Recall that for an elliptic distribution of circulation F_R is equal to 1.

F_1 represents the basic free surface effect, including effects of sweep and taper:

$$F_1 = \left[\cos\Lambda/\{\pi\kappa A_k(1+TR)\}\right]\left\{1 - e^{-\pi k A_k (1+TR)/\cos\Lambda}\right\} \tag{E.46}$$

F_2 is an additional term due to taper:

$$F_2 = \left[\cos^2\Lambda(TR-1)/\left\{\pi\kappa A_k\left(1+TR\right)\right\}^2\right]$$

$$\times\left[1 - \{1 + \pi\kappa A_k(1+TR)/\cos\Lambda\}\, e^{-\pi k A_k (1+TR)/\cos\Lambda}\right] \tag{E.47}$$

Dividing (E.44) by $\frac{1}{2}\,\rho V_b^2\, S$ leads to the following expression for the side force coefficient

$$C_S = C_{S_0}\left[1 - (4/\pi)\, F_R\,(F_1 + F_2)\right] \tag{E.48},$$

with C_{S_0} given by Eq. (E.37).

Effect of the Hull
Although there are many 'classical' theoretical models for the effect of a hull on the side force produced by a keel (see, e.g., Oossanen 1981), there is, to the author's knowledge, none that takes into account the effect of the free surface. However, it seems possible to extend the model of the free surface effect for a surface piercing fin as described above to the case of a fin attached to a surface piercing body.

As discussed in Sects. 5.17, 6.3, and in more detail in Appendix C, the main effect of the hull is an increase of the *active span* from the value b of the fin or keel proper to $b + D_h$, where D_h is the draught of the hull, plus a corresponding increase of the *effective span* b'_e (See Fig. 6.3.8). Due to the mechanism of *lift carry-over* the hull carries circulation of a constant value, equal to the circulation Γ_R at the root of the fin, over its full depth. In terms of the integral (E.28) expressing the lift or side force, this means that the range of integration must be extended and split. Without free surface effects the integral can be written as

$$I_{1.0} = \rho\, V_b\left\{\int_{z_0=-(b+D_h)}^{-D_h} \Gamma(z_0)\,dz_0 + \int_{z_0=-D_h}^{0} \Gamma(0)\,dz_0\right\} \tag{E.50},$$

Evaluating the integral in the same way as before gives the following result:

$$I_{1.0} = \rho\, V_b b \Gamma_{\text{avw}} \left[1 + \{\Gamma_R/\Gamma_{\text{avw}}\}\, D_h/b \right] \tag{E.51}$$

The average circulation Γ_{avw} should, as in Appendix C, be taken as the average circulation on the fin only. The expressions (C.6) and (C.7) still hold except for the value of the aspect ratio A, which should be replaced by $2A_k$ and the efficiency factor e_k or effective span b_e of the fin proper, which should also be based on $2A_k$ rather than A. It then follows (see also Appendix C) that Γ_{avw} can be expressed as

$$\Gamma_{\text{avw}}/V = \{2\pi E b_k /(2A_k)\}\ (\lambda - \lambda_0) F_\alpha$$
$$/\left[1 + 2E\{1 + (\Gamma_R/\Gamma_{\text{avw}})\}\, D_h/b/\{2A_k(e_k + D_{he}/b)\} \right] \tag{E.52}$$

where

$$D_{he}/b = (D/b)/\left\{ 1 + (0.1\ R/b)^2 \right\} \tag{E.53}$$

It also follows from Appendix C.1 that the edge factor E has to be based also on $2A_k$ rather than A. It is further noted that the upwash factor F_α is given by (C.14) and the cross-sectional radius of curvature of the hull R in Eq. (E.53) by Eq. (C.15). The ratio $\Gamma_R/\Gamma_{\text{avw}}$ is to be taken as in Eq. (C.9).

The question is (again) how to model the pressure relief effect of the free surface. Considering the basic integral (E.1), the first question, after splitting the z-range of the integration, is what to take for the chord length c. It can be argued that, because the fin itself is still the 'generator' of the lift force, the root chord c_R, together with the circulation Γ_R at the root are still the governing quantities. This suggests that the lift on the submerged part of the hull can be modeled as if the keel were extended upto the waterline with a segment of constant circulation, constant chord and zero sweep. It suggests also that the power in the exponential function in, for example Eq. (E.41), should be taken as $2\pi\kappa z_0/c_R$ and TR $= 1$ for $-D_h \le z_0 \le 0$. For $z_0 \le -D_h$ the integrant of (E.41) would remain what it is. With this, the part $(I_{1.1})$ of (E.41) that represents the free surface effect takes the following form:

$$I_{1.1} = -\rho V_b b \int\limits_{z_0 = -(b + D_h)}^{-D_h} \left\{ \Gamma_R \left[1 - (z_0/b)(TR - 1) \right] e^{\pi\kappa A_k (1 + TR)\, z_0/(b\, \cos\Lambda)} \right\} d(z_0/b) +$$

$$-\rho V_b c_R \int\limits_{z_0 = -D_h}^{0} \{\Gamma_R e^{2\pi\kappa z_0/c_R}\} d(z_0/c_R) \tag{E.54}$$

This can be integrated to give:

$$I_{1.1} = -\rho\, V_b b \Gamma_R [F_{1.1} + F_{2.1} + F_{1.2}] \tag{E.55},$$

In Eq. (E.55) Γ_R is defined by Eq. (E.45), (E.46) and Eq. (E.54). The terms between square brackets are found to take the following forms

$$F_{1.1} = \cos\Lambda/\{\pi\kappa A_k(1+TR)\} \times \{ e^{-\pi\kappa A_k(1+TR)\,(D_h/b)/\cos\Lambda} - e^{-\pi\kappa A_k(1+TR)\,(1+D_h/b)/\cos\Lambda} \}$$

$$(E.56),$$

$$F_{2.1} = \cos^2\Lambda(TR-1)/\{\pi\kappa A_k(1+TR)\}^2$$
$$\times \left[\{1+\pi\kappa A_k(1+TR)(D_h/b)/\cos\Lambda\} \{e^{-\pi\kappa A_k(1+TR)(D_h/b)/\cos\Lambda} + \right.$$
$$\left. -e^{-\pi\kappa A_k(1+TR)(1+D_h/b)/\cos\Lambda}\} \right]$$

$$F_{1.2} = \left[1/\{\pi\kappa A_k(1+TR)\} \right]\left[1-e^{-\pi\kappa A_k(1+TR)(1+D_h/b)} \right] \qquad (E.57)$$

The total side force can then be written as

$$S = I_{1.0} + I_{1.1} = \rho V_b b \Gamma_{avw}[1+(4/\pi)\,F_R\{D_h/b - F_{1.1} - F_{2.1} - F_{1.2}\}] \qquad (E.58)$$

The side force coefficient C_S is obtained by dividing by $\frac{1}{2}\rho V_b^2 S$:

$$C_S = 2\pi E(\lambda - \lambda_0)\,[1+(4/\pi)\,F_R\{D_h/b - (F_{1.1} + F_{2.1} + F_{1.2})\}]$$
$$/[1+2E\{1+(4/\pi)\,F_R D_h/b\}/(1+D_h/b)/\{2A_k(e_k + D_{he}/b)\}] \qquad (E.59)$$

This can also be written as

$$C_S = F_{FS}C_{S_0} \qquad (E.60),$$

where

$$F_{FS} = [1+(4/\pi)F_R\{D_h/b - (F_{1.1} + F_{2.1} + F_{1.2})\}]/[1+(4/\pi)F_R D_h/b] \qquad (E.61)$$

and C_{S_0} is the side force coefficient without the free surface effect, i.e.

$$C_{S_0} = 2\pi E(\lambda - \lambda_0)\,[1+(4/\pi)\,F_R D_h/b\,]$$
$$/[1+2E\{1+(4/\pi)\,F_R D_h/b\}/(1+D_h/b)/\{2A_k e_k(1+D_{he}/b)\}] \qquad (E.62)$$

We will call F_{FS} the free surface factor.

It is emphasized, that although fairly complex, the model described above is necessarily also fairly crude. For example, it does not model the effects of the length of the hull or Froude number on the side force, nor the effect of the hull on the span-wise distribution of circulation. It is believed, nevertheless, that it is sufficiently representative to model the main trends. This is supported by comparisons with the results of towing tank tests; see Fig. E.1.

Figure E.1 compares lift (side force) curve slopes determined according to the method described above with the results of towing tank tests for several yacht keel configurations. Configuration 0 is a surface piercing fin (A=3, Λ=0, TR=1) with-

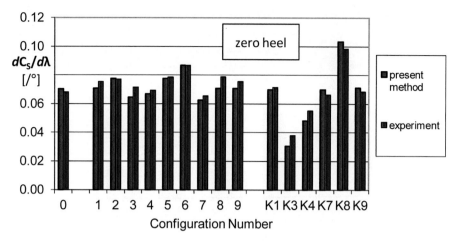

Fig. E.1 *Comparison of estimated and measured lift (side force) curve slopes for several keel-hull configurations*

out a hull (Ba and Guilbaund 1990). Configurations 1–9 represent 9 different hull forms, all with the same keel (Gerritsma et al. 1977) ($A_k = 0.65$, $\Lambda = 36°$, TR$=0.63$). Configurations K1 to K9 represent keels of different shape, all attached to the same hull (Gerritsma and Keuning 1985).

As described in Sect. 6.5 and, utilizing the derivations in Appendices C.1 and C.2, the coefficient of the induced hydrodynamic resistance of a sailing yacht can be expressed as

$$C_{D_i} = C_S \alpha_i = C_S C_{S0} \frac{\{1 + (4/\pi)\, F_R D_h / b_k\}}{\{1 + D_h / b_k\}\, \pi A_k \{e + (D_h / b_k)\}^0} \qquad (E.63)$$

With (E.60) this can also be written as

$$C_{D_i} = K_i C_S^{\,2}$$

where

$$K_i = (1/F_{FS}) \frac{\{1 + (4/\pi)\, F_R D_h / b_k\}}{\{1 + D_h / b_k\}\, \pi\, 2 A_k \{e + (D_h / b_k)\}} \qquad (E.64),$$

and F_{FS} is the free surface factor introduced above.

The total drag due to lift or side force can be expressed in a similar way:

$$K = (R - R_0)/S^2 \qquad (E.65),$$

where R is the resistance or drag and S is the side force or lift. R_0 is the resistance at zero side force.

Figure E.2 provides a comparison of drag-due-to-lift factors K for the configurations K1 to K9.

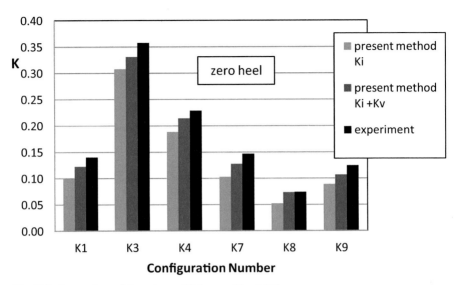

Fig. E.2 *Comparison of drag-due-to-lift factors (Fr = 0.35)*

In considering Fig. E.2 it should be realized that the experimental results contain all components of drag-due-to-lift, i.e. the induced drag, the viscous drag-due-to-lift as well as any wave-making resistance due to side force or leeway. The theoretical results of the present method contain the induced drag factor K_i (Sect. 6.5) and the viscous drag-due-to-lift (expressed in terms of a viscous drag-due-to-lift factor K_v) as determined according to the form factor method of Appendix H, but not the wave-drag-due to-side-force (Appendix J). Because of the latter one would expect the theoretical results for the drag-due-to-side-force to be a little smaller than in the experiment. This, indeed, appears to be the case.

Because the experiments were performed with rudder, the effect of the rudder on the lift curve slope and the induced drag has also been taken into account in the theoretical results. The effect of the downwash induced by the keel at the position of the rudder and the forces on the rudder were found to be significant.

Effects of Heel

As discussed in Sect. 6.7 the effects of heel on the hydrodynamic side force acting on a sailing yacht are two-fold. First, there is a reduction of the lift curve slope of the keel and, secondly, there is already a side force, 'acting the wrong way', at zero angle of leeway.

The most important effect on the lift curve slope is caused by the reduction, by a factor $\cos\varphi$, of the angle of attack of the keel as compared with the angle of leeway ($\alpha \approx \lambda \cos\varphi$, see Sect. 6.7). For the horizontal component C_S of the side force this was seen to imply

$$\Delta C_S(\varphi)/\Delta\lambda = \{\Delta C_L(\varphi)/\Delta\alpha\} \cos^2(\varphi) \qquad (E.66)$$

Fig. E.3 *Illustrating the reduction of effective span due to heel*

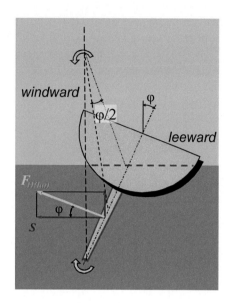

There is also a secondary effect in the sense that the lift curve slope $\Delta C_L(\varphi)/\Delta\alpha$ depends on the angle of heel. As mentioned this is caused by the fact that, under heel, the (virtual) tip vortex of the mirror image of the keel with respect to the water surface is closer to the real keel. This implies a reduction of the effective span. Another effect is due to the smaller distance, by a factor $\cos\varphi$, between the keel and the water surface (Fig. E.3). This causes a larger loss of lift due to the free surface effect.

A crude approximation for the effect of heel on the effective span is given by

$$b_e(\varphi)/b_e(0) \cong (4/3) \cos(\varphi/2) / \{1 + \cos(\varphi/2)/(3 \cos\varphi)\} \qquad (E.67)$$

This expression can be derived by considering the downwash normal to the plane of the keel as induced by the real and virtual tip vortices at the semi-span position. Figure E.4 illustrates that the effect of heel on the effective span is modest.

The effect of heel on the loss of lift due to the proximity of the free surface can be approximated by replacing the factor κ, in the exponential functions in the integral I_1, (see Eq. (E.54), by $\kappa\cos\varphi$. This comes down to replacing the quantity κ in the

Fig. E.4 *Effective span as a function of heel angle*

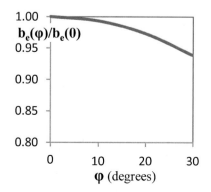

functions F_1, (Eq. (E.56)), and F_2, (Eq. (E.57)), by $\kappa\cos\varphi$. The author has found that the effect is negligible for most if not all practical purposes.

The side force at zero leeway is caused by the fact that, under heel, the flow about the hull is, in general, asymmetric. There is one exception and that is when the under-water part of the cross-section of the hull is a segment of a circle. The latter implies a section coefficient C_m $(= C_p/C_B)$ equal to

$$C_m = (R/D_h)^2 (D_h/B_{WL}) \arcsin(B_{WL}/2R) - \tfrac{1}{2}(R/D_h - 1) \equiv C_m^* \qquad (E.68),$$

where R is the radius of the circle segment. The latter can be approximated by Eq. (C.15) with $D=2D_h$. Figure E.5 gives C_m^* as a function of B_{WL}/D_h.

Based on this, the side force at zero leeway can be chosen to be a function of a factor of the type

$$\left\{1-(C_m/C_m^*) B_{WL}/D_h/2\right\}$$

Note that this factor takes the value zero for a half-circle.

On the other hand, it is also clear that the asymmetry becomes insignificant when $B_{WL}/D_h \to 0$. This suggests that the side force at zero leeway is proportional to a factor

$$(B_{WL}/D_h)\left\{1-(C_m/C_m^*) B_{WL}/D_h/2\right\}$$

It is also reasonable to assume that the lift induced by the asymmetric flow about the hull on the upper part of the keel is proportional to the length of the root chord of the keel. This leads to the following tentative expression

$$\delta C_L(\varphi) = c_k F_{FS}(B_{WL}/D_h)\left\{1-(C_m/C_m^*)B_{WL}/D_h/2\right\} \times (c_R D_h/S_k)\tan\varphi/\cos\varphi$$
$$(E.69)$$

Fig. E.5 *Section coefficient of circle segment shaped hull cross-sections as a function of beam/draft ratio*

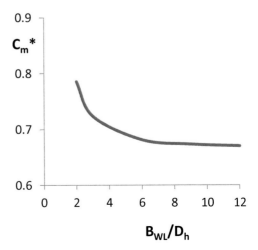

The factor F_{FS} (Eq. (E.69) models the effect of the presence of the free surface. The factor $\tan\varphi/\cos\varphi$ is suggested by the fact that $\delta C_L(\varphi)$ must be an anti-symmetric function of φ and by the progressive dependence on φ that is found in towing tank tests (Gerritsma et al. 1977).

It has been found by the author that a reasonable correlation with results of towing tank tests (Gerritsma and Keuning 1984, 1985;) is obtained when the value of the constant c_k is taken as $c_k = 0.023$.

There is also an important effect of heel on the induced resistance, in particular through the lift $\delta C_L(\varphi)$ due to heel at zero leeway. As discussed above and in sub-Section 6.7.2, the lift $\delta C_L(\varphi), < 0$, due to heel at zero leeway is caused by asymmetry of the pressure field and/or by cross-flow induced at the trailing edge of the keel by the hull. In either case, the loss of lift due to heel must be compensated by an increase of the angle of leeway.

When the (negative) lift due to heel at zero leeway is caused by cross-flow, the keel experiences a 'hydrodynamic twist' that causes a certain spanwise distribution of circulation and associated lift. As a consequence there is also a spanwise distribution of circulation at zero lift with an associated amount of 'zero-lift induced drag'. In this situation the keel does not have to produce more lift but 'only' needs more leeway to produce the required amount of lift. It follows from Abbott et al. (1945) that the induced drag can then be expressed as

$$C_{D_i} = K_i C_L^2 + C_L f_1(A, TR)\delta C_L(\varphi) + f_2(A, TR)(\delta C_L(\varphi))^2 \qquad (E.70)$$

Here, $C_L = C_L(\varphi)$ is the actual lift under heel. The factors $f_1(A, TR)$ and $f_2(A, TR)$ depend on the aspect ratio and the taper ratio of the keel. It is found in Abbott et al. (1945) that f_1 and f_2 are, in general, small ($< \cong 0.003$), so that the second and third terms of Eq. (E.70) are only of (relative) significance for high aspect ratios and/or low levels of lift when the basic term $K_i C_L^2$ is also small.

It is a different matter when the lift $\delta C_L(\varphi)$ due to heel at zero leeway is caused by asymmetry of the pressure field induced by the hull. In that case the keel must fully compensate the negative lift under heel at zero leeway. That is, it must produce a total amount of lift due to leeway $C_L(\lambda,\varphi)$ equal to $C_L - \delta C_L(\varphi)$, (see also Sect. Effects of Heel on Vertical Force and Side Force). The induced drag is then given by

$$C_{D_i} = K_i\{C_L(\varphi) - \delta C_L(\varphi)\}^2 \qquad (E.71)$$

Because of the small values of the factors f_1 and f_2 this is, in general, much larger than the value implied by Eq. (E.70).

It is the author's impression, from comparisons with experimental results (Gerritsma and Keuning 1985) that about 75 % of the additional induced resistance due to heel is caused by the asymmetric pressure field induced by the hull and about 25 % by cross-flow. A suitable, but crude, approximation for the induced resistance under heel is therefore

$$C_{D_i} = K_i\{C_L(\varphi) - 0.75\delta C_L(\varphi)\}^2 \qquad (E.72)$$

Additional, but smaller effects of heel on the induced resistance are cause by the reduction of the effective span as illustrated by Fig. E.4 and the increased effect of the free surface mentioned above. In terms of the expressions (E.63/64) for the induced drag coefficient the effect of the latter two is to increase the induced drag factor K_i by an increase of the free surface factor F_{FS} and multiplication of the denominator of Eq. (E.64) by the ratio $b_e(\varphi)/b_e(0)$ as given by Eq. (E.67).

References

Abbott IH, Von Doenhoff AE, Stivers LS (1945) Summary of airfoil data, NACA Report No. 824

Gerritsma J, Keuning JA (1984) Experimental analysis of five keel-hull combinations, Delft University of Technology, Ship Hydromechanics Lab, Report no. 637

Gerritsma J, Keuning JA (1985) Further experiments with keel-hull combinations. Delft University of Technology, Ship Hydromechanics Laboratory, Rept. Nr 699-P

Gerritsma J, Moeyes G, Onnink R (1977) Test results of a systematic yacht hull series. 5th HISWA symposium yacht architecture

Oossanen Peter van (1981) Method for the calculation of the resistance and side force of sailing yachts, paper presented at Conference on 'calculator and computer aided design for small craft—the way ahead', R.I.N.A., 1981

Prandtl L, Tietjens OG (1934a) Fundamentals of aero- & hydromechanics. McGraw-Hill, New York (also: Dover Publications, NY, 1957)

Prandtl L, Tietjens OG (1934b) Applied aero- & hydromechanics. McGraw-Hill, New York (also: Dover Publications, NY, 1957)

Appendix F

The Downwash (or Side Wash) Behind a Keel

The downwash behind a lifting surface like the keel of a sailing yacht is caused by both the bound vorticity and the trailing vorticity (Sects. 5.13, 5.15; see also Fig. F.1). In principle it varies in both the streamwise and spanwise directions, except in the case of an elliptical spanwise distribution of circulation when the downwash is constant in the spanwise direction (at least far downstream).

Qualitatively the downwash behaves as indicated in Fig. F.2. This figure gives the downwash parameter $d\alpha_\varepsilon/d\alpha$ as a function of the normalized distance $\Delta x / b_L$. Here, α_ε is the flow angle caused by the downwash and b_k is half the (effective) span of the lifting surface (the geometric span in case of a sailing yacht keel). Note that the downwash parameter $d\alpha_\varepsilon/d\alpha$ takes the value 1 at the trailing edge. This is a consequence of the fact that, at least in attached flow, the flow direction at the trailing edge is equal to the direction of the bisector of the trailing edge angle ("Kutta condition", see Sect. 5.14).

The part of the downwash induced by the bound vorticity decreases rapidly in the streamwise direction aft of the trailing edge (positioned at 'TE' in Fig. F.2) of the lifting surface. It goes to zero far downstream.

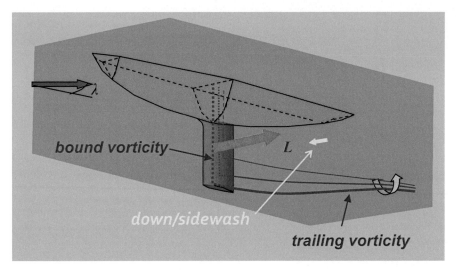

Fig. F.1 *Downwash/sidewash behind a lifting surface like the keel of a sailing yacht*

Fig. F.2 *Variation of down-wash behind the trailing edge of the root section of a lifting surface (like the keel of a sailing yacht)*

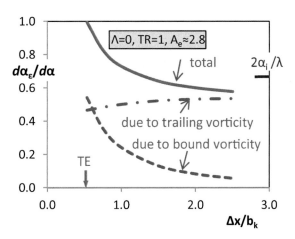

The part due to the trailing vorticity has the same value as the induced angle of attack (in radians) at the location of the keel and, in theory, for inviscid flow, twice that value far downstream. The latter is due to the fact that the trailing vortices extend in the downstream direction only at the location of the keel but both very far upstream and very far downstream at a large distance behind the keel. It follows that, in theory, $d\alpha_\varepsilon/d\alpha \to 2\alpha_i/\alpha$ for $x \to \infty$, at least for an elliptical spanwise distribution of circulation. For keels with a taper ratio $> \cong 0.5$, the downwash far downstream behind the root section is a little smaller due to the shift of circulation from the root towards the tip. The opposite is the case for taper ratios $< \cong 0.5$.

Quantitative estimates of the downwash behind lifting surfaces like sailing yacht keels can be obtained on the basis of methods and data for aircraft wings. Silberstein and Katzoff (1939) presents charts for straight wings of aspect ratios ≥ 6, based on a theoretical-empirical model. An empirical formula for the downwash behind wings

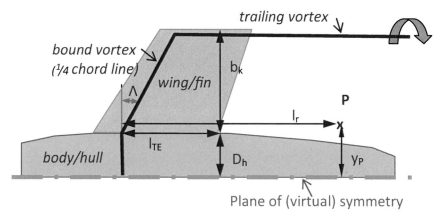

Fig. F.3 *Simple line vortex model for estimating the downwash behind a lifting surface like the keel of a sailing yacht*

of more arbitrary planform can be found in Etkin and Reid (1996). In Gerritsma (1971) the downwash behind sailing yacht keels at the position of the rudder is taken at 0.8 times the theoretical downwash at infinity downstream, irrespective of the longitudinal position of the rudder.

Because the charts of Silberstein and Katzoff (1939) are applicable only to lifting surfaces of high aspect ratio and zero sweep, they are not applicable to many sailing yacht keels. The more general empirical formulae of Etkin and Reid (1996), on the other hand, seem to underestimate the downwash for small taper ratios and small aspect ratios and do not seem to be very accurate close to the trailing edge. The theoretical-empirical model described below is believed to be a little more versatile and more accurate than those of Silberstein and Katzoff (1939) and Etkin and Reid (1996).

The lifting surface is modeled by a simple, swept 'horsehoe' vortex system as in Fig. F.3. Note that the bound vortex of the wing or fin is continued inside the body or hull to model lift carry-over. The dowwash induced by the bound (or 'lifting') vortex of the wing or fin in a point P in the plane of the fin, can, using the so-called Biot and Savart law of vortex theory (Prandtl and Tietjens 1957), be approximated as

$$\alpha_{\varepsilon bw} = w_w (w_{1w} + w_{2w}) \tag{F.1},$$

with the factor w_w given by

$$w_w = (\Gamma_{avw}/V)/[4\pi \{l_r' \cos\Lambda - (y_P - D_h) \sin\Lambda\}] \tag{F.2}$$

The term w_{1w} is associated with the root end of the bound vortex of the fin/keel (see Fig. F.3) and can be written as

$$w_{1w} = \sin(\arctan[\{l_r' \sin\Lambda + (y_P - D_h)\cos\Lambda\}/\{l_r' \cos\Lambda - (y_P - D_h)\sin\Lambda\}]) \tag{F.3}$$

The term w_{2w} is associated with the tip end of the bound vortex and can be written as

$$w_{2w} = \sin \arctan[\{b_k /\cos\Lambda - (l_r' \sin\Lambda + (y_P - D_h)\cos\Lambda)\}$$
$$/\{l_r' \cos\Lambda - (y_P - D_h) \sin\Lambda\}] \tag{F.4}$$

A similar contribution to the downwash is due to the mirror image of the bound vortex of the fin. An expression for this contribution is obtained by replacing the quantity y_P in Eqs. (F.2)–(F.4) by $-y_P$.

The contribution of the extension of the bound vortex inside the body or hull (including its mirror image) can be written as

$$\alpha_{\varepsilon bb} = w_b(w_{1b} + w_{2b}) \tag{F.5},$$

where

$$w_b = (\Gamma_R /V)/(4\pi l_r') \tag{F.6},$$

$$w_{1b} = \sin[\arctan\{(y_P + D_h)/l_r'\}] \tag{F.7}$$

and

$$w_{2b} = -\sin[\arctan\{(y_P - D_h)/l_r'\}] \tag{F.8}$$

With respect to the formulae given above it is to be mentioned that

- Γ_{avw} is the average circulation on the wing/fin or keel. It can (see Appendix C) be expressed as

$$\Gamma_{avw}/V = C_{Lw0}S_k/(2b_k) = C_{Lw0}b_k/(2A_k)$$
$$= C_{L0}b_k/[2A_k\{1+(\Gamma_R /\Gamma_{avw}) D_h/b_k\}] \tag{F.9},$$

  where it is emphasized that C_{L0} is to be taken as the lift without free surface effect (see also Appendix E).

- Γ_R is the circulation at the root of the wing/fin or keel. As described in Appendix C.1 it can be expressed as

$$\Gamma_R /V = (4/\pi) F_R\Gamma_{avw}/V \tag{F.10},$$

  with the factor F_R given in Appendix C.1 (Eq. (C.9).

- The quantity l_r' is related to the longitudinal distance l_r between the point P and the trailing edge of the root section:

$$l_r' = l_r + \Delta l_r \tag{F.11}$$

The correction term Δl_r is chosen such that the total downwash at the trailing edge of the root section is equal to the angle of attack α so that the downwash factor $d\alpha_\varepsilon /d\alpha$ is equal to 1, as it should.

The average downwash induced by the trailing vorticity plus its mirror image can, in agreement with Appendices B and C, be written as

$$\alpha_{\varepsilon iav} = C_{L0}\left[1+\sin\left(\arctan\{(l_r'-b_k\sin\Lambda)/(b_k+D_h)\}\right)\right]$$
$$/\left\{\pi 2A_k\left(e_k+D_h/b_k\right)\left(1+D_h/b_k\right)\right\} \tag{F.12}$$

The total downwash is obtained by summing the contributions of the bound and trailing vorticity. It is further noted that (the contributions to) the downwash factor $d\alpha_\varepsilon/d\alpha$ are obtained if the lift coefficient C_{L0} in Eq. (F.12) is replaced by the lift curve slope $dC_{L0}/d\alpha$.

Using the data given in Silberstein and Katzoff (1939), the author has found that the spanwise variation of the downwash induced by the trailing vorticity can be approximated through multiplication of Eq. (F.12) with the function

$$f(y_P) = f(F_R) + 4\{y_P/(b_k+D_h)\}^2\{1-f(F_R)\} \tag{F.13},$$

where

$$f(F_R) = 1 + 3(F_R - 1.01) \tag{F.14}$$

and

$$\alpha_{\varepsilon i}(y_P) = \alpha_{\varepsilon iav}f(y_P) \tag{F.15}$$

While the distance between a point in the plane of the wing and the trailing vortex sheet is zero at zero lift, this distance increases with increasing angle of attack. The effect of this is that the downwash in the plane of the wing is a little smaller than that in the plane of the trailing vortex sheet. It can be derived from the Biot and Savart law of vortex theory (Prandtl and Tietjens 1957) that this can be compensated by introducing a second multiplication factor, also to be applied to Eq. (F.12), of the form

$$f_i = 1/(h_0 - h_i)^2 \tag{F.16}$$

In this expression h_0 is the geometrical displacement of the vortex sheet due to angle of attack. This is given by

$$h_0 = \{(1 - l_{TE})/b_k\}\tan\alpha \tag{F.17}$$

'h_i' is the additional displacement due to the downwash of the keel. Using the data of Silberstein and Katzoff (1939) the author has found that this can be approximated by

$$h_i = 1.3\tan\alpha_{\varepsilon i0}\{(1 - l_{TE})/b_k\}^{0.95} \tag{F.18},$$

where $\alpha_{\varepsilon i0}$ is the downwash in the plane of the wing or fin as given by Eqs. (F.12)–(F.15).

In principle, similar multiplication factors have to be applied to the contributions (F.5) of the bound vorticity. It appears, however, that this is hardly significant, except for very high angles of attack and at a relatively close distance behind the trailing edge of the wing.

References

Etkin B, Reid LD (1996) Dynamics of flight: stability and control. John Wiley and Sons, New York

Gerritsma J (1971) Course-keeping qualities and motions in waves of a sailing yacht. 3rd AIAA symposium on the aerodynamics and hydrodynamics of sailing

Prandtl L, Tietjens OG (1934a) Fundamentals of aero- & hydromechanics. McGraw-Hill, New York (also: Dover Publications, NY, 1957)

Prandtl L, Tietjens OG (1934b) Applied aero- & hydromechanics. McGraw-Hill, New York (also: Dover Publications, NY, 1957)

Silberstein A, Katzoff S (1939) Design charts for predicting downwash angles and wake characteristics of plain and flapped wings. NACA Report No. 648

Appendix G

Estimation of the Yawing Moment of a Sailing Yacht Hull

As indicated in Sect. 5.16, the yawing moment of a fully submerged, ellipsoidal body of revolution is given by

$$M_{Hz} \cong \tfrac{1}{2}\rho V^2 2\lambda \ (1 - D/L)\nabla \qquad (G.1),$$

where D/L is the diameter/length ratio. A corresponding dimensionless moment coefficient is defined by

$$C_{MHz} \equiv M_{Hz}/\left(\tfrac{1}{2}\rho V_b^{\,2}\nabla\right) \qquad (G.2)$$

which gives

$$C_{MHz} \cong 2\lambda \ (1 - D/L) \qquad (G.3)$$

According to Lewis (1988) a more general approximation for hulls of surface ships at zero heel can be written as

$$C_{MHz} \cong 2\lambda(k_2 f_s - k_1) \qquad (G.4)$$

where

$$k_2 \cong 1 - D_h/L_{WL} \qquad (G.5)$$

and

$$k_1 \cong D_h/L_{WL} \qquad (G.6)$$

The factor 'f$_s$' in (G.4) is a function of the geometry of the hull. It follows from Lewis (1988) that

$$f_s \div C_s D_h^2 L_{WL} = C_s / (C_B D_h / B_{WL})$$ (G.7)

The coefficient C$_s$ is a 'sectional-inertia coefficient' that can be expressed in terms of the sectional area coefficient C$_m$. In Lewis (1988) the relation between C$_s$ and C$_m$ is given graphically. The author has found that, algebraically, it can be approximated as

$$C_m \cong C_{m0} + (C_s - 1) D_h / B_{WL}$$ (G.8)

where

$$C_{m0} \cong -3.6\, C_s^2 + 7.09\, C_s - 2.705$$ (G.9)

In principle 'f$_s$' also depends on the longitudinal variation of C$_s$ and the local draft of the hull. A simple approximation, adopted here, is to ignore the longitudinal variation and to link C$_s$ to the area coefficient of the mid-ship section.

It is further clear, that for a fully submerged ellipsoidal body of revolution one should have f$_s$ = 1. Is This is obtained by taking 'f$_s$' as

$$f_s^* = \left(C_s / C_s^*\right) / \left\{\left(C_B / C_B^*\right)\left(D_h / B_{WL}\right) / \left(D_h / B_{WL}\right)^*\right\}$$ (G.10)

Here, the quantities C$_s^*$ = 1, C$_B^*$ = 0.523 and $(D_h / B_{WL})^*$ = 0.5 are those of (half) an ellipsoidal body of revolution.

In Keuning and Vermeulen (2003) an expression similar to Eqs. (G.4)–(G.7) is used for the yawing moment of sailing yacht hulls. The main difference is that Keuning and Vermeulen (2003) uses the approximation k$_2$ ≈ 1 and k$_1$ ≈ 0, which is valid for slender bodies.

A drawback of the formulations given above is that they do not, at least explicitly, take into account free surface effects. Since the formulation (G.4) to (G.10) is meant for surface ships (Lewis 1989), one might assume that free surface effects are, implicitly, incorporated. This, however, does not hold for the expression (G.3) for a fully submerged ellipsoidal body of revolution.

On the basis of Appendix E one would expect that a factor of the type

$$\left\{1 - e^{-2\pi\kappa D_h / (L_{WL}/2)}\right\}$$

should be applicable to the yawing moment (as well as to all lateral pressure forces and associated moments) of semi-submerged ellipsoidal bodies of revolution. In other words one may assume that an expression of the type

$$C_{MHz} \cong 2\lambda\, (1 - 2D_h / L_{WL})\left\{1 - e^{-2\pi\kappa D_h / (L_{WL}/2)}\right\}$$ (G.11a)

or

$$C_{MHz} \cong 2\lambda\, \left\{(1 - D_h / L_{WL}) f' - D_h / L_{WL}\right\}$$ (G.11b)

with

$$f' = 1 - e^{-2\pi\kappa D_h/(L_{WL}/2)} \times \{(1 - 2D_h/L_{WL})/(1 - D_h/L_{WL})\} \equiv f_{sf} \qquad (G.12)$$

should be applicable to a semi-submerged ellipsoidal body of revolution. (Note that for a semi-submerged ellipsoidal body of revolution D_h would be equal to half the maximum diameter of the ellipsoid).

Comparing the expressions given above suggests that the yawing moment coefficient of a surface ship can be approximated by Eq. (G.11b) with f' given by the linear combination

$$f' = \{1 - (f_s^*)p\} f_s^* + (f_s^*)^p f_{sf} \qquad (G.13)$$

Unfortunately there is hardly any information in the literature on the yawing moment of sailing yacht hulls. The correlation of results of the estimation method given above with the limited amount of experimental data given in Keuning and Vermeulen (2003) is only modest but suggests that the power 'p' in (G.13) should be taken as ½.

Effect of Heel
The effect of heel on the yawing moment of a hull can be split in two parts; a yawing moment coefficient $\delta C_{MHz}(\varphi)$ at zero leeway and the effect of heel on the rate of change $\dfrac{dC_{MHz}(\varphi)}{d\lambda}$. The total yawing moment coefficient is then written as

$$C_{MHz}(\varphi) = \delta C_{MHz}(\varphi) + \lambda \frac{dC_{MHz}(\varphi)}{d\lambda} \qquad (G.14)$$

The yawing moment coefficient $\delta C_{MHz}(\varphi)$ due to heel at zero leeway is, like the side force due to heel at zero leeway, caused by the asymmetry that a hull adopts under heel (see the last paragraph of Appendix E and recall that the effect is zero when the hull has a cross-section with the shape of a circle segment). This means that $\delta C_{MHz}(\varphi)$, like $\delta C_L(\varphi)$, can be expected to be proportional to a term $\{(C_m/C_m^*) B_{WL}/D_h/2 - 1\}$. A tentative formula for estimating the magnitude of $\delta C_{MHz}(\varphi)$ is therefore

$$\delta C_{MHz}(\varphi) = c_{M\varphi}(B_{WL}/D_h)\{(C_m/C_m^*) B_{WL}/D_h/2 - 1\}$$
$$\times \{1 - e^{-2\pi\kappa D_h/(L_{WL}/2)}\} \tan\varphi/\cos\varphi \qquad (G.15)$$

Correlation with experimental data summarized in Keuning and Vermeulen (2003) is rather modest but suggests that the constant '$c_{M\varphi}$' in (G.15) should be chosen as $c_{M\varphi} \cong 0.4$

It appears from the limited amount of experimental data that is available in the literature (Keuning and Vermeulen 2003) that the effect of heel on the rate of change

$\dfrac{dC_{\text{MHz}}(\varphi)}{d\lambda}$ with leeway is not insignificant. We have seen above that the yawing moment due to leeway at zero heel can be modeled like (see Eq. (G.11))

$$\frac{dC_{\text{MHz}}(\varphi=0)}{d\lambda} = 2\{(1-D_h/L_{\text{WL}})f' - D_h/L_{\text{WL}}\} \qquad (G.16)$$

with f' given by Eqs. (G.13),(G.12), (G.10). It is further clear, that there will be no effect of heel when the cross-sectional shape of the under-water part of a hull is part of a circle segment. This suggests (see Appendix E) that $\dfrac{dC_{\text{MHz}}(\varphi)}{d\lambda}$ is also approximately proportional to a term of the type

$$1 + c_{M\lambda\varphi}(B_{\text{WL}}/D_h)\{\left(C_m/C_m^{\;*}\right)B_{\text{WL}}/D_h/2 - 1\}\tan\varphi/\cos\varphi \qquad (G.17)$$

Combining (G.16) and (G.17) gives

$$\frac{dC_{\text{MHz}}(\varphi)}{d\lambda} = \frac{dC_{\text{MHz}}(\varphi=0)}{d\lambda}$$
$$\times [1 + c_{M\lambda\varphi}(B_{\text{WL}}/D_h)\{\left(C_m/C_m^{\;*}\right)B_{\text{WL}}/D_h/2 - 1\}\tan\varphi/\cos\varphi] \qquad (G.18)$$

Equation (G.18) has been found to give fair correlation with experimental data (Keuning and Vermeulen 2003) when the constant $c_{M\lambda\varphi}$ is chosen to be $\cong 0.04$.

References

Keuning JA, Vermeulen KJ (2003) The yaw balance of sailing yachts upright and heeled, Report no. 1363-P, Ship Hydromechanics Laboratory, Delft University of Technology

Lewis EV (ed) (1988) Principles of naval architecture, vol II. SNAME, Jersey City. ISBN 0-9397703-01-5

Lewis EV (ed) (1989) Principles of naval architecture, vol III—motions in waves and controllability. SNAME, Jersey City. ISBN 0-939773-02-3

Appendix H

Form Factors for the Viscous Resistance of Sailing Yacht Components

Introduction
As discussed in Sect. 5.11 the 'viscous' or profile drag coefficient of a body can be expressed as

$$C_{D_v} = (1+k)C_{F_0} \qquad (H.1)$$

Here, C_{F_0} is the friction drag coefficient of an equivalent, non-lifting flat plate with the same length and Reynolds number as the body. 'k' is a form factor, usually representing the combined effects of the increase of the fluid velocity ('supervelocity') due to the thickness or volume of the body on the friction drag and the displacement thickness of the boundary layer (form drag).

Alternatively, one can separate the effects due to supervelocity and displacement thickness by writing

$$k = k_0 + k_1 \qquad (H.2),$$

so that

$$C_{D_v} = (1+k_0+k_1)C_{F_0} \qquad (H.3)$$

Then,

$$C_{D_F} = (1+k_0)C_{F_0} \qquad (H.4)$$

is the friction drag coefficient and

$$C_{D_f} = k_1 C_{F_0} \qquad (H.5)$$

is the form drag or 'viscous pressure drag' coefficient due to displacement thickness. An advantage of this formulation is that it provides better insight in the physical mechanisms involved.

Because, as mentioned in Sect. 5.11 and 6.5, the literature contains several different empirical 'laws' for the friction drag of turbulent boundary layers, there is also some ambiguity in the choice of the level of the form factor. This is the case in particular when the form factor is based on correlations with model experiments (towing tank tests) at low Reynolds number. Differences between different turbulent friction laws may be as large as 10% at (very) low Reynolds numbers (see Fig. H.1).

Because of the relatively high values of the ITTC correlation line at low Reynolds numbers, form factors based on the latter tend to have low values. The latter is even more so when the Reynolds number is chosen to be based on 70% of the length of the waterline, as is sometimes the case (Keuning and Sonnenberg 1998a). Another problem is that unknown positions of transition in the experiment can lead to erroneous values of experimentally determined form factors.

Bodies of Revolution

A generally accepted approximation for the form factor(s) of sailing yacht hulls can, to the author's knowledge, not be found in the literature. However, for bodies

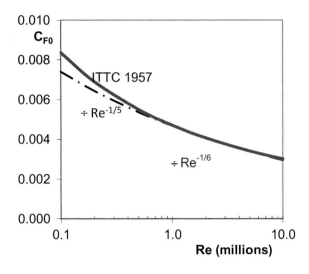

Fig. H.1 *Flat plate friction coefficient as a function of Reynolds number according to different formulae*

of revolution at zero lift there are some (Hoerner 1965; Torenbeek 1982) that are widely used in the aeronautics community. In this author's opinion a form factor k_0 for the friction drag of slender bodies of revolution should be of the form

$$k_0 = c_0 \left(D_{max}/L \right)^2 \tag{H.6}$$

where c_0 is a constant. Such a formula is in agreement with the theoretical result (Ashley and Landahl 1965), that the flow velocity on a slender body of revolution at zero angle of attack is proportional to the maximum cross-sectional area.

A useful result for the boundary layer pressure drag or form drag of streamlined bodies of revolution with attached flow, based on calculations by Young (1989), is

$$C_{D_f}/C_{D_v} \cong 0.4 \left(D_{max}/L \right) \tag{H.7}$$

From this it follows that

$$k_1/\left(1 + k_0 + k_1 \right) \approx 0.4 \left(D_{max}/L \right) \tag{H.8}$$

or

$$k_1 \approx \left(1 + k_0 \right) 0.4 \left(D_{max}/L \right) / \{ 1 - 0.4 \left(D_{max}/L \right) \} \tag{H.9}$$

It can be verified that, for $c_0 = 3.0$, results of Eq. (H.2) with (H.6), (H.9) are in reasonable agreement with the commonly used form factors of Hoerner (1965) and Torenbeek (1982), which are based on correlations with experimental (wind tunnel) results (see Fig. H.2).

It will be clear that a form factor of the type discussed above can only provide a rather crude approximation of the viscous resistance. The reason is, of course, that it does not take into account the geometry of the body in sufficient detail to

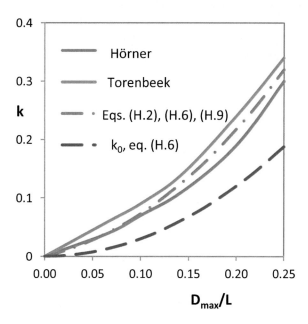

Fig. H.2 *Form factors for slender bodies of revolution*

model phenomena like boundary layer convergence or divergence (Sect. 5.16) and/or separation (except for the convergence and divergence of boundary layer flow that occurs on streamlined bodies of revolution).

Equations (H.3), (H.6), (H.9) should be applicable to keel bulbs with a near-circular cross-section. Sailing yacht hulls are, however, another matter because of the free surface effect and because they usually have a non-circular cross-section. We will return to this later in this Appendix.

Keels and Rudders

It is first of all noted that it is common practice for lifting surfaces like keels and rudders to choose the projected area S_{proj} as the reference area for the lift and drag coefficients. For this reason Eq. (H.3) is replaced by

$$C_{D_v} = (1 + k) 2 \, C_{F_0} \qquad (H.10)$$

The factor 2 appears because the equivalent flat plate is now two-sided. It also implies that 'k' now includes the effect of the increase of the wetted area due to thickness.

Because the geometry and associated fluid dynamic behaviour of lifting surfaces is less three-dimensional than that of hulls, the dependence of the form factor on volume or thickness is also different. Commonly used form factors for keels and rudders are those due to Hoerner (1965):

$$k = 2\,t/c + 60\left(t/c\right)^{4} \tag{H.11}$$

A similar formula proposed by Torenbeek (1982) reads

$$k = 2.7\,t/c + 100\left(t/c\right)^{4} \tag{H.12}$$

Note that Eqs. (H.11), (H.12) represent the effect of thickness only. More general formulae, involving the effects of sweep angle and lift can be found in the ESDU Data Sheets (ESDU Data Sheets) and have been proposed by Kroo (www.adg.stanford.edu).

The form factor proposed by Kroo for the friction drag due to thickness, including the effect of sweep, is well founded theoretically and can be written as

$$k_0 = 2C_0\left(t/c\right)\cos\Lambda + 0.5C_0^{\,2}\left(t/c\right)^{2}\left(1 + 5\,\cos^2\Lambda\right) \tag{H.13}$$

where C_0 is a coefficient that, in principle, depends on the shape of the section.

There is also a weak dependence of C_0 on Reynolds number (Young 1989; Kroo www.adg.stanford.edu). This in the sense that C_0 increases slightly with Reynolds number. It is caused by the fact that the streamwise distribution of the displacement thickness of the boundary layer causes a negative supervelocity that vanishes when the Reynolds number goes to infinity. This effect is, in general, neglected.

It has also been found by Young (1989) that the form drag (coefficient) C_{Df} of two-dimensional airfoils approximately satisfies the relation

$$C_{D_f}/C_{D_v} \cong t/c \tag{H.14}$$

This means, that in analogy with the situation for bodies of revolution, described above, one can write, for $\Lambda = 0$,

$$k_1/\left(1 + k_0 + k_1\right) \cong t/c \tag{H.15}$$

This gives

$$k_1 \cong \left(1 + k_0\right)\left(t/c\right)/\left(1 - t/c\right) \tag{H.16}$$

From the theory of swept wings (Küchemann 1978) it can be derived that for lifting surfaces with sweep Eq. (H.16) should be expanded to read

$$k_1 \cong \left(1 + k_0\right)\left(t/c\right)\cos\Lambda/\{1 - \left(t/c\right)\cos\Lambda\} \tag{H.17}$$

As before, the total form factor 'k' is then given by (see (H.2)–(H.5))

$$k = k_0 + k_1$$

Figure H.3 compares the different form factors given above as a function of the thickness/chord ratio t/c for $\Lambda = 0$.

Fig. H.3 *Form factors for viscous drag due to thickness of two-dimensional airfoils*

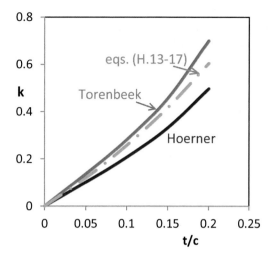

The constant C_0 in Eq. (H.13) has been taken equal to 0.6. In the author's experience this leads to results that correlate well with the experimental data given in McWilkinson et al. (1974) for NACA 6-series sections with the position of maximum thickness at about 30–40 % of the chord. This is a type of foil section that is often used in sailing yacht keels. It also leads to good agreement with results given in Kroo (www.adg.stanford.edu). The level of 'k' is a little larger than Hoerner's (Hoerner 1965), which is based on NACA experimental data (Abbott and Von Doenhoff 1949; Abbott et al. 1945). Values of 'k' which are almost identical with Hoerner's (1965) are obtained for $C_0 = 0.4$. Torenbeek's (1982) values are obtained for $C_0 = 0.7$.

Because the form factors given above are based on data for two-dimensional flow, they are valid only for very high aspect ratios. The effect of aspect ratio can, tentatively, be taken into account by realizing that when the span of a foil becomes very small, that is of the order of the thickness/chord ratio t/c, the foil begins, fluid-dynamically, to behave, as a slender body. Form factors for the latter were given above (see Eq. (H.6–H.9)).

Rewritten in terms of lifting surface quantities like t/c and the projected area S_{proj}, the viscous drag of a slender body of revolution can be expressed as

$$C_{D_v} = \left(1 + k_{0bw} + k_{1bw}\right)\left(S_{wet}/S_{proj}\right)C_{F_0} \qquad (H.18),$$

with

$$k_{0bw} = 3\left(t/c\right)^2 \qquad (H.19)$$

and

$$k_{1bw} = \left(1 + k_{0bw}\right)0.4\left(t/c\right)/(1 - 0.4\ t/c) \qquad (H.20)$$

This means, that the viscous drag of keels and rudders of vanishing aspect ratio can also be approximated as (H.18) to (H.20). It also means that the viscous drag of keels and rudders of arbitrary aspect ratio can be written as a combination of the ex-pressures (H.10), (H.13), (H.17) and (H.18), (H.19), (H.20). A suitable interpolation function is the 'edge correction factor' E introduced in Sect. 5.15, Eq. (5.15.13). One can write

$$C_{D_v} = \{1 + k_{0w} + k_{1w}\} 2C_{F_0} \tag{H.21},$$

with

$$k_{0w} = E_0 k_{0ww} + (1 - E_0) k_{0bw} \left(S_{wet}/S_{proj}\right)/2 \tag{H.22}$$

and

$$k_{1w} = E_0 k_{1ww} + (1 - E_0) k_{1bw} \left(S_{wet}/S_{proj}\right)/2 \tag{H.23}$$

Here, k_{0ww} and k_{1ww} are given by the expressions (H.13)and (H.17) for the form factors of lifting surfaces and k_{0bw}, k_{1bw} are given by the expressions (H.19), (H.20) for the form factors of bodies of revolution. The edge correction factor E_0 is taken as

$$E_0 = (A/2)/\sqrt{\{1 + (A/2)^2\}} \tag{H.24}$$

Note that Eq. (H.24) does not contain the angle of sweep Λ like Eq. (5.15.14). The reason is that the effect of sweep is already contained in the expressions (H.13), (H.17). It is further noted that the aspect ratio 'A' in Eq. (H.24) should be taken as twice the geometrical aspect ratio of the keel or rudder. This because of the reflection plane effect of the free water surface and/or the hull (see Appendix E). The ratio S_{wet}/S_{proj} in Eqs. (H.22, 23) can, for most practical purposes, be taken as $\approx 2(1 + t/c)$.

As already indicated in Sects. 5.11 and 5.14 the viscous drag of a lifting surface also depends on the lift, albeit that the effect of lift on the friction drag is usually negligible. There is, however, a significant effect of lift on the pressure drag and, hence, on the form factor k_1. Under lifting conditions the displacement thickness on the suction side of a foil increases with lift while that on the pressure side decreases. This difference in displacement thickness causes a reduction $\delta_v \alpha_e$ of the effective angle of attack that is proportional with the lift L of the foil. As in the case of the induced drag (Sect. 5.15) this causes a resistance equal to the product of the lift L and the reduction $\delta_v \alpha_e$ of the effective angle of attack. Because $\delta_v \alpha_e$ is proportional with L, the 'viscous drag due to lift' is proportional to L^2. This means that the viscous drag coefficient contains a term proportional to C_L^2, which can be represented by a form factor k_{1L}:

$$k_{1L} = C_{1L} C_L^2 \tag{H.25}$$

Using the data of Abbott and Von Doenhoff (1949), the author has found that $C_{1L} \approx 1.0$ represents an average value as long as there is attached flow. Foil sections with a large leading-edge radius have a smaller value of c_{1L} (down to 0.6) and foil sections with a small leading edge radius a higher value (up to 1.4). C_L is the lift coefficient of the foil, given by

$$C_L = L/(\tfrac{1}{2}\rho V^2 S_{proj}) \qquad (H.26)$$

The form factor k_{1L} is to be added to Eq. (H.23):

$$k_{1w} = E_0 k_{1ww} + k_{1L} + (1-E_0)\, k_{1bw}\left(S_{wet}/S_{proj}\right)/2 \qquad (H.27)$$

Note that the 'edge factor' E_0 is not applied to k_{1L} because the effect of aspect ratio as well as that of sweep) is already contained in the level of the lift.

Graphical representations of the form factors introduced above are given in Sect. 6.5.

As already discussed in Sect. 5.14, camber, as sometimes applied in dagger boards and flap deflection, as sometimes applied on keels, are also known to increase the viscous resistance, at least at zero lift. Because camber causes lift or side force, in proportion to the amount of camber, its effect on the viscous resistance is similar to that of lift. This means that, in the case of two-dimensional flow, its effect on the viscous resistance can be taken into account through a form factor of the type

$$k_{1c} = C_{1c}\left(f_c/c\right)^2 \qquad (H.28a),$$

Here, f_c/c is the maximum camber/chord ratio. Using the data of Abbott and Von Doenhoff (1949), the author has found that an average value of the coefficient C_{1c} for foils with the position of maximum camber at about 35–50% of the chord is about 35. For foils with a large amount of camber (Milgram 1971) but attached flow this should be refined to read

$$k_{1c} = C_{1c}\left(f_c/c\right)^2 + 0.05\left\{1+25\ \left(x_f/c - 0.3\right)^2\right\}\left(10\ f_c/c\right)^4 \qquad (H.28b),$$

For lifting surfaces with sweep it is found that (H.28a) should be extended to

$$k_{1c} = C_{1c}\left(f_c/c\right)^2 \cos^2\Lambda \qquad (H.29)$$

The form factor k_{1c} is to be added to the term k_{1ww} in Eq. (H.27), giving

$$k_{1w} = E_0\left(k_{1ww} + k_{1c}\right) + k_{1L} + (1-E_0)\, k_{1bw}\left(S_{wet}/S_{proj}\right)/2 \qquad (H.30)$$

Camber also causes the minimum of the viscous drag to occur at a lift coefficient (C_{LDmin}) different from zero. Lifting surface theory suggests that the magnitude of the lift coefficient for minimum drag is proportional to the amount of camber, i.e.

$$C_{LDmin} = C_{minD} E_0\left(f_c/c\right)\cos\Lambda \qquad (H.31)$$

Using the experimental data of Abbott and Von Doenhoff (1949), the author has found that a representative value of the constant C_{minD}, for airfoils with a maximum camber/chord ratio $f_c/c < 0.05$ at (effective) Reynolds numbers of about 2×10^6, is about 8. For large amounts of camber (Milgram 1971) this should be taken as

$$C_{minD} = 8\left\{1 + 0.4\,\left(x_f/c\right)^2\left(21\,f_c/c\right)^2\right\}$$

A further implication of camber is that Eq. (H.25), expressing the viscous drag due to lift, should be replaced by

$$k_{1L} = C_{1L}\left(C_L - C_{LDmin}\right)^2 \tag{H.32}$$

Figures H.4 and H.5 give an impression of the magnitude of the dependence on camber of the form factor for viscous resistance and the lift coefficient for minimum viscous resistance. Because the additional viscous resistance due to camber

Fig. H.4 *Form factor for minimum viscous resistance of lifting surfaces as a function of camber/chord ratio*

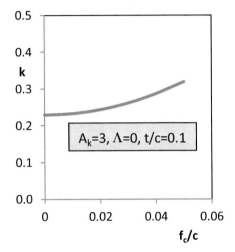

Fig. H.5 *Lift coefficient for minimum viscous resistance of lifting surfaces as a function of camber/chord ratio*

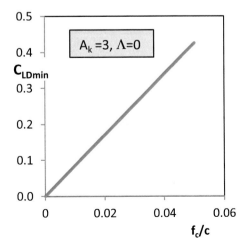

is proportional to the square of the camber/chord ratio it is, in general, negligible for small values of f_c/c. This is less the case for the shift of the lift coefficient for minimum viscous drag because of its linear dependence on f_c/c (see Eq. (H.31)) and the relatively strong dependence of the viscous resistance on lift (see Fig. 6.5.6).

The effect of a flap or trimtab on the viscous resistance is similar to that of camber. That is, a (small) flap deflection causes additional viscous drag that is proportional to the square of the camber induced by the flap deflection. Because the induced camber is proportional to the chord of the flap and the angle of the flap deflection there holds:

$$k_{1f} = C_{1f}\{\delta_f\left(c_f/c\right)(1-c_f/c)\}^2\cos^2\Lambda \qquad (H.33)$$

In (H.33) δ_f is the angle of flap deflection (in radians) and c_f is the flap chord.

Flap deflection also causes a shift of the angle of attack and lift at which the viscous drag has a minimum value:

$$C_{LDminf} = C_{minDf}E_0\{\delta_f\left(c_f/c\right)(1-c_f/c)\}\cos\Lambda \qquad (H.34)$$

Unfortunately, little is known about the value of the coefficients C_{1f} and C_{minDf}. However, there is reason to believe that the drag penalty of a trailing-edge flap for a given shift of the lift coefficient for minimum viscous drag is much larger than in the case of camber. This is due to the fact that a plain trailing-edge flap causes a less favourable pressure distribution when deflected (Abbott and Von Doenhoff 1949). A tentative estimate based on the scarce data on plain flaps available in Abbott et al. (1945) is $C_{1f}=9$ and $C_{minDf}=350$.

Sailing Yacht Hulls

The form factor formulae for bodies of revolution, given above as Eqs. (H.6), (H.9), can be generalized, to some extent, for application to bodies with non-circular cross-sections through replacing the diameter/length ratio D_{max}/L by the maximum cross-sectional area S_{Xmax}. Using the identity

$$S_{Xmax} = (\pi/4)D_{max}^2 \qquad (H.35)$$

for circular cross-sections one can write

$$\begin{aligned}
k &= k_0 + k_1 \\
&= 3.0\,(4/\pi)\left(S_{Xmax}/L^2\right) + \{1+3.0(4/\pi)\left(S_{Xmax}/L^2\right)\} \\
&\quad \times 0.4\sqrt{\{(4/\pi)\left(S_{Xmax}/L^2\right)\}}/\{1-0.4\sqrt{\{(4/\pi)\left(S_{Xmax}/L^2\right)\}} \\
&= 3.8\,S_{Xmax}/L^2 + \left(1+3.8\,S_{Xmax}/L^2\right) \\
&\quad \times 0.45\sqrt{\left(S_{Xmax}/L^2\right)}\,/\,\{1-0.45\sqrt{\left(S_{Xmax}/L^2\right)}\}
\end{aligned} \qquad (H.36)$$

Equation (H.36) should be applicable to (keel)bulbs. For ship hulls one would have to take into account that the water surface acts as a reflection plane (Sect. 6.3). This means that if S_{Xmax} is taken to be the maximum cross-sectional area of the submerged part of the hull, its value has to be doubled to obtain the effective cross-sectional area. In other words, a version of (H.36) that may be applied to sailing yacht hulls is

$$k = 7.6\, S_{Xmax}/L_{WL}^{2} + (1 + 7.6\, S_{Xmax}/L_{WL}^{2})$$
$$\times 0.64\, \sqrt{(S_{Xmax}/L_{WL}^{2})}/\left\{1 - 0.64\, \sqrt{(S_{Xmax}/L_{WL}^{2})}\right\} \qquad \text{(H.37),}$$

where L_{WL} is the length of the waterline.

At this point, it is useful to note that while k_0 is proportional to S_{Xmax}, k_1 contains terms proportional to $S_{Xmax}^{1/2}$, S_{Xmax} and $S_{Xmax}^{3/2}$, respectively. It is also noted that the quantity S_{Xmax}/L_{WL}^{2} can be replaced by

$$S_{Xmax}/L_{WL}^{2} = (\nabla/L_{WL}^{3})/C_P \qquad \text{(H.38)}$$

The quantity ∇/L_{WL}^{3} is known as the displacement/length ratio and C_P is the prismatic coefficient of the hull (See Chap. 2).

The accurate experimental determination of form factors for sailing yacht hulls is, unfortunately, a general and recognized problem (Oossanen 1981; Keuning and Sonnenberg 1998a). As already mentioned, the reason seems to be that in model (towing tank) tests there is usually a significant but arbitrary amount of laminar flow that makes the accurate determination of the equivalent flat plate friction drag coefficient C_{F_0} in the experiment very difficult if not impossible. As a consequence reliable values cannot be found in the literature and it has not been possible to verify the applicability of Eq. (H.37) to sailing yachts extensively.

Although a generally accepted form factor for sailing yacht hulls can, to the author's knowledge, not be found in the literature, there are a number of form factors that have been proposed for other types of ship. A rather complex example, given by Holtrop and Mennen (Holtrop and Mennen 1978), based on statistical analyses of experimental data, is of the type

$$1 + k_h \approx 0.93 + 0.49(B_{WL}/L_{WL})(D_h/L_{WL})^{0.5}(L_{WL}^{3}/\nabla)^{1/3}(1 - C_P)^{-0.5} \qquad \text{(H.39)[1]}$$

An even more complex form, also involving the position of the longitudinal centre of buoyancy (LCB) is given in Holtrop and Mennen (1982).

In Eq. (H.39) C_P is the prismatic coefficient, given by

$$C_P = \nabla/\left(L_{WL} S_{Xmax}\right) \qquad \text{(H.40)}$$

---

[1] constants and powers have been rounded off to one or two significant figures

Other examples are (Gross and Watanabe 1972)

$$k_h = 23\, C_B / \left\{ (L_{WL}/B_{WL})^2 \sqrt{(B_{WL}/D_h)} \right\} \tag{H.41}$$

and (USNA (Vroman 2003))

$$k_h = 19 \left(C_B B_{WL}/L_{WL} \right)^2 \tag{H.42}$$

In (H.41), (H.42) C_B is the so-called block coefficient of the hull defined as

$$C_B = \nabla / \left(L_{WL} B_{WL} D_h \right) = C_P\, S_{Xmax} / \left(B_{WL} D_h \right) \tag{H.43},$$

Equations (H.39), (H.41) and (H.42) can be rewritten as, respectively,

$$k_h \approx -0.07 + 0.49 \left(B_{WL}/L_{WL} \right)^{3/2} \left(D_h/B_{WL} \right)^{1/2} \left(C_P\, S_{Xmax}/L_{WL}{}^2 \right)^{-1/3} \left(1 - C_P \right)^{-0.5} \tag{H.44}$$

$$k_h = 23\, C_P \left(S_{Xmax}/L_{WL}{}^2 \right) \sqrt{ \left(B_{WL}/D_h \right) } \tag{H.45}$$

$$k_h = 19 \left\{ C_P \left(S_{Xmax}/L_{WL}{}^2 \right) \left(B_{WL}/D_h \right) \left(L_{WL}/B_{WL} \right) \right\}^2 \tag{H.46}$$

Since L_{WL}/B_{WL} can be approximated by

$$L_{WL}/B_{WL} \approx \sqrt{ \left\{ (\pi/4)/(B_{WL}/D_h) / (S_{Xmax}/L_{WL}{}^2) \right\} } \tag{H.47},$$

Equation (H.44) can also be written as

$$k_h \approx -0.07 + 0.49\, (\pi/4)^{-3/4} (S_{Xmax}/L_{WL}{}^2)^{5/12} C_P{}^{-1/3} (1 - C_P)^{-0.5} \left(B_{WL}/D_h \right)^{1/4} \tag{H.48}$$

and Eq. (46) as

$$k_h \approx 19 (\pi/4) C_P{}^2 \left(S_{Xmax}/L_{WL}{}^2 \right) \sqrt{ \left(B_{WL}/D_h \right) } \tag{H.49}$$

Note that, unlike (H.36/37), none of these expressions distinguishes between friction drag and form drag components and that the dependence of k_h on the various parameters is totally different.

It is interesting to apply Eqs. (H.48), (H.45), (H.49) to the case where the hull is a half-body of revolution and compare the results with those of (H.36/37). One should then multiply S_{Xmax} by 0.5 and put $D_h/B_{WL} = 0.5$. This gives

$$k_h \approx -0.07 + 0.49\, (4/\pi)^{3/4} 0.5^{1/6} (S_{Xmax}/L^2)^{5/12} C_P{}^{-1/3} (1 - C_P)^{-0.5} \tag{H.50},$$

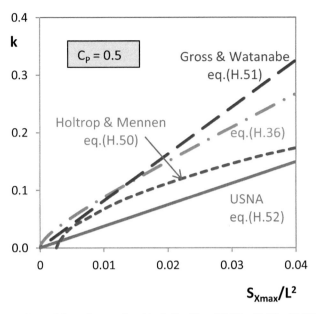

Fig. H.6 *Comparison of form factors for ship hulls (Eqs. (H.50), (H.51), (H.52)) and slender bodies of revolution (Eq. (H.36)) applied to slender bodies of revolution*

$$k_h = 16 \, C_P S_{Xmax} / L_{WL}^{\ 2} \tag{H.51}$$

$$k_h = 15 \, C_P^{\ 2} \, S_{Xmax} / L_{WL}^{\ 2} \tag{H.52}$$

The result, for C_P=0.5, a representative value for streamlined bodies of revolution, is shown in Fig. H.6. It can be noted that there are considerable differences between the various form factors. Obviously those given by Eq. (H.41/51) and Eq. (H.42/52) are both linear in S_{max} but differ by a factor 2 in slope. In terms of general level, those given by Eq. (H.41/51) and Eq. (H.36) are in reasonable agreement.

Because there is little doubt about the applicability of Eq. (H.36/37) to slender bodies of revolution, it is desirable to extend Eq. (H.36/37) so as to render it better applicability for sailing yachts. For this purpose the composite expressions (H.21) to (H.23) for slender lifting surfaces can be rewritten in terms of (shallow) hull related quantities:

$$C_{D_v} = \left(1 + k_{0b} + k_{1b}\right) C_{F_0} \tag{H.53},$$

with

$$k_{0b} = E_0 k_{0wb} (S_{WA}/S_{wet}) + (1 - E_0) \, k_{0bb} \tag{H.54}$$

and

$$k_{1b} = E_0 k_{1wb} (S_{WA}/S_{wet}) + (1 - E_0) \, k_{1bb} \tag{H.55}$$

The expressions (H.13) and (H.17) for k_{0w} and k_{1w} can, for $\Lambda = 0$, be rewritten as

$$
\begin{aligned}
k_{0wb} &= 2\,C_0(2D_h/L_{WL}) + 0.5\,C_0^2(2D_h/L_{WL})^2(1+5) \\
&= 4\,C_0(D_h/B_{WL})(B_{WL}/L_{WL}) + 12\,C_0^2\left\{(D_h/B_{WL})(B_{WL}/L_{WL})\right\}^2
\end{aligned}
\tag{H.56},
$$

and

$$
\begin{aligned}
k_{1wb} &= (1+k_{0wb})(2D_h/L_{WL})/(1-2D_h/L_{WL}) \\
&= 2(1+k_{0wb})(D_h/B_{WL})(B_{WL}/L_{WL})/\{1-2(D_h/B_{WL})(B_{WL}/L_{WL})\}
\end{aligned}
\tag{H.57},
$$

with the beam/length ratio B_{WL}/L_{WL} given by Eq. (H.47).

It follows from Eqs. (H.36), (H.37), that the coefficients k_{0bb} and k_{1bb} are given by

$$
k_{0bb} = 7.6\,S_{Xmax}/L_{WL}^2
\tag{H.58}
$$

and

$$
k_{1bb} = (1+k_{0bb})0.64\,\sqrt{\left(S_{Xmax}/L_{WL}^2\right)}/\left\{1-0.64\,\sqrt{\left(S_{Xmax}/L_{WL}^2\right)}\right\}
\tag{H.59}
$$

It is further noted that the quantity S_{WA} in Eqs. (H.54,55) represents the waterplane area. The ratio S_{WA}/S_{wet} can, for most practical purposes, be taken as

$$
S_{WA}/S_{wet} \approx 1/(1+2D_h/L_{WL})
\tag{H.60}
$$

The 'edge factor' E_0 is given by Eq. (H.24) with the aspect ratio given by

$$
A = B_{WL}^2/S_{WA} = \left(B_{WL}^2/L_{WL}^2\right)/\left\{\left(S_{WA}/S_{wet}\right)\left(S_{wet}/L_{WL}^2\right)\right\}
\tag{H.61}
$$

The quantity S_{wet}/L_{WL}^2 can, according to Keuning and Sonnenberg (1998a), be approximated empirically by

$$
S_{wet}/L_{WL}^2 \approx (1.97+0.171\,B_{WL}/D_h)\sqrt{\left\{C_P\left(S_{Xmax}/L_{WL}^2\right)\right\}}
\tag{H.62}
$$

The formulation given above implies a form factor that depends primarily on the maximum cross-sectional area and the beam/draft ratio of the hull. Vroman (2003); Van Driest and Blumer (1960); Prandtl and Schlichting (1934) and Candries et al. suggest that the block coefficient C_B, the prismatic coefficient C_P, and, possibly, the LCB position are also parameters of some importance. However, there is, apparently, considerable divergence of opinion on **how** the viscous drag depends on these parameters. This should, perhaps, not be surprising. Hull parameters like the block and prismatic coefficients and LCB are typically associated with wave-making resistance but are not necessarily of much importance for the viscous resistance. For the latter, there is little doubt that the cross-sectional area S_{Xmax}/L^2 (or the displacement/length ratio ∇/L_{WL}^3) is the most important.

Nevertheless, it can be argued that the prismatic coefficient (as well as the block coefficient) represents some crude measure of the convergence and divergence of the boundary layer flow on the hull. In general, the boundary layer flow on a sailing yacht hull will be diverging on the part of the hull forward of the maximum section and converging on the rear part. When the prismatic coefficient is large, the divergence and convergence of the boundary layer flow will be concentrated near, respectively, the bow and stern. For the stern region this means a more rapid growth of the displacement thickness (and possibly flow separation) leading to a higher form drag. In other words one should expect a higher value of the form factor 'k_1', associated with displacement thickness, for high values of C_p.

An exception is possibly the case where the front half of the hull has a low prismatic coefficient and the rear half has a very large one with a wide transom type stern and a relatively flat underbody (like many modern yachts). In this case the flow would detach along the whole width of the stern and there would be very little convergence of boundary layer flow. This would mean a little more friction drag but less form drag[2].

There is possibly also a significant dependence on the longitudinal position of the maximum section. With a rearward position, the convergence of the boundary layer flow towards the stern will be larger, which leads to higher form drag, than with a more forward position of the maximum section. This suggest, that the form drag will increase when the LCB moves aft. A possible exception is, again, the case of a wide transom type stern with a relatively flat underbody.

It is not very likely that the prismatic coefficient has a large effect on the form factor 'k_0', associated with friction drag due to supervelocity, at least for moderate values of C_p. The reason is that an increase of C_p causes larger longitudinal curvature and associated higher velocities near the bow and stern, but smaller curvature and lower velocities amidships. As a result there will probably not be much change in the average supervelocity level.

The dependence of the friction drag on the beam/draft ratio B_{WL}/D_h is another matter. For large B_{WL}/D_h (or small D_h/B_{WL}) the hull begins to behave as a slender wing with relative thickness $2D_h/L_{WL}$ and aspect ratio B_{WL}^2/S_{WA} where S_{WA} is the waterplane area. This case is already covered by the formulation (H.53) to (H.62) given above.

For $B_{WL}/D_h \rightarrow 0$, or $D_h/B_{WL} \rightarrow \infty$, the hull would take the form of a vertical, two-sided flat plate, implying $k_0 \rightarrow 0$. Since a flat plate at zero angle of attack cannot have form drag this should also imply $k_1 \rightarrow 0$. This case is, however, not of practical importance, because, for most sailing yachts, $B_{WL}/D_h > 2$.

As already mentioned, it is not very likely that the viscous resistance of sailing yacht hulls depends strongly on the prismatic coefficient. Exceptions are probably cases with extremely small or extremely large prismatic coefficients. Such configurations are, however, not of practical significance for sailing yachts.

---

[2] The discussion suggests that there might be a point in distinguishes prismatic coefficients for the front and rear halves of the hull.

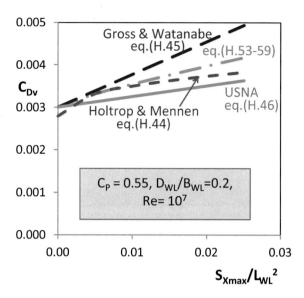

Fig. H.7 *Viscous drag coefficient of sailing yacht hulls as a function of cross-sectional area according to different formulae*

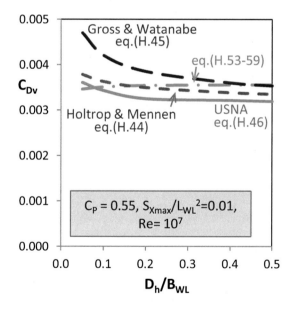

Fig. H.8 *Viscous drag coefficient of sailing yacht hulls as a function of beam/draft ratio according to different formulae*

Figure H.7 presents the viscous drag coefficient according to different formulae as a function of the relative cross-sectional area for hulls with a prismatic coefficient of 0.55 and a beam/draft ratio of 5. The Reynolds number is 10^7. It can be noticed that the differences between the various formulae are significant but that the agreement between the results of Eq. (H.44) and (H.53–59) is quite reasonable.

Figure H.8 presents the variation with the draft/beam ratio D_h/B_{WL}. Here also, the differences between the results of the various formulae are significant. A point of particular concern is the difference in character between the results of Eq. (H.53–

Fig. H.9 *Viscous drag coefficient of sailing yacht hulls as a function of prismatic coefficient according to different formulae*

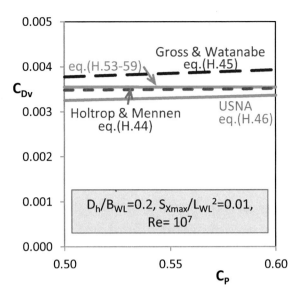

59) and the other curves. The behaviour of the latter does not seem to be in agreement with the requirement that, for $D_h/B_{WL} \to 0$, the viscous drag should become equal to that of a (one-sided) flat plate.

Figure H.9 presents the variation of the viscous drag as a function of the prismatic coefficient. The figure confirms that the variation with C_p is small. The difference in level between the 'classical' formulae is, again, significant.

A firm conclusion about the applicability of the various formula can, of course, not be drawn. It is clear however, that the proposed formula Eqs. (H.53–59) gives the best results when applied to bodies of revolution. It appears also that the results of Eqs. (H.53–59), based on analysis, and those of Eq. (H.44), based on statistics, are in reasonable agreement for an average sailing yacht hull characterized by $S_{Xmax}/L_{WL}^2 \approx 0.01$, $D_h/B_{WL} \approx 0.2$ and $C_p \approx 0.55$. This in spite of the fact that the backgrounds of the formulae are totally different.

Effects of Fluid Dynamic Interference

Additional viscous resistance is caused by mutual fluid dynamic interference between a hull and its appendages. This is caused by additional supervelocity generated on the hull by the appendages and vice versa.

Most, but not all, of the additional viscous resistance caused by the hull on the appendages is covered by the reflection plane model described in Sect. 6.3. The model implies that the effective aspect ratio is twice the geometrical onH. Another part is caused by additional supervelocity on the appendages induced by the hull. This part should be roughly proportional to the form factor k_h of the hull, given in Fig. 6.5.3. The ratio D_h/b, between the draft of the hull and the span of the appendage, is another. The reason is that the supervelocity induced by a body decreases with increasing (lateral) distance from the body. Obviously, the total effect of the additional supervelocity induced by the hull must go to zero when $D_h/b \to 0$. The appendage experiences the full effect only in the case of very small span, that is $D_h/b \to \infty$.

The discussion just given suggests that the additional viscous resistance of an appendage due to interference with the hull could be modelled by an additional form factor δk_h, to be added in Eq. (H.21),which is given by

$$\delta k_h = f(D_h/b)k_h \tag{H.63}$$

A suitable form of the interpolation function $f(D_h/b)$ is possibly

$$f(D_h/b) = (D_h/b)^2 / \left\{ 1 + (D_h/b)^2 \right\} \tag{H.64}$$

Equations (H.63), (H.64) indicate that the additonal friction drag of an appendage due to interference with the hull can be a significant fraction (upto, say, 10%) of the total viscous resistance of the appendage, in praticular for shallow draft, low aspect ratio appendages.

Similarly, the appendages also induce additional supervelocities on the hull, which lead to increased viscous resistance of the hull. A part of this is due to the thickness of the appendage and is felt as additional friction drag. Another part is due to the lift carry-over (Sect. 5.17) and is felt as pressure (form) drag, like the effect of lift on the viscous drag of a lifting surface.

Unfortunately there are no simple methods available in the literature for estimating the magnitude of the additional resistance of local velocity disturbances in threedimensional (hull) boundary layers. Nevertheless, a few statements can be made about it:

- The addional friction drag of a hull due to an attached appendage must be proportional to some positive power, probably close to 1, of the form factor k_{0w}, (see Eq. (H.21)), of the appendage. It is also clear that it must be a function of the relative dimensions of appendage and hull, such as the ratio c_R/L_{WL} between the root chord of the appendage and the length of the waterline of the hull. The additional friction drag of the hull due to interference can then be modelled by an additional form factor, to be added to Eq. (H.53), of the type

$$\delta k_{0bw} = C_{0bw} f(c_R/L_{WL})k_{0w} \tag{H.65}$$

Here, $f(c_R/L_{WL}) \to 0$ for $c_R/L_{WL} \to 0$ and $f(c_R/L_{WL})$ becomes of the order of 1 for $c_R/L_{WL} \to 1$. Unfortunately nothing can be said of the precise form of the function $f(c_R/L_{WL})$. The same applies to the constant C_{0bw}, except for that it is probably also of the order of 1.

- The additional form drag (pressure drag) due to the lift carry-over from an appendage on the hull should, like the form drag due to lift of a lifting surface, also be proportional to L^2, or, perhaps, rather L_R^2, where L_R is the local lift per unit span at the root section of the appendage. It should also be a function of the ratio L_{WL}/c_R, like the function $f(c_R/L_{WL})$ introduced above. Tentatively, one can postulate that the form drag due to lift carry-over from an appendage to a hull can be modelled in a similar way as for a lifting fin, by another additional form factor, also to be added to Eq. (H.53), of the type

$$\delta k_{1Lbw} = 2C_{1Lbw} f\left(L_{WL}/c_R\right)\left(C_{LR}c_R/L_{WL}\right)^2 \left(S_{hlat}/S_h\right) \qquad (H.66)$$

In Eq. (H.66) S_{hlat} is the lateral projected area of the hull and S_h the wetted area. Again, nothing can be said about the precise form of the function $f\left(L_{WL}/c_R\right)$ and the constant C_{1Lbw} except that both are probably also of the order of 1.

It is clear that the the precise magnitude of the additional viscous resistance of the hull due to interference with an appendage cannot be estimated properly with any certainty along the lines given above. Again, we can conclude only that the additional viscous drag of the hull due to interference with an appendage can be a significant fraction (upto, say, 10%) of the total viscous resistance of a sailing yacht hull. Computational Fluid Dynamics is required to obtain more accurate answers.

References

Abbott IH, Von Doenhoff AE (1949) Theory of wing sections. McGraw Hill, New York

Abbott IH, Von Doenhoff AE, Stivers LS (1945) Summary of airfoil data, NACA Report No. 824

Ashley H, Landahl M (1985/1965) Aerodynamics of wings and bodies. Dover Publications, New York. ISBN 0-486-64899-0

Gross A, Watanabe K (1972) Form factor, report of performance committee. App. 4, 13 ITTC

Hoerner SF (1965) Fluid dynamic drag. Hoerner Fluid Dynamics, Bakersfield

Holtrop J, Mennen GGJ (1978) A statistical power prediction method. ISP 25

Holtrop J, Mennen GGJ (1982) An approximate power prediction method. ISP 29

Keuning JA, Sonnenberg UB (1998a) Approximation of the hydrodynamic forces on a sailing yacht based on the delft systematic yacht hull series. Proceedings of the 15th HISWA international symposium on yacht design and yacht construction, Amsterdam, Nov 1998

Küchemann D (1978) The aerodynamic design of aircraft. Pergamon Press, Oxford

Kroo I (n.d.) Form factor. http://www.adg.stanford.edu

McWilkinson DG, Blackerby WT, Paterson JH (1974) Correlation of full-scale drag predictions with flight measurements on the C-141aircraft-phase II, wind tunnel test, analysis and prediction techniques. Vol 1-drag predictions, wind tunnel data analysis and correlation. NASA CR-2333, Feb 1974

Oossanen Peter van (1981) Method for the calculation of the resistance and side force of sailing yachts, paper presented at Conference on 'calculator and computer aided design for small craft—the way ahead', R.I.N.A., 1981

Prandtl L, Schlichting H (1934) Das Widerstandsgesetz rauher Platten, Werft-Reederei-Hafen, 1–4

Prandtl L, Tietjens OG (1934a) Fundamentals of aero- & hydromechanics. McGraw-Hill, New York (also: Dover Publications, NY, 1957)

Torenbeek E (1982) Synthesis of airplane design. Kluwer, Netherlands

Van Driest ER, Blumer CB (1960) Effect of roughness on transition in supersonic flow. AGARD Rept. 255

Vroman R (2003) Principles of ship performance, EN200 courses, Chap 7, resistance and power-
    ing of ships. http://www.usna.edu
Young AD (1989) Boundary layers, AIAA Education Series. ISBN 0-930403-57-6

Appendix I

Kelvin Ship Waves

A ship, or any object for that matter, moving through the water surface with a speed
V_b causes wave-like disturbances at each point in time. We can think of this in
terms of circular wavelets caused by pebbles thrown in a pond along a straight line
at regular intervals in space and time. When the ship is at the position indicated by
'O', the wave emitted 't' seconds ago, when the ship was at the position Q, will have
reached the position indicated by the circle with radius $c_w t$ around Q, where c_w is the
propagation speed of the wavelet. It is recalled from Sect. 5.19 that this propagation
speed is a function of the wavelength:

$$c_w = \sqrt{(g\,\lambda_w/2\pi)} \tag{I.1}$$

When the ship emits waves of only one wavelength, that is waves with one and the
same propagation speed, the circular wavelets generated at different points in time
form a common, straight wave front. In Fig. I.1 this is indicated as envelope for

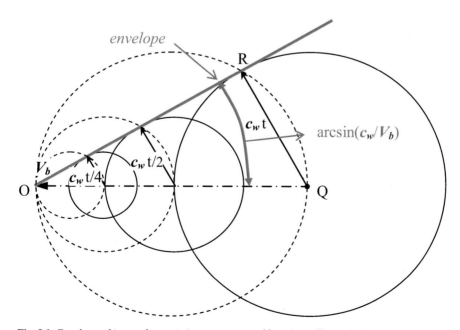

Fig. I.1 Envelope of 'monochromatic' waves generated by a 'travelling point'

circular wavelets generated at points in time corresponding with t, t/2 and t/4. This envelope makes an angle arcsin (c_w/V_b) with the direction of motion of the ship.

In principle a ship emits wavelets of all wave lengths. However, only a certain group of wavelets can contribute to the formation of a steady wave pattern that moves with the ship. For the wavelets emitted from the point Q, this group is characterized by the dotted circle with the OQ as the middle line. Obviously, all the points 'R' of this group, that move with the ship, are positioned on this circle. Similar 'dotted circles' can be constructed for the points on OQ corresponding with t/2 and t/4.

One conclusion that can be drawn from this is that only wavelets with a propagation speed between zero and V_b can contribute to a steady wave pattern that moves with the ship. This means also that, at least in principle, there is an infinitely large number of envelope-type wave fronts, each corresponding with a certain propagation speed between zero and V_b.

An important mechanism is further that waves can intensify each other at certain locations due to interference. These are the locations where the crests or troughs of different waves coincide. At other locations they may tend to cancel each other, such as where a crest meets a trough or vice versa. Hence, interfering waves form new, so-called composite waves. The latter are the waves that are visible to an observer's eye.

It has been found[3] that composite waves formed by interfering elementary waves of almost the same wavelength travel at a speed that differs from the propagation speed of the elementary waves when the propagation speed is a function of the wavelength. When the propagation speed c_w of the elementary waves is proportional to the square root of the wavelength, as is the case for water surface waves (see Eq. (I.1)), the propagation speed c_g of the composite waves (the 'group velocity') is found to be half that of the elementary waves:

$$c_g = \tfrac{1}{2}c_w \tag{I.2}$$

This means that all points R′ in the water surface, that belong to composite waves formed by elementary waves that travel at the same speed as the ship and that are generated at the point Q, must be positioned at a circle with diameter MQ, where M is the point halfway between Q and O (See Fig. I.2). As indicated in Fig. I.2 the composite waves also form an oblique wave front, that, unlike the front of the elementary waves, is visible to an observer's eye.

It is easily verified that the angle ψ' between the envelope and the boat speed is equal to about 19.5°: It follows from Fig. I.2 that

$$\sin \psi' = M'R'/M'O = (V_b/4)/(3V_b/4) = 1/3 \tag{I.3}$$

which gives $\psi' \approx 19.5°$.

It is important to note that ψ' is independent of boat speed. It means that we must expect that a ship always generates an oblique wave front radiating out at an angle of about 19.5°, irrespective of boat speed. The earliest explanation of this

---

[3] See, e.g. H. Lamb, *'Hydrodynamics'*, Cambridge University Press, 1916.

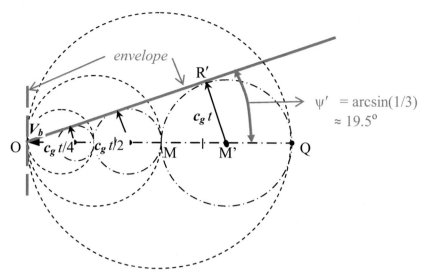

Fig. I.2 *Envelope of composite waves travelling at half the speed of elementary waves generated by a 'travelling point'*

phenomenon is believed to have been given by Lord Kelvin (1887, 1904). For this reason wave patterns of this kind are known as Kelvin ship waves or Kelvin wave patterns.

Two further remarks are in order at this point.

The first is that, since the ship was considered as a moving point in the description given above, the associated wave pattern is representative only for an observer at a large distance from the ship.

The second remark is that, as one may see in Fig. I.2, the (large) circles representing the loci of points on elementary waves that travel with the ship, also tend to form a common envelope at the point O where the ship is. This is already a hint, that, near the ship, there will also be waves with an orientation normal to the direction of motion.

References

Kelvin L (1904) Deep-sea ship-waves. Proceedings of the royal society of Edinburgh

Appendix J

Wave-Making Resistance

Wave-making resistance manifests itself as pressure drag. We have seen in Sect. 5.2 that the pressure drag D_p of a body can be expressed as

$$D_p = f_{np}S_Z\sin\alpha + f_{tp}S_X\cos\alpha \tag{J.1},$$

where S_X is the (maximum) cross-sectional area, S_Z the area of the longitudinal section through the main plane of the body and α is the angle of attack. The quantities f_{np} and f_{tp} are functions of the distribution of pressure forces around the body. In sailing yacht terminology Eq. (J.1) is written as

$$D_p = f_{np}S_{lat}\sin\lambda + f_{tp}S_{Xmax}\cos\lambda \tag{J.2},$$

where S_{lat} is the lateral area of (the under-water part of) the hull and appendages and λ is the leeway angle.

For small values of λ Eq. (J.2) reduces to

$$D_p = f_{np}S_{lat}\lambda + f_{tp}S_{Xmax} \qquad (\lambda \text{ in radians}) \tag{J.3}$$

In Eq. (J.3) one can distinguish a symmetric part, that is the term proportional to S_{Xmax}, and an anti-symmetric part, that is the term involving the angle of leeway λ. Obviously, the wave-making resistance due to displacement volume (at zero leeway) is related to the symmetric part of Eq. (J.3). The term involving λ contains any wave-making resistance due to leeway and/or side force.

We will first consider the wave-making resistance due to displacement volume.

Wave-Making Resistance Due to Displacement Volume
Because the wave-making resistance at zero leeway is a longitudinally directed pressure drag, that is caused by differences in water level, it must scale like the product of the pressure due to the deformation of the water surface (Sect. 5.19) and the cross-sectional area. This means that the wave-making resistance due to displacement volume $R_{w\nabla}$ (also) scales like

$$R_{w\nabla} \div \rho g Z_h S_{Xmax} \tag{J.4},$$

Here, Z_h represents the elevation of the water surface and S_{Xmax} is the maximum cross-sectional area. We have seen in Sect. 5.19 that the surface elevation Z_h can be expressed in terms of local flow velocity (see Eqs. (5.19.5),(5.19.9)):

$$\rho g Z_h \approx \tfrac{1}{2}\rho(V_b^2 - V_{WL}^2) = \tfrac{1}{2}\rho V_b^2\left(1 - V_{WL}^2/V_b^2\right) \tag{J.5},$$

where V_{WL} is the flow velocity at the waterline.

This means that, also using the identity $S_{Xmax} = \nabla/(L_{WL}C_P)$, Eq. (J.4) can be rewritten as

$$R_{w\nabla} \div \tfrac{1}{2}\rho V_b^2 S_{Xmax} = \tfrac{1}{2}\rho g\, Fr^2 L_{WL}\nabla/(L_{WL}C_P), \tag{J.6a},$$

or

$$R_{w\nabla} \doteq \tfrac{1}{2}\, \mathrm{Fr}^2\, \nabla/C_P \tag{J.6b}$$

Here, as before, ∇ is the displacement volume, ∇ is the displacement weight, L_{WL} is the length of the waterline and C_p is the prismatic coefficient.

Note that Eq. (J.6) includes the general scaling (6.56) with displacement. At the same time it is more specific because of the inclusion of the dependence on boat speed, length of the waterline and prismatic coefficient.

Taking Eq. (J.6b) as a basis, an alternative definition of the coefficient of the wave-making resistance due to volume would be:

$$C_{Rw\nabla} \equiv R_{w\nabla}/(\tfrac{1}{2}\mathrm{Fr}^2 \nabla/C_P) \tag{(J.7)}$$

Because of the wave interference phenomena mentioned in Sect. Wave Making Resistance one should still expect the wave-making resistance coefficient to be a function of Froude number. At the same time one would also expect $C_{Rw\nabla}$ to vary less rapidly with Froude number than in the case of the classical definition (6.58).

It follows from Eq. (J.5) that

$$C_{Rw\nabla} \doteq \left\{1 - (V_b + v_{WL})^2/V_b)^2\right\} \approx -2v_{WL}/V_b \tag{J.8},$$

where $v_{WL} = V_{WL} - V_b$ is the perturbation velocity, or supervelocity, at the waterline.

We have seen in Appendix H that the average supervelocity of the hull of a sailing yacht is, primarily, a function of the cross-sectional area/length ratio S_{Xmax}/L^2 or the displacement/length ratio $(\nabla/L_{WL}{}^3)/C_p$. From this it follows that

$$v_{WL}/V_b \doteq S_{Xmax}/L^2 \equiv (\nabla/L_{WL}{}^3)/C_P \tag{J.9}$$

It is further clear that the supervelocity on the hull at the waterline must also be a function of the beam/draft ratio B_{WL}/D_h of the hull. Because the hull becomes a vertical flat plate for $B_{WL}/D_h \to 0$, v_{WL} will also tend to zero under these conditions. When the cross-section of the hull is a semi-circle ($B_{WL}/D_h = 2$), the super-velocity at the waterline will be the same as on a body of revolution. For larger values of B_{WL}/D_h, the super-velocity at the waterline will be higher, (except, possibly, for very high values of B_{WL}/D_h when the shape of the hull tends to a horizontal flat plate).

The discussion just given and inclusion of the dependence on Froude number, to model the effect of wave interference, leads to the following tentative expression for the dependence of the wave-making resistance coefficient:

$$C_{Rw\nabla} = f_\nabla(\mathrm{Fr})f(B_{WL}/D_h)S_{Xmax}/L^2 \tag{J.10}$$

The function $f_\nabla(\mathrm{Fr})$ models the dependence on Froude number. It is, in general (Lewis 1988), of the type of a 4th degree polynomial in Fr. The function $f(B_{WL}/D_h)$ is possibly of the type $(B_{WL}/2D_h)^p$, where p is > 0.

Equation (J.10) reflects, that for a given Froude number, the wave-making resistance coefficient is primarily a function of the cross-sectional area/length ratio S_{Xmax}/L_{WL}^2 or the displacement/length ratio and the beam/draft ratio B_{WL}/D_h. In addition there will be a dependence on the prismatic coefficient and the position of the longitudinal centre of buoyancy, in particular because of the variation with Froude number of sinkage and pitch angle, as discussed in Sect. Wave Making Resistance.

Effect of Leeway on the Wave-Making Resistance Due to Volume
According to Eq. (J.3) the anti-symmetric part (due to leeway) of the pressure drag of a body is given by

$$D_p = f_{np} S_{lat} \lambda, \quad (\lambda \text{ in radians}) \tag{J.11}$$

The function f_{np} is now related to the differences in the elevation of the free surface caused by the angle of leeway of the hull. This means that the effect of leeway on the wave-making resistance due to volume can be written as

$$R_{wV\lambda} \div \rho\, g Z_h S_{lath} \lambda \tag{J.12},$$

Because the factor $\rho\, g\, Z_h$ represents a pressure difference due to leeway or yaw, it can, when multiplied with a characteristic area $(D_h L_{WL})$ and arm length (L_{WL}) be related to the hydrodynamic moment M_{Hz}:

$$\rho g Z_h L_{WL} D_h L_{WL} \div M_{Hz} \tag{J.13}$$

In Sect. 5.16 we have seen that the fluid dynamic moment of a body is given by

$$M = \tfrac{1}{2}\rho V^2 \nabla \sin 2\alpha \tag{J.14}$$

Hence, it follows from Eq. (J.13) that

$$\rho g Z_h \div \tfrac{1}{2}\rho V_b^2 \nabla \sin 2\lambda \,/\!\left(L_{WL}^2 D_h\right)$$

or, for small λ

$$\rho g Z_h \div \tfrac{1}{2}\rho V_b^2 \left(\nabla/L_{WL}^3\right)/\left(D_h/L_{WL}\right) 2\lambda \tag{J.15}$$

Substitution in Eq. (J.12) gives

$$R_{wV\lambda} \div \tfrac{1}{2}\rho V_b^2 \left(\nabla/L_{WL}^3\right)/\left(D_h/L_{WL}\right) S_{lath}\, 2\lambda^2 \tag{J.16},$$

This suggests that a coefficient for the effect of leeway or yaw on the wave-making resistance due to volume can be defined as

$$C_{\mathrm{Rw}\nabla\lambda} \div R_{w\nabla\lambda} / \left(\tfrac{1}{2}\rho V_b^2 S_{\mathrm{lath}} \right) \tag{J.17}$$

It then follows from Eq. (J.16) that

$$C_{\mathrm{Rw}\nabla\lambda} \div \left\{ \left(\nabla / L_{\mathrm{WL}}^3 \right) / (D_h / L_{\mathrm{WL}}) \right\} \lambda^2 \tag{J.18}$$

Equation (J.18) implies that the effect of leeway or yaw on the wave-making resistance due to volume of a sailing yacht is proportional to the product of the displacement length ratio and the square of the angle of leeway and inversely proportional to the draft/length ratio of the hull. This, of course, in addition to the general scaling factors $\tfrac{1}{2}\rho V_b^2$ and S_{lath} and a dependence on Froude number. Unfortunately, there is, to the author's knowledge, no information on the absolute magnitude of this part of the wave-making resistance and its dependence on Froude number. However, it is probably negligible for most if not all practical purposes, except, perhaps, for large angles of yaw or leeway and high displacement/length ratios.

Wave-Making Resistance Due to Side Force
Because the side force produced by the keel (and carried-over to the hull) causes deformation of the water surface it also causes wave-making resistance (R_{wS}). Like the effect of leeway on the wave-making resistance due to volume the wave-making resistance due to side force also scales like Eq. (J.12). I.e.

$$R_{wS} \div \rho g Z_h S_{\mathrm{lat}} \lambda \tag{J.19},$$

with the lateral area written as

$$S_{\mathrm{lat}} = S_{\mathrm{lath}} + S_k \tag{J.20},$$

Replacing λ by $C_S / \dfrac{dC_S}{d\lambda}$, Eq. (J.19) can be rewritten as

$$R_{wS} \div \rho g Z_h S_k (1 + S_{\mathrm{lath}} / S_k) C_S / \frac{dC_S}{d\lambda} \tag{J.21}$$

The elevation Z_h of the water surface should now be expressed in terms of the side force or circulation of the keel and hull. It follows from Appendix E, Eq. (E.7), that

$$\rho g\, Z_{h+/-} = \tfrac{1}{2}\rho \left(V_b^2 - V_{+/-}^2 \right) = \tfrac{1}{2}\rho V_b^2 \left\{ 1 - (V_{+/-} / V_b)^2 \right\}, \tag{J.22}$$

where the subscript $+/-$ refers to the pressure and the suction side of the keel/hull, respectively. One can write

$$V_{+/-} = V_b +/- v_\Gamma \tag{J.23},$$

where v_Γ is the perturbation velocity at the water line due to the circulation about the keel/hull. This can be expressed as

$$v_\Gamma \doteq \Gamma_R / L_{CWL} \qquad (J.24),$$

in which Γ_R is the circulation around the root section of the keel (which is the same as the circulation around the waterline) and L_{CWL} is the circumference of the water-line. The latter can be approximated as

$$L_{CWL} \approx 2L_{WL} + B_{WL} = 2L_{WL}(1 + 0.5\,B_{WL}/L_{WL}) \qquad (J.25)$$

For small perturbations $(v_\Gamma/V_b \ll 1)$ Eq. (J.22) can be approximated as

$$\rho g Z_h \approx \rho V_b^2 v_\Gamma / V_b \qquad (J.26)$$

With Eqs. (J.24) and (J.25) this gives

$$\rho g Z_h \doteq \rho V_b^2 (\Gamma_R/V_b)\{2\,L_{WL}(1 + 0.5\,B_{WL}/L_{WL})\} \qquad (J.27a)$$

or

$$\rho g Z_h \div \tfrac{1}{2}\rho V_b^2 (\Gamma_R/V_b)/\{L_{WL}(1 + 0.5\,B_{WL}/L_{WL})\} \qquad (J.27b)$$

It follows from the discussion given above that the wave-making resistance due to the side force can be expressed as

$$R_{wS} = \tfrac{1}{2}\rho V_b^2 S_k C_{RwS} \qquad (J.28)$$

Then, there holds for the non-dimensional coefficient C_{RwS}:

$$C_{RwS} \div (\Gamma_R/V_b)(1 + S_{lath}/S_k)\,C_S / \left(\frac{dC_S}{d\lambda}\right) / \{L_{WL}(1 + 0.5 B_{WL}/L_{WL})\} \quad (J.29)$$

It can be derived from Appendices D and E that the factor Γ_R/V_b in Eq. (J.29) can be written as

$$\Gamma_R/V_b = \left[(2/\pi)F_R / \left[1 + (4/\pi)F_R \{D_h/b_k - (F_{1.1} + F_{2.1} + F_{1.2})\} \right] (b_k/A_k)C_S \right] \quad (J.30)$$

The various quantities in Eq. (J.30) are described in Appendix E:

- C_S is the side force coefficient
- F_R is a function involving the ratio between the circulation at the root and the average circulation of the keel. It depends on taper ratio, sweep angle and aspect ratio
- D_h/b_k is the ratio between the draught of the hull and the span of the keel
- b_k is the span of the keel

- A_k is the aspect ratio of the keel
- The functions $F_{1.1}$, $F_{2.1}$ and $F_{1.2}$ model the free surface effect on the side force

Equations (J.29), (J.30) suggest that the coefficient C_{RwS} for the wave-making resistance due to side force can be written as

$$C_{RwS} = f_S(Fr)K_{w0}C_S^2 \qquad (J.31),$$

or

$$C_{RwS} = K_w C_S^2 \qquad (J.32),$$

where

$$K_w = f_S(Fr)K_{w0} \qquad (J.33)$$

The factor K_{w0} is given by

$$K_{w0} = (2/\pi)F_R / \left(\frac{dC_S}{d\lambda}\right) / [1 + (4/\pi)F_R \{D_h/b_k - (F_{1.1} + F_{2.1} + F_{1.2})\}]$$
$$\times (b_k/L_{WL})(1 + S_{lath}/S_k)/\{A_k(1 + 0.5B_{WL}/L_{WL})\} \qquad (J.34)$$

The ratio S_{lath}/S_k can be approximated by

$$S_{lath}/S_k \approx 0.65 A_k(D_h/b_k)(L_{WL}/b_k) \qquad (J.35)$$

Note that Eq. (J.32) implies that the coefficient for the wave-making resistance due to side force, like the induced resistance (Sect. Wave Making Resistance and Appendix E), is proportional to the square of the side force coefficient. The wave-making resistance due to side force is further, again like the induced drag factor K_i, a function of the sweep angle, taper ratio and aspect ratio of the keel as well as the relative dimensions of keel and hull and the beam/length ratio of the hull. However, it is a stronger function of aspect ratio because the lift curve slope $\frac{dC_S}{d\lambda}$ also depends on aspect ratio.

Figures J.1, J.2, J.3, and J.4 illustrate that the keel aspect ratio and the relative dimensions of keel and hull (b_k/L_{WL}) are the most important parameters for the drag-due-to-side-force factor K_{w0}. The taper ratio of the keel is only important for very small aspect ratios. The effect of the angle of sweep of the keel and the beam/length ratio of the hull is quite small.

The function $f_S(Fr)$ in Eqs. (J.31), (J.33) models the effect of Froude number (wave interference) but there is, unfortunately, hardly any information about it. The empirical/statistical method of Keuning and Sonnenberg (1998a) for estimating the (total) resistance due to side force suggests that the latter is approximately proportional to $(1.23 - 0.73 \, Fr)^{-2}$. Assuming that the viscous resistance due to side force and

Fig. J.1 Wave drag due to side force
factor for keel-hull configurations as a
function of keel-span/hull-length ratio

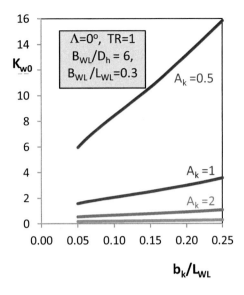

Fig. J.2 Wave drag due to side force
factor of keel-hull configurations, as
a function of the taper ratio of the
keel

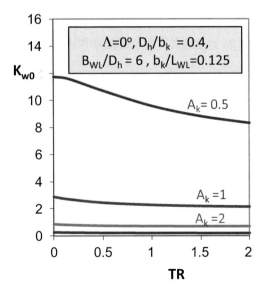

Fig. J.3 Wave drag due to side force
factor of keel-hull configurations as
a function of the angle of sweep of
the keel

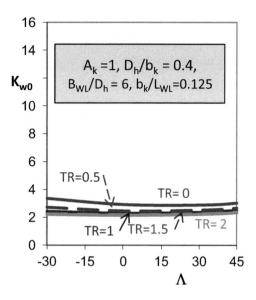

Fig. J.4 Wave drag due to side force
factor of keel-hull configurations, as
a function of the beam/length ratio
of the hull

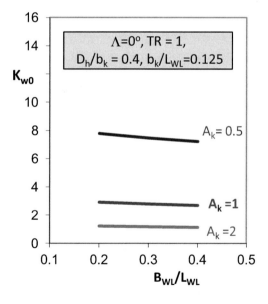

the induced resistance are independent of Froude number, this would mean, that, at a Froude number of 0.35, the wave-making resistance due to side force is about 35 % of the total resistance due to side force.

On the other hand, Fig. E.2 suggests that it can only be of the order of 10–15 % of the total resistance due to side force (at Fr = 0.35). In any case, the analysis given above suggests that the wave-making resistance due to side force is, in general, quite small (only a fraction of the induced resistance) and probably negligible for many practical purposes. Exceptions are, possibly, yachts with a (big) keel of very low aspect ratio at high Froude numbers.

Effects of Heel

The most important effect of heel on the wave-making resistance is that on the wave-making resistance due to the volume of the hull. It appears that there is considerable divergence of opinion in the literature (Oossanen 1981; Keuning and Sonnenberg 1998a, b) on the way in which the wave-making resistance due to volume varies with heel. In Oossanen (1981) the variation is taken as $1/\cos^2\varphi$. Keuning and Sonnenberg (1998a, b) describe empirical-statistical approximation formulae that, in the author's experience, produce significantly different results.

It would seem, however, that some characteristics of the variation of the wave-making resistance with heel are clear. Firstly, the wave-making resistance is, obviously, a symmetric function of the heel angle φ (such as $\cos\varphi$, or even powers of φ), because it does not matter for the resistance whether a yacht heels to starboard or to port. Secondly, it is also clear that there will be no change due to heel in wave-making resistance of the hull if the cross-sectional shape of the hull is that of a circle segment. The reason is that the shape of the submersed part of the hull is not changed by heel, provided that the angle covered by the circle segment is sufficiently large in relation to the angle of heel. We have seen in Appendix E that for a circle segment there holds

$$\{1-(C_m/C_m^*)\} = 0 \tag{J.36},$$

where C_m is the maximum-section coefficient and C_m^* is the section coefficient of a circle segment with beam/draft ratio B_{WL}/D_h (see Eq. (E.68)).

It would seem further that the variation would be inversely proportional to (some power of) the vertical position under heel of the centre of buoyancy. The wave-making resistance can be expected to increase when the centre of buoyancy comes closer to the water surface and to decrease when VCB goes away from the water surface. This effect can be approximated by a factor of the type

$$\{1+(B_{WL}/D_h)^2\}/\{\cos\varphi+(B_{WL}/D_h)\sin\varphi\}^2 \tag{J.37}$$

It is easily verified that this factor takes the value $1/\cos^2\varphi$ for $B_{WL}/D_h \to 0$ and the value $1/\sin^2\varphi$ for $B_{WL}/D_h \to \infty$.

It also appears (Keuning and Sonnenberg 1998a) that the variation with heel of the wave-making resistance is influenced significantly by the position of the longitudinal centre of buoyancy. This in the sense that the variation with heel of the wave-making resistance is smaller when the LCB position is further aft.

Based on the discussion just given a tentative expression for the variation with heel of the wave-making resistance is

$$C_{Rw\nabla}(\varphi) = C_{Rw\nabla}(\varphi = 0) \left[1 + c_1 \{ 1 - c_2 (C_m / C_m^*) + c_3 LCB^2 \} \right.$$
$$\left. \times \{ 1 + (B_{WL}/D_h)^2 \} / \{ \cos\varphi + (B_{WL}/D_h) \sin\varphi \}^2 \right]$$

$$(J.38)$$

Tentative values for the constants c_1, c_2, c_3 are 0.3, 0.9 and -10, respectively.

In the author's experience results of this formula are in reasonable agreement with the scarce amount of towing test results published in the literature (Keuning and Sonnenberg 1998a).

References

Keuning JA, Sonnenberg UB (1998a) Approximation of the hydrodynamic forces on a sailing yacht based on the delft systematic yacht hull series. Proceedings of the 15th HISWA international symposium on yacht design and yacht construction, Amsterdam, Nov 1998

Keuning JA, Sonnenberg UB (1998b) Developments in the velocity prediction based on the Delft systematic yacht hull series, Report no. 1132-P, Ship Hydromechanics Laboratory, Delft University of Technology, March 1998

Oossanen Peter van (1981) Method for the calculation of the resistance and side force of sailing yachts, paper presented at Conference on 'calculator and computer aided design for small craft—the way ahead', R.I.N.A., 1981

Appendix K

Natural Frequencies of Heaving, Pitching and Rolling Motions

It is shown in Lewis (1989) that the natural frequency f_{0z} of the heaving motion of a ship can be expressed as

$$f_{0z} \approx \frac{1}{2\pi} \sqrt{\frac{gSwp}{(\nabla + \nabla')}} = \frac{1}{2\pi} \frac{\sqrt{(g/L_{WL})}}{\sqrt{[\{C_{wp}(B_{WL}/L_{WL})\}/\{(\nabla + \nabla')/L_{WL}^3\}]}}$$

$$(K.1)$$

Here, B_{WL} and L_{WL} are, as before, the maximum beam and the length of the waterline, respectively, ∇ is the displacement volume, g the constant of gravitational acceleration and S_{WP} is the waterplane area (see Chap. 2). The quantity ∇' is the volume of the so-called added mass, that is the virtual, additional mass involved with the motion due to the periodic displacement of the volume of water occupied by the ship. ∇' depends on the frequency but, for sailing yachts, can be taken equal to about 1.8∇ as a first approximation (Keuning 1985, 1998, 2006).

A similar expression (Lewis 1989) for the natural frequency of the pitching motion reads

$$f_{00} \approx \frac{1}{2\pi}\sqrt{\{(\nabla\,gm_L)/(I_{yy}+I'_{yy})\}} = \frac{1}{2\pi}\sqrt{(g/L_{WL})}\sqrt{\left\{(gm_L/L_{WL})/\left(C_{ky}{}^2+A'\right)\right\}} \qquad (K.2)$$

In Eq. (K.2) gm_L is the longitudinal metacentric height (Sect. 6.2). The author has found that this can be approximated by

$$gm_L \approx \{1+0.1(1+C_P)\}\,L_{WL} \qquad (K.3)$$

C_{ky} is a coefficient related to the mass moment of inertia I_{yy} (Sect. 2.3) around the pitching axis. It is defined by

$$I_{yy} = \rho\nabla C_{ky}{}^2 L_{WL}{}^2 \qquad (K.4)$$

The product $C_{ky}L_{WL} = k_y$ is called the longitudinal radius of gyration (Sect. 2.3). For sailing yachts C_{ky} is of the order of 0.27, i.e. the longitudinal radius of gyration is about 27% of the length of the waterline, or about 25% of the overall length (Keuning 1985, 1998, 2006).

The quantities I'_{yy} and A' in Eq. (K.2) represent the effect of the moment of inertia around the pitching axis of the added mass. For sailing yachts their values can be taken as

$$I'_{yy} \cong \rho\nabla A'\,L_{WL}{}^2 \qquad (K.5)$$

where A' can be taken as

$$A' \cong 0.05 \qquad (K.6)$$

for most practical purposes (Lewis 1989; Keuning 1985, 1998, 2006). This means (Keuning 1985, 1998, 2006) that $I'_{yy} \cong 0.7\,I_{yy}$.

For the rolling motion a similar expression for the natural frequency $f_{0\varphi}$ is given by

$$f_{0\varphi} \approx \frac{1}{2\pi}\sqrt{\{(\nabla\,gm_T)/(I_{xx}+I'_{xx})\}} = \frac{1}{2\pi}\sqrt{(g/L_{WL})}\sqrt{\left\{(gm_T/L_{WL})/\left(C_{kx}{}^2+\hat{A}\right)\right\}} \qquad (K.7)$$

Here, gm_T is the lateral or transverse metacentric height and I_{xx} is the mass moment of inertia around the rolling axis. C_{kx}, when multiplied by L_{WL}, represents the transverse radius of gyration.

It is the author's impression that for sailing yachts gm_T/L_{WL} is of the order of 0.45 B_{WL}/L_{WL} and that C_{kx} is of the order of 0.18.

The quantities I'_{xx} and $\hat{A}$ represent the effect of the moment of inertia around the pitching axis of the added mass. They are related in the same way as Eq. (K.5):

$$I'_{xx} = \rho \nabla \; \hat{A} L_{WL}^{\;2} \qquad\qquad (K.8)$$

For most ship hulls the moment of inertia around the rolling axis of the added mass is much smaller than that of the real mass. For ships with (semi-) circular cross-sections without appendages of significant volume it should, in fact, be zero. This suggests that $\hat{A}$ can be taken equal to zero for many practical purposes. For yachts with highly non-circular cross-sections and a keel of large volume, $\hat{A}$ is probably still much smaller than 0.05 (the corresponding value for the longitudinal moment of inertia of the added mass, Eq. (K.6)).

References

Keuning JA (1985) Resistance of sailing yacht propellers. Delft University of Technology, Ship Hydromechanics Laboratory, Report nr. 656

Keuning JA (1998) Dynamic behaviour of sailing yachts in waves, Chap 6 of Reference Claughton et al. (1999 in Chap. 5)

Keuning JA (2006) An approximate method for the added resistance in waves of a sailing yacht, Report 1481-P, Ship Hydromechanics Laboratory, Delft University of Technology

Lewis EV (ed) (1989) Principles of naval architecture, vol III—motions in waves and controllability. SNAME, Jersey City. ISBN 0-939773-02-3

Appendix L

Forces and Moments Due to Wave Encounter

L.1 Excitation Forces

As discussed in detail in Lewis (1989), an essential role in the mechanism of motion excitation by waves is played by the orbital motion of the water particles in waves (Sect. 5.19). For the excitation of the pitching and rolling motions this is illustrated by Fig. L.1.

The figure illustrates that, when sailing upwind on a port tack, a yacht at a wave crest experiences a bow down pitching moment and a rolling moment to port due to the orbital motion of the water particles in the wave acting on the hull. On the same port tack a yacht experiences a bow up pitching moment and a rolling moment to starboard when in a wave trough. It is easily verified that the opposite is the case when a yacht sails with the waves. It will also be clear from the figure that the pitching moment induced by the waves will be most pronounced for $\lambda_w \approx 2L_{WL} \, | \cos\gamma_w |$. For the rolling moment this will be the case when $\lambda_w \approx 2B_{WL} \, | \sin\gamma_w |$.

Considering the theory and experimental data contained by Lewis (1989), the author has found that the amplitude of the total excitation moment around an axis

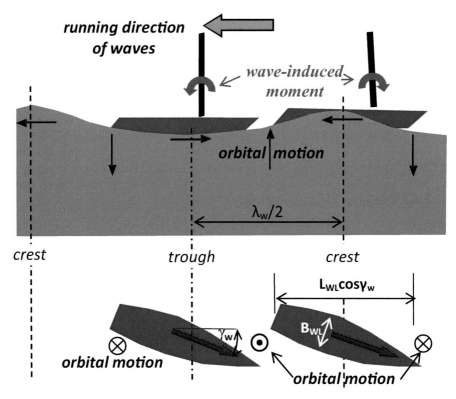

Fig. L.1 *Effect of orbital motion of water particles on pitching and rolling moments due to waves*

in the water plane, caused by the orbital motion of the water particles acting on the hull is approximately proportional to a factor of the type

$$F_w = f(\omega_e) e^{-2.5[(L_{WL}/\lambda_w)\{|\cos\gamma_w| + \sin^2\gamma_w (B_{WL}/L_{WL})\} - 0.55]^2} \tag{L.1}$$

Then, for the pitching component of this moment there holds

$$F_{ex\theta} = |\cos\gamma_w| F_w \tag{L.2}$$

and for the rolling component

$$F_{ex\varphi} = |\sin\gamma_w| F_w \tag{L.3}$$

In the expression (L.1), $f(\omega_e)$ is a (weak) function of the circular frequency of encounter of the yacht with the waves. The experimental data given in Lewis (1989) suggests that

$$f(\omega_e) \cong (\omega_e/\omega_e')^{0.15} \tag{L.4}$$

where

$$\omega_e = \omega_0 + 2\pi\sqrt{(g/L_{WL})(L_{WL}/\lambda_w)}Fr\cos\gamma_w \tag{L.5}$$

$$\omega_0 = \sqrt{(2\pi g/\lambda_w)} \tag{L.6}$$

and ω_e' is the value of ω_e for the condition under which the exponent in (L.1) becomes 0 and ω_0 is the natural circular frequency.

The factors $F_{ex\theta}$ and $F_{ex\varphi}$ can be considered as dimensionless amplitudes of the external pitching and rolling moments induced by the waves on the hull, normalized by their respective maximum values. For the pitching motion the latter occurs when the exponent in (L.1) and γ_w are zero. For the rolling motion this occurs when the exponent is zero and γ_w is 90°.

For the heaving motion it can be argued that the maximum occurs for $\gamma_w = 90°$ and that the amplitude of the motion will increase with increasing wavelength. A corresponding factor for the heaving motion is therefore

$$F_{exh} = f(\omega_e)e^{-[(L_{WL}/\lambda_w)\{|\cos\gamma_w|+\sin^2\gamma_w(B_{WL}/L_{WL})\}]^2} \tag{L.7}$$

Note that these are the factors that are displayed in Figs. 6.5.34, 6.5.35, and 6.5.36.

The mechanism for the excitation of the yawing motion is similar and illustrated by Fig. 6.8.11. The maximum excitation through the hull now occurs when the bow is at the crest of a wave and the stern is in a trough, or the other way around. This behavior can be described qualitatively by a factor $F_{ex\psi}$ of the type

$$F_{ex\psi} = \sin\gamma_w f(\omega_e)e^{-2.5[(L_{WL}/\lambda_w)\{|\cos\gamma_w|+\sin^2\gamma_w(B_{WL}/L_{WL})\}-0.55]^2} \tag{L.8}$$

Note that it is this factor that is displayed in Fig. 6.8.10.

Two further remarks are in order here.

The first is that the discussion given above is applicable only to hulls without appendages. While keel and rudders will not have a significant influence on the pitching and heaving excitation, at least at zero heel, there is no doubt that there will be a significant effect on the excitation in roll and yaw. The excitation in roll will be enhanced by the effect of the lateral orbital motion of the water particles on the side force on keel and rudder (Fig. L.2). Because the orbital motion decreases

Fig. L.2 *Orbital motion and wave forces in beam seas*

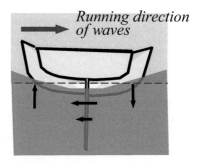

exponentially with depth the wave-induced side force will be larger for low aspect ratio appendages than for high aspect ratio keels and rudders with the same area. On the other hand the 'lever arm' is smaller for low draft appendages so it is not immediately obvious what the consequences are for the rolling moment.

The excitation in yaw will also be enhanced by the wave-induced side force on the rudder. Because the keel is usually positioned relatively close to the centre of gravity and there is little horizontal orbital motion at the position of the keel when the stern is on a wave crest, the effect of the keel on the wave-induced yawing moment will probably be small.

The second remark is that there will be an effect of heel on the wave excitation, in particular in the presence of appendages. The pitching excitation, for example, will, under heel, be enhanced by a vertical force component on the rudder. Similarly, heel will also cause a rolling excitation through keel and rudder for head and stern waves. Heel will also cause a horizontal, wave–induced force on the rudder and as-sociated excitation in yaw in head and stern waves.

It is the author's impression that, in the quantitative sense, very little is known about the phenomena just described. Literature on the subject seems almost non-existent.

L.2 The Scaling of Added Resistance

The added resistance in waves is, for by far the greater part, a pressure drag due to wave-making. This means that it scales in a way similar to Eq. (J.4), i.e.;

$$R_{aw} \div \rho g Z_h S_{Xmax} \tag{L.9}$$

There is however, a difference. That is, the height Z_h of the waves generated by the yacht should now be interpreted as the time-averaged height or amplitude of the waves that are generated by the periodic heaving and pitching motion. While the wave height due to motion in flat water was seen to be proportional to $\frac{1}{2}\rho V_b^2$ (see Eq. (J.5)), the height of waves generated by pitching and plunging can be shown to be a function of the relative vertical velocity V_{wz} of the water particles relative to the moving body (Lewis 1989). V_{wz} is determined by, in the first place, the orbital motion of the water particles in the waves.

From dimensional analysis it follows that

$$\rho g Z_h \div \rho V_{wz}^2 \tag{L.10},$$

so that

$$R_{aw} \div \rho V_{wz}^2 S_{Xmax} \tag{L.11}$$

In regular, sinusoidal waves, the vertical velocity V_{wz} of the water particles relative to a moving body can be shown (Lewis 1989) to be proportional to the ratio of the wave height (h_w) or wave amplitude ($\zeta_w = \frac{1}{2}h_w$) and the period T_e (or frequency $f_e = 1/T_e$) of encounter:

$$V_{wz} \div \zeta_w/T_e = \zeta_w \{1/T_w + (V_b/\lambda_w)\cos\gamma_w\} \tag{L.12}$$

Hence, Eq. (L.11) can also be written as

$$R_{aw} \div \rho(\zeta_w/T_e)^2 S_{Xmax} \tag{L.13},$$

or, using the identity $S_{Xmax} \equiv \nabla/(C_P L_{WL})$, as

$$R_{aw} \div \rho(\zeta_w/T_e)^2 \nabla/(C_P L_{WL}) \tag{L.14}$$

This can also be written as

$$R_{aw} \div \rho \, \zeta_w^2 \{1/T_w + (V_b/\lambda_w)\cos\gamma_w\}^2 \nabla/(C_P L_{WL}) \tag{L.15a},$$

or

$$R_{aw} \div \rho(\zeta_w/T_w)^2 \{1 + 2\pi(Fr/T_w')\cos\gamma_w\}^2 \nabla/(C_P L_{WL}) \tag{L.15b},$$

or

$$R_{aw} \div \rho g (h_w/T_w')^2 \{1 + 2\pi(Fr/T_w')\cos\gamma_w\}^2 \nabla/(C_P L_{WL}^3) \tag{L.15c}$$

Here, T_w' is a dimensionless wave period defined by

$$T_w' = T_w \sqrt{(g/L_{WL})} \tag{L.16}$$

Note that for low Froude numbers and/or long wave length and/or beam seas ($\gamma_w \approx 90°$) the factor $\{1 + Fr\sqrt{(2\pi L_{WL}/\lambda_w)}\cos\gamma_w\}$ can be replaced by 1.

It is further clear that R_{aw} must also be a function of the wave length/boat length ratio $\lambda_w/(L_{WL}\cos\gamma_w)$ and the difference or ratio between the frequency of encounter with waves and the natural frequencies of the pitching and heaving motions of the yacht. For very short wave lengths $(\lambda_w/(L_{WL}\cos\gamma_w) \to 0)$ a yacht cannot be triggered into pitching or heaving and the added resistance becomes negligible. This is also the case for beam seas and for waves that are very long with respect to the boat length $(\lambda_w/(L_{WL}\cos\gamma_w) \to \infty)$. A behavior of this kind can be modeled by applying a factor of the type

$$\lambda_w/(L_{WL}\cos\gamma_w)/\left[1 + \{\lambda_w/(L_{WL}\cos\gamma_w)\}^2\right] \tag{L.17a}$$

or

$$(T_w')^2/\cos\gamma_w/\left[1 + a\{(T_w')^2/\cos\gamma_w\}^2\right] \tag{L.17b},$$

where 'a' is a constant, to the expression (L.12) for the vertical velocity.

Synchronisms leading to a (relative) maximum of the added resistance will occur when the frequency of encounter with waves coincides with the natural frequency of the pitching and/or heaving motion of the yacht. A behavior of this kind can be modeled by a factor of the type

$$1/\left[1 + b \left\{p\,(T_e' - T_{0p}')^2 + (1-p)(T_e' - T_{0h}')^2\right\}\right] \qquad (L.18),$$

where 'b' is another constant and 'p' is a parameter expressing the relative importance of the pitching and heaving motions.

For irregular, composite waves one may expect a scaling like (L.15) based on the significant wave height $h_{w1/3}$ and the average wave period T_{wave} (see Sect. 5.19). Because a seaway consists of waves of different wave lengths, the effects of wavelength and synchronism or resonance as expressed by Eqs. (L.17) and L.16) will be less explicit in a seaway than in regular waves of one wavelength.

However, it appears (Keuning and Sonnenberg 1998a), perhaps somewhat surprisingly, to be common practice to use, instead of an expression of the type of Eq. (L.15), the following, simplified scaling:

$$\boldsymbol{R_{aw}} \div \rho\, g h_{w1/3}{}^2 L_{WL} \qquad (L.19)$$

The latter implies a non-dimensional coefficient for the added resistance in waves defined by

$$C_{Raw} = \boldsymbol{R_{aw}} / (\rho g h_{w1/3}{}^2 L_{WL}) \qquad (L.20)$$

Comparing Eqs. (L.15) and (L.19) suggests that C_{Raw} as defined by (L.20) will be proportional to a factor

$$(1/T_{wave}')^2 \left\{1 + 2\pi(Fr/T_w')\cos\gamma_w\right\}^2 C_{WP}(B_{WL}/L_{WL}) \qquad (L.21)$$

C_{Raw} should also be expected to be a function of the quantities determining the natural heaving and pitching periods, such as the displacement and length, beam/length ratio and gyradius.

References

Keuning JA, Sonnenberg UB (1998a) Approximation of the hydrodynamic forces on a sailing yacht based on the delft systematic yacht hull series. Proceedings of the 15th HISWA international symposium on yacht design and yacht construction, Amsterdam, Nov 1998
Lewis EV (ed) (1989) Principles of naval architecture, vol III—motions in waves and controllability. SNAME, Jersey City. ISBN 0-939773-02-3

Appendix M

A Model for the Effect of a Mast on the Aerodynamics of a Thin Sail Section

As demonstrated in literature on the subject (Refs. Marchai 2000; Milgram 1971, 2000; Abbott and Von Doenhoff 1949; Liebeck 1973; Olejniczak and Lyrintzis 1994; Goovaerts 2000; Chapin et al. 2004, 2005, 2008; Collie et al. 2004; Larsson and Eliasson 1996; Wilkinson 1984; Chapin et al. 2004; Haddad and Lepine 2003; Bethwaite 2010; Gentry 1976; Marchai 1964, 2000), the effects of a mast on the aerodynamics of a sail are a rather complex function of several parameters, such as mast diameter to sail chord ratio d/c, the camber and lift coefficient of the sail and the Reynolds number. There is no doubt that the only rigorous ways to study and model these effects is by physical (wind tunnel or full scale) or numerical (CFD) experiments. It is useful, however, for example for the application in Velocity Prediction Programs, to be able to model the main effects of a mast on the aerodynamics of a sail section also by a relatively simple set of formulae in the spirit of the 'form factor' methods described in Sects. 5.14, Viscous Resistance and Appendix H. This appendix reflects an attempt to construct a model of this nature.

A logical starting point for the formulation of such a model is to consider the flow about a circular cylinder. This type of flow is known to be strongly dependent on Reynolds number (Prandtl and Tietjens 1957). For Reynolds numbers $\left(Re = VD/v \right)$ between 2000 and 3000 the flow is unsteady, producing the Von Kármán vortex trail introduced in Sect. 5.20. For a medium size sailing yacht the Reynolds number based on the chord length of the sail was found to be of the order of 10^6 (See Fig. 5.9.3). This means that for a mast section it is of the order of $(D/c) \times 10^6$, where D is the diameter of the mast and 'c' the chord length of the sail. Because for most yachts D/c is of the order of 0.1, the mast section Reynolds number Re_D is of the order of 1×10^5. This means that the flow will, in general, be steady, except at very low wind speeds and/or for small mast diameters.

While the lift of a circular cylinder in steady flow is zero because of the symmetry of the flow field, the drag depends strongly on Reynolds number. Figure M.1 gives the drag coefficient C_D of a circular cylinder as a function of Reynolds number for the range of Reynolds numbers that is applicable to sailing yachts. For $Re_D < 1.5 \times 10^5$ the drag coefficient is practically constant with a value of about 1.2. For $Re_D > 5 \times 10^5$ it is also constant, with a value of about 0.3. In between there is a transition region.

It appears (Prandtl and Tietjens 1957) that for $Re_D < 1.5 \times 10^5$ the flow is characterized by laminar separation of the boundary layer while for $Re_D > 5 \times 10^5$ the boundary layer is turbulent at separation. Because the turbulent boundary layer separates further downstream (Sect. 5.12) the wake and 'dead water' region behind the cylinder is considerably thinner than in the laminar case. As a result there is less form drag (pressure drag).

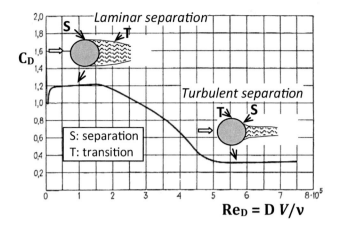

Fig. M.1 *Drag coefficient of a circular cylinder as a function of Reynolds number (Prandtl and Tietjens 1957)*

When a sail section is attached to a circular cylinder there is mutual interaction between the two. First of all the sail will be fully submerged in the wake of the cylinder. Because there is a velocity defect in the wake (see Sect. 5.11), the effective velocity of the onset-flow of the sail section will be lower than the velocity of the undisturbed flow. This tends to reduce both the lift and the drag of the sail section. Together with the drag D_{cyl} of the mast section the effect on the total drag D_{m+s} of the mast-sail configuration can be expressed as

$$D_{m+s} = D_{cyl} + \{1 - f_0(D/c)\} D_s \tag{M.1},$$

where D_s is the drag of the sail section without the presence of the mast and $f_0(D/c)$ is a factor accounting for the reduced effective flow velocity experienced by the sail in the wake of the mast. Obviously, we must have $f_0(D/c) \to 0$ for $D/c \to 0$ and $f_0(D/c) \to 1$ for $D/c \to \infty$. An example of a simple function exhibiting this behaviour is

$$f_0(D/c) = \frac{k_0(D/c)^p}{1 + k_0(D/c)^p} \tag{M.2},$$

where k_0 is a constant and the power $p > 0$. Since the effective flow velocity in the wake of the mast/cylinder will depend on the drag of the cylinder the constant k_0 will also be a function of the drag of the cylinder. Figure M.1 teaches that this means that k_0 will have significantly different values for laminar and turbulent boundary layer separation on the cylinder.

Equation (M.1) can also be written as

$$D_{m+s} = D_s + \delta D \tag{M.3}$$

with

$$\delta D = D_{cyl} - f_0(D/c)D_s \tag{M.4}$$

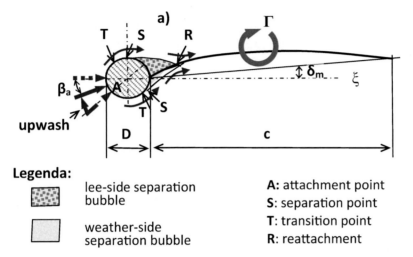

Legenda:

	lee-side separation bubble	**A:** attachment point
	weather-side separation bubble	**S:** separation point
		T: transition point
		R: reattachment

Fig. M.2 *Main parameters of the flow about a mast-sail section*

In dimensionless quantities, that is divided by $\frac{1}{2}\rho V^2 c$, Eq. (M.4) can be written as

$$\delta_0 C_D = C_{Dcyl}\frac{D}{c} - f_0(D/c)C_{Ds} \qquad (M.5)$$

The effect of the defect velocity in the wake of the cylinder on the lift of the sail will be similar. However, because the mast cylinder does, basically, not produce any lift by itself, the effect of the velocity defect on the lift (coefficient) of the sail can be expressed as

$$\delta_0 C_{Ls} = -f_0(D/c)C_{Ls} \qquad (M.6)$$

But, there is more to it. The flow about the sail affects the flow about the mast also and this bounces back to the sail. Three different, but related, mechanisms can be distinguished when considering the effects of the sail on the flow about the mast.

The first is that the sail splits the wake of the mast/cylinder in two parts. Under conditions of symmetrical flow, that is a sail without camber at zero angle of attack, this will not have a significant effect on the flow about mast.

Secondly, there are effects of the apparent wind angle. Even when the sail does not carry any load or lift, the flow about the mast-cylinder will become asymmetric because the stagnation or attachment point ('A', see Fig. M.2) will move to a point on the cylinder which is no longer opposite to the point where the sail is attached to the mast-cylinder. As a consequence the mast-cylinder will develop some lift. The amount of lift will be proportional to the apparent wind angle 'β_a'. Airfoil theory (Ashley and Landahl 1965) teaches that the factor of proportionality will be $<4\pi$ (per radian) when the lift of the cylinder is made non-dimensional by its diameter. When made non-dimensional by the chord length of the sail it will also be proportional to the mast-diameter/sail-chord ratio D/c. It is further clear that it should go to zero for $D/c \to \infty$ when the sail does no longer have any influence on the flow about the mast. In terms of lift coefficient based on the chord length 'c' of the sail the lift coefficient of the mast-cylinder can be expressed as

$$\delta C_{Lm} = k_{1m} f_1 (D/c) \beta_a \tag{M.7}$$

where $4\pi > k_{1m} > 0$ and

$$f_1 (D/c) = \frac{D/c}{1 + (D/c)^2} \tag{M.8}$$

or

$$f_1 (D/c) \approx D/c \tag{M.8a}$$

for small values of D/c.

The third mechanism is due to the lift of the sail. When the sail section carries lift, through angle of attack and/or camber, the mast-cylinder will experience additional upwash by the bound vorticity/circulation of the sail section. From airfoil theory it follows that this induces an additional angle of attack $\delta \alpha_m$ on the mast/cylinder given by

$$\delta \alpha_m \cong \frac{C_{Ls} + \delta_0 C_{Ls}}{2\pi \left(k_c + \frac{1}{2} \dfrac{D}{c} \right)} \quad \text{(radians)} \tag{M.9}$$

The factor k_c represents the position of the lift vector of the sail section in parts of the chord length, measured from the leading edge. This means that it is a function of the lift and moment coefficients of the sail section and, hence, of the camber and the angle of attack (see also Sect. 5.14). For zero camber and attached flow k_c takes the value ¼. For large amounts of camber and attached flow k_c can be as large as ½. For separated flow around and beyond maximum lift, the lift vector moves further backward with increasing angle of attack implying that near maximum lift k_c increases with angle of attack.

The dependence of k_c on the amount of camber and angle of attack implies that the sail-induced angle of attack of the mast/cylinder decreases with increasing camber of the sail and with increasing amount of separated flow on the suction side of the sail.

Other geometrical factors that will induce asymmetry in the flow about the mast are (the amount of) camber and the sheeting angle of the sail. The effect of this, apart from the effect through the lift-induced upwash, is difficult to estimate. It would seem, however, that both camber and sheeting angle will tend to reduce the asymmetry of the flow about the mast/cylinder. The reason for this is that both camber and sheeting angle tend to reduce the inclination of the flow relative to the local direction of the surface of the cylinder in the front part of the sail. The effect, if significant, can be accounted for, in principle, through the constant k_1 in Eq. (M.8), with less asymmetry implying a smaller value of k_1.

The total amount of lift of the mast/cylinder, induced by the sail, can thus be approximated as

$$\delta C_{Lm} = k_{1m} f_1 (D/c)(\beta_a + \delta\alpha_m) \tag{M.10},$$

where the function $f_1(D/c)$ is given by Eq. (M.8) and the sail-induced angle of attack $\delta\alpha_m$ given by Eq. (M.9). Note that, in general, δC_{Lm} will be >0.

When the mast/cylinder carries lift, there is, in all probability, also an effect on its drag. The additional form drag due to lift of the mast will be proportional to $(\delta C_{Lm})^2$, like the form drag due to lift of a two-dimensional airfoil (Sect. 5.15 and Appendix H). This means that the additional form drag of the mast/cylinder due to sail-induced lift can, when non-dimensionalised by the sail chord, be approximated as

$$\delta C_{Dm} = k_m [\delta C_{Lm}/\{f_1(D/c)\}]^2 C_{Dcyl} D/c \tag{M.11},$$

with δC_{Lm} given by (M.10). The form factor k_m is known to be of order [1] for attached flow (Appendix H), but possibly much larger for the type of separated flow that occurs on a circular cylinder. In this context it should be mentioned that the corresponding formula in Appendix H for the form drag due to lift of an airfoil with attached flow contains a factor $2C_{F0}$ rather than the factor C_{Dcyl} in Eq. (M.11). The latter is meant to take into account that the flow about the mast-cylinder is always separated, with a much higher drag level than a flat plate or thin airfoil.

As already indicated above, the asymmetry in the flow about the mast-cylinder influences, in turn, the flow about the sail. One mechanism is that the circulation around the mast/cylinder section induces downwash at the position of the sail. Airfoil theory teaches that this induces a negative angle of attack on the sail:

$$\delta\alpha_s \cong -\frac{\delta C_{Lm}}{2\pi\left(3/4 + \frac{1}{2}\frac{D}{c}\right)} \tag{M.12}$$

This causes a corresponding reduction $\delta_1 C_{Ls}$ (<0) of the lift as well as the drag ($\delta_1 C_{Ds} < 0$) of the sail section that can be expressed as

$$\delta_1 C_{Ls} = C_{Ls}(\alpha + \delta\alpha_s) - C_{Ls}(\alpha) \tag{M.13}$$

$$\delta_1 C_{Ds} = C_{Ds}(\alpha + \delta\alpha_s) - C_{Ds}(\alpha) \tag{M.14}$$

It is further known (Wilkinson 1984) that an important factor for the loss of lift and the increase of the drag of the sail is the condition (thickness) of the boundary layer on the suction side (lee side) of the sail. The larger the difference between the displacement thickness of the boundary layer on the suction side and that on the pressure side, the larger will be the loss of lift due to viscous effects and the increase of the form drag of the sail section. This is probably influenced by the asymmetry in the flow about the mast, the size of the separation bubbles in particular. However, with leading edge separation bubbles already present without a mast, it is not immediately clear to what extent the mast increases the loss of lift and the associated form drag of the sail. More asymmetry and more lift in the flow about the mast-cylinder will possibly cause a larger difference between the displacement thickness of the boundary layer on the suction side and that on the pressure side and, hence, a larger viscous loss of lift and more form drag of the sail section than in the case without mast. If this is the case, the effect can possibly be represented by introducing a multiplication factor in Eqs. (M.13) and (M.14):

$$\delta_1 C_{Ls} := k_{1sL} \delta_1 C_{Ls} \tag{M.15}$$

$$\delta_1 C_{Ds} := k_{1sD} \delta_1 C_{Ds} \tag{M.16}$$

With $\delta_1 C_{Lsand} < 0$ and $\delta_1 C_{Ds} < 0$ it is to be expected that $k_{1sL} < 1$ and that $k_{1sD} < k_{1sL}$. Collecting the various terms we get for the effect of the mast on lift:

$$\delta C_L = \delta_0 C_{Ls} + \delta C_{Lm} + \delta_1 C_{Ls} \tag{M.17},$$

with the specific terms given by Eqs. (M.6), (M.10) and (M.15).

The constants k_0, k_{1m} and k_{1s} as well as the power p are to be determined through correlation with experiment. The constant k_c is, as already indicated above, of the order of 0.4.

Collecting the various terms for the effect of the mast on drag one can write:

$$\delta C_D = C_{Dcyl} + \delta_0 C_{Ds} + \delta C_{Dm} + \delta_1 C_{Ds} \tag{M.18},$$

with the various terms given by Eqs. (M.5), (M.11) and (M.16) and with the additional constant k_m also to be determined through correlation with experiment.

The author has found from correlation with wind tunnel data (Chapin et al 2005; Wilkinson 1984), that for the powers p in Eq. (M.2) there holds p $\cong$ 1. The constant k_0 was taken as

$$k_0 = k_0' C_{Dcyl} \tag{M.19}$$

The factor k_0' was found to be of the order of $\cong 12$, k_{1sL} of the order of $\cong 0.5$ and $k_{1sD} \cong 0$. The values found for 'k_{1m}' and 'k_m' were $k_{1m} \cong 13$ and $k_m \cong 0.02$.

Figure M.3 illustrates the correlation with wind tunnel data for the lift loss due to a cylindrical mast section as a function of the angle of attack or apparent heading β_a (sheeting angle $\delta = 0$) according to Eq. (M.17). Results are given for sail sections with different amounts of camber ($f_c/c = 12\%$, 8% and 3%) and two different mast diameters ($D/c = 0.053$ and 0.08).

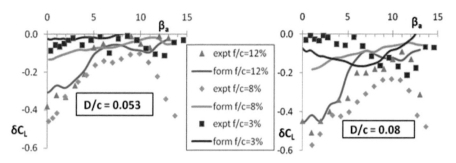

Fig. M.3 *Lift loss due to the presence of a mast/cylinder of sail sections with different amounts of camber* $(Re_c \cong 0.2 \times 10^6)$

It can be noted that the correlation shown in Fig. M.3 is not very good. This is believed to be due, at least partly, to the fact that the experimental data given in Refs. Chapin et al. (2005) and Haddad and Lepine (2003) are subject to errors[4].

It turned out that the constants k_{1s} and k_c play an important role in this correlation. For k_{1s} different values had to be adopted for lift and drag. Figure M.4 gives the factor k_c as a function of the angle of attack α (or the apparent heading β_a, sheeting angle $\delta=0$) for different values of f_c/c. It is useful to note that for small angles of attack the upwash factor k_c is approximately proportional with the amount of camber. For angles of attack beyond about 7.5 degrees, when, at the given Reynolds number, there is massive boundary layer separation on the suction side of the sail, k_c increases almost linearly with α. The angle of attack at which the transition in the behavior of k_c takes place may be dependent on Reynolds number, but to what extent is unknown.

Figure M.5 shows that the correlation for drag is equally modest. This, presumably, is also caused, at least partly, by problems with the experiment.

Fig. M.4 *Factor modelling the sail-induced upwash at the position of the mast as a function of angle of attack and camber/chord ratio*

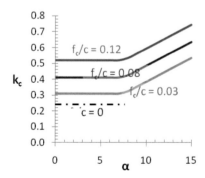

---

[4] The author has the impression, from a comparison with other wind tunnel data for sail sections without mast, that the data contained by Refs. Chapin (2005), Haddad and Lepine (2003) have not been (properly) corrected for systematic errors associated with the open jet type of test section. For this reason the author has applied corrections to the data that bring the data in agreement with the results of other wind tunnel experiments (Milgram 1971; Marchai 2000) for the same sail sections (without mast).

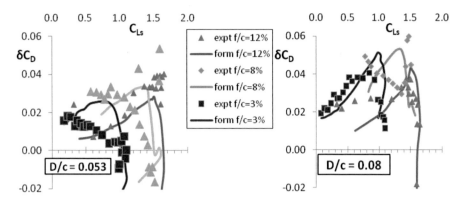

Fig. M.5 *Effect on drag of the presence of a mast/cylinder attached to sail sections with different amounts of camber (Re$_c$ $\cong 0.2 \times 10^6$)*

A remarkable fact is further that the best correlation is obtained when the drag coefficient of the cylinder based on its diameter is taken as $C_{Dcyl} = 0.4$. Considering Fig. M.1 this would mean that the flow would be almost fully turbulent. Given the fact that in the experiment the Reynolds number is as low as $Re_D \cong .15 \times 10^5$ this is quite remarkable.

References

Abbott IH, Von Doenhoff AE (1949) Theory of wing sections. McGraw Hill, New York

Ashley H, Landahl M (1985/1965) Aerodynamics of wings and bodies. Dover Publications, New York. ISBN 0-486-64899-0

Bethwaite F (2010) High performance sailing, 2nd edn. Adlard Coles Nautical, London, pp 311–314

Chapin VG, Jamme S, Neyhousser R (2004) Aérodynamique grand voile—mât', 2nd workshop Ecole Navale

Chapin VG et al (2005) Sailing yacht rig improvements through viscous computational fluid dynamics. The 17th Chesapeake sailing yacht symposium, Anapolis

Chapin VG et al (2008) Design optimization of interacting sails through viscous CFD. International conference on innovation in high performance sailing yachts, Lorient, RINA, May 2008

Collie SJ, Jackson PS, Gerritsen M, Fallow JB (2004) Two-dimensional CFD-based parametric analysis of downwind sail designs. RINA

Gentry AE (1976) Studies of mast section aerodynamics. Presented at The Ancient Interface VII, 7th AIAA symposium on sailing, Long Beach, Jan 1976

Goovaerts FEG (2000) Development of a high power sail with the aid of gurney flaps. Master's thesis Delft University of Technology, Laboratory of Aero- and Hydrodynamics, Sub-faculty of Applied Physics, Rept. No. MEAH-199

Haddad B, Lepine B (2003) Etude expérimentale de l'interaction mât-voile. Rapport de stage de fin d'étude, Ecole Navale

Larsson L, Eliasson RE (1996) Principles of yacht design. Adlard Coles Nautical, London, pp 141–142. ISBN 0-7136-3855-9

Liebeck RH (1973) A class of airfoils designed for high lift in incompressible flow. J Aircr 10(10)

Olejniczak J, Lyrintzis AS (1994) Design of optimized airfoils in subcritical flow. J Aircr 31(3):680–687

Marchai CA (1964) Sailing theory and practice. Adlard Coles Ltd, Southampton

Marchai CA (2000) Aero-hydrodynamics of sailing. Adlard Coles Nautical, London, p 240, 444,553, 587, 652, pp 302–325, 328–338, 510–516, 604–606, 658–675

Milgram JH (1971) Section data for thin, highly cambered airfoils in incompressible flow, NASA CR-1767

Prandtl L, Tietjens OG (1934a) Fundamentals of aero- & hydromechanics. McGraw-Hill, New York (also: Dover Publications, NY, 1957)

Wilkinson S (1984) Partially separated flow around 2D masts and sails. PhD Thesis, University of Southampton

Appendix N

Optimum Angle of Heel

For a given type of yacht, sailing at a given apparent wind angle it is possible to determine an 'optimum' angle of heel that is defined as the angle of heel for which the driving force, under influence of increasing wind speed and/or sail area and increasing heel attains a maximum. A mathematical expression for this 'optimum' angel of heel can be derived through application of the calculus of variations.

We first recall that the driving force T of a sailing yacht can be expressed as

$$T = \tfrac{1}{2}\rho V_a^2 S_s C_T \tag{N.1},$$

or, setting $q = \tfrac{1}{2}\rho V_a^2$, where q is the dynamic pressure, as

$$T = qS_s C_T \tag{N.2}$$

The product $q\,S_s$ can be called the dynamic pressure force of the sails.

The calculus of variations (http://en.wikipedia.org/wiki/Calculus_of_variations) teaches that a variation dT of the driving force as the result of a variation $d(qS_s)$ of the dynamic pressure force is given by

$$dT = C_T d(qS_s) + qS_s dC_T \tag{N.3}$$

The second term $qS_s\,dC_T$ appears because the driving force coefficient C_T is a function of the angle of heel φ and the latter depends also on wind speed and sail area and, hence, on qS_s. This can be expressed as

$$dC_T = \frac{\partial C_T}{\partial \varphi}d\varphi = \frac{\partial C_T}{\partial \varphi}\frac{d\varphi}{d(qS)}d(qS_s) \tag{N.4},$$

where the derivatives $\dfrac{\partial C_T}{\partial \varphi}$ and $\dfrac{d\varphi}{d(qS)}$ represent the dependence of the driving force on heel and the dependence of heel on the dynamic pressure force, respectively.

The derivative $\dfrac{d\varphi}{d(qS)}$ is coupled to the heeling moment and the righting moment of the yacht. We recall that these neutralize each other in equilibrium conditions. This means that the heeling moment

$$M_x = H(\zeta_{CE} - \zeta_{CLR}) = qS_sC_H(\zeta_{CE} - \zeta_{CLR}) \qquad (N.5),$$

where H and C_H are the heeling force and heeling force coefficient, respectively, and ζ_{CE}, ζ_{CLR} the (vertical) positions of the centre of effort of the sails and the centre of lateral resistance of hull plus keel, respectively. In equilibrium conditions the heeling moment M_x is equal to the righting moment M_{rx}. We have seen in Sect. 6.2 that the latter can be expressed as

$$M_{rx} = \Delta gm \sin\varphi \qquad (N.6),$$

where Δ is the displacement force (weight) of the yacht and gm the metacentric height. This means that, in equilibrium conditions, there holds

$$\Delta gm \sin\varphi = qS_sC_H(\zeta_{CE} - \zeta_{CLR}) \qquad (N.7)$$

From Eq. (N.7) it can be derived that, for a given apparent wind angle, a variation $d\varphi$ of the angle of heel due to a change $d(qS_s)$ of the dynamic pressure force on the sails satisfies the following relation

$$\Delta gm \cos\varphi \, d\varphi = \left\{ C_H + qS_s \frac{\partial C_H}{\partial \varphi} \frac{d\varphi}{d(qS)} \right\} (\zeta_{CE} - \zeta_{CLR}) \, d(qS_s) \qquad (N.8)$$

This can be rewritten as

$$\frac{d\varphi}{d(qS)} = \left\{ C_H + qS_s \frac{\partial C_H}{\partial \varphi} \frac{d\varphi}{d(qS)} \right\} \frac{(\zeta_{CE} - \zeta_{CLR})}{\cos\varphi \Delta \, gm}$$

or

$$\frac{d\varphi}{d(qS)} = \frac{\dfrac{C_H}{\cos\varphi} \dfrac{(\zeta_{CE} - \zeta_{CLR})}{\Delta \, gm}}{1 - \dfrac{qS_s}{\cos\varphi} \dfrac{(\zeta_{CE} - \zeta_{CLR})}{\Delta \, gm} \dfrac{\partial C_H}{\partial \varphi}} \qquad (N.9)$$

Introducing the heel forcing parameter

$$K_\varphi = qS_s \frac{(\zeta_{CE} - \zeta_{CLR})}{\Delta \, gm} \tag{N.10}$$

this can be simplified as

$$qS_s \frac{d\varphi}{d(qS)} = \frac{\dfrac{K_\varphi C_H}{\cos\varphi}}{1 - \dfrac{K_\varphi}{\cos\varphi} \dfrac{\partial C_H}{\partial\varphi}} \tag{N.11}$$

Substituting Eqs. (N.4), (N.5) and (N.11) into (N.3) gives

$$dT = \left[C_T - \frac{\partial C_T}{\partial\varphi} \frac{\dfrac{K_\varphi}{\cos\varphi} C_H}{1 - \dfrac{K_\varphi}{\cos\varphi} \dfrac{\partial C_H}{\partial\varphi}} \right] d(qS_s) \tag{N.12}$$

According to the calculus of variation the driving force T attains its maximum when its variation dT is equal to zero. This is the case when the quantity between the square brackets is zero, i.e. when

$$C_T(\varphi) - \frac{\partial C_T}{\partial\varphi} \frac{\dfrac{K_\varphi}{\cos\varphi} C_H(\varphi)}{1 - \dfrac{K_\varphi}{\cos\varphi} \dfrac{\partial C_H}{\partial\varphi}} = 0 \tag{N.13}$$

When, for the apparent wind angle considered, the driving force coefficient C_T, the heeling force coefficient C_H and the heel forcing parameter K_φ are known as a function of the heel angle φ, Eq. (N.13) can be used to determine the heel angle φ beyond which an increase in wind speed or sail area does no longer lead to an increase of the driving force.

Note that equation Eq. (N.13) implies that, for a given apparent wind angle and a given level of C_T, the resulting 'optimum' (effective) heel angle will be smaller for larger values of the heel forcing parameter K_φ, that is for tender yachts, than for stiff yachts, when K_φ is small.

References

http://en.wikipedia.org/wiki/Calculus_of_variations

Index

© Springer International Publishing Switzerland 2015
J. W. Slooff, *The Aero- and Hydromechanics of Keel Yachts*,
DOI 10.1007/978-3-319-13275-4

Printed in the United States
By Bookmasters